T0202726

Sequential and Parallel
Algorithms and Data Structures

Peter Sanders • Kurt Mehlhorn
Martin Dietzfelbinger • Roman Dementiev

Sequential and Parallel Algorithms and Data Structures

The Basic Toolbox

 Springer

Peter Sanders
Karlsruhe Institute of Technology
Karlsruhe, Germany

Kurt Mehlhorn
Max Planck Institute for Informatics
Saarbrücken, Germany

Martin Dietzfelbinger
Technische Universität Ilmenau
Ilmenau, Germany

Roman Dementiev
Intel Deutschland GmbH
Feldkirchen, Germany

Intel material (including descriptions of algorithms and their implementations, text, figures, and tables describing their experimental evaluation and experimental results on Intel processors, and sample code) is reproduced by permission of Intel Corporation.

ISBN 978-3-030-25211-3   ISBN 978-3-030-25209-0   (eBook)
https://doi.org/10.1007/978-3-030-25209-0

© Springer Nature Switzerland AG 2019
This work is subject to copyright. All rights are reserved by the Publisher, whether the whole or part of the material is concerned, specifically the rights of translation, reprinting, reuse of illustrations, recitation, broadcasting, reproduction on microfilms or in any other physical way, and transmission or information storage and retrieval, electronic adaptation, computer software, or by similar or dissimilar methodology now known or hereafter developed.
The use of general descriptive names, registered names, trademarks, service marks, etc. in this publication does not imply, even in the absence of a specific statement, that such names are exempt from the relevant protective laws and regulations and therefore free for general use.
The publisher, the authors, and the editors are safe to assume that the advice and information in this book are believed to be true and accurate at the date of publication. Neither the publisher nor the authors or the editors give a warranty, express or implied, with respect to the material contained herein or for any errors or omissions that may have been made. The publisher remains neutral with regard to jurisdictional claims in published maps and institutional affiliations.

This Springer imprint is published by the registered company Springer Nature Switzerland AG
The registered company address is: Gewerbestrasse 11, 6330 Cham, Switzerland

To all algorithmicists

Preface

An algorithm is a precise and unambiguous recipe for solving a class of problems. If formulated as programs, they can be executed by a machine. Algorithms are at the heart of every nontrivial computer application. Therefore, every computer scientist and every professional programmer should know about the basic algorithmic toolbox: structures that allow efficient organization and retrieval of data, key algorithms for problems on graphs, and generic techniques for modeling, understanding, and solving algorithmic problems. This book is a concise introduction to this basic toolbox, intended for students and professionals familiar with programming and basic mathematical language. Although the presentation is concise, we have included many examples, pictures, informal explanations, and exercises, and some linkage to the real world. The book covers the material typically taught in undergraduate and first-year graduate courses on algorithms.

Most chapters have the same basic structure. We begin by discussing a problem as it occurs in a real-life situation. We illustrate the most important applications and then introduce simple solutions *as informally as possible and as formally as necessary* to really understand the issues at hand. When we move to more advanced and optional issues, this approach gradually leads to a more mathematical treatment, including theorems and proofs. Thus, the book should work for readers with a wide range of mathematical expertise. There are also advanced sections (marked with a *) where we *recommend* that readers should skip them on first reading. Exercises provide additional examples, alternative approaches and opportunities to think about the problems. It is highly recommended to take a look at the exercises even if there is no time to solve them during the first reading. In order to be able to concentrate on ideas rather than programming details, we use pictures, words, and high-level pseudocode to explain our algorithms. A section "Implementation Notes" links these abstract ideas to clean, efficient implementations in real programming languages such as C++ and Java. Each chapter ends with a section on further findings that provides a glimpse at the state of the art, generalizations, and advanced solutions.

The first edition of this book treated only sequential algorithms and data structures. Today, almost every computer, be it a desktop, a notebook, or a smartphone, has multiple cores, and sequential machines have become an exotic species. The reason for this change is that sequential processors have ceased to get proportional performance improvements from increased circuit complexity. Although the number of transistors in an integrated circuit (still) doubles every two years (Moore's law), the only reasonable way to use this transistor budget is to put multiple processor cores on a chip. The consequence is that nowadays every performance-critical application has to be parallelized. Moreover, big data – the explosion of data set sizes in many applications – has produced an enormous demand for algorithms that scale to a large number of processors. *This paradigm shift has profound effects on teaching algorithms.* Parallel algorithms are no longer a specialized topic reserved for a small percentage of students. Rather, every student needs some exposure to parallel algorithms, and parallel solution paradigms need to be taught early on. As a consequence, parallel algorithms should be integrated tightly and early into algorithms courses. We therefore decided to include parallel algorithms in the second edition of the book. Each chapter now has some sections on parallel algorithms. The goals remain the same as for the first edition: a careful balance between simplicity and efficiency, between theory and practice, and between classical results and the forefront of research. We use a slightly different style for the sections on parallel computing. We include concrete programming examples because parallel programming is still more difficult than sequential programming (the programs are available at github.com/basic-toolbox-sample-code/basic-toolbox-sample-code/). We also reference original work directly in the text instead of in the section on history because the parallel part is closer to current research.

Algorithmics is a modern and active area of computer science, even at the level of the basic toolbox. We have made sure that we present algorithms in a modern way, including explicitly formulated invariants. We also discuss important further aspects, such as algorithm engineering, memory hierarchies, algorithm libraries, and certifying algorithms.

We have chosen to arrange most of the material by problem domain and not by solution technique. The chapter on optimization techniques is an exception. We find that an organization by problem domain allows a more concise presentation. However, it is also important that readers and students obtain a good grasp of the available techniques. Therefore, we have structured the optimization chapter by techniques, and an extensive index provides cross-references between different applications of the same technique. Bold page numbers in the index indicate the pages where concepts are defined.

This book can be used in multiple ways in teaching. We have used the first edition of the book and a draft version of the second edition in undergraduate and graduate courses on algorithmics. In a first year undergraduate course, we concentrated on the sequential part of the book. In a second algorithms course at an advanced bachelor level or master's level and with students with some experience in parallel programming we made most of the sequential part of the book a prerequisite and concentrated on the more advanced material such as the starred sections and the parts on external memory and parallel algorithms. If a first algorithms course is taught later in the undergraduate curriculum, the book can be used to teach parallel and sequential algorithms in an integrated way. Although we have included some material about parallel programming and several concrete programming examples, the parallel part of the book works best for readers who already have some background in parallel programming. Another approach is to use the book to provide concrete algorithmic content for a parallel programming course that uses another book for the programming part. Last but not least, the book should be useful for independent study and to professionals who have a basic knowledge of algorithms but less experience with parallelism.

Follow us on an exciting tour through the world of algorithms.

Ilmenau, Saarbrücken, Karlsruhe, Heidelberg   *Martin Dietzfelbinger*
August 2018   *Kurt Mehlhorn*
                                                                   Peter Sanders
                                                               Roman Dementiev

A first edition of this book was published in 2008. Since then the book has been translated into Chinese, Greek, Japanese, and German. Martin Dietzfelbinger translated the book into German. Actually, he did much more than a translation. He thoroughly revised the book and improved the presentation at many places. He also corrected a number of mistakes. Thus, the book gained through the translation, and we decided to make the German edition the reference for any future editions. It is only natural that we asked Martin to become an author of the German edition and any future editions of the book.

Saarbrücken, Karlsruhe   *Kurt Mehlhorn*
March 2014   *Peter Sanders*

Soon after the publication of the German edition, we started working on the revised English edition. We decided to expand the book into the parallel world for the reasons indicated in the preface. Twenty years ago, parallel machines were exotic, nowadays, sequential machines are exotic. However, the parallel world is much more diverse and complex than the sequential world, and therefore algorithm-engineering issues become more important. We concluded that we had to go all the way to implementations and experimental evaluations for some of the parallel algorithms. We invited Roman Dementiev to work with us on the algorithm engineering aspects of parallel computing. Roman received his PhD in 2006 for a thesis on "Algorithm Engineering for Large Data Sets". He now works for Intel, where he is responsible for performance engineering of a major database system.

Ilmenau, Saarbrücken, Karlsruhe   *Martin Dietzfelbinger*
November 2017   *Kurt Mehlhorn*
                                                                   Peter Sanders

Contents

1	**Appetizer: Integer Arithmetic**	1
	1.1 Addition	2
	1.2 Multiplication: The School Method	6
	1.3 Result Checking	10
	1.4 A Recursive Version of the School Method	11
	1.5 Karatsuba Multiplication	13
	1.6 Parallel Multiplication	15
	1.7 Algorithm Engineering	17
	1.8 The Programs	18
	1.9 Proofs of Lemma 1.7 and Theorem 1.9	21
	1.10 Implementation Notes	23
	1.11 Historical Notes and Further Findings	23
2	**Introduction**	25
	2.1 Asymptotic Notation	26
	2.2 The Sequential Machine Model	29
	2.3 Pseudocode	32
	2.4 Parallel Machine Models	38
	2.5 Parallel Pseudocode	44
	2.6 Designing Correct Algorithms and Programs	46
	2.7 An Example – Binary Search	49
	2.8 Basic Algorithm Analysis	52
	2.9 Average-Case Analysis	57
	2.10 Parallel-Algorithm Analysis	62
	2.11 Randomized Algorithms	63
	2.12 Graphs	68
	2.13 **P** and **NP**	73
	2.14 Implementation Notes	77
	2.15 Historical Notes and Further Findings	79

3   Representing Sequences by Arrays and Linked Lists 81
   3.1   Processing Arrays in Parallel 82
   3.2   Linked Lists 86
   3.3   Processing Linked Lists in Parallel 91
   3.4   Unbounded Arrays 97
   3.5   *Amortized Analysis 102
   3.6   Stacks and Queues 106
   3.7   Parallel Queue-Like Data Structures........................ 109
   3.8   Lists versus Arrays 113
   3.9   Implementation Notes..................................... 114
   3.10 Historical Notes and Further Findings 116

4   Hash Tables and Associative Arrays 117
   4.1   Hashing with Chaining 120
   4.2   Universal Hashing... 122
   4.3   Hashing with Linear Probing 128
   4.4   Chaining versus Linear Probing 130
   4.5   *Perfect Hashing 131
   4.6   Parallel Hashing 134
   4.7   Implementation Notes..................................... 147
   4.8   Historical Notes and Further Findings 149

5   Sorting and Selection 153
   5.1   Simple Sorters 156
   5.2   Simple, Fast, and Inefficient Parallel Sorting 158
   5.3   Mergesort – an $O(n \log n)$ Sorting Algorithm 160
   5.4   Parallel Mergesort.. 162
   5.5   A Lower Bound 165
   5.6   Quicksort 168
   5.7   Parallel Quicksort 173
   5.8   Selection... 178
   5.9   Parallel Selection... 180
   5.10 Breaking the Lower Bound 182
   5.11 *Parallel Bucket Sort and Radix Sort 186
   5.12 *External Sorting 187
   5.13 Parallel Sample Sort with Implementations 191
   5.14 *Parallel Multiway Mergesort 201
   5.15 Parallel Sorting with Logarithmic Latency..................... 205
   5.16 Implementation Notes..................................... 207
   5.17 Historical Notes and Further Findings 208

6   Priority Queues 211
   6.1   Binary Heaps 213
   6.2   Addressable Priority Queues 218
   6.3   *External Memory 224
   6.4   Parallel Priority Queues 226
   6.5   Implementation Notes....................................... 229
   6.6   Historical Notes and Further Findings 230

7   Sorted Sequences 233
   7.1   Binary Search Trees 235
   7.2 (a,b)-Trees and Red–Black Trees 238
   7.3   More Operations 245
   7.4   Amortized Analysis of Update Operations..................... 248
   7.5   Augmented Search Trees 250
   7.6   Parallel Sorted Sequences 252
   7.7   Implementation Notes....................................... 254
   7.8   Historical Notes and Further Findings 256

8   Graph Representation 259
   8.1   Unordered Edge Sequences 260
   8.2   Adjacency Arrays – Static Graphs 261
   8.3   Adjacency Lists – Dynamic Graphs 261
   8.4   The Adjacency Matrix Representation 263
   8.5   Implicit Representations.................................... 264
   8.6   Parallel Graph Representation 265
   8.7   Implementation Notes....................................... 267
   8.8   Historical Notes and Further Findings 268

9   Graph Traversal 271
   9.1   Breadth-First Search 272
   9.2   Parallel Breadth-First Search 274
   9.3   Depth-First Search 282
   9.4   Parallel Traversal of DAGs.................................. 294
   9.5   Implementation Notes....................................... 298
   9.6   Historical Notes and Further Findings 299

10   Shortest Paths 301
   10.1   From Basic Concepts to a Generic Algorithm 302
   10.2   Directed Acyclic Graphs 306
   10.3   Nonnegative Edge Costs (Dijkstra's Algorithm) 307
   10.4   *Average-Case Analysis of Dijkstra's Algorithm 311
   10.5   Monotone Integer Priority Queues 313
   10.6   Arbitrary Edge Costs (Bellman–Ford Algorithm) 318
   10.7   All-Pairs Shortest Paths and Node Potentials................ 320
   10.8   Shortest-Path Queries 322

10.9   Parallel Shortest Paths   328
10.10 Implementation Notes.   330
10.11 Historical Notes and Further Findings   331

11 Minimum Spanning Trees   333
11.1   Cut and Cycle Properties   334
11.2   The Jarník–Prim Algorithm   337
11.3   Kruskal's Algorithm   338
11.4   The Union–Find Data Structure.   340
11.5   *External Memory   344
11.6   *Parallel Algorithms   348
11.7   Applications.   351
11.8   Implementation Notes.   353
11.9   Historical Notes and Further Findings   354

12 Generic Approaches to Optimization   357
12.1   Linear Programming – Use a Black-Box Solver   359
12.2   Greedy Algorithms – Never Look Back   365
12.3   Dynamic Programming – Build It Piece by Piece   368
12.4   Systematic Search – When in Doubt, Use Brute Force   373
12.5   Local Search – Think Globally, Act Locally   379
12.6   Evolutionary Algorithms   389
12.7   Implementation Notes.   391
12.8   Historical Notes and Further Findings   392

13 Collective Communication and Computation   393
13.1   Broadcast   396
13.2   Reduction   402
13.3   Prefix Sums   404
13.4   Synchronization.   406
13.5   (All)-Gather/Scatter   412
13.6   All-to-All Message Exchange   413
13.7   Asynchronous Collective Communication   418

14 Load Balancing   419
14.1   Overview and Basic Assumptions   420
14.2   Prefix Sums – Independent Tasks with Known Size   422
14.3   The Master–Worker Scheme   424
14.4   (Randomized) Static Load Balancing   426
14.5   Work Stealing   427
14.6   Handling Dependencies   433

A   Mathematical Background 435
    A.1   Mathematical Symbols 435
    A.2   Mathematical Concepts 436
    A.3   Basic Probability Theory 438
    A.4   Useful Formulae 443

B   Computer Architecture Aspects 447
    B.1   Cores and Hardware Threads 447
    B.2   The Memory Hierarchy 448
    B.3   Cache Coherence Protocols 449
    B.4   Atomic Operations 451
    B.5   Hardware Transactional Memory 451
    B.6   Memory Management 451
    B.7   The Interconnection Network 452
    B.8   CPU Performance Analysis 453
    B.9   Compiler 454

C   Support for Parallelism in C++ 455
    C.1   "Hello World" C++11 Program with Threads 455
    C.2   Locks 456
    C.3   Asynchronous Operations 457
    C.4   Atomic Operations 457
    C.5   Useful Tools 458
    C.6   Memory Management 458
    C.7   Thread Scheduling 459

D   The Message Passing Interface (MPI) 461
    D.1   "Hello World" and What Is an MPI Program? 461
    D.2   Point-to-Point Communication 463
    D.3   Collective Communication 464

E   List of Commercial Products, Trademarks and Software Licenses ... 465
    E.1   BSD 3-Clause License 465

References 467

Index 487

1

Appetizer: Integer Arithmetic

An appetizer is supposed to stimulate the appetite at the beginning of a meal. This is exactly the purpose of this chapter. We want to stimulate your interest in algorithmic[1] techniques by showing you a surprising result. Although, the school method for multiplying integers has been familiar to all of us since our school days and seems to be the natural way to multiply integers, it is not the best multiplication algorithm; there are much faster ways to multiply large integers, i.e., integers with thousands or even millions of digits, and we shall teach you one of them.

Algorithms for arithmetic are not only among the oldest algorithms, but are also essential in areas such as cryptography, geometric computing, computer algebra, and computer architecture. The three former areas need software algorithms for arithmetic on long integers, i.e., numbers with up to millions of digits. We shall present the algorithms learned in school, but also an improved algorithm for multiplication. The improved multiplication algorithm is not just an intellectual gem but is also extensively used in applications. Computer architecture needs very fast algorithms for moderate-length (32 to 128 bits) integers. We shall present fast parallel algorithms suitable for hardware implementation for addition and multiplication. On the way, we shall learn basic algorithm analysis and basic algorithm engineering techniques in a simple setting. We shall also see the interplay of theory and experiment.

We assume that integers[2] are represented as digit strings. In the base B number system, where B is an integer larger than 1, there are digits $0, 1, \ldots, B-1$, and a digit string $a_{n-1}a_{n-2}\ldots a_1 a_0$ represents the number $\sum_{0 \le i < n} a_i B^i$. The most important systems with a small value of B are base 2, with digits 0 and 1, base 10, with digits 0

---

[1] The Soviet stamp on this page shows Muhammad ibn Musa al-Khwarizmi (born approximately 780; died between 835 and 850), Persian mathematician and astronomer from the Khorasan province of present-day Uzbekistan. The word "algorithm" is derived from his name.

[2] Throughout this chapter, we use "integer" to mean a nonnegative integer.

© Springer Nature Switzerland AG 2019
P. Sanders et al., *Sequential and Parallel Algorithms and Data Structures*,
https://doi.org/10.1007/978-3-030-25209-0_1

to 9, and base 16, with digits 0 to 15 (frequently written as 0 to 9, A, B, C, D, E, and F). Larger bases, such as 2^8, 2^{16}, 2^{32}, and 2^{64}, are also useful. For example,

$$\text{"10101" in base 2 represents} \quad 1\cdot 2^4 + 0\cdot 2^3 + 1\cdot 2^2 + 0\cdot 2^1 + 1\cdot 2^0 = 21,$$
$$\text{"924" in base 10 represents} \quad 9\cdot 10^2 + 2\cdot 10^1 + 4\cdot 10^0 = 924.$$

We assume that we have two primitive operations at our disposal: the addition of three digits with a two-digit result (this is sometimes called a full adder), and the multiplication of two digits with a two-digit result.[3] For example, in base 10, we have

$$\begin{array}{r} 3 \\ 5 \\ 5 \\ \hline 13 \end{array} \quad \text{and} \quad 6\cdot 7 = 42.$$

We shall measure the efficiency of our algorithms by the number of primitive operations executed.

We can artificially turn any n-digit integer into an m-digit integer for any $m \geq n$ by adding additional leading 0's. Concretely, "425" and "000425" represent the same integer. We shall use a and b for the two operands of an addition or multiplication and assume throughout this chapter that a and b are n-digit integers. The assumption that both operands have the same length simplifies the presentation without changing the key message of the chapter. We shall come back to this remark at the end of the chapter. We refer to the digits of a as a_{n-1} to a_0, with a_{n-1} being the most significant digit (also called leading digit) and a_0 being the least significant digit, and write $a = (a_{n-1}\ldots a_0)$. The leading digit may be 0. Similarly, we use b_{n-1} to b_0 to denote the digits of b, and write $b = (b_{n-1}\ldots b_0)$.

1.1 Addition

We all know how to add two integers $a = (a_{n-1}\ldots a_0)$ and $b = (b_{n-1}\ldots b_0)$. We simply write one under the other with the least significant digits aligned, and sum the integers digitwise, carrying a single digit from one position to the next. This digit is called a *carry*. The result will be an $(n+1)$-digit integer $s = (s_n\ldots s_0)$. Graphically,

a_{n-1} ... $a_1\ a_0$			first operand
b_{n-1} ... $b_1\ b_0$			second operand
$c_n\ c_{n-1}$ ... $c_1\ \ 0$			carries
$s_n\ s_{n-1}$ ... $s_1\ s_0$			sum

---

[3] Observe that the sum of three digits is at most $3(B-1)$ and the product of two digits is at most $(B-1)^2$, and that both expressions are bounded by $(B-1)\cdot B^1 + (B-1)\cdot B^0 = B^2 - 1$, the largest integer that can be written with two digits. This clearly holds for $B = 2$. Increasing B by 1 increases the first expression by 3, the second expression by $2B-1$, and the third expression by $2B+1$. Hence the claim holds for all B.

where $c_0,\ldots,c_n$ is the sequence of carries and $s = (s_n \ldots s_0)$ is the sum. We have $c_0 = 0$, $c_{i+1} \cdot B + s_i = a_i + b_i + c_i$ for $0 \le i < n$, and $s_n = c_n$. As a program, this is written as

```
c = 0 : Digit                                    // Variable for the carry digit
for i:=0 to n − 1 do  add aᵢ, bᵢ, and c to form sᵢ and a new carry c
sₙ:=c
```

We need one primitive operation for each position, and hence a total of n primitive operations.

Theorem 1.1. *The addition of two n-digit integers requires exactly n primitive operations. The result is an $(n+1)$-digit integer.*

1.1.1 Parallel Addition

Our addition algorithm produces the result digits one after the other, from least significant to most significant. In fact, the ith carry depends on the $(i-1)$th carry, which in turn depends on the $(i-2)$nd carry, and so on. So, it seems natural to compute the result digits sequentially. Is there a parallel algorithm for addition that produces all digits in time less than linear in the number of digits? Parallel addition is not an academic exercise but crucial for microprocessor technology, because processors need fast, hardware-implemented algorithms for arithmetic. Suppose that engineers had insisted on using serial addition algorithms for microprocessors. In that case it is likely that we would still be using 8-bit processors wherever possible, since they would be up to eight times faster than 64-bit processors.

Our strategy for parallel addition is simple and ambitious – we want to perform all digit additions $a_i + b_i + c_i$ in parallel. Of course, the problem is how to compute the carries c_i in parallel. Consider the following example of the addition of two 8-digit binary numbers a and b:

position	8 7 6 5 4 3 2 1 0
a	1 0 0 0 0 1 1 1
b	0 1 0 0 1 0 1 0
c	0 0 0 0 1 1 1 0 0
x	p p s s p p g p
y	s s s s g g g p

Row c indicates the carries. Why is there a carry into position 4, why is $c_4 = 1$? Because position 1 generates a carry, since $a_1 + b_1 = 2$, and positions 2 and 3 propagate it, since $a_2 + b_2 = a_3 + b_3 = 1$. Why is there no carry into position 8, why is $c_8 = 0$? Positions 6 and 7 would propagate a carry, since $a_6 + b_6 = a_7 + b_7 = 1$, but there is nothing to propagate, since no carry is generated in position 5 since $a_5 + b_5 = 0$. The general rule is: We have a carry into a certain position if a carry is generated somewhere to the right and then propagated through all intermediate positions.

Rows x and y implement this rule. Row x indicates whether a position generates (g), propagates (p), or stops (s) a carry. We have

$$x_i = \begin{cases} \text{g} & \text{if } a_i + b_i = 2, \\ \text{p} & \text{if } a_i + b_i = 1, \\ \text{s} & \text{if } a_i + b_i = 0. \end{cases}$$

The different x_i's are independent and can be computed in parallel. Row y gives information whether we have a carry into a certain position. For $i \geq 1$, we have a carry into position i, i.e., $c_i = 1$, if and only if $y_{i-1} = \text{g}$. We never have a carry into position 0, i.e., $c_0 = 0$. We give two rules for determining the sequence of y's. Consider the maximal subsequences of the x-sequence consisting of a string of p's followed by either an s or a g or the right end. Within each subsequence, turn all symbols into the last symbol; leave the sequence of trailing p's unchanged. In our example, the maximal subsequences are $x_7 x_6 x_5$, x_4, and $x_3 x_2 x_1$, and the sequence of trailing p's consists of x_0. Therefore $y_7 y_6 y_5$ become s's, y_4 becomes s, $y_3 y_2 y_1$ become g's, and y_0 becomes p. It would be equally fine to turn the trailing p's into s's. However, this would make the formal treatment slightly more cumbersome. Alternatively, we may state the rule for the y_i's as follows:

$$y_i = \begin{cases} x_i & \text{if } i = 0 \text{ or } x_i \in \{\text{s}, \text{g}\}, \\ y_{i-1} & \text{otherwise, i.e., } i > 0 \text{ and } x_i = \text{p}. \end{cases}$$

Our task is now to compute the y_i's in parallel. The first step is to rewrite the definition of y_i as

$$y_i = \begin{cases} x_0 & \text{if } i = 0, \\ x_i \otimes y_{i-1} & \text{if } i > 0, \end{cases} \quad \text{where}$$

$\otimes$	s	p	g
s	s	s	s
p	s	p	g
g	g	g	g

The operator $\otimes$[4] returns its left argument if this argument is s or g and returns its right argument when the left argument is p. We next expand the definition of y_i and obtain

$$y_i = x_i \otimes y_{i-1} = x_i \otimes (x_{i-1} \otimes y_{i-2}) = x_i \otimes (x_{i-1} \otimes (x_{i-2} \ldots (x_1 \otimes x_0) \ldots)).$$

This formula corresponds to the sequential computation of y_i. We first compute $x_1 \otimes x_0$, then left-multiply by x_2, then by x_3, and finally by x_i. The operator $\otimes$ is associative (we shall prove this below) and hence we can change the evaluation order without changing the result. Compare the following formulae for y_6:

$$y_6 = x_6 \otimes (x_5 \otimes (x_4 \otimes (x_3 \otimes (x_2 \otimes (x_1 \otimes x_0)))))$$

and

$$y_6 = (x_6 \otimes ((x_5 \otimes x_4) \otimes ((x_3 \otimes x_2) \otimes (x_1 \otimes x_0))).$$

---

[4] Pronounced "otimes".

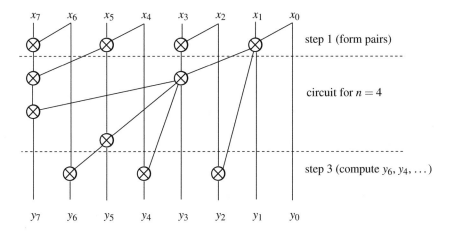

Fig. 1.1. The computation of y_7 to y_0 from x_7 to x_0. The horizontal partition of the computation corresponds to the general description of the algorithm. We first form pairs. In the second row of gates, we combine pairs, and in the third row, we combine pairs of pairs. Then we start to fill in gaps. In row four, we compute y_5, and in the last row, we compute y_6, y_4, and y_2.

The latter formula corresponds to a parallel evaluation of y_6. We compute $x_5 \otimes x_4$, $x_3 \otimes x_2$, and $x_1 \otimes x_0$ in parallel, then $x_6 \otimes (x_5 \otimes x_4)$ and $(x_3 \otimes x_2) \otimes (x_1 \otimes x_0)$, and finally y_6. Thus three rounds of computation suffice instead of the six rounds for the sequential evaluation. Generally, we can compute y_i in $\lceil \log i \rceil$ rounds of computation. The computation of the different y_i's can be intertwined, as Fig. 1.1 shows.

We next give a more formal treatment for a general base B. We start with the simple observation that the carry digit is either 0 or 1.

Lemma 1.2. $c_i \in \{0,1\}$ *for all* i.

Proof. We have $c_0 = 0$ by definition. Assume inductively, that $c_i \in \{0,1\}$. Then $a_i + b_i + c_i \leq 2(B-1) + 1 = 2B - 1 = 1 \cdot B + B - 1$ and hence $c_{i+1} \in \{0,1\}$. □

Two input digits a_i and b_i will *generate* a carry no matter what happens to the right of them if $a_i + b_i > B - 1$. On the other hand, the addition involving a_i and b_i will *stop* any carry if $a_i + b_i < B - 1$; this holds because an incoming carry is at most 1. If $a_i + b_i = B - 1$, an incoming carry will be *propagated* to the left. Hence, we define

$$x_i := \begin{cases} \mathrm{s} & \text{if } a_i + b_i < B - 1 \text{ (stop)}, \\ \mathrm{p} & \text{if } a_i + b_i = B - 1 \text{ (propagate)}, \\ \mathrm{g} & \text{if } a_i + b_i > B - 1 \text{ (generate)}. \end{cases} \qquad (1.1)$$

Lemma 1.3. *The operator* $\otimes$ *is associative, i.e.,* $(u \otimes v) \otimes w = u \otimes (v \otimes w)$ *for any* u, v, *and* w. *Let* $y_i = \bigotimes_{0 \leq j \leq i} x_j$. *Then* $c_i = 1$ *if and only if* $y_{i-1} = \mathrm{g}$.

Proof. If $u \neq \mathrm{p}$, then $(u \otimes v) \otimes w = u \otimes w = u = u \otimes (v \otimes w)$. If $u = \mathrm{p}$, then $(u \otimes v) \otimes w = v \otimes w = u \otimes (v \otimes w)$.

We have a carry into position i if and only if there is a $k < i$ such that a carry is generated in position k and propagated by positions $k+1$ to $i-1$, i.e., $x_k = \text{g}$ and $x_{k+1} = \ldots x_{i-1} = \text{p}$. Then $y_k = x_k = \text{g}$. This completes the argument when $i-1 = k$. Otherwise, $y_{k+1} = \text{p} \otimes y_k = \text{g}$, $y_{k+2} = \text{p} \otimes y_{k+1} = \text{g}$, $\ldots$, $y_{i-1} = \text{p} \otimes y_{i-2} = \text{g}$. □

We now come to the parallel computation of the y_i's. For simplicity, we assume n to be a power of 2. If $n = 1 = 2^0$, we simply return x_0. Otherwise, we do the following:

(a) Combine consecutive pairs of inputs: $z_0 = x_1 \otimes x_0$, $z_1 = x_3 \otimes x_2$, $\ldots$, $z_{n/2-1} = x_{n-1} \otimes x_{n-2}$.
(b) Apply the algorithm recursively to the sequence $z_{n/2-1}, \ldots, z_0$ to obtain its sequence of prefix sums $w_{n/2-1}, \ldots, w_0$.
(c) Assemble the output as $y_{2i+1} = w_i$ for $i \in 0..n/2 - 1$ and $y_0 = x_0$ and $y_{2i} = x_{2i} \otimes w_{i-1}$ for $i \in 1..n/2 - 1$.

Figure 1.1 illustrates the computation for $n = 8$. The number $N(n)$ of gates satisfies $N(1) = 0$ and $N(n) = n/2 + N(n/2) + n/2 - 1 \leq N(n/2) + n$. Note that we use $n/2$ gates in step (a), $n/2 - 1$ gates in step (c), and $N(n/2)$ gates in step (b). By repeated substitution, we obtain $N(n) \leq n + N(n/2) \leq n + n/2 + n/4 + n/8 + \ldots = O(n)$. The number of rounds of computation is 0 for $n = 1$, is 1 for $n = 2$, and grows by 2 whenever n doubles. Thus the number of rounds is $2 \log n - 1$.

The algorithm for computing the y_i's in parallel is an example of a *prefix sum* computation – a frequently useful collective computation described in more detail in Sect. 13.3.

1.2 Multiplication: The School Method

We all know how to multiply two integers. In this section, we shall review the "school method". In a later section, we shall get to know a method which is significantly faster for large integers.

We shall proceed slowly. We first review how to multiply an n-digit integer a by a one-digit integer b_j. We use b_j for the one-digit integer, since this is how we need it below. For any digit a_i of a, we form the product $a_i \cdot b_j$. The result is a two-digit integer $(c_i d_i)$, i.e.,

$$a_i \cdot b_j = c_i \cdot B + d_i.$$

We form two integers, $c = (c_{n-1} \ldots c_0 0)$ and $d = (d_{n-1} \ldots d_0)$ from the c_i's and d_i's, respectively. Since the c's are the higher-order digits in the products, we add a 0 digit at the end. We add c and d to obtain the product $p_j = a \cdot b_j$. Graphically,

$$
(a_{n-1} \ldots a_i \ldots a_0) \cdot b_j \quad \longrightarrow \quad
\frac{\begin{array}{l} c_{n-1}\ c_{n-2}\ \ldots\ c_i \quad c_{i-1}\ \ldots\ c_0\ 0 \\ \qquad\quad d_{n-1}\ \ldots\ d_{i+1}\ d_i \quad \ldots\ d_1\ d_0 \end{array}}{\text{sum of } c \text{ and } d}
$$

Let us determine the number of primitive operations. For each i, we need one primitive operation to form the product $a_i \cdot b_j$, for a total of n primitive operations. Then we add two $(n+1)$-digit numbers $(c_{n-1}\ldots c_0 0)$ and $(0 d_{n-1}\ldots d_0)$. However, we may simply copy the digit d_0 to the result and hence effectively add two n-digit numbers. This requires n primitive operations. So the total number of primitive operations is $2n$. The leftmost addition of c_{n-1} and the carry into this position generates carry 0 as $a \cdot b_j \leq (B^n - 1) \cdot (B - 1) < B^{n+1}$, and hence the result can be written with $n+1$ digits.

Lemma 1.4. *We can multiply an n-digit number by a one-digit number with $2n$ primitive operations. The result is an $(n + 1)$-digit number.*

When you multiply an n-digit number by a one-digit number, you will probably proceed slightly differently. You combine[5] the generation of the products $a_i \cdot b_j$ with the summation of c and d into a single phase, i.e., you create the digits of c and d when they are needed in the final addition. We have chosen to generate them in a separate phase because this simplifies the description of the algorithm.

Exercise 1.1. Give a program for the multiplication of a and b_j that operates in a single phase.

We can now turn to the multiplication of two n-digit integers. The school method for integer multiplication works as follows: We first form partial products p_j by multiplying a by the jth digit b_j of b, and then sum the suitably aligned products $p_j \cdot B^j$ to obtain the product of a and b. Graphically,

$$
\begin{array}{cccccccc}
 & & p_{0,n} & p_{0,n-1} & \cdots & p_{0,2} & p_{0,1} & p_{0,0} \\
 & p_{1,n} & p_{1,n-1} & p_{1,n-2} & \cdots & p_{1,1} & p_{1,0} & \\
p_{2,n} & p_{2,n-1} & p_{2,n-2} & p_{2,n-3} & \cdots & p_{2,0} & & \\
 & & & \cdots & & & & \\
p_{n-1,n} \cdots & p_{n-1,3} & p_{n-1,2} & p_{n-1,1} & p_{n-1,0} & & &
\end{array}
$$

sum of the n partial products

The description in pseudocode is more compact. We initialize the product p to 0 and then add to it the partial products $a \cdot b_j \cdot B^j$ one by one:

$p = 0 : \mathbb{N}$
for $j := 0$ **to** $n - 1$ **do** $p := p + a \cdot b_j \cdot B^j$

Let us analyze the number of primitive operations required by the school method. Each partial product p_j requires $2n$ primitive operations, and hence all partial products together require $2n^2$ primitive operations. The product $a \cdot b$ is a $2n$-digit number, and hence all summations $p + a \cdot b_j \cdot B^j$ are summations of $2n$-digit integers. Each such addition requires at most $2n$ primitive operations, and hence all additions together require at most $2n^2$ primitive operations. Thus, we need no more than $4n^2$ primitive operations in total.

---

[5] In the literature on compiler construction and performance optimization, this transformation is known as *loop fusion*.

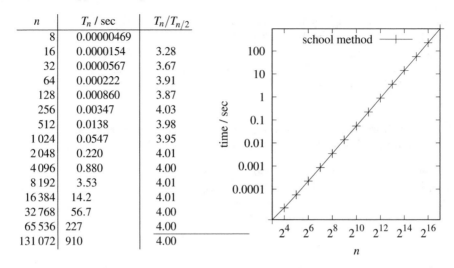

n	T_n / sec	$T_n/T_{n/2}$
8	0.00000469	
16	0.0000154	3.28
32	0.0000567	3.67
64	0.000222	3.91
128	0.000860	3.87
256	0.00347	4.03
512	0.0138	3.98
1 024	0.0547	3.95
2 048	0.220	4.01
4 096	0.880	4.00
8 192	3.53	4.01
16 384	14.2	4.01
32 768	56.7	4.00
65 536	227	4.00
131 072	910	4.00

Fig. 1.2. The running time of the school method for the multiplication of n-digit integers. The three columns of the table on the *left* give n, the running time T_n in seconds of the C++ implementation given in Sect. 1.8, and the ratio $T_n/T_{n/2}$. The plot on the *right* shows $\log T_n$ versus $\log n$, and we see essentially a line. Observe that if $T_n = \alpha n^\beta$ for some constants α and β, then $T_n/T_{n/2} = 2^\beta$ and $\log T_n = \beta \log n + \log \alpha$, i.e., $\log T_n$ depends linearly on $\log n$ with slope β. In our case, the slope is 2. Please use a ruler to check.

A simple observation allows us to improve this bound. The number $a \cdot b_j \cdot B^j$ has $n + 1 + j$ digits, the last j of which are 0. We can therefore start the addition in the $(j+1)$th position. Also, when we add $a \cdot b_j \cdot B^j$ to p, we have $p = a \cdot (b_{j-1} \ldots b_0)$, i.e., p has $n + j$ digits. Thus, the addition of p and $a \cdot b_j \cdot B^j$ amounts to the addition of two $n + 1$-digit numbers and requires only $n + 1$ primitive operations. Therefore, all $n - 1$ additions together require no more than $(n - 1)(n + 1) < n^2$ primitive operations. We have thus shown the following result.

Theorem 1.5. *The school method multiplies two n-digit integers with $3n^2$ primitive operations.*

We have now analyzed the numbers of primitive operations required by the school methods for integer addition and integer multiplication. The number M_n of primitive operations for the school method for integer multiplication is essentially $3n^2$. We say that M_n *grows quadratically*. Observe also that

$$\frac{M_n}{M_{n/2}} = \frac{3n^2}{3(n/2)^2} = 4,$$

i.e., quadratic growth has the consequence of essentially quadrupling the number of primitive operations when the size of the instance is doubled.

Assume now that we actually implement the multiplication algorithm in our favorite programming language (we shall do so later in the chapter), and then time the program on our favorite machine for various n-digit integers a and b and various n. What should we expect? We want to argue that we shall see quadratic growth. The reason is that *primitive operations are representative of the running time of the algorithm*. Consider the addition of two n-digit integers first. What happens when the program is executed? For each position i, the digits a_i and b_i have to be moved to the processing unit, the sum $a_i + b_i + c$ has to be formed, the digit s_i of the result needs to be stored in memory, the carry c is updated, the index i is incremented, and a test for loop exit needs to be performed. Thus, for each i, the same number of machine cycles is executed. We have counted one primitive operation for each i, and hence the number of primitive operations is representative of the number of machine cycles executed. Of course, there are additional effects, for example pipelining and the complex transport mechanism for data between memory and the processing unit, but they will have a similar effect for all i, and hence the number of primitive operations is also representative of the running time of an actual implementation on an actual machine. The argument extends to multiplication, since multiplication of a number by a one-digit number is a process similar to addition and the second phase of the school method for multiplication amounts to a series of additions.

Let us confirm the above argument by an experiment. Figure 1.2 shows execution times of a C++ implementation of the school method; the program can be found in Sect. 1.8. For each n, we performed a large number[6] of multiplications of n-digit random integers and then determined the average running time T_n; T_n is listed in the second column of the table. We also show the ratio $T_n/T_{n/2}$. Figure 1.2 also shows a plot of the data points[7] $(\log n, \log T_n)$. The data exhibits approximately quadratic growth, as we can deduce in various ways. The ratio $T_n/T_{n/2}$ is always close to four, and the double logarithmic plot shows essentially a line of slope 2. The experiments are quite encouraging: *Our theoretical analysis has predictive value. Our theoretical analysis showed quadratic growth of the number of primitive operations, we argued above that the running time should be related to the number of primitive operations, and the actual running time essentially grows quadratically.* However, we also see systematic deviations. For small n, the growth factor from one row to the next is by less than a factor of four, as linear and constant terms in the running time still play a substantial role. For larger n, the ratio is very close to four. For very large n (too large to be timed conveniently), we would probably see a factor larger than four, since the access time to memory depends on the size of the data. We shall come back to this point in Sect. 2.2.

Exercise 1.2. Write programs for the addition and multiplication of long integers. Represent integers as sequences (arrays or lists or whatever your programming language offers) of decimal digits and use the built-in arithmetic to implement the primitive operations. Then write ADD, MULTIPLY1, and MULTIPLY functions that add

---

[6] The internal clock that measures CPU time returns its timings in some units, say milliseconds, and hence the rounding required introduces an error of up to one-half of this unit. It is therefore important that the experiment whose duration is to be timed takes much longer than this unit, in order to reduce the effect of rounding.

[7] Throughout this book, we use $\log x$ to denote the logarithm to base 2, $\log_2 x$.

integers, multiply an integer by a one-digit number, and multiply integers, respectively. Use your implementation to produce your own version of Fig. 1.2. Experiment with using a larger base than 10, say base 2^{16}.

Exercise 1.3. Describe and analyze the school method for division.

1.3 Result Checking

Our algorithms for addition and multiplication are quite simple, and hence it is fair to assume that we can implement them correctly in the programming language of our choice. However, writing software[8] is an error-prone activity, and hence we should always ask ourselves whether we can check the results of a computation. For multiplication, the authors were taught the following technique in elementary school. The method is known as *Neunerprobe* in German, "casting out nines" in English, and *preuve par neuf* in French.

Add the digits of a. If the sum is a number with more than one digit, sum its digits. Repeat until you arrive at a one-digit number, called the checksum of a. We use s_a to denote this checksum. Here is an example:

$$4528 \rightarrow 19 \rightarrow 10 \rightarrow 1.$$

Do the same for b and the result c of the computation. This gives the checksums s_b and s_c. All checksums are single-digit numbers. Compute $s_a \cdot s_b$ and form its checksum s. If s differs from s_c, c is not equal to $a \cdot b$. This test was described by al-Khwarizmi in his book on algebra.

Let us go through a simple example. Let $a = 429$, $b = 357$, and $c = 154153$. Then $s_a = 6$, $s_b = 6$, and $s_c = 1$. Also, $s_a \cdot s_b = 36$ and hence $s = 9$. So $s_c \neq s$ and hence c is not the product of a and b. Indeed, the correct product is $c = 153153$. Its checksum is 9, and hence the correct product passes the test. The test is not foolproof, as $c = 135153$ also passes the test. However, the test is quite useful and detects many mistakes.

What is the mathematics behind this test? We shall explain a more general method. Let q be any positive integer; in the method described above, $q = 9$. Let s_a be the remainder, or residue, in the integer division of a by q, i.e., $s_a = a - \lfloor a/q \rfloor \cdot q$. Then $0 \leq s_a < q$. In mathematical notation, $s_a = a \bmod q$.[9] Similarly, $s_b = b \bmod q$ and $s_c = c \bmod q$. Finally, $s = (s_a \cdot s_b) \bmod q$. If $c = a \cdot b$, then it must be the case that $s = s_c$. Thus $s \neq s_c$ proves $c \neq a \cdot b$ and uncovers a mistake in the multiplication (or a mistake in carrying out casting out nines). What do we know if $s = s_c$? We know that q divides the difference of c and $a \cdot b$. If this difference is nonzero, the mistake will be detected by any q which does not divide the difference.

---

[8] The bug in the division algorithm of the floating-point unit of the original Pentium chip became infamous. It was caused by a few missing entries in a lookup table used by the algorithm.

[9] The method taught in school uses residues in the range 1 to 9 instead of 0 to 8 according to the definition $s_a = a - (\lceil a/q \rceil - 1) \cdot q$.

Let us continue with our example and take $q = 7$. Then $a \bmod 7 = 2$, $b \bmod 7 = 0$ and hence $s = (2 \cdot 0) \bmod 7 = 0$. But $135153 \bmod 7 = 4$, and we have uncovered the fact that $135153 \neq 429 \cdot 357$.

Exercise 1.4. Explain why casting out nines corresponds to the case $q = 9$. Hint: $10^k \bmod 9 = 1$ for all $k \geq 0$.

Exercise 1.5 (*Elferprobe*, **casting out elevens**)**.** Powers of ten have very simple remainders modulo 11, namely $10^k \bmod 11 = (-1)^k$ for all $k \geq 0$, i.e., $1 \bmod 11 = +1$, $10 \bmod 11 = -1$, $100 \bmod 11 = +1$, $1\,000 \bmod 11 = -1$, etc. Describe a simple test to check the correctness of a multiplication modulo 11.

1.4 A Recursive Version of the School Method

We shall now derive a recursive version of the school method. This will be our first encounter with the *divide-and-conquer* paradigm, one of the fundamental paradigms in algorithm design.

Let a and b be our two n-digit integers which we want to multiply. Let $k = \lfloor n/2 \rfloor$. We split a into two numbers a_1 and a_0; a_0 consists of the k least significant digits and a_1 consists of the $n - k$ most significant digits.[10] We split b analogously. Then

$$a = a_1 \cdot B^k + a_0 \quad \text{and} \quad b = b_1 \cdot B^k + b_0,$$

and hence

$$a \cdot b = a_1 \cdot b_1 \cdot B^{2k} + (a_1 \cdot b_0 + a_0 \cdot b_1) \cdot B^k + a_0 \cdot b_0.$$

This formula suggests the following algorithm for computing $a \cdot b$:

(a) Split a and b into a_1, a_0, b_1, and b_0.
(b) Compute the four products $a_1 \cdot b_1$, $a_1 \cdot b_0$, $a_0 \cdot b_1$, and $a_0 \cdot b_0$.
(c) Add the suitably aligned products to obtain $a \cdot b$.

Observe that the numbers a_1, a_0, b_1, and b_0 are $\lceil n/2 \rceil$-digit numbers and hence the multiplications in step (b) are simpler than the original multiplication if $\lceil n/2 \rceil < n$, i.e., $n > 1$. The complete algorithm is now as follows. To multiply one-digit numbers, use the multiplication primitive. To multiply n-digit numbers for $n \geq 2$, use the three-step approach above.

It is clear why this approach is called *divide-and-conquer*. We reduce the problem of multiplying a and b to some number of *simpler* problems of the same kind. A divide-and-conquer algorithm always consists of three parts: In the first part, we split the original problem into simpler problems of the same kind (our step (a)); in the second part, we solve the simpler problems using the same method (our step (b)); and, in the third part, we obtain the solution to the original problem from the solutions to the subproblems (our step (c)). Instead of "we solve the simpler problems using

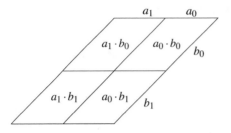

Fig. 1.3. Visualization of the school method and its recursive variant. The rhombus-shaped area indicates the partial products in the multiplication $a \cdot b$. The four subareas correspond to the partial products $a_1 \cdot b_1$, $a_1 \cdot b_0$, $a_0 \cdot b_1$, and $a_0 \cdot b_0$. In the recursive scheme, we first sum the partial products in the four subareas and then, in a second step, add the four resulting sums.

the same method", one usually says more elegantly "we solve the simpler problems recursively".

What is the connection of our recursive integer multiplication to the school method? The two methods are strongly related. Both methods compute all n^2 digit products $a_i b_j$ and then sum the resulting two-digit results (appropriately shifted). They differ in the order in which these summations are performed. Figure 1.3 visualizes the digit products and their place values as rhombus-shaped regions. The school method adds the digit products row by row. The recursive method first computes the partial results corresponding to the four subregions, which it then combines to obtain the final result with three additions. We may almost say that our recursive integer multiplication is just the school method in disguise. In particular, the recursive algorithm also uses a quadratic number of primitive operations.

We next derive the quadratic behavior without appealing to the connection to the school method. This will allow us to introduce recurrence relations, a powerful concept for the analysis of recursive algorithms.

Lemma 1.6. *Let $T(n)$ be the maximum number of primitive operations required by our recursive multiplication algorithm when applied to n-digit integers. Then*

$$T(n) \leq \begin{cases} 1 & \text{if } n = 1, \\ 4 \cdot T(\lceil n/2 \rceil) + 2 \cdot 2 \cdot n & \text{if } n \geq 2. \end{cases}$$

Proof. Multiplying two one-digit numbers requires one primitive multiplication. This justifies the case $n = 1$. So, assume $n \geq 2$. Splitting a and b into the four pieces a_1, a_0, b_1, and b_0 requires no primitive operations.[11] Each piece has at most $\lceil n/2 \rceil$ digits and hence the four recursive multiplications require at most $4 \cdot T(\lceil n/2 \rceil)$ primitive operations. The products $a_0 \cdot b_0$ and $a_1 \cdot b_1 \cdot B^{2k}$ can be combined into a single number without any cost as the former number is a $2k$-digit number and the latter number ends with $2k$ many 0's. Finally, we need two additions to assemble the final result. Each addition involves two numbers of at most $2n$ digits and hence requires at most $2n$ primitive operations. This justifies the inequality for $n \geq 2$. □

[10] Observe that we have changed notation; a_0 and a_1 now denote the two parts of a and are no longer single digits.

[11] It will require work, but it is work that we do not account for in our analysis.

In Sect. 2.8, we shall learn that such recurrences are easy to solve and yield the already conjectured quadratic execution time of the recursive algorithm.

Lemma 1.7. *Let $T(n)$ be the maximum number of primitive operations required by our recursive multiplication algorithm when applied to n-digit integers. Then $T(n) \leq 5n^2$ if n is a power of 2, and $T(n) \leq 20n^2$ for all n.*

Proof. We refer the reader to Sect. 1.9 for a proof. □

1.5 Karatsuba Multiplication

In 1962, the Soviet mathematician Karatsuba [175] discovered a faster way of multiplying large integers. The running time of his algorithm grows like $n^{\log 3} \approx n^{1.58}$. The method is surprisingly simple. Karatsuba observed that a simple algebraic identity allows one multiplication to be eliminated in the divide-and-conquer implementation, i.e., one can multiply n-digit numbers using only *three* multiplications of integers half the size.

The details are as follows. Let a and b be our two n-digit integers which we want to multiply. Let $k = \lfloor n/2 \rfloor$. As above, we split a into two numbers a_1 and a_0; a_0 consists of the k least significant digits and a_1 consists of the $n-k$ most significant digits. We split b in the same way. Then

$$a = a_1 \cdot B^k + a_0 \quad \text{and} \quad b = b_1 \cdot B^k + b_0$$

and hence (the magic is in the second equality)

$$\begin{aligned} a \cdot b &= a_1 \cdot b_1 \cdot B^{2k} + (a_1 \cdot b_0 + a_0 \cdot b_1) \cdot B^k + a_0 \cdot b_0 \\ &= a_1 \cdot b_1 \cdot B^{2k} + ((a_1 + a_0) \cdot (b_1 + b_0) - (a_1 \cdot b_1 + a_0 \cdot b_0)) \cdot B^k + a_0 \cdot b_0. \end{aligned}$$

At first sight, we have only made things more complicated. A second look, however, shows that the last formula can be evaluated with only three multiplications, namely, $a_1 \cdot b_1$, $a_0 \cdot b_0$, and $(a_1 + a_0) \cdot (b_1 + b_0)$. We also need six additions.[12] That is three more than in the recursive implementation of the school method. The key is that additions are cheap compared with multiplications, and hence saving a multiplication more than outweighs the additional additions. We obtain the following algorithm for computing $a \cdot b$:

(a) Split a and b into a_1, a_0, b_1, and b_0.
(b) Compute the three products

$$p_2 = a_1 \cdot b_1, \quad p_0 = a_0 \cdot b_0, \quad p_1 = (a_1 + a_0) \cdot (b_1 + b_0).$$

---

[12] Actually, five additions and one subtraction. We leave it to readers to convince themselves that subtractions are no harder than additions.

(c) Add the suitably aligned products to obtain $a \cdot b$, i.e., compute $a \cdot b$ according to the formula

$$a \cdot b = (p_2 \cdot B^{2k} + p_0) + (p_1 - (p_2 + p_0)) \cdot B^k.$$

The first addition can be performed by concatenating the corresponding digit strings and requires no primitive operation.

The numbers a_1, a_0, b_1, b_0, $a_1 + a_0$, and $b_1 + b_0$ are $(\lceil n/2 \rceil + 1)$-digit numbers and hence the multiplications in step (b) are simpler than the original multiplication if $\lceil n/2 \rceil + 1 < n$, i.e., $n \geq 4$. The complete algorithm is now as follows: To multiply three-digit numbers, use the school method, and to multiply n-digit numbers for $n \geq 4$, use the three-step approach above.

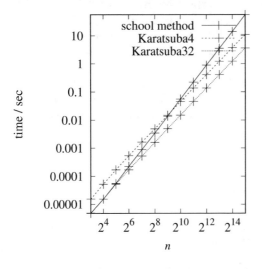

Fig. 1.4. The running times of implementations of the Karatsuba and school methods for integer multiplication. The running times of two versions of Karatsuba's method are shown: Karatsuba4 switches to the school method for integers with fewer than four digits, and Karatsuba32 switches to the school method for integers with fewer than 32 digits. The slopes of the lines for the Karatsuba variants are approximately 1.58. The running time of Karatsuba32 is approximately one-third the running time of Karatsuba4.

Figure 1.4 shows the running times $T_S(n)$, $T_{K4}(n)$, and $T_{K32}(n)$ of C++ implementations of the school method and of two variants of the Karatsuba method for the multiplication of n-digit numbers. Karatsuba4 (running time $T_{K4}(n)$) uses the school method for numbers with fewer than four digits and Karatsuba32 (running time $T_{K32}(n)$) uses the school method for numbers with fewer than 32 digits; we discuss the rationale for this variant in Sect. 1.7. The scales on both axes are logarithmic. We see, essentially, straight lines of different slope. The running time of the school method grows like n^2, and hence the slope is 2 in the case of the school method. The slope is smaller in the case of the Karatsuba method, and this suggests that its running time grows like n^β with $\beta < 2$. In fact, the ratios[13] $T_{K4}(n)/T_{K4}(n/2)$ and $T_{K32}(n)/T_{K32}(n/2)$ are close to three, and this suggests that β is such that $2^\beta = 3$ or $\beta = \log 3 \approx 1.58$. Alternatively, you may determine the slope from Fig. 1.4. We shall prove below that $T_K(n)$ grows like $n^{\log 3}$. We say that *the Karatsuba method has*

---

[13] $T_{K4}(1024) = 0.0455$, $T_{K4}(2048) = 0.1375$, and $T_{K4}(4096) = 0.41$.

better asymptotic behavior than the school method. We also see that the inputs have to be quite big before the superior asymptotic behavior of the Karatsuba method actually results in a smaller running time. Observe that for $n = 2^8$, the school method is still faster, that for $n = 2^9$, the two methods have about the same running time, and that the Karatsuba method wins for $n = 2^{10}$. The lessons to remember are:

* Better asymptotic behavior ultimately wins.
* An asymptotically slower algorithm can be faster on small inputs.

In the next section, we shall learn how to improve the behavior of the Karatsuba method for small inputs. The resulting algorithm will always be at least as good as the school method. It is time to derive the asymptotics of the Karatsuba method.

Lemma 1.8. *Let $T_K(n)$ be the maximum number of primitive operations required by the Karatsuba algorithm when applied to n-digit integers. Then*

$$T_K(n) \leq \begin{cases} 3n^2 & \text{if } n \leq 3, \\ 3 \cdot T_K(\lceil n/2 \rceil + 1) + 8n & \text{if } n \geq 4. \end{cases}$$

Proof. Multiplying two n-digit numbers using the school method requires no more than $3n^2$ primitive operations, according to Theorem 1.5. This justifies the first line. So, assume $n \geq 4$. Splitting a and b into the four pieces a_1, a_0, b_1, and b_0 requires no primitive operations.[14] Each piece and the sums $a_0 + a_1$ and $b_0 + b_1$ have at most $\lceil n/2 \rceil + 1$ digits, and hence the three recursive multiplications require at most $3 \cdot T_K(\lceil n/2 \rceil + 1)$ primitive operations. We need two additions to form $a_0 + a_1$ and $b_0 + b_1$. The results of these additions have fewer than n digits and hence the additions need no more than n elementary operations each. Finally, we need three additions in order to compute the final result from the results of the multiplications. These are additions of numbers with at most $2n$ digits. Thus these additions require at most $3 \cdot 2n$ primitive operations. Altogether, we obtain the bound stated in the second line of the recurrence.   □

In Sect. 2.8, we shall learn some general techniques for solving recurrences of this kind.

Theorem 1.9. *Let $T_K(n)$ be the maximum number of primitive operations required by the Karatsuba algorithm when applied to n-digit integers. Then $T_K(n) \leq 153 n^{\log 3}$ for all n.*

Proof. We refer the reader to Sect. 1.9 for a proof.   □

1.6 Parallel Multiplication

Both the recursive version of the school method and the Karatsuba algorithm are good starting points for parallel algorithms. For simplicity, we focus on the school

---

[14] It will require work, but remember that we are counting primitive operations.

method. Recall that the bulk of the work is done in the recursive multiplications a_0b_0, a_1b_1, a_0b_1, and a_1b_0. These four multiplications can be done independently and in parallel. Hence, in the first level of recursion, up to four processors can work in parallel. In the second level of recursion, the parallelism is already $4 \cdot 4 = 16$. In the ith level of recursion, 4^i processors can work in parallel. Fig. 1.5 shows a graphical representation of the resulting computation for multiplying two numbers $a_{11}a_{10}a_{01}a_{00}$ and $b_{11}b_{10}b_{01}b_{00}$.

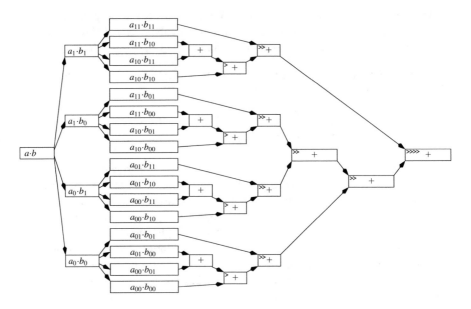

Fig. 1.5. Task graph for parallel recursive multiplication with two levels of parallel recursion and using the school method on two-digit numbers as the base case. A ">" stands for a shift by two digits.

What is the running time of the algorithm if an unlimited number of processors are available? This quantity is known as the *span* of a parallel computation; see Sect. 2.10. For simplicity, we assume a sequential addition algorithm. As before, we shall count only arithmetic operations on digits. For the span $S(n)$ for multiplying two n-digit integers, we get

$$S(n) \le \begin{cases} 1 & \text{if } n = 1, \\ S(\lceil n/2 \rceil) + 3 \cdot 2 \cdot n & \text{if } n \ge 2. \end{cases}$$

This recurrence has the solution $S(n) \le 12n$. Note that this is much less than the quadratic number of operations performed. Hence, we can hope for considerable speedup by parallel processing – even without parallel addition.

Exercise 1.6. Show that by using parallel addition (see Sect. 1.1.1), we can achieve a span $O(\log^2 n)$ for parallel school multiplication.

Parallel programming environments for multicore processors make it relatively easy to exploit parallelism in the kind of divide-and-conquer computations described above, see Sect. 2.14 for details. Roughly, they create a *task* for recursive calls and automatically assign cores to tasks. They ensure that enough tasks are created to keep all available cores busy, but also that tasks stay reasonably large. Section 14.5 explains the *load-balancing* algorithms behind this programming model.

Exercise 1.7. Describe a parallel divide-and-conquer algorithm for the Karatsuba method. Show that its span is linear in the number of input digits.

1.7 Algorithm Engineering

Karatsuba integer multiplication is superior to the school method for large inputs. In our implementation, the superiority only shows for integers with more than 1000 digits. However, a simple refinement improves the performance significantly. Since the school method is superior to the Karatsuba method for short integers, we should stop the recursion earlier and switch to the school method for numbers which have fewer than n_0 digits for some yet to be determined n_0. We call this approach the *refined Karatsuba method*. It is never worse than either the school method or the original Karatsuba algorithm as long as n_0 is not chosen too large.

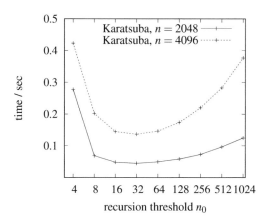

Fig. 1.6. The running time of the Karatsuba method as a function of the recursion threshold n_0. The times consumed for multiplying 2048-digit and 4096-digit integers are shown. The minimum is at $n_0 = 32$.

What is a good choice for n_0? We shall answer this question both experimentally and analytically. Let us discuss the experimental approach first. We simply time the refined Karatsuba algorithm for different values of n_0 and then adopt the value giving the smallest running time. For our implementation, the best results were obtained for $n_0 = 32$ (see Fig. 1.6). The asymptotic behavior of the refined Karatsuba method is shown in Fig. 1.4. We see that the running time of the refined method still grows like $n^{\log 3}$, that the refined method is about three times faster than the basic Karatsuba method and hence the refinement is highly effective, and that the refined method is never slower than the school method.

Exercise 1.8. Derive a recurrence for the worst-case number $T_R(n)$ of primitive operations performed by the refined Karatsuba method.

We can also approach the question analytically. If we use the school method to multiply n-digit numbers, we need $3n^2$ primitive operations. If we use one Karatsuba step and then multiply the resulting numbers of length $\lceil n/2 \rceil + 1$ using the school method, we need about $3(3(n/2 + 1)^2) + 7n$ primitive operations. The latter is smaller for $n \geq 23$, and hence a recursive step saves primitive operations as long as the number of digits is more than 23. You should not take this as an indication that an actual implementation should switch at integers of approximately 23 digits, as the argument concentrates solely on primitive operations. You should take it as an argument that it is wise to have a nontrivial recursion threshold n_0 and then determine the threshold experimentally.

Exercise 1.9. Throughout this chapter, we have assumed that both arguments of a multiplication are n-digit integers. What can you say about the complexity of multiplying n-digit and m-digit integers? (a) Show that the school method requires no more than $\alpha \cdot nm$ primitive operations for some constant α. (b) Assume $n \geq m$ and divide a into $\lceil n/m \rceil$ numbers of m digits each. Multiply each of the fragments by b using Karatsuba's method and combine the results. What is the running time of this approach?

1.8 The Programs

We give C++ programs for the school and Karatsuba methods below. These programs were used for the timing experiments described in this chapter. The programs were executed on a machine with a 2 GHz dual-core Intel Core 2 Duo T7200 processor with 4 MB of cache memory and 2 GB of main memory. The programs were compiled with GNU C++ version 3.3.5 using optimization level −O2.

A digit is simply an unsigned int and an integer is a vector of digits; here, "vector" is the vector type of the standard template library. A declaration *integer a*(*n*) declares an integer with n digits, *a.size*() returns the size of a, and $a[i]$ returns a reference to the ith digit of a. Digits are numbered starting at 0. The global variable B stores the base. The functions *fullAdder* and *digitMult* implement the primitive operations on digits. We sometimes need to access digits beyond the size of an integer; the function *getDigit*(*a*, *i*) returns $a[i]$ if i is a legal index for a and returns 0 otherwise:

```
typedef unsigned int digit;
typedef vector<digit> integer;
unsigned int B = 10;                    // Base, 2 <= B <= 2^16

void fullAdder( digit a, digit b, digit c, digit & s, digit & carry)
{ unsigned int sum = a + b + c; carry = sum/B; s = sum - carry*B; }
```

```
void digitMult ( digit  a,  digit  b,  digit & s,  digit & carry)
{ unsigned int prod = a*b; carry = prod/B; s = prod - carry*B; }
```

```
digit  getDigit (const integer& a, int  i)
{ return ( i < a.size()? a[i] : 0 ); }
```

We want to run our programs on random integers: *randDigit* is a simple random
generator for digits, and *randInteger* fills its argument with random digits.

```
unsigned int X = 542351;
```

```
digit  randDigit () { X = 443143*X + 6412431; return X % B ; }
```

```
void randInteger(integer& a)
{ int n = a.size (); for (int i=0; i<n; i++) a[i] = randDigit ();}
```

We now come to the school method of multiplication. We start with a routine that
multiplies an integer *a* by a digit *b* and returns the result in *atimesb*. We need a carry-
digit *carry*, which we initialize to 0. In each iteration, we compute *d* and *c* such that
$c * B + d = a[i] * b$. We then add *d*, the *c* from the previous iteration, and the *carry*
from the previous iteration, store the result in *atimesb*[i], and remember the *carry*.
The school method (the function *mult*) multiplies *a* by each digit of *b* and then adds
it at the appropriate position to the partial result (the function *addAt*):

```
void mult(const integer& a, const digit& b, integer& atimesb)
{ int n = a.size (); assert(atimesb.size() == n+1);
  digit  carry = 0, c, d, cprev = 0;
  for (int i = 0; i < n; i++)
  { digitMult (a[i],b,d,c);
    fullAdder (d, cprev, carry, atimesb[i], carry);
    cprev = c;
  }
  d = 0;
  fullAdder (d, cprev, carry, atimesb[n], carry);  assert(carry == 0);
}
```

```
void addAt(integer& p, const integer& atimesbj, int  j)
{ // p has length n+m,
  digit  carry = 0; int L = p.size ();
  for (int i = j; i < L; i++)
    fullAdder (p[i], getDigit (atimesbj, i-j), carry, p[i], carry);
  assert(carry == 0);
}
```

```
integer  mult(const integer& a, const integer& b)
{ int n = a.size (); int m = b.size ();
  integer p(n + m,0); integer atimesbj(n+1);
  for (int j = 0; j < m; j++)
  { mult(a, b[j], atimesbj); addAt(p, atimesbj, j); }
  return p;
}
```

For Karatsuba's method, we also need algorithms for general addition and subtraction. The subtraction method may assume that the first argument is no smaller than the second. It computes its result in the first argument:

```
integer add(const integer& a, const integer& b)
{ int  n = max(a.size(),b.size ());
  integer s(n+1); digit  carry = 0;
  for (int i = 0; i < n; i++)
    fullAdder (getDigit (a,i ),  getDigit (b,i ),  carry,  s[i ],  carry);
  s[n] = carry;
  return s;
}
void sub(integer& a, const integer& b) //  requires  a >= b
{ digit  carry = 0;
  for (int i = 0; i < a.size (); i++)
    if ( a[i] >= ( getDigit (b,i ) + carry ))
      { a[i] = a[i] - getDigit (b,i ) - carry; carry = 0; }
    else { a[i] = a[i] + B - getDigit (b,i ) - carry; carry = 1;}
  assert(carry == 0);
}
```

The function *split* splits an integer into two integers of half the size:

```
void split (const integer& a,integer& a1,  integer& a0)
{ int n = a.size (); int  k = n/2;
  for (int i = 0; i < k; i++) a0[i] = a[i ];
  for (int i = 0; i < n - k; i++) a1[i] = a[k+ i ];
}
```

The function *Karatsuba* works exactly as described in the text. If the inputs have fewer than $n0$ digits, the school method is employed. Otherwise, the inputs are split into numbers of half the size and the products $p0$, $p1$, and $p2$ are formed. Then $p0$ and $p2$ are written into the output vector and subtracted from $p1$. Finally, the modified $p1$ is added to the result:

```
integer Karatsuba(const integer& a, const integer& b, int n0)
{ int n = a.size (); int  m = b.size (); assert(n == m); assert(n0 >= 4);
  integer p(2*n);
  if (n < n0) return mult(a,b);
  int k = n/2; integer a0(k), a1(n - k), b0(k), b1(n - k);
  split (a,a1,a0); split (b,b1,b0);
  integer p2 = Karatsuba(a1,b1,n0),
          p1 = Karatsuba(add(a1,a0),add(b1,b0),n0),
          p0 = Karatsuba(a0,b0,n0);
  for (int i = 0; i < 2*k; i++) p[i] = p0[i ];
  for (int i = 2*k; i < n+m; i++) p[i] = p2[i - 2*k];
  sub(p1,p0); sub(p1,p2); addAt(p,p1,k);
  return p;
}
```

The following program generated the data for Fig. 1.4:

```
inline double cpuTime() { return double(clock())/CLOCKS_PER_SEC; }

int main(){

for (int n = 8; n <= 131072; n *= 2)
{ integer a(n), b(n); randInteger(a); randInteger(b);

    double T = cpuTime(); int k = 0;
    while (cpuTime() - T < 1) {   mult(a,b); k++; }
    cout << "\n" << n << "␣school␣=␣" << (cpuTime() -T)/k;

    T = cpuTime(); k = 0;
    while (cpuTime() - T < 1) {   Karatsuba(a,b,4); k++; }
    cout << "␣Karatsuba4␣=␣" << (cpuTime() -T) /k; cout.flush ();

    T = cpuTime(); k = 0;
    while (cpuTime() - T < 1) {   Karatsuba(a,b,32); k++; }
    cout << "␣Karatsuba32␣=␣" << (cpuTime() -T) /k; cout.flush ();
}
return 0;
}
```

1.9 Proofs of Lemma 1.7 and Theorem 1.9

To make this chapter self-contained, we include proofs of Lemma 1.7 and Theorem 1.9. We start with an analysis of the recursive version of the school method. Recall that $T(n)$, the maximum number of primitive operations required by our recursive multiplication algorithm when applied to n-digit integers, satisfies the recurrence relation

$$T(n) \leq \begin{cases} 1 & \text{if } n = 1, \\ 4 \cdot T(\lceil n/2 \rceil) + 4n & \text{if } n \geq 2. \end{cases}$$

We use induction on n to show $T(n) \leq 5n^2 - 4n$ when n is a power of 2. For $n = 1$, we have $T(1) \leq 1 = 5n^2 - 4n$. For $n > 1$, we have

$$T(n) \leq 4T(n/2) + 4n \leq 4(5(n/2)^2 - 4n/2) + 4n = 5n^2 - 4n,$$

where the second inequality follows from the induction hypothesis. For general n, we observe that multiplying n-digit integers is certainly no more costly than multiplying $2^{\lceil \log n \rceil}$-digit integers and hence $T(n) \leq T(2^{\lceil \log n \rceil})$. Since $2^{\lceil \log n \rceil} \leq 2n$, we conclude $T(n) \leq 20n^2$ for all n.

Exercise 1.10. Prove a bound on the recurrence $T(1) \leq 1$ and $T(n) \leq 4T(n/2) + 9n$ when n is a power of 2.

How did we know that "$5n^2 - 4n$" is the bound to be shown? There is no magic here. For $n = 2^k$, repeated substitution yields

$$T(2^k) \le 4 \cdot T(2^{k-1}) + 4 \cdot 2^k \le 4 \cdot (4 \cdot T^{k-2} + 4 \cdot 2^{k-1}) + 4 \cdot 2^k$$
$$\le 4 \cdot (4 \cdot (4 \cdot T(2^{k-3}) + 4 \cdot 2^{k-2}) + 4 \cdot 2^{k-1}) + 4 \cdot 2^k$$
$$\le 4^3 T(2^{k-3}) + 4 \cdot (4^2 \cdot 2^{k-2} + 4^1 \cdot 2^{k-1} + 2^k) \le \cdots$$
$$\le 4^k T(1) + 4 \sum_{0 \le i \le k-1} 4^i 2^{k-i} \le 4^k + 4 \cdot 2^k \sum_{0 \le i \le k-1} 2^i$$
$$\le 4^k + 4 \cdot 2^k (2^k - 1) = n^2 + 4n(n-1) = 5n^2 - 4n.$$

We now turn to the proof of Theorem 1.9. Recall that T_K satisfies the recurrence

$$T_K(n) \le \begin{cases} 3n^2 & \text{if } n \le 3, \\ 3 \cdot T_K(\lceil n/2 \rceil + 1) + 8n & \text{if } n \ge 4. \end{cases}$$

The recurrence for the school method has the nice property that if n is a power of 2, the arguments of T on the right-hand side are again powers of two. This is not true for T_K. However, if $n = 2^k + 2$ and $k \ge 1$, then $\lceil n/2 \rceil + 1 = 2^{k-1} + 2$, and hence we should now use numbers of the form $n = 2^k + 2$, $k \ge 0$, as the basis of the inductive argument. We shall show that

$$T_K(2^k + 2) \le 51 \cdot 3^k - 16 \cdot 2^k - 8$$

for $k \ge 0$. For $k = 0$, we have

$$T_K(2^0 + 2) = T_K(3) \le 3 \cdot 3^2 = 27 = 51 \cdot 3^0 - 16 \cdot 2^0 - 8.$$

For $k \ge 1$, we have

$$T_K(2^k + 2) \le 3 T_K(2^{k-1} + 2) + 8 \cdot (2^k + 2)$$
$$\le 3 \cdot \left(51 \cdot 3^{k-1} - 16 \cdot 2^{k-1} - 8\right) + 8 \cdot (2^k + 2)$$
$$= 51 \cdot 3^k - 16 \cdot 2^k - 8.$$

Again, there is no magic in coming up with the right induction hypothesis. It is obtained by repeated substitution. Namely,

$$T_K(2^k + 2) \le 3 T_K(2^{k-1} + 2) + 8 \cdot (2^k + 2)$$
$$\le 3^k T_K(2^0 + 2) + 8 \cdot \left(3^0(2^k + 2) + 3^1(2^{k-1} + 2) + \ldots + 3^{k-1}(2^1 + 2)\right)$$
$$\le 27 \cdot 3^k + 8 \cdot \left(2^k \frac{(3/2)^k - 1}{3/2 - 1} + 2 \frac{3^k - 1}{3 - 1}\right)$$
$$\le 51 \cdot 3^k - 16 \cdot 2^k - 8,$$

where the first inequality uses the fact that $2^k + 2$ is even, the second inequality follows from repeated substitution, the third inequality uses $T_K(3) = 27$, and the last inequality follows by a simple computation.

It remains to extend the bound to all n. Let k be the minimum integer such that $n \leq 2^k + 2$. Then $k \leq 1 + \log n$. Also, multiplying n-digit numbers is no more costly than multiplying $(2^k + 2)$-digit numbers, and hence

$$T_K(n) \leq 51 \cdot 3^k - 16 \cdot 2^k - 8 \leq 153 \cdot 3^{\log n} \leq 153 \cdot n^{\log 3},$$

where we have used the equality $3^{\log n} = 2^{(\log 3) \cdot (\log n)} = n^{\log 3}$.

Exercise 1.11. Solve the recurrence

$$T_R(n) \leq \begin{cases} 3n^2 + 2n & \text{if } n < n_0, \\ 3 \cdot T_R(\lceil n/2 \rceil + 1) + 8n & \text{if } n \geq n_0, \end{cases}$$

where n_0 is a positive integer. Optimize n_0.

1.10 Implementation Notes

The programs given in Sect. 1.8 are not optimized. The base of the number system should be a power of 2 so that sums and carries can be extracted by bit operations. Also, the size of a digit should agree with the word size of the machine and a little more work should be invested in implementing the primitive operations on digits.

1.10.1 C++

GMP [127] and LEDA [195] offer exact arithmetic on integers and rational numbers of arbitrary size, and arbitrary-precision floating-point arithmetic. Highly optimized implementations of Karatsuba's method are used for multiplication here.

1.10.2 Java

java.math implements exact arithmetic on integers of arbitrary size and arbitrary-precision floating-point numbers.

1.11 Historical Notes and Further Findings

Is the Karatsuba method the fastest known method for integer multiplication? No, much faster methods are known. Karatsuba's method splits an integer into two parts and requires three multiplications of integers of half the length. The natural extension is to split integers into k parts of length n/k each. If the recursive step requires ℓ multiplications of numbers of length n/k, the running time of the resulting algorithm grows like $n^{\log_k \ell}$. In this way, Toom [316] and Cook [78] reduced the running time to[15] $O(n^{1+\varepsilon})$ for arbitrary positive ε. Asymptotically even more efficient are the

---

[15] The $O(\cdot)$ notation is defined in Sect. 2.1.

algorithms of Schönhage and Strassen [286] and Schönhage [285]. The former multiplies n-bit integers with $O(n \log n \log \log n)$ bit operations, and can be implemented to run in this time bound on a Turing machine. The latter runs in linear time $O(n)$ and requires the machine model discussed in Sect. 2.2. In this model, integers with $\log n$ bits can be multiplied in constant time. The former algorithm was improved by Fürer [117] and De et al. [84] to $O((n \log n) 2^{c \log^*(n)})$ bit operations. Here, $\log^* n$ is the smallest integer $k \geq 0$ such that $\log(\log(\ldots \log(n) \ldots)) \leq 1$ (k-fold application of the logarithm function). The function $\log^* n$ grows extremely slowly. In March 2019 an algorithm for integer multiplication was announced that requires $O(n \log n)$ bit operations [146]. This bound is widely believed to be best possible.

Modern microprocessors use a base $B = 2$ multiplication algorithm that does quadratic work but requires only $O(\log n)$ wire delays [149]. The basic idea is to first compute n^2 digit products in parallel. Bits with the same position in the output are combined together using full adders, which reduce three values at position i to one value at position i and one value at position $i + 1$. This way, $\log_{3/2} n$ layers of full adders suffice to reduce up to n initial bits at position i to two bits. The remaining bits can be fed into an adder with logarithmic delay as in Sect. 1.1.1.

2

Introduction

When you want to become a sculptor,[1] you have to learn some basic techniques: where to get the right stones, how to move them, how to handle the chisel, how to erect scaffolding, … . Knowing these techniques will not make you a famous artist, but even if you have an exceptional talent, it will be very difficult to develop into a successful artist without knowing them. It is not necessary to master all of the basic techniques before sculpting the first piece. But you always have to be willing to go back to improve your basic techniques.

This introductory chapter plays a similar role in this book. We introduce basic concepts that make it simpler to discuss and analyze algorithms in the subsequent chapters. There is no need for you to read this chapter from beginning to end before you proceed to later chapters. You can also skip the parts on parallel processing when you are only considering sequential algorithms. On the first reading, we recommend that you should read carefully to the end of Sect. 2.3 and skim through the remaining sections. We begin in Sect. 2.1 by introducing some notation and terminology that allow us to argue about the complexity of algorithms in a concise way. We then introduce machine models in Sect. 2.2 that allow us to abstract from the highly variable complications introduced by real hardware. The models are concrete enough to have predictive value and abstract enough to permit elegant arguments. Section 2.3 then introduces a high-level pseudocode notation for algorithms that is much more convenient for expressing algorithms than the machine code of our abstract machine. Pseudocode is also more convenient than actual programming languages, since we can use high-level concepts borrowed from mathematics without having to worry about exactly how they can be compiled to run on actual hardware. We frequently annotate programs to make algorithms more readable and easier to prove correct. This is the subject of Sect. 2.6. Section 2.7 gives the first comprehensive example: binary search in a sorted array. In Sect. 2.8, we introduce mathematical techniques

---

[1] The above illustration of Stonehenge is from [255].

© Springer Nature Switzerland AG 2019
P. Sanders et al., *Sequential and Parallel Algorithms and Data Structures*,
https://doi.org/10.1007/978-3-030-25209-0_2

for analyzing the complexity of programs, in particular, for analyzing nested loops and recursive procedure calls. Additional analysis techniques are needed for average-case analysis and parallel algorithm analysis; these are covered in Sects. 2.9 and 2.10, respectively. Randomized algorithms, discussed in Sect. 2.11, use coin tosses in their execution. Section 2.12 is devoted to graphs, a concept that will play an important role throughout the book. In Sect. 2.13, we discuss the question of when an algorithm should be called efficient, and introduce the complexity classes **P** and **NP** and the concept of **NP**-completeness. Finally, as in most chapters of this book, we close with implementation notes (Sect. 2.14) and historical notes and further findings (Sect. 2.15).

2.1 Asymptotic Notation

The main purpose of algorithm analysis is to give performance guarantees, for example bounds on running time, that are at the same time accurate, concise, general, and easy to understand. It is difficult to meet all these criteria simultaneously. For example, the most accurate way to characterize the running time T of an algorithm is to view T as a mapping from the set I of all inputs to the set of nonnegative numbers $\mathbb{R}_+$. For any problem instance i, $T(i)$ is the running time on i. This level of detail is so overwhelming that we could not possibly derive a theory about it. A useful theory needs a more global view of the performance of an algorithm.

Hence, we group the set of all inputs into classes of "similar" inputs and summarize the performance on all instances in the same class in a single number. The most useful grouping is by *size*. Usually, there is a natural way to assign a size to each problem instance. The size of an integer is the number of digits in its representation, and the size of a set is the number of elements in that set. The size of an instance is always a natural number. Sometimes we use more than one parameter to measure the size of an instance; for example, it is customary to measure the size of a graph by its number of nodes and its number of edges. We ignore this complication for now. We use $\text{size}(i)$ to denote the size of instance i, and I_n to denote the set of instances of size n for $n \in \mathbb{N}$. For the inputs of size n, we are interested in the maximum, minimum, and average execution times:[2]

$$
\begin{aligned}
\text{worst case:} \quad & T(n) = \max\{T(i) : i \in I_n\}; \\
\text{best case:} \quad & T(n) = \min\{T(i) : i \in I_n\}; \\
\text{average case:} \quad & T(n) = \frac{1}{|I_n|} \sum_{i \in I_n} T(i).
\end{aligned}
$$

We are most interested in the worst-case execution time, since it gives us the strongest performance guarantee. A comparison of the best and the worst case tells us how much the execution time varies for different inputs in the same class. If the

---

[2] We shall make sure that $\{T(i) : i \in I_n\}$ always has a proper minimum and maximum, and that I_n is finite when we consider averages.

discrepancy is big, the average case may give more insight into the true performance of the algorithm. Section 2.9 gives an example.

We shall perform one more step of data reduction: We shall concentrate on *growth rate* or *asymptotic analysis*. Functions $f(n)$ and $g(n)$ have the *same growth rate* if there are positive constants c and d such that $c \leq f(n)/g(n) \leq d$ for all sufficiently large n, and $f(n)$ *grows faster* than $g(n)$ if, for all positive constants c, we have $f(n) \geq c \cdot g(n)$ for all sufficiently large n. For example, the functions n^2, $n^2 + 7n$, $5n^2 - 7n$, and $n^2/10 + 10^6 n$ all have the same growth rate. Also, they grow faster than $n^{3/2}$, which in turn grows faster than $n \log n$. The growth rate talks about the behavior for large n. The word "asymptotic" in "asymptotic analysis" also stresses the fact that we are interested in the behavior for large n.

Why are we interested only in growth rates and the behavior for large n? We are interested in the behavior for large n because the whole purpose of designing efficient algorithms is to be able to solve large instances. For large n, an algorithm whose running time has a smaller growth rate than the running time of another algorithm will be superior. Also, our machine model is an abstraction of real machines and hence can predict actual running times only up to a constant factor. A pleasing side effect of concentrating on growth rate is that we can characterize the running times of algorithms by simple functions. However, in the sections on implementation, we shall frequently take a closer look and go beyond asymptotic analysis. Also, when using one of the algorithms described in this book, you should always ask yourself whether the asymptotic view is justified.

The following definitions allow us to argue precisely about *asymptotic behavior*. Let $f(n)$ and $g(n)$ denote functions that map nonnegative integers to nonnegative real numbers:

$$O(f(n)) = \{g(n) : \exists c > 0 : \exists n_0 \in \mathbb{N}_+ : \forall n \geq n_0 : g(n) \leq c \cdot f(n)\},$$
$$\Omega(f(n)) = \{g(n) : \exists c > 0 : \exists n_0 \in \mathbb{N}_+ : \forall n \geq n_0 : g(n) \geq c \cdot f(n)\},$$
$$\Theta(f(n)) = O(f(n)) \cap \Omega(f(n)),$$
$$o(f(n)) = \{g(n) : \forall c > 0 : \exists n_0 \in \mathbb{N}_+ : \forall n \geq n_0 : g(n) \leq c \cdot f(n)\},$$
$$\omega(f(n)) = \{g(n) : \forall c > 0 : \exists n_0 \in \mathbb{N}_+ : \forall n \geq n_0 : g(n) \geq c \cdot f(n)\}.$$

The left-hand sides should be read as "big O of f", "big omega of f", "theta of f", "little o of f", and "little omega of f", respectively. A remark about notation is in order here. In the definitions above, we use "$f(n)$" and "$g(n)$" with two different meanings. In "$O(f(n))$" and "$\{g(n) : \ldots\}$", they denote the functions f and g and the "n" emphasizes that these are functions of the argument n, and in "$g(n) \leq c \cdot f(n)$", they denote the values of the functions at the argument n.

Let us see some examples. $O(n^2)$ is the set of all functions that grow at most quadratically, $o(n^2)$ is the set of functions that grow less than quadratically, and $o(1)$ is the set of functions that go to 0 as n goes to infinity. Here "1" stands for the function $n \mapsto 1$, which is 1 everywhere, and hence $f \in o(1)$ if $f(n) \leq c \cdot 1$ for any positive c and sufficiently large n, i.e., $f(n)$ goes to zero as n goes to infinity. Generally, $O(f(n))$ is the set of all functions that "grow no faster than" $f(n)$. Similarly,

$\Omega(f(n))$ is the set of all functions that "grow at least as fast as" $f(n)$. For example, the Karatsuba algorithm for integer multiplication has a worst-case running time in $O(n^{1.58})$, whereas the school algorithm has a worst-case running time in $\Omega(n^2)$, so that we can say that the Karatsuba algorithm is asymptotically faster than the school algorithm. The "little o" notation $o(f(n))$ denotes the set of all functions that "grow strictly more slowly than" $f(n)$. Its twin $\omega(f(n))$ is rarely used, and is only shown for completeness.

The growth rate of most algorithms discussed in this book is either a polynomial or a logarithmic function, or the product of a polynomial and a logarithmic function. We use polynomials to introduce our readers to some basic manipulations of asymptotic notation.

Lemma 2.1. *Let* $p(n) = \sum_{i=0}^{k} a_i n^i$ *denote any polynomial and assume* $a_k > 0$. *Then* $p(n) \in \Theta(n^k)$.

Proof. It suffices to show that $p(n) \in O(n^k)$ and $p(n) \in \Omega(n^k)$. First observe that for $n \geq 1$,

$$p(n) \leq \sum_{i=0}^{k} |a_i| n^i \leq n^k \sum_{i=0}^{k} |a_i|,$$

and hence $p(n) \leq (\sum_{i=0}^{k} |a_i|) n^k$ for all positive n. Thus $p(n) \in O(n^k)$.

Let $A = \sum_{i=0}^{k-1} |a_i|$. For positive n, we have

$$p(n) \geq a_k n^k - A n^{k-1} = \frac{a_k}{2} n^k + n^{k-1} \left(\frac{a_k}{2} n - A \right)$$

and hence $p(n) \geq (a_k/2) n^k$ for $n > 2A/a_k$. We choose $c = a_k/2$ and $n_0 = 2A/a_k$ in the definition of $\Omega(n^k)$, and obtain $p(n) \in \Omega(n^k)$.   □

Exercise 2.1. Right or wrong? (a) $n^2 + 10^6 n \in O(n^2)$, (b) $n \log n \in O(n)$, (c) $n \log n \in \Omega(n)$, (d) $\log n \in o(n)$.

Asymptotic notation is used a lot in algorithm analysis, and it is convenient to stretch mathematical notation a little in order to allow sets of functions (such as $O(n^2)$) to be treated similarly to ordinary functions. In particular, we shall always write $h = O(f)$ instead of $h \in O(f)$, and $O(h) = O(f)$ instead of $O(h) \subseteq O(f)$. For example,

$$3n^2 + 7n = O(n^2) = O(n^3).$$

Never forget that sequences of "equalities" involving O-notation are really membership and inclusion relations and, as such, can only be read from left to right.

If h is a function, F and G are sets of functions, and $\circ$ is an operator such as $+$, $\cdot$, or $/$, then $F \circ G$ is a shorthand for $\{f \circ g : f \in F, g \in G\}$, and $h \circ F$ stands for $\{h\} \circ F$. So $f(n) + o(f(n))$ denotes the set of all functions $f(n) + g(n)$ where $g(n)$ grows strictly more slowly than $f(n)$, i.e., the ratio $(f(n) + g(n))/f(n)$ goes to 1 as n goes to infinity. Equivalently, we can write $(1 + o(1)) f(n)$. We use this notation whenever we care about the constant in the leading term but want to ignore *lower-order terms*.

Lemma 2.2. *The following rules hold for* O*-notation:*

$$cf(n) = \Theta(f(n)) \text{ for any positive constant } c,$$
$$f(n) + g(n) = \Omega(f(n)),$$
$$f(n) + g(n) = O(f(n)) \text{ if } g(n) = O(f(n)),$$
$$O(f(n)) \cdot O(g(n)) = O(f(n) \cdot g(n)).$$

Exercise 2.2. Prove Lemma 2.2.

Exercise 2.3. Sharpen Lemma 2.1 and show that $p(n) = a_k n^k + o(n^k)$.

Exercise 2.4. Prove that $n^k = o(c^n)$ for any integer k and any $c > 1$. How does $n^{\log\log n}$ compare with n^k and c^n?

2.2 The Sequential Machine Model

In 1945, John von Neumann (Fig. 2.1) introduced a computer architecture [243] which was simple, yet powerful. The limited hardware technology of the time forced him to come up with a design that concentrated on the essentials; otherwise, realization would have been impossible. Hardware technology has developed tremendously since 1945. However, the programming model resulting from von Neumann's design is so elegant and powerful that it is still the basis for most of modern programming. Usually, programs written with von Neumann's model in mind also work well on the vastly more complex hardware of today's machines.

Fig. 2.1. John von Neumann, born Dec. 28, 1903 in Budapest, died Feb. 8, 1957 in Washington, DC.

The variant of von Neumann's model used in algorithmic analysis is called the *RAM* (random access machine) model. It was introduced by Shepherdson and Sturgis [294] in 1963. It is a *sequential* machine with uniform memory, i.e., there is a single processing unit, and all memory accesses take the same amount of time. The (main) memory, or *store*, consists of infinitely many cells $S[0]$, $S[1]$, $S[2]$, …; at any point in time, only a finite number of them will be in use. In addition to the main memory, there are a small number of *registers* $R_1, \ldots, R_k$.

The memory cells store "small" integers, also called *words*. In our discussion of integer arithmetic in Chap. 1, we assumed that "small" meant one-digit. It is more reasonable and convenient to assume that the interpretation of "small" depends on the size of the input. Our default assumption is that integers whose absolute value is bounded by a polynomial in the size of the input can be stored in a single cell. Such integers can be represented by a number of bits that is logarithmic in the size of the input. This assumption is reasonable because we could always spread out the contents of a single cell over logarithmically many cells with a logarithmic overhead in time and space and obtain constant-size cells. The assumption is convenient because we want to be able to store array indices in a single cell. The assumption is necessary because allowing cells to store arbitrary numbers would lead to absurdly overoptimistic algorithms. For example, by repeated squaring, we could generate a number with 2^n bits in n steps. Namely, if we start with the number $2 = 2^1$, squaring it once gives $4 = 2^2 = 2^{2^1}$, squaring it twice gives $16 = 2^4 = 2^{2^2}$, and squaring it n times gives 2^{2^n}.

Our model supports a limited form of parallelism. We can perform simple operations on a logarithmic number of bits in constant time.

A RAM can execute (machine) programs. A program is simply a sequence of machine instructions, numbered from 0 to some number ℓ. The elements of the sequence are called *program lines*. The program is stored in a program store. Our RAM supports the following *machine instructions*:

- $R_i := S[R_j]$ *loads* the contents of the memory cell indexed by the contents of R_j into register R_i.
- $S[R_j] := R_i$ *stores* the contents of register R_i in the memory cell indexed by the contents of R_j.
- $R_i := R_j \odot R_h$ executes the binary operation $\odot$ on the contents of registers R_j and R_h and stores the result in register R_i. Here, $\odot$ is a placeholder for a variety of operations. The *arithmetic* operations are the usual $+$, $-$, and $*$; they interpret the contents of the registers as integers. The operations **div** and **mod** stand for integer division and the remainder, respectively. The *comparison* operations $\leq$, $<$, $>$, and $\geq$ for integers return *truth values*, i.e., either *true* ($= 1$) or *false* ($= 0$). The *logical* operations $\wedge$ and $\vee$ manipulate the truth values 0 and 1. We also have bitwise Boolean operations | (OR), & (AND), and $\oplus$ (exclusive OR, XOR). They interpret contents as bit strings. The shift operators $>>$ (shift right) and $<<$ (shift left) interpret the first argument as a bit string and the second argument as a nonnegative integer. We may also assume that there are operations which interpret the bits stored in a register as a floating-point number, i.e., a finite-precision approximation of a real number.
- $R_i := \odot R_j$ executes the *unary* operation $\odot$ on the contents of register R_j and stores the result in register R_i. The operators $-$, $\neg$ (logical NOT), and $\sim$ (bitwise NOT) are available.
- $R_i := C$ assigns the *constant* value C to R_i.
- JZ k, R_i continues execution at program line k, if register R_i is 0, and at the next program line otherwise (*conditional branch*). There is also the variant JZ R_j, R_i, where the target of the jump is the program line stored in R_j.
- J k continues execution at program line k (*unconditional branch*). Similarly to JZ, the program line can also be specified by the content of a register.

A program is executed on a given input step by step. The input for a computation is stored in memory cells $S[1]$ to $S[R_1]$ and execution starts with program line 1. With the exception of the branch instructions JZ and J, the next instruction to be executed is always the instruction in the next program line. The execution of a program terminates if a program line is to be executed whose number is outside the range $1..\ell$. Recall that ℓ is the number of the last program line.

We define the execution time of a program on an input in the most simple way: *Each instruction takes one time step to execute. The total execution time of a program is the number of instructions executed.*

It is important to remember that the RAM model is an abstraction. One should not confuse it with physically existing machines. In particular, real machines have a finite memory and a fixed number of bits per register (e.g., 32 or 64). In contrast, the word size and memory of a RAM scale with input size. This can be viewed as an abstraction of the historical development. Microprocessors have had words of 4, 8, 16, and 32 bits in succession, and now often have 64-bit words. Words of 64 bits can index a memory of size 2^{64}. Thus, at current prices, memory size is limited by cost and not by physical limitations. This statement was also true when 32-bit words were introduced.

Our complexity model is a gross oversimplification: Modern processors attempt to execute many instructions in parallel. How well they succeed depends on factors such as data dependencies between successive operations. As a consequence, an operation does not have a fixed cost. This effect is particularly pronounced for memory accesses. The worst-case time for a memory access to the main memory can be hundreds of times higher than the best-case time. The reason is that modern processors attempt to keep frequently used data in *caches* – small, fast memories close to the processors. How well caches work depends a lot on their architecture, the program, and the particular input. App. B discusses hardware architecture in more detail.

We could attempt to introduce a very accurate cost model, but this would miss the point. We would end up with a complex model that would be difficult to handle. Even a successful complexity analysis would lead to a monstrous formula depending on many parameters that change with every new processor generation. Although such a formula would contain detailed information, the very complexity of the formula would make it useless. We therefore go to the other extreme and eliminate all model parameters by assuming that each instruction takes exactly one unit of time. The result is that constant factors in our model are quite meaningless – one more reason to stick to asymptotic analysis most of the time. We compensate for this drawback by providing implementation notes, in which we discuss implementation choices and shortcomings of the model. Two important shortcomings of the RAM model, namely the lack of a memory hierarchy and the limited parallelism ,are discussed in the next two subsections.

2.2.1 External Memory

The organization of the memory is a major difference between an RAM and a real machine: a uniform flat memory in a RAM and a complex memory hierarchy in a real machine. In Sects. 5.12, 6.3, 7.7, and 11.5 we shall discuss algorithms that have been specifically designed for huge data sets which have to be stored on slow memory, such as disks. We shall use the *external-memory model* to study these algorithms.

The external-memory model is like the RAM model except that the fast memory is limited to M words. Additionally, there is an external memory with unlimited size. There are special *I/O operations*, which transfer B consecutive words between slow and fast memory. The reason for transferring a block of B words instead of a single word is that the memory access time is large for a slow memory in comparison with the transfer time for a single word. The value of B is chosen such that the transfer time for B words is approximately equal to the access time. For example, the external memory could be a hard disk; M would then be the size of the main memory, and B would be a block size that is a good compromise between low latency and high bandwidth. With current technology, $M = 8\,\text{GB}$ and $B = 2\,\text{MB}$ are realistic values. One I/O step would then take around $10\,\text{ms}$, which is $2 \cdot 10^7$ clock cycles of a $2\,\text{GHz}$ machine. With another setting of the parameters M and B, one can model the smaller access time difference between a hardware cache and main memory.

2.3 Pseudocode

Our RAM model is an abstraction and simplification of the machine programs executed on microprocessors. The purpose of the model is to provide a precise definition of running time. However, the model is much too low-level for formulating complex algorithms. Our programs would become too long and too hard to read. Instead, we formulate our algorithms in *pseudocode*, which is an abstraction and simplification of imperative programming languages such as C, C++, Java, C#, Rust, Swift, Python, and Pascal, combined with liberal use of mathematical notation. We now describe the conventions used in this book, and derive a timing model for pseudocode programs. The timing model is quite simple: *Basic pseudocode instructions take constant time, and procedure and function calls take constant time plus the time to execute their body.* We justify the timing model by outlining how pseudocode can be translated into equivalent RAM code. We do this only to the extent necessary for understanding the timing model. There is no need to worry about compiler optimization techniques, since constant factors are ignored in asymptotic analysis anyway. The reader may decide to skip the paragraphs describing the translation and adopt the timing model as an axiom. The syntax of our pseudocode is akin to that of Pascal [166], because we find this notation typographically nicer for a book than the more widely known syntax of C and its descendants C++ and Java.

2.3.1 Variables and Elementary Data Types

A *variable declaration* "$v = x : T$" introduces a variable v of type T and initializes it to the value x. For example, "*answer* $= 42 : \mathbb{N}$" introduces a variable *answer* assuming nonnegative integer values and initializes it to the value 42. When the type of a variable is clear from the context, we shall sometimes omit it from the declaration. A type is either a basic type (e.g., integer, Boolean value, or pointer) or a composite type. We have predefined composite types such as arrays, and application-specific classes (see below). When the type of a variable is irrelevant to the discussion, we use the unspecified type *Element* as a placeholder for an arbitrary type. We take the liberty of extending numeric types by the values $-\infty$ and ∞ whenever this is convenient. Similarly, we sometimes extend types by an undefined value (denoted by the symbol $\bot$), which we assume to be distinguishable from any "proper" element of the type T. In particular, for pointer types it is useful to have an undefined value. The values of the pointer type "**Pointer to** T" are handles to objects of type T. In the RAM model, this is the index of the first cell in a region of storage holding an object of type T.

A declaration "$a : Array\ [i..j]$ **of** T" introduces an *array a* consisting of $j - i + 1$ *elements* of type T, stored in $a[i]$, $a[i+1]$, ..., $a[j]$. Arrays are implemented as contiguous pieces of memory. To find an element $a[k]$, it suffices to know the starting address of a and the size of an object of type T. For example, if register R_a stores the starting address of an array $a[0..k]$, the elements have unit size, and R_i contains the integer 42, the instruction sequence "$R_1 := R_a + R_i$; $R_2 := S[R_1]$" loads $a[42]$ into register R_2. The size of an array is fixed at the time of declaration; such arrays are called *static*. In Sect. 3.4, we show how to implement *unbounded arrays* that can grow and shrink during execution.

A declaration "$c :$ **Class** *age* : $\mathbb{N}$, *income* : $\mathbb{N}$ **end**" introduces a variable c whose values are pairs of integers. The components of c are denoted by *c.age* and *c.income*. For a variable c, **addressof** c returns a handle to c, i.e., the address of c. If p is an appropriate pointer type, $p :=$ **addressof** c stores a handle to c in p and $*p$ gives us back c. The fields of c can then also be accessed through $p \rightarrow age$ and $p \rightarrow income$. Alternatively, one may write (but nobody ever does) $(*p).age$ and $(*p).income$.

Arrays and objects referenced by pointers can be allocated and deallocated by the commands **allocate** and **dispose**. For example, $p :=$ **allocate** $Array\ [1..n]$ **of** T allocates an array of n objects of type T. That is, the statement allocates a contiguous chunk of memory of size n times the size of an object of type T, and assigns a handle to this chunk (= the starting address of the chunk) to p. The statement **dispose** p frees this memory and makes it available for reuse. With **allocate** and **dispose**, we can cut our memory array S into disjoint pieces that can be referred to separately. These functions can be implemented to run in constant time. The simplest implementation is as follows. We keep track of the used portion of S by storing the index of the first free cell of S in a special variable, say *free*. A call of **allocate** reserves a chunk of memory starting at *free* and increases *free* by the size of the allocated chunk. A call of **dispose** does nothing. This implementation is time-efficient, but not space-efficient. Any call of **allocate** or **dispose** takes constant time. However, the total space consumption is the total space that has ever been allocated and not the maximum space simultaneously used, i.e., allocated but not yet freed, at any one time. It is not known whether an arbitrary sequence of **allocate** and **dispose** operations can be realized space-efficiently and with constant time per operation. However, for all algorithms presented in this book, **allocate** and **dispose** can be realized in a time- and space-efficient way.

We borrow some composite data structures from mathematics. In particular, we use tuples, sequences, and sets. *Pairs*, *triples*, and other *tuples* are written in round brackets, for example $(3, 1)$, $(3, 1, 4)$, and $(3, 1, 4, 1, 5)$. Since tuples contain only a constant number of elements, operations on them can be broken into operations on their constituents in an obvious way. *Sequences* store elements in a specified order; for example, "$s = \langle 3,1,4,1 \rangle$: *Sequence* of $\mathbb{Z}$" declares a sequence s of integers and initializes it to contain the numbers 3, 1, 4, and 1 in that order. Sequences are a natural abstraction of many data structures, such as files, strings, lists, stacks, and queues. In Chap. 3, we shall study many ways of representing sequences. In later chapters, we shall make extensive use of sequences as a mathematical abstraction with little further reference to implementation details. The empty sequence is written as $\langle \rangle$.

Sets play an important role in mathematical arguments, and we shall also use them in our pseudocode. In particular, you will see declarations such as "$M = \{3,1,4\}$: *Set* of $\mathbb{N}$" that are analogous to declarations of arrays or sequences. Sets are usually implemented as sequences.

2.3.2 Statements

The simplest statement is an assignment $x := E$, where x is a variable and E is an expression. An assignment is easily transformed into a constant number of RAM instructions. For example, the statement $a := a + bc$ is translated into "$R_1 := R_b * R_c$; $R_a := R_a + R_1$", where R_a, R_b, and R_c stand for the registers storing a, b, and c, respectively. From C, we borrow the shorthands $++$ and $--$ for incrementing and decrementing variables. We also use parallel assignment to several variables. For example, if a and b are variables of the same type, "$(a,b) := (b,a)$" swaps the contents of a and b.

The conditional statement "**if** C **then** I **else** J", where C is a Boolean expression and I and J are statements, translates into the instruction sequence

$$eval(C);\ \text{JZ}\ sElse,\ R_c;\ trans(I);\ \text{J}\ sEnd;\ trans(J),$$

where $eval(C)$ is a sequence of instructions that evaluate the expression C and leave its value in register R_c, $trans(I)$ is a sequence of instructions that implement statement I, $trans(J)$ implements J, $sElse$ is the address of the first instruction in $trans(J)$, and $sEnd$ is the address of the first instruction after $trans(J)$. The sequence above first evaluates C. If C evaluates to false (= 0), the program jumps to the first instruction of the translation of J. If C evaluates to true (= 1), the program continues with the translation of I and then jumps to the instruction after the translation of J. The statement "**if** C **then** I" is a shorthand for "**if** C **then** I **else** ;", i.e., an if–then–else with an empty "else" part.

Our written representation of programs is intended for humans and uses less strict syntax than do programming languages. In particular, we usually group statements by indentation and in this way avoid the proliferation of brackets observed in programming languages such as C that are designed as a compromise between readability for humans and for computers. We use brackets only if the program would be ambiguous otherwise. For the same reason, a line break can replace a semicolon for the purpose of separating statements.

The loop "**repeat** I **until** C" translates into $trans(I);\ eval(C);\ \text{JZ}\ sI,\ R_c$, where sI is the address of the first instruction in $trans(I)$. We shall also use many other types of loops that can be viewed as shorthands for various repeat loops. In the following list, the shorthand on the left expands into the statements on the right:

while C **do** I	**if** C **then repeat** I **until** $\neg C$
for $i := a$ **to** b **do** I	$i := a;$ **while** $i \leq b$ **do** $I; i{+}{+}$
for $i := a$ **to** ∞ **while** C **do** I	$i := a;$ **while** C **do** $I; i{+}{+}$
foreach $e \in s$ **do** I	**for** $i := 1$ **to** $|s|$ **do** $e := s[i]; I$

Many low-level optimizations are possible when loops are translated into RAM code. These optimizations are of no concern to us. For us, it is only important that the execution time of a loop can be bounded by summing the execution times of each of its iterations, including the time needed for evaluating conditions.

2.3.3 Procedures and Functions

A subroutine with the name *foo* is declared in the form "**Procedure** *foo*(D) *I*", where *I* is the body of the procedure and D is a sequence of variable declarations specifying the parameters of *foo*. A call of *foo* has the form *foo*(P), where P is a parameter list. The parameter list has the same length as the variable declaration list. Parameter passing is either "by value" or "by reference". Our default assumption is that basic objects such as integers and Booleans are passed by value and that complex objects such as arrays are passed by reference. These conventions are similar to the conventions used by C and guarantee that parameter passing takes constant time. The semantics of parameter passing is defined as follows. For a value parameter x of type T, the actual parameter must be an expression E of the same type. Parameter passing is equivalent to the declaration of a local variable x of type T initialized to E. For a reference parameter x of type T, the actual parameter must be a variable of the same type and the formal parameter is simply an alternative name for the actual parameter.

As with variable declarations, we sometimes omit type declarations for parameters if they are unimportant or clear from the context. Sometimes we also declare parameters implicitly using mathematical notation. For example, the declaration **Procedure** $bar(\langle a_1, \ldots, a_n \rangle)$ introduces a procedure whose argument is a sequence of n elements of unspecified type.

Most procedure calls can be compiled into machine code by simply substituting the procedure body for the procedure call and making provisions for parameter passing; this is called *inlining*. Value passing is implemented by making appropriate assignments to copy the parameter values into the local variables of the procedure. Reference passing to a formal parameter $x : T$ is implemented by changing the type of x to **Pointer to** T, replacing all occurrences of x in the body of the procedure by $(*x)$ and initializing x by the assignment $x := $ **addressof** y, where y is the actual parameter. Inlining gives the compiler many opportunities for optimization, so that inlining is the most efficient approach for small procedures and for procedures that are called from only a single place.

Functions are similar to procedures, except that they allow the return statement to return a value. Figure 2.2 shows the declaration of a recursive function that returns $n!$ and its translation into RAM code. The substitution approach

Function *factorial*(n) : $\mathbb{Z}$
  if $n = 1$ **then return** 1 **else return** $n \cdot factorial(n-1)$

```
factorial :                                          // the first instruction of factorial
```
$R_n := S[R_r - 1]$ // load n into register R_n. Abbreviation of $R_{tmp} := R_r - 1$; $R_n := S[R_{tmp}]$
```
JZ thenCase, Rn                                      // jump to then case, if n is 0
```
$S[R_r] = \text{aRecCall}$ // else case; return address for recursive call
$S[R_r + 1] := R_n - 1$ // parameter is $n - 1$
$R_r := R_r + 2$ // increase stack pointer
```
J factorial                                         // start recursive call
aRecCall :                                          // return address for recursive call
```
$R_{result} := S[R_r - 1] * R_{result}$ // store $n * factorial(n-1)$ in result register
```
J return                                            // goto return
thenCase :                                          // code for then case
```
$R_{result} := 1$ // put 1 into result register
```
return :                                            // code for return
```
$R_r := R_r - 2$ // free activation record
```
J  S[Rr]                                            // jump to return address
```

Fig. 2.2. A recursive function *factorial* and the corresponding RAM code. The RAM code returns the function value in the register R_{result}. To keep the presentation short, we take the liberty of directly using subexpressions, where, strictly speaking, sequences of assignments using temporary registers would be needed.

fails for *recursive* procedures and functions that directly or indirectly call themselves – substitution would never terminate. Realizing recursive procedures in RAM code requires the concept of a *recursion stack*. Explicit subroutine calls over a stack are also used for large procedures that are called multiple times where inlining would unduly increase the code size. The recursion stack is a reserved part of the memory. Register R_r always points to the first free entry in this stack.

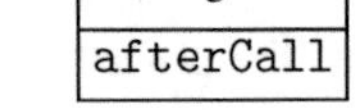

Fig. 2.3. The recursion stack of a call *factorial*(5) when the recursion has reached *factorial*(3).

The stack contains a sequence of *activation records*, one for each active procedure call. The activation record for a procedure with k parameters and ℓ local variables has size $1 + k + \ell$. The first location contains the return address, i.e., the address of the instruction where execution is to be continued after the call has terminated, the next k locations are reserved for the parameters, and the final ℓ locations are for the local variables. A procedure call is now implemented as follows. First, the calling procedure *caller* pushes the return address and the actual parameters onto the stack, increases R_r accordingly, and jumps to the first instruction of the called routine *called*. The called routine reserves space for its local variables by increasing R_r appropriately. Then the body of *called* is executed. During execution of the body, any access to the ith formal parameter $(0 \le i < k)$ is an access to $S[R_r - \ell - k + i]$ and any access to the ith local variable $(0 \le i < \ell)$ is an access to $S[R_r - \ell + i]$. When *called* executes a **return** statement, it decreases R_r by $1 + k + \ell$ (observe that *called* knows k and ℓ) and execution continues at the return address (which can be found at $S[R_r]$). Thus control is returned to *caller*. Note that recursion is no problem with this scheme, since each incarnation of a routine will have its own stack area for its parameters and local variables. Figure 2.3 shows the contents of the recursion stack of a call *factorial*(5) when the recursion has reached *factorial*(3). The label `afterCall` is the address of the instruction following the call *factorial*(5), and `aRecCall` is defined in Fig. 2.2.

Exercise 2.5 (sieve of Eratosthenes). Translate the following pseudocode for finding all prime numbers up to n into RAM machine code. There is no need to translate the output command, in which the value in the box is output as a number. Argue correctness first.

$a = \langle 1, \ldots, 1 \rangle : Array\ [2..n]\ \textbf{of}\ \{0,1\}$ // if $a[i]$ is false, i is known to be nonprime
for $i := 2$ **to** $\lfloor \sqrt{n} \rfloor$ **do**   // nonprimes $\le n$ have a factor $\le \lfloor \sqrt{n} \rfloor$
    if $a[i]$ **then**   // i is prime
        for $j := 2i$ **to** n **step** i **do** $a[j] := 0$ // all multiples of i are nonprime
    for $i := 2$ **to** n **do** **if** $a[i]$ **then** output("$\boxed{i}$ is prime")

2.3.4 Object Orientation

We also need a simple form of object-oriented programming so that we can separate the interface and the implementation of data structures. We introduce our notation by way of example. The definition

Class *Complex*$(x, y : Number)$ **of** *Number*
    $re = x : Number$
    $im = y : Number$
    Function *abs* : *Number* **return** $\sqrt{re^2 + im^2}$
    Function *add*$(c' : Complex) : Complex$
                return *Complex*$(re + c'.re, im + c'.im)$

gives a (partial) implementation of a complex number type that can use arbitrary numeric types such as $\mathbb{Z}$, $\mathbb{Q}$, and $\mathbb{R}$ for the real and imaginary parts. Our class names (here "*Complex*") will usually begin with capital letters. The real and imaginary parts are stored in the *member variables* re and im, respectively. Now, the declaration "$c : Complex(2,3)$ **of** $\mathbb{R}$" declares a complex number c initialized to $2 + 3i$, where i is the imaginary unit. The expression $c.im$ evaluates to the imaginary part of c, and $c.abs$ returns the absolute value of c, a real number.

The type after the **of** allows us to parameterize classes with types in a way similar to the template mechanism of C++ or the generic types of Java. Note that in the light of this notation, the types "*Set* **of** *Element*" and "*Sequence* **of** *Element*" mentioned earlier are ordinary classes. Objects of a class are initialized by setting the member variables as specified in the class definition.

2.4 Parallel Machine Models

We classify parallel machine models into two broad classes: shared-memory machines and distributed-memory machines. In both cases, we have *p processing elements* (PEs). In the former case, these PEs share a common memory and all communication between PEs is through the shared memory. In the latter case, each PE has
its own private memory, the PEs are connected by a communication network, and all
communication is through the network. We introduce shared-memory machines in
Sect. 2.4.1 and discuss distribute- memory machines in Sect. 2.4.2.

2.4.1 Shared-Memory Parallel Computing

In a shared-memory machine, the PEs share a common memory (Fig. 2.4). Each PE
knows its number i_{proc} (usually from $1..p$ or $0..p-1$). The theoretical variant of this
model is known as the *PRAM (parallel random access machine)*. PRAMs come in
several flavors. The main distinction is whether concurrent access to the same memory cell is allowed. This leads to the submodels *EREW-PRAM*, *CREW-PRAM*, and
CRCW-PRAM where "C"stands for "concurrent" (concurrent access allowed), "E"
stands for "exclusive" (concurrent access forbidden), "R" stands for "read" and "W"
stands for "write". Thus a CREW-PRAM supports concurrent reads but forbids concurrent writes. Real-world shared-memory machines support something resembling
concurrent read, so that we do not need to bother with the complications introduced
by exclusive reads. We therefore concentrate on the CREW and CRCW. Concurrent
writing makes the model more powerful, but we have be careful with the semantics
of concurrent writing. To illustrate the pitfalls of concurrent memory access, let us
consider a simple example: Two PEs a and b share the same counter variable c, say
because they want to count how often a certain event happens. Suppose the current
value of c is 41 and both a and b want to increment c at the same time. Incrementing
means first loading the old value into a register, and then incrementing the register
and storing it back in memory. Suppose both PEs read the value 41, increment this
value to 42 and then write it back – all at the same time. Afterwards, $c = 42$, although
the programmer probably intended $c = 43$. Different semantics of concurrent writing lead to several subflavors of CRCW-PRAMs. The two most widely used ones

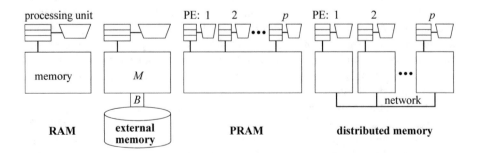

Fig. 2.4. Machine models used in this book.

are "common" and "arbitrary". The *common CRCW-PRAM* only allows concurrent write access if all PEs writing to the same memory cell are writing the same value. There is no such restriction in *arbitrary CRCW-PRAMs*. If several different values are written to the same cell in the same time step, one of these values is chosen arbitrarily. We shall try to avoid algorithms requiring the CRCW model, since concurrent write access to the same memory cells causes problems in practice. In particular, it becomes difficult to ensure correctness, and performance may suffer.

PRAM models assume globally synchronized time steps – every PE executes exactly one machine instruction in every time step. This makes it relatively easy to write and understand PRAM programs. Unfortunately, this assumption is untenable in practice – the execution of instructions (in particular memory access instructions) happens in several stages and it depends on the current state of the computation how long it takes to finish the instruction. Moreover, even instructions accessing memory at the same time may see different values in the same memory cell owing to the behavior of caches. More realistic models of shared memory therefore introduce additional mechanisms for explicitly controlling concurrent memory access and for synchronizing PEs.

A very general such mechanism is *transactions* (or *critical sections*). A transaction t consists of a piece of code that is executed *atomically*. Atomic means indivisible – during the execution of t, no other PE writes to the memory cells accessed by t. For example, if the PEs a and b in the example above were to execute transactions

begin transaction $c := c + 1$ **end transaction**

then the hardware or runtime system would make sure that the two transactions were executed one after the other, resulting in the correct value $c = 43$ after executing both transactions. Some processor architectures support transactions in hardware, while others only support certain atomically executed instructions that can be used to implement higher-level concepts. In principle, these can be used to support general transactions in software. However, this is often considered too expensive, so one often works with less general concepts. See Sect. B.5 for some details of actual hardware implementations of transactional memory.

Perhaps the most important *atomic instruction* for concurrent memory access is *compare-and-swap* (CAS).

Function $CAS(i, expected, desired) : \{1, 0\}$
    begin transaction
        if $S[i] = expected$ **then** $S[i] := desired;$ **return** 1   // success
        else $expected := S[i];$ **return** 0   // failure
    end transaction

A call of CAS specifies a value expected to be present in memory cell $S[i]$ and the value that it wants to write. If the expectation is true, the operation writes the desired value into the memory cell and succeeds (returns 1). If not, usually because some other PE has modified $S[i]$ in the mean time, the operation writes the actual value of $S[i]$ into the variable *expected* and fails (returns 0). CAS can be used to implement transactions acting on a single memory cell. For example, atomically adding a value to a memory cell can be done as follows:

Function *fetchAndAdd*(i, Δ)
    expected := $S[i]$
    repeat *desired* := *expected* + Δ **until** $CAS(i, expected, desired)$
    return *desired*

The function reads the value of $S[i]$ and stores the old and the incremented value in *expected* and *desired*, respectively. It then calls $CAS(i, expected, desired)$. If the value of $S[i]$ has not changed since it was stored in *expected*, the call succeeds, and the incremented value is stored in $S[i]$ and returned. Otherwise, the current value of $S[i]$ is stored in *expected* and the call fails. Then another attempt to increment the variable is made.

Regardless of whether hardware transactions or atomic instructions are used, when many PEs try to write to the same memory cell at once, some kind of serialization will take place and performance will suffer. This effect is called *write contention*. Asymptotically speaking, it will take time $\Omega(p)$ if all PEs are involved. Note that this is far from the behavior of a CRCW-PRAM, where concurrent writing is assumed to work in constant time.

Let us look at a simple example where this makes a difference: Assume each PE has a Boolean value and we want to compute the logical OR of all these values. On a common CRCW-PRAM, we simply initialize a global variable g to false, and each PE with a local value of true writes true to g. Within the theoretical model, this works in constant time. However, when we try to do this on a real-world machine using transactions, we may observe time $\Omega(p)$ when all local values are true. We shall later see more complicated algorithms achieving time $O(\log p)$ on CREW machines.

In order to get closer to the real world, additional models of PRAMs have been proposed that assume that the cost of memory access is proportional to the number of PEs concurrently accessing the same memory cell. For example, QRQW (queue-read-queue-write) means that contention has to be taken into account for both reading and writing [126]. Since modern machines support concurrent reading by placing copies of the accessed data in the machine caches, it also makes sense to consider CRQW (concurrent-read-queue-write) models. In this book, we shall sometimes use the aCRQW-PRAM model, where the "a" stands for "asynchronous", i.e., there is no step-by-step synchronization between the PEs.[3] For the cost model, this means that all instructions take constant time, except for write operations, whose execution time is proportional to the number of PEs trying to access that memory cell concurrently.

---

[3] There is previous work on asynchronous PRAMs [125] that is somewhat different, however, in that it subdivides computations into synchronized phases. Our aCRQW model performs only local synchronization. Global synchronization requires a separate subroutine that can be implemented using $O(\log p)$ local synchronizations (see Sect. 13.4.2).

2.4.2 Distributed-Memory Parallel Computing

Another simple way to extend the RAM model is to connect several RAMs with a communication network (Fig. 2.4). The network is used to exchange messages. We assume that messages have exactly one sender and one receiver (*point-to-point communication*) and that exchanging a message of length ℓ takes time $\alpha + \ell\beta$ regardless which PEs are communicating. In particular, several messages can be exchanged at once except that no PE may send several messages at the same time or receive several messages at the same time. However, a PE is allowed to send one message and to receive another message at the same time. This mode of communication is called (full-duplex) single-ported communication. The function call $send(i,m)$ sends a message m to PE i and $receive(i,m)$ receives a message from PE i. When the parameter i is dropped, a message from any PE can be received and the number of the actual sender is the return value of the *receive* function. For every call $send(i,m)$, PE i must eventually execute a matching receive operation. The *send* operation will only complete when the matching *receive* has completed, i.e., PE i now has a copy of message m.[4] Thus sends and receives *synchronize* the sender and the receiver. The integration of data exchange and synchronization is an important difference with respect to shared-memory parallel programming. We will see that the message-passing style of parallel programming often leads to more transparent programs despite the fact that data exchange is so simple in shared-memory programming. Moreover, message-passing programs are often easier to debug than shared-memory programs.

Let us consider a concrete example. Suppose each PE has stored a number x in a local variable and we want to compute the sum of all these values so that afterwards PE 0 knows this sum. The basic idea is to build a binary tree on top of the PEs and to add up the values by layer. Assume that PEs are numbered from 0 to $p-1$ and that p is a power of two. In the first round, each odd numbered PE sends his value to the PE numbered one smaller and then stops. The even numbered PEs sum the number received to the number they hold. In the second round, the even numbered PEs whose number is not divisible by four send to the PE whose number is smaller by two. Continuing in this way, the total sum is formed in time $O(\log p)$. The following lines of pseudocode formalize this idea. They work for arbitrary p.

```
Function reduceAdd(x)                    // let i denote the local PE number
    for (d := 1;  d < p;  d *= 2)                                    1
        if (i bitand d) = 0 then                                    2
            if i + d < p then receive(i+d,x');   x += x'            3
        else send(i−d,x);   return                                  4
    return x                             // only reached by PE 0     5
```

Initially all PEs are active. The layer counter d is a power of two and is also interpreted as a bit string. It starts at $d = 1 = 2^0$. Let us first assume that p is a power of two. PEs with an odd PE number (i **bitand** $d \neq 0$) exit from the for-loop and send

---

[4] Most algorithms in this book use synchronous communication. However, *asynchronous* send operations can also have advantages since they allow us do decouple communication and cooperation. An example can be found in Sect. 6.4 on parallel priority queues.

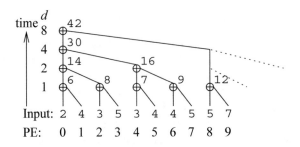

Fig. 2.5. Parallel summation using a tree

their value to PE $i-1$. PEs with an even PE number (i **bitand** $d = 0$) issue a receive request $receive(i+d, x')$ (note that $i+d < p$ for all even PEs) and add the value received to their own value. At this point, the odd-numbered PEs have terminated and the even-numbered PEs have increased the value of d to 2. In the second round, exactly the same reduction happens on the next to last bit, i.e., PEs whose number ends with 00 issue a *receive* and PEs whose number ends with 10 issue a *send* and terminate. Generally, in a round with $d = 2^k$, $k \geq 0$, the PEs whose PE numbers end with 10^k send their value to PE $i-d$ and terminate. The PEs whose PE numbers end with 0^{k+1} receive and add the received value to their current value. When $d = p$, the for-loop terminates and only PE 0 remains active. It returns x, which now contains the desired global sum.

The program is also correct when p is not a power of two since we have made sure that no PE tries to receive from a nonexistent PE and all sending PEs have a matching receiving PE. Figure 2.5 shows an example computation. In this example, processor 8 issues a $receive(9, x')$ in round 1, sits idle in rounds 2 and 3, and sends its value to PE 0 in round 4.

On a (synchronous) PRAM, we can do something very similar – it is even slightly easier. Assume the inputs are stored in an array $x[0..p-1]$. We replace "$receive(i+d, x'); x += x'$" by "$x[i] += x[i+d]$", drop lines 4 and 5, and obtain the final result in $x[0]$. However, if we use this code on an asynchronous shared-memory machine (e.g., in our aCRQW PRAM model), it is incorrect. We have to add additional synchronization code to make sure that an addition is only performed when its input values are available. The resulting code will be at least as complex as the message-passing code. Of course, we could use the function *fetchAndAdd* mentioned in from the preceding section to realize correct concurrent access to a global counter. However, this will take time $O(p)$ owing to write contention, and, indeed, it is likely that on a large machine it would be faster to perform one global synchronization followed by adding up the values in $x[0..p-1]$ sequentially.

2.4.3 Parallel Memory Hierarchies

The models presented in Sects. 2.2.1–2.4.2 each address an important aspect of real-world machines not present in the RAM model. However, they are still a gross simpli-

fication of reality. Modern machines have a memory hierarchy with multiple levels
and use many forms of parallel processing (see Fig. 2.6). Appendix B describes a
concrete machine that we used for the experiments reported in this book. We next
briefly discuss some important features found in real-world machines.

Many processors have 128–512-bit *SIMD* registers that allow the parallel execution of a *single instruction* on *multiple data* objects (*SIMD*).

They are *superscalar*, i.e., they can execute multiple independent instructions
from a sequential instruction stream in parallel.

Simultaneous multithreading allows processors to better utilize parallel execution
units by running multiple threads of activity on a single processor core sharing the
same first-level (L1) cache.

Even mobile devices nowadays have *multicore* processors, i.e., multiple processor cores, that can independently execute programs. There are further levels of on-
chip cache. The further up in this hierarchy, the more PEs share the same cache. For
example, each PE may have its own L1 and L2 caches but eight cores on one chip
might share a rather large L3 cache. Most servers have several multicore processors
accessing the same shared memory. Accessing the memory chips directly connected
to the processor chip is usually faster than accessing memory connected to other
processors. This effect is called *nonuniform memory access* (NUMA).

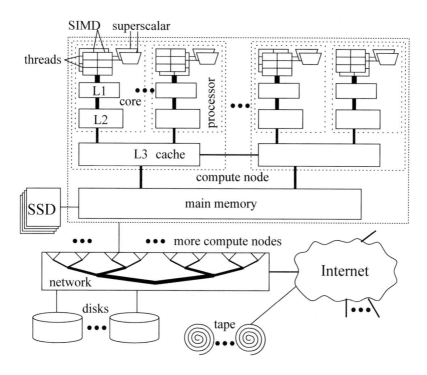

Fig. 2.6. Example of a parallel memory hierarchy

Coprocessors, in particular *graphics processing units* (GPUs), have even more parallelism on a single chip. GPUs have their own peculiarities complicating the model. In particular, large groups of threads running on the same piece of hardware can only execute in SIMD mode, i.e., in the same time step, they all have to execute the same instruction.

High-performance computers consist of multiple server-type systems interconnected by a fast, dedicated network.

Finally, more loosely connected computers of all types interact through various kinds of network (the internet, radio networks, . . .) in *distributed systems* that may consist of millions of nodes.

Storage devices such as solid state disks (SSDs), hard disks, or tapes may be connected to different levels of this hierarchy of processors. For example, a high-performance computer might have solid state disks connected to each multiprocessor board while its main communication network has ports connected to a large array of hard disks. Over the internet, it might do data archiving to a remote tape library.

Attempts to grasp this complex situation using a single overarching model can easily fall into a complexity trap. It is not very difficult to define more general and accurate models. However, the more accurate the model gets, the more complicated it gets to design and analyze algorithms for it. A specialist may succeed in doing all this (at least for selected simple problems), but it would still then be very difficult to understand and apply the results. In this book, we avoid this trap by flexibly and informally combining simple models such as those shown in Fig. 2.4 in a case-by-case fashion. For example, suppose we have many shared-memory machines connected by a fast network (as is typical for a high-performance computer). Then we could first design a distributed-memory algorithm and then replace its local computations by shared-memory parallel algorithms. The sequential computations of a thread could then be made cache-efficient by viewing them as a computation in the external-memory model. These isolated considerations can be made to work on the overall system by making appropriate modifications to the model parameters. For example, the external-memory algorithm may use only $1/16$ of the L3 cache if there are 16 threads running in parallel on each processor sharing the same L3 cache.

2.5 Parallel Pseudocode

We now introduce two different ways of writing parallel pseudocode.

2.5.1 Single-Program Multiple-Data (SPMD)

Most of our parallel algorithms will look very similar to sequential algorithms. We design a single program that is executed on all PEs. This does not mean that all PEs execute the same sequence of instructions. The program may refer to the local ID i_{proc} and the total number p of PEs. In this way, each incarnation of the program may behave differently. In particular, the different PEs may work on different parts of the data. In the simplest case, each PE performs the same sequence of instructions on a

disjoint part of the data and produces a part of the overall result. No synchronization is needed between the PEs. Such an algorithm is called *embarrasingly parallel*.
Making large parts of a computation embarassingly parallel is a major design goal.
Of course, some kind of interaction between the PEs is needed in most algorithms.
The PEs may interact using primitives from the shared memory model, such as CAS,
or from the distributed-memory model such as *send/receive*. Often, we are even able
to abstract from the concrete model by using constructs that can be implemented in
both models.

In our SPMD programs each PE has its private, local version of every variable.
We write $v@j$ to express a remote access to the copy of v at PE j. We shall even use
this notation for distributed memory algorithms when it is clear how to translate it
into a pair of procedure calls $send(i,v)$ on PE j and $receive(j,\ldots)$ on PE i.

2.5.2 Explicitly Parallel Programs

Sometimes it is more elegant to augment a sequential program with explicit parallel
constructs. In particular, writing **do‖** in a for-loop indicates that the computations
for each loop index can be done independently and in parallel. The runtime system
or the programming language can then distribute the loop iterations over the PEs.
If all these operations take the same time, each PE will be responsible for n/p loop
iterations. Otherwise, one of the load-balancing algorithms presented in Chap. 14
can be used to distribute the loop iterations. This declarative style of parallelization
also extends to initializing arrays.

Threads and Tasks. So far, we have implicitly assumed that there is a one-to-one
correspondence between computational activities (*threads*) and pieces of hardware
(processor cores or hardware-supported threads). Sometimes, however, a more flexible approach is appropriate. For example, allowing more threads than PEs can increase efficiency when some of the threads are waiting for resources. Moreover,
when threads represent pieces of work (*tasks*) of different size, we can leave it to
the runtime system of our programming language to do the load balancing.

In our pseudocode, we separate statements by the separator " ‖ " to indicate that
all these statements can be executed in parallel; it is up to the runtime system how
it exploits this parallelism. We refer to Sect. 14.5 for load balancing algorithms that
make this decision. A parallel statement finishes when all its constituent subtasks
have finished. For example, for a parallel implementation of the recursive multiplication algorithm presented in Sect. 1.4, we could write

$$a0b0 := a_0 \cdot b_0 \parallel a0b1 := a_0 \cdot b_1 \parallel a1b0 := a_1 \cdot b_0 \parallel a1b1 := a_1 \cdot b_1$$

to compute all the partial products in parallel. Note that in a recursive multiplication
algorithm for n-digit numbers, this would result in up to n^2 tasks. Our most frequent
use of the ‖ operator is of the form $send(i,a) \parallel receive(j,b)$ to indicate concurrent
send and receive operations exploiting the full-duplex capability of our distributed-
memory model (see Sect. 2.4.2).

Locks. These are an easy way to manage concurrent data access. A lock protects a set S of memory cells from concurrent access. Typically S is the state of a data structure or a piece of it. We discuss the most simple *binary lock* first. A lock can be held by at most one thread, and only the thread holding the lock can access the memory cells in S. For each lock, there is a corresponding global *lock variable* $\ell_S \in \{0,1\}$; it has the value 1 if some thread holds the lock and has the value 0 otherwise. This variable can be modified only by the procedures *lock* and *unlock*. A thread u can set (or *acquire*) lock ℓ_S by calling $lock(\ell_S)$. This call will wait for any other thread currently holding the lock and returns when it has managed to acquire the lock for thread u. A thread holding a lock may *release* it by calling $unlock(\ell_S)$, which resets ℓ_S to 0. We refer to Sect. 13.4.1 for possible implementations. There are also more general locks distinguishing between read and write access – multiple readers are allowed but a writer needs exclusive access. Section C.2 gives more information.

The simplicity of locks is deceptive. It is easy to create situations of cyclic waiting. Consider a situation with two threads. Thread 1 first locks variable A and then variable B. Thread 2 first locks variable B and then variable A. Suppose both threads reach their first lock operation at the same time. Then thread 1 acquires the lock on A and thread 2 acquires the lock on B. Then both of them proceed to their second lock operation and both will start to wait. Thread 1 waits for thread 2 to release the lock on B, and thread 2 waits for thread 1 to release the lock on A. Hence, they will both wait forever. This situation of cyclic waiting is called a *deadlock*.

Indeed, an interesting area of research are algorithms and data structures that do *not* need locks at all (non-blocking, lock-free, wait-free) and thus avoid many problems connected to locks. The hash table presented in Sects. 4.6.2–4.6.3 is lock-free as long as it does not need to grow.

2.6 Designing Correct Algorithms and Programs

An algorithm is a general method for solving problems of a certain kind. We describe algorithms using natural language and mathematical notation. Algorithms, as such, cannot be executed by a computer. The formulation of an algorithm in a programming language is called a program. Designing correct algorithms and translating a correct algorithm into a correct program are nontrivial and error-prone tasks. In this section, we learn about assertions and invariants, two useful concepts in the design of correct algorithms and programs.

2.6.1 Assertions and Invariants

Assertions and *invariants* describe properties of the program state, i.e., properties of single variables and relations between the values of several variables. Typical properties are that a pointer has a defined value, an integer is nonnegative, a list is nonempty, or the value of an integer variable *length* is equal to the length of a certain list L. Figure 2.7 shows an example of the use of assertions and invariants

Function $power(a : \mathbb{R};\ n_0 : \mathbb{N}) : \mathbb{R}$
    assert $n_0 \geq 0$ *and* $\neg(a = 0 \wedge n_0 = 0)$ // It is not so clear what 0^0 should be
    $p = a : \mathbb{R};\quad r = 1 : \mathbb{R};\quad n = n_0 : \mathbb{N}$ // we have $p^n r = a^{n_0}$
    while $n > 0$ **do**
        invariant $p^n r = a^{n_0}$
        if *n is odd* **then** $n--;\ r := r \cdot p$ // invariant violated between assignments
        else $(n, p) := (n/2, p \cdot p)$ // parallel assignment maintains invariant
    assert $r = a^{n_0}$ // This is a consequence of the invariant and $n = 0$
    return r

Fig. 2.7. An algorithm that computes integer powers of real numbers

in a function $power(a, n_0)$ that computes a^{n_0} for a real number a and a nonnegative integer n_0.

We start with the assertion **assert** $n_0 \geq 0$ and $\neg(a = 0 \wedge n_0 = 0)$. This states that the program expects a nonnegative integer n_0 and that a and n_0 are not allowed to be both 0.[5] We make no claim about the behavior of our program for inputs that violate this assertion. This assertion is therefore called the *precondition* of the program. It is good programming practice to check the precondition of a program, i.e., to write code which checks the precondition and signals an error if it is violated. When the precondition holds (and the program is correct), a *postcondition* holds at the termination of the program. In our example, we assert that $r = a^{n_0}$. It is also good programming practice to verify the postcondition before returning from a program. We shall come back to this point at the end of this section.

One can view preconditions and postconditions as a *contract* between the caller and the called routine: If the caller passes parameters satisfying the precondition, the routine produces a result satisfying the postcondition.

For conciseness, we shall use assertions sparingly, assuming that certain "obvious" conditions are implicit from the textual description of the algorithm. Much more elaborate assertions may be required for safety-critical programs or for formal verification.

Preconditions and postconditions are assertions that describe the initial and the final state of a program or function. We also need to describe properties of intermediate states. A property that holds whenever control passes a certain location in the program is called an *invariant*. Loop invariants and data structure invariants are of particular importance.

2.6.2 Loop Invariants

A *loop invariant* holds before and after each loop iteration. In our example, we claim that $p^n r = a^{n_0}$ before each iteration. This is true before the first iteration. The initialization of the program variables takes care of this. In fact, an invariant frequently

---

[5] The usual convention is $0^0 = 1$. The program is then also correct for $a = 0$ and $n_0 = 0$.

tells us how to initialize the variables. Assume that the invariant holds before execution of the loop body, and $n > 0$. If n is odd, we decrement n and multiply r by p. This reestablishes the invariant (note that the invariant is violated between the assignments). If n is even, we halve n and square p, and again reestablish the invariant. When the loop terminates, we have $p^n r = a^{n_0}$ by the invariant, and $n = 0$ by the condition of the loop. Thus $r = a^{n_0}$ and we have established the postcondition.

The algorithm in Fig. 2.7 and many more algorithms described in this book have a quite simple structure. A few variables are declared and initialized to establish the loop invariant. Then, a main loop manipulates the state of the program. When the loop terminates, the loop invariant together with the termination condition of the loop implies that the correct result has been computed. The loop invariant therefore plays a pivotal role in understanding why a program works correctly. Once we understand the loop invariant, it suffices to check that the loop invariant is true initially and after each loop iteration. This is particularly easy if the loop body consists of only a small number of statements, as in the example above.

2.6.3 Data Structure Invariants

More complex programs encapsulate their state in objects and offer the user an abstract view of the state. The connection between the abstraction and the concrete representation is made by an invariant. Such *data structure invariants* are declared together with the data type. They are true after an object is constructed, and they are preconditions and postconditions of all methods of a class.

For example, we shall discuss the representation of sets by sorted arrays. Here, set is the abstraction and sorted array is the concrete representation. The data structure invariant will state that the data structure uses an array a and an integer n, that n is the size of the set stored, that the set S stored in the data structure is equal to $\{a[1], \ldots, a[n]\}$, and that $a[1] < a[2] < \ldots < a[n]$. The methods of the class have to maintain this invariant, and they are allowed to leverage the invariant; for example, the search method may make use of the fact that the array is sorted.

2.6.4 Certifying Algorithms

We mentioned above that it is good programming practice to check assertions. It is not always clear how to do this efficiently; in our example program, it is easy to check the precondition, but there seems to be no easy way to check the postcondition. In many situations, however, *the task of checking assertions can be simplified by computing additional information*. This additional information is called a *certificate* or *witness*, and its purpose is to simplify the check of an assertion. When an algorithm computes a certificate for the postcondition, we call the algorithm a *certifying algorithm*. We shall illustrate the idea by an example. Consider a function whose input is a graph $G = (V, E)$. Graphs are defined in Sect. 2.12. The task is to test whether the graph is bipartite, i.e., whether there is a labeling of the nodes of G with the colors blue and red such that any edge of G connects nodes of different colors. As specified so far, the function returns true or false – true if G is bipartite, and false otherwise.

With this rudimentary output, the postcondition cannot be checked. However, we may augment the program as follows. When the program declares G bipartite, it also returns a two-coloring of the graph. When the program declares G nonbipartite, it also returns a cycle of odd length in the graph (as a sequence e_1 to e_k of edges). For the augmented program, the postcondition is easy to check. In the first case, we simply check whether all edges connect nodes of different colors, and in the second case, we check that the returned sequence of edges is indeed an odd-length cycle in G. An odd-length cycle proves that the graph is nonbipartite. Most algorithms in this book can be made certifying without increasing the asymptotic running time.

2.7 An Example – Binary Search

Binary search is a very useful technique for searching in an ordered set of elements. We shall use it over and over again in later chapters.

The simplest scenario is as follows. We are given a sorted array $a[1..n]$ of pairwise distinct elements, i.e., $a[1] < a[2] < \ldots < a[n]$, and an element x. We want to find the index k with $a[k-1] < x \leq a[k]$; here, $a[0]$ and $a[n+1]$ should be interpreted as virtual elements with values $-\infty$ and $+\infty$, respectively. We can use these virtual elements in the invariants and the proofs, but cannot access them in the program.

Binary search is based on the principle of divide-and-conquer. We choose an index $m \in [1..n]$ and compare x with $a[m]$. If $x = a[m]$, we are done and return $k = m$. If $x < a[m]$, we restrict the search to the part of the array before $a[m]$, and if $x > a[m]$, we restrict the search to the part of the array after $a[m]$. We need to say more clearly what it means to restrict the search to a subarray. We have two indices ℓ and r and have restricted the search for x to the subarray $a[\ell+1]$ to $a[r-1]$. More precisely, we maintain the invariant

$$\text{(I)} \qquad 0 \leq \ell < r \leq n+1 \quad \text{and} \quad a[\ell] < x < a[r].$$

This is true initially, with $\ell = 0$ and $r = n+1$. Once ℓ and r become consecutive indices, we may conclude that x is not contained in the array. Figure 2.8 shows the complete program.

We now prove correctness of the program. We shall first show that the loop invariant holds whenever the loop condition "$\ell + 1 < r$" is checked. We do so by induction on the number of iterations. We have already established that the invariant holds initially, i.e., if $\ell = 0$ and $r = n+1$. This is the basis of the induction. For the induction step, we have to show that if the invariant holds before the loop condition is checked and the loop condition evaluates to true, then the invariant holds at the end of the loop body. So, assume that the invariant holds before the loop condition is checked and that $\ell + 1 < r$. Then we enter the loop, and $\ell + 2 \leq r$ since ℓ and r are integral. We compute m as $\lfloor (r+\ell)/2 \rfloor$. Since $\ell + 2 \leq r$, we have $\ell < m < r$. Thus m is a legal array index, and we can access $a[m]$. If $x = a[m]$, we stop. Otherwise, we set either $r = m$ or $\ell = m$ and hence have $\ell < r$ and $a[\ell] < x < a[r]$. Thus the invariant holds at the end of the loop body and therefore before the next test of the loop condition. Hence (I) holds whenever the loop condition is checked.

Function $binarySearch(x : Element, a : Array\ [1..n]$ **of** $Element) : 1..n+1$
$\quad(\ell,r) := (0, n+1)$
$\quad$**assert** (I)   // (I) holds here.
$\quad$**while** $\ell + 1 < r$ **do**
$\quad\quad$**invariant** (I): $0 \le \ell < r \le n+1 \wedge a[\ell] < x < a[r]$ // (I) is the loop invariant.
$\quad\quad$**assert** (I) and $\ell + 1 < r$ // Invariant (I) holds here. Also $\ell + 1 < r$.
$\quad\quad m := \lfloor (r+\ell)/2 \rfloor$ // $\ell < m < r$
$\quad\quad s := compare(x, a[m])$ // -1 if $x < a[m]$, 0 if $x = a[m]$, $+1$ if $x > a[m]$
$\quad\quad$**if** $s = 0$ **then return** m // $x = a[m]$
$\quad\quad$**if** $s < 0$ **then** $r := m$ // $a[\ell] < x < a[m] = a[r]$
$\quad\quad$**if** $s > 0$ **then** $\ell := m$ // $a[\ell] = a[m] < x < a[r]$
$\quad\quad$**assert** (I)   // Invariant (I) holds here.
$\quad$**assert** (I) and $\ell + 1 = r$ // Invariant (I) holds here. Also $\ell + 1 = r$.
$\quad$**return** r // $a[r-1] < x < a[r]$

Fig. 2.8. Binary search for x in a sorted array a. Returns an index k with $a[k-1] < x \le a[k]$.

It is now easy to complete the correctness proof. If we do not enter the loop, we have $\ell + 1 \ge r$. Since $\ell < r$ by the invariant and ℓ and r are integral, we have $\ell + 1 = r$. Thus $a[r-1] < x < a[r]$ by the second part of the invariant. We have now established correctness: The program returns either an index k with $a[k] = x$ or an index k with $a[k-1] < x < a[k]$.

We next argue termination. We observe first that if an iteration is not the last one, then we either increase ℓ or decrease r. Hence, $r - \ell$ decreases. Thus the search terminates. We want to show more. We want to show that the search terminates in a logarithmic number of steps. We therefore study the quantity $r - \ell - 1$. This is the number of indices i with $\ell < i < r$, and hence a natural measure of the size of the current subproblem. We shall show that each iteration at least halves the size of the problem. Indeed, in a round $r - \ell - 1$ decreases to something less than or equal to

$$\max\{r - \lfloor (r+\ell)/2 \rfloor - 1, \lfloor (r+\ell)/2 \rfloor - \ell - 1\}$$
$$\le \max\{r - ((r+\ell)/2 - 1/2) - 1, (r+\ell)/2 - \ell - 1\}$$
$$= \max\{(r - \ell - 1)/2, (r-\ell)/2 - 1\} = (r - \ell - 1)/2,$$

and hence is at least halved. We start with $r - \ell - 1 = n + 1 - 0 - 1 = n$, and hence have $r - \ell - 1 \le \lfloor n/2^h \rfloor$ after h iterations.

Let us use k to denote the number of times the comparison between x and $a[m]$ is performed If x occurs in the array, the k-th comparison yields that $x = a[m]$, which ends the search. Otherwise testing the loop condition in the $k + 1$-th iteration yields that $r \le \ell + 1$, and the search ends with this test. So when the loop condition is tested for the k-th time we must have $\ell + 1 < r$. Thus $r - \ell - 1 \ge 1$ after the $k - 1$-th iteration, and hence $1 \le n/2^{k-1}$, which means $k \le 1 + \log n$. We conclude that, at most, $1 + \log n$ comparisons are performed. Since the number of comparisons is a natural number, we can sharpen the bound to $1 + \lfloor \log n \rfloor$.

Theorem 2.3. *Binary search locates an element in a sorted array of size n in at most* $1 + \lfloor \log n \rfloor$ *comparisons between elements. The computation time is* $O(\log n)$.

Exercise 2.6. Show that the above bound is sharp, i.e., for every n, there are instances where exactly $1 + \lfloor \log n \rfloor$ comparisons are needed.

Exercise 2.7. Formulate binary search with two-way comparisons, i.e., distinguish between the cases $x \leq a[m]$ and $x > a[m]$.

We next discuss two important extensions of binary search. First, there is no need for the values $a[i]$ to be stored in an array. We only need the capability to compute $a[i]$, given i. For example, if we have a strictly increasing function f and arguments i and j with $f(i) < x \leq f(j)$, we can use binary search to find $k \in i + 1 .. j$ such that $f(k-1) < x \leq f(k)$. In this context, binary search is often referred to as the *bisection method*.

Second, we can extend binary search to the case where the array is infinite. Assume we have an infinite array $a[1..\infty]$ and some x, and we want to find the smallest k such that $x \leq a[k]$. If x is larger than all elements in the array, the procedure is allowed to diverge. We proceed as follows. We compare x with $a[2^0]$, $a[2^1]$, $a[2^2]$, $a[2^3]$, ..., until the first i with $x \leq a[2^i]$ is found. This is called an *exponential search*. If $x = a[2^i]$ or $i \leq 1$ (note that in the latter case, either $x \leq a[1]$ or $a[1] < x < a[2]$ or $x = a[2]$), we are done. Otherwise, $i > 1$ and $a[2^{i-1}] < x < a[2^i]$, and we complete the task by binary search on the subarray $a[2^{i-1} + 1 .. 2^i - 1]$. This subarray contains $2^i - 2^{i-1} - 1 = 2^{i-1} - 1$ elements. Note that one comparison is carried out if $x \leq a[1]$.

Theorem 2.4. *The combination of exponential and binary search finds* $x > a[1]$ *in an unbounded sorted array in at most* $2 \lceil \log k \rceil$ *comparisons, where* $a[k-1] < x \leq a[k]$.

Proof. If $a[1] < x \leq a[2]$, two comparisons are needed. So, we may assume $k > 2$, and hence the exponential search ends with $i > 1$. We need $i + 1$ comparisons to find the smallest i such that $x \leq a[2^i]$ and $\lfloor \log(2^i - 2^{i-1} - 1) \rfloor + 1 = \lfloor \log(2^{i-1} - 1) \rfloor + 1 = i - 1$ comparisons for the binary search. This gives a total of $2i$ comparisons. Since $k > 2^{i-1}$, we have $i < 1 + \log k$, and the claim follows. Note that $i \leq \lceil \log k \rceil$ since i is integral. $\qquad\square$

Binary search is certifying. It returns an index k with $a[k-1] < x \leq a[k]$. If $x = a[k]$, the index proves that x is stored in the array. If $a[k-1] < x < a[k]$ and the array is sorted, the index proves that x is not stored in the array. Of course, if the array violates the precondition and is not sorted, we know nothing. There is no way to check the precondition in logarithmic time.

We have described binary search as an iterative program. It can also, and maybe even more naturally, be described as a recursive procedure; see Fig. 2.9. As above, we assume that we have two indices ℓ and r into an array a with index set $1..n$ such that $0 \leq \ell < r \leq n + 1$ and $a[\ell] < x < a[r]$. If $r = \ell + 1$, we stop. This is correct by the assertion $a[\ell] < x < a[r]$. Otherwise, we compute $m = \lfloor (\ell + r)/2 \rfloor$. Then $\ell < m < r$. Hence we may access $a[m]$ and compare x with this entry (in a three-way fashion). If $x = a[m]$, we found x and return m. This is obviously correct. If $x < a[m]$, we make

Function *binSearch(x : Element, $\ell, r : 0..n+1$, a : Array $[1..n]$ **of** Element) : $1..n+1$*
    assert $0 \leq \ell < r \leq n+1 \wedge a[\ell] < x < a[r]$ // The precondition
    if $\ell+1 = r$ **then return** r // x is not in the array and $a[r-1] < x < a[r]$
    $m := \lfloor (r+\ell)/2 \rfloor$ // $\ell < m < r$
    $s := compare(x, a[m])$ // -1 if $x < a[m]$, 0 if $x = a[m]$, $+1$ if $x > a[m]$
    if $s = 0$ **then return** m // $x = a[m]$
    if $s < 0$ **then return** $binSearch(x, l, m, a)$
    if $s > 0$ **then return** $binSearch(x, m, r, a)$

Fig. 2.9. A recursive function for binary search

the recursive call for the index pair (ℓ, m). Note that $a[\ell] < x < a[m]$ and hence the precondition of the recursive call is satisfied. If $x > a[m]$, we make the recursive call for the index pair (m, r).

Observe that at most one recursive call is generated, and that the answer to the recursive call is also the overall answer. This situation is called *tail recursion*. Tail recursive procedures are easily turned into loops. The body of the recursive procedure becomes the loop body. Each iteration of the loop corresponds to a recursive call; going to the next recursive call with new parameters is realized by going to the next round in the loop, after changing variables. The resulting program is our iterative version of binary search.

2.8 Basic Algorithm Analysis

In this section, we introduce a set of simple rules for determining the running time of pseudocode. We start with a summary of the principles of algorithm analysis as we established them in the preceding sections. We abstract from the complications of a real machine to the simplified RAM model. In the RAM model, running time is measured by the number of instructions executed. We simplify the analysis further by grouping inputs by size and focusing on the worst case. The use of asymptotic notation allows us to ignore constant factors and lower-order terms. This coarsening of our view also allows us to look at upper bounds on the execution time rather than the exact worst case, as long as the asymptotic result remains unchanged. The total effect of these simplifications is that the running time of pseudocode can be analyzed directly. There is no need to translate the program into machine code first.

We now come to the set of rules for analyzing pseudocode. Let $T(I)$ denote the worst-case execution time of a piece of program I. The following rules then tell us how to estimate the running time for larger programs, given that we know the running times of their constituents:

- *Sequential composition*: $T(I; I') = T(I) + T(I')$.
- *Conditional instructions*: $T(\text{if } C \text{ then } I \text{ else } I') = O(T(C) + \max(T(I), T(I')))$.

- *Loops*: $T(\textbf{repeat } I \textbf{ until } C) = \mathrm{O}\left(\sum_{i=1}^{k(n)} T(I,C,i)\right)$, where $k(n)$ is the maximum number of loop iterations on inputs of length n, and $T(I,C,i)$ is the time needed in the ith iteration of the loop, including the test C.

We postpone the treatment of subroutine calls to Sect. 2.8.2. Of the rules above, only the rule for loops is nontrivial to apply; it requires evaluating sums.

2.8.1 "Doing Sums"

We introduce some basic techniques for evaluating sums. Sums arise in the analysis of loops, in average-case analysis, and also in the analysis of randomized algorithms.

For example, the insertion sort algorithm introduced in Sect. 5.1 has two nested loops. The loop variable i of the outer loop runs from from 2 to n. For any i, the inner loop performs at most $i - 1$ iterations. Hence, the total number of iterations of the inner loop is at most

$$\sum_{i=2}^{n} (i-1) = \sum_{i=1}^{n-1} i = \frac{n(n-1)}{2} = \Theta(n^2),$$

where the second equality comes from (A.12). Since the time for one execution of the inner loop is $\Theta(1)$, we get a worst-case execution time of $\Theta(n^2)$.

All nested loops with an easily predictable number of iterations can be analyzed in an analogous fashion: Work your way outwards by repeatedly finding a closed-form expression for the currently innermost loop. Using simple manipulations such as $\sum_i c a_i = c \sum_i a_i$, $\sum_i (a_i + b_i) = \sum_i a_i + \sum_i b_i$, or $\sum_{i=2}^{n} a_i = -a_1 + \sum_{i=1}^{n} a_i$, one can often reduce the sums to simple forms that can be looked up in a catalog of sums. A small sample of such formulae can be found in Appendix A.4. Since we are usually interested only in the asymptotic behavior, we can frequently avoid doing sums exactly and resort to estimates. For example, instead of evaluating the sum above exactly, we may argue more simply as follows:

$$\sum_{i=2}^{n} (i-1) \le \sum_{i=1}^{n} n = n^2 = \mathrm{O}(n^2),$$

$$\sum_{i=2}^{n} (i-1) \ge \sum_{i=\lceil n/2 \rceil}^{n} n/2 = \lfloor n/2 \rfloor \cdot n/2 = \Omega(n^2).$$

2.8.2 Recurrences

In our rules for analyzing programs, we have so far neglected subroutine calls. Non-recursive subroutines are easy to handle, since we can analyze the subroutine separately and then substitute the bound obtained into the expression for the running time of the calling routine. For recursive programs, however, this approach does not lead to a closed formula, but to a recurrence relation.

For example, for the recursive variant of the school method of multiplication, we obtained $T(1) = 1$ and $T(n) = 4T(\lceil n/2 \rceil) + 4n$ for the number of primitive operations. For the Karatsuba algorithm, the corresponding equations were $T(n) = 3n^2$ for $n \leq 3$ and $T(n) = 3T(\lceil n/2 \rceil + 1) + 8n$ for $n > 3$. In general, a *recurrence relation* (or recurrence) defines a function in terms of the values of the same function on smaller arguments. Explicit definitions for small parameter values (the base case) complete the definition. Solving recurrences, i.e., finding nonrecursive, closed-form expressions for the functions defined by them, is an interesting subject in mathematics. Here we focus on the recurrence relations that typically emerge from divide-and-conquer algorithms. We begin with a simple case that will suffice for the purpose of understanding the main ideas. We have a problem of size $n = b^k$ for some integer k. If $k \geq 1$, we invest linear work cn dividing the problem into d subproblems of size n/b and in combining the results. If $k = 0$, there are no recursive calls, we invest work a in computing the result directly, and are done.

Theorem 2.5 (master theorem (simple form)). *For positive constants a, b, c, and d, and integers n that are nonnegative powers of b, consider the recurrence*

$$r(n) = \begin{cases} a & \text{if } n = 1, \\ cn + d \cdot r(n/b) & \text{if } n > 1. \end{cases}$$

Then,

$$r(n) = \begin{cases} \Theta(n) & \text{if } d < b, \\ \Theta(n \log n) & \text{if } d = b, \\ \Theta(n^{\log_b d}) & \text{if } d > b. \end{cases}$$

Figure 2.10 illustrates the main insight behind Theorem 2.5. We consider the amount of work done at each level of recursion. We start with a problem of size n, i.e., at the zeroth level of the recursion we have one problem of size n. At the first level of the recursion, we have d subproblems of size n/b. From one level of the recursion to the next, the number of subproblems is multiplied by a factor of d and the size of the subproblems shrinks by a factor of b. Therefore, at the ith level of the recursion, we have d^i problems, each of size n/b^i. Thus the total size of the problems at the ith level is equal to

$$d^i \frac{n}{b^i} = n \left(\frac{d}{b} \right)^i.$$

The work performed for a problem (excluding the time spent in recursive calls) is c times the problem size, and hence the work performed at any level of the recursion is proportional to the total problem size at that level. Depending on whether d/b is less than, equal to, or larger than 1, we have different kinds of behavior.

If $d < b$, the work *decreases geometrically* with the level of recursion and the *topmost* level of recursion accounts for a constant fraction of the total execution time. If $d = b$, we have the same amount of work at *every* level of recursion. Since there are logarithmically many levels, the total amount of work is $\Theta(n \log n)$. Finally, if $d > b$, we have a geometrically *growing* amount of work at each level of recursion so that the *last* level accounts for a constant fraction of the total running time. We formalize this reasoning next.

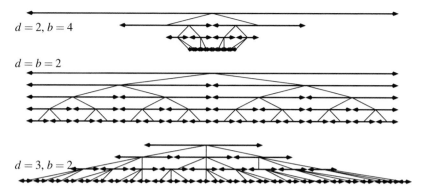

Fig. 2.10. Examples of the three cases of the master theorem. Problems are indicated by horizontal line segments with arrows at both ends. The length of a segment represents the size of the problem, and the subproblems resulting from a problem are shown in the line below it. The topmost part of figure corresponds to the case $d = 2$ and $b = 4$, i.e., each problem generates two subproblems of one-fourth the size. Thus the total size of the subproblems is only half of the original size. The middle part of the figure illustrates the case $d = b = 2$, and the bottommost part illustrates the case $d = 3$ and $b = 2$.

Proof. We start with a single problem of size $n = b^k$. We call this level zero of the recursion.[6] At level 1, we have d problems, each of size $n/b = b^{k-1}$. At level 2, we have d^2 problems, each of size $n/b^2 = b^{k-2}$. At level i, we have d^i problems, each of size $n/b^i = b^{k-i}$. At level k, we have d^k problems, each of size $n/b^k = b^{k-k} = 1$. Each such problem has a cost of a, and hence the total cost at level k is ad^k.

Let us next compute the total cost of the divide-and-conquer steps at levels 0 to $k - 1$. At level i, we have d^i recursive calls each for subproblems of size b^{k-i}. Each call contributes a cost of $c \cdot b^{k-i}$, and hence the cost at level i is $d^i \cdot c \cdot b^{k-i}$. Thus the combined cost over all levels is

$$\sum_{i=0}^{k-1} d^i \cdot c \cdot b^{k-i} = c \cdot b^k \cdot \sum_{i=0}^{k-1} \left(\frac{d}{b}\right)^i = cn \cdot \sum_{i=0}^{k-1} \left(\frac{d}{b}\right)^i.$$

We now distinguish cases according to the relative sizes of d and b.

Case $d = b$. We have a cost of $ad^k = ab^k = an = \Theta(n)$ for the bottom of the recursion and cost of $cnk = cn\log_b n = \Theta(n\log n)$ for the divide-and-conquer steps.

Case $d < b$. We have a cost of $ad^k < ab^k = an = O(n)$ for the bottom of the recursion. For the cost of the divide-and-conquer steps, we use the summation formula (A.14) for a geometric series, namely $\sum_{0 \le i < k} q^i = (1 - q^k)/(1 - q)$ for $q > 0$ and $q \ne 1$, and obtain

---

[6] In this proof, we use the terminology of recursive programs in order to provide intuition. However, our mathematical arguments apply to any recurrence relation of the right form, even if it does not stem from a recursive program.

$$cn \cdot \sum_{i=0}^{k-1} \left(\frac{d}{b}\right)^i = cn \cdot \frac{1-(d/b)^k}{1-d/b} < cn \cdot \frac{1}{1-d/b} = O(n)$$

and

$$cn \cdot \sum_{i=0}^{k-1} \left(\frac{d}{b}\right)^i = cn \cdot \frac{1-(d/b)^k}{1-d/b} > cn = \Omega(n).$$

Case $d > b$. First, note that

$$d^k = (b^{\log_b d})^k = b^{k\log_b d} = (b^k)^{\log_b d} = n^{\log_b d}.$$

Hence the bottom of the recursion has a cost of $an^{\log_b d} = \Theta(n^{\log_b d})$. For the divide-and-conquer steps we use the geometric series again and obtain

$$cb^k \frac{(d/b)^k - 1}{d/b - 1} = c\frac{d^k - b^k}{d/b - 1} = cd^k \frac{1-(b/d)^k}{d/b - 1} = \Theta(d^k) = \Theta(n^{\log_b d}). \qquad \square$$

There are many generalizations of the master theorem: We might break the recursion earlier, the cost of dividing and conquering might be nonlinear, the size of the subproblems might vary within certain bounds or vary stochastically, the number of subproblems might depend on the input size, etc. We refer the reader to the books [137, 289] and the papers [13, 100, 267] for further information. The recurrence $T(n) = 3n^2$ for $n \le 3$ and $T(n) \le 3T(\lceil n/2 \rceil + 1) + 8n$ for $n \ge 4$, governing Karatsuba's algorithm, is not covered by our master theorem, which neglects rounding issues. We shall now state, without proof, a more general version of the master theorem. Let $r(n)$ satisfy

$$r(n) \le \begin{cases} a & \text{if } n \le n_0, \\ cn^s + d \cdot r(\lceil n/b \rceil + e_n) & \text{if } n > n_0, \end{cases} \qquad (2.1)$$

where $a > 0$, $b > 1$, $c > 0$, $d > 0$, and $s \ge 0$ are real constants, and the e_n, for $n > n_0$, are integers such that $-\lceil n/b \rceil < e_n \le e$ for some integer $e \ge 0$. In the recurrence governing Karatsuba's algorithm, we have $n_0 = 3$, $a = 27$, $c = 8$, $s = 1$, $d = 3$, $b = 2$, and $e_n = 1$ for $n \ge 4$.

Theorem 2.6 (master theorem (general form)).
If $r(n)$ satisfies the recurrence (2.1), then

$$r(n) = \begin{cases} O(n^s) & \text{for } d < b^s, \text{ i.e., } \log_b d < s, \\ O(n^s \log n) & \text{for } d = b^s, \text{ i.e., } \log_b d = s, \\ O(n^{\log_b d}) & \text{for } d > b^s, \text{ i.e., } \log_b d > s. \end{cases} \qquad (2.2)$$

Exercise 2.8. Consider the recurrence

$$C(n) = \begin{cases} 1 & \text{if } n = 1, \\ C(\lfloor n/2 \rfloor) + C(\lceil n/2 \rceil) + cn & \text{if } n > 1. \end{cases}$$

Show that $C(n) = O(n \log n)$.

*Exercise 2.9. Suppose you have a divide-and-conquer algorithm whose running time is governed by the recurrence $T(1) = a$, $T(n) \leq cn + \lceil \sqrt{n} \, \rceil \, T(\lceil n / \lceil \sqrt{n} \, \rceil \rceil)$. Show that the running time of the program is $O(n \log \log n)$. Hint: Define a function $S(n)$ by $S(n) = T(n)/n$ and study the recurrence for $S(n)$.

Exercise 2.10. Access to data structures is often governed by the following recurrence: $T(1) = a$, $T(n) = c + T(n/2)$. Show that $T(n) = O(\log n)$.

2.8.3 Global Arguments

The algorithm analysis techniques introduced so far are syntax-oriented in the following sense: In order to analyze a large program, we first analyze its parts and then combine the analyses of the parts into an analysis of the large program. The combination step involves sums and recurrences.

We shall also use a quite different approach, which one might call semantics-oriented. In this approach we associate parts of the execution with parts of a combinatorial structure and then argue about the combinatorial structure. For example, we might argue that a certain piece of program is executed at most once for each edge of a graph or that the execution of a certain piece of program at least doubles the size of a certain structure, that the size is 1 initially and at most n at termination, and hence the number of executions is bounded logarithmically.

2.9 Average-Case Analysis

In this section we shall introduce you to average-case analysis. We shall do so by way of three examples of increasing complexity. We assume that you are familiar with basic concepts of probability theory such as discrete probability distributions, expected values, indicator variables, and the linearity of expectations. The use of the language and tools of probability theory suggests the following approach to average case analysis. We view the inputs as coming from a probability space, e.g., all inputs from a certain size with the uniform distribution, and determine the expexted complexity for an instance sampled randomly from this space. Section A.3 reviews the basic probability theory.

2.9.1 Incrementing a Counter

We begin with a very simple example. Our input is an array $a[0..n-1]$ filled with digits 0 and 1. We want to increment the number represented by the array by 1:

```
i := 0
while (i < n and a[i] = 1) do a[i] = 0; i++;
if i < n then a[i] = 1
```

How often is the body of the while-loop executed? Clearly, n times in the worst case and 0 times in the best case. What is the average case? The first step in an average-case analysis is always to define the model of randomness, i.e., to define the underlying probability space. We postulate the following model of randomness: Each digit is 0 or 1 with probability $1/2$, and different digits are independent. Alternatively, we may say that all bit strings of length n are equally likely. The loop body is executed k times, if either $k < n$ and $a[0] = a[1] = \ldots = a[k-1] = 1$ and $a[k] = 0$ or if $k = n$ and all digits of a are equal to 1. The former event has probability $2^{-(k+1)}$, and the latter event has probability 2^{-n}. Therefore, the average number of executions is equal to

$$\sum_{0 \le k < n} k 2^{-(k+1)} + n 2^{-n} \le \sum_{k \ge 0} k 2^{-k} = 2,$$

where the last equality is the same as (A.15).

2.9.2 Left-to-Right Maxima

Our second example is slightly more demanding. Consider the following simple program that determines the maximum element in an array $a[1..n]$:

$m := a[1];$ **for** $i := 2$ **to** n **do if** $a[i] > m$ **then** $m := a[i]$

How often is the assignment $m := a[i]$ executed? In the worst case, it is executed in every iteration of the loop and hence $n - 1$ times. In the best case, it is not executed at all. What is the average case? Again, we start by defining the probability space. We assume that the array contains n distinct elements and that any order of these elements is equally likely. In other words, our probability space consists of the $n!$ permutations of the array elements. Each permutation is equally likely and therefore has probability $1/n!$. Since the exact nature of the array elements is unimportant, we may assume that the array contains the numbers 1 to n in some order. We are interested in the average number of *left-to-right maxima*. A left-to-right maximum in a sequence is an element which is larger than all preceding elements. So, $(1, 2, 4, 3)$ has three left-to-right-maxima and $(3, 1, 2, 4)$ has two left-to-right-maxima. For a permutation π of the integers 1 to n, let $M_n(\pi)$ be the number of left-to-right-maxima. What is $E[M_n]$? We shall describe two ways to determine the expectation. For small n, it is easy to determine $E[M_n]$ by direct calculation. For $n = 1$, there is only one permutation, namely (1), and it has one maximum. So $E[M_1] = 1$. For $n = 2$, there are two permutations, namely $(1, 2)$ and $(2, 1)$. The former has two maxima and the latter has one maximum. So $E[M_2] = 1.5$. For larger n, we argue as follows.

We write M_n as a sum of indicator variables I_1 to I_n, i.e., $M_n = I_1 + \ldots + I_n$, where I_k is equal to 1 for a permutation π if the kth element of π is a left-to-right maximum. For example, $I_3((3, 1, 2, 4)) = 0$ and $I_4((3, 1, 2, 4)) = 1$. We have

$$\begin{aligned}
E[M_n] &= E[I_1 + I_2 + \ldots + I_n] \\
&= E[I_1] + E[I_2] + \ldots + E[I_n] \\
&= \text{prob}(I_1 = 1) + \text{prob}(I_2 = 1) + \ldots + \text{prob}(I_n = 1),
\end{aligned}$$

where the second equality is the linearity of expectations (A.3) and the third equality follows from the I_k's being indicator variables. It remains to determine the probability that $I_k = 1$. The kth element of a random permutation is a left-to-right maximum if and only if the kth element is the largest of the first k elements. In a random permutation, any position is equally likely to hold the maximum, so that the probability we are looking for is $\mathrm{prob}(I_k = 1) = 1/k$ and hence

$$E[M_n] = \sum_{1 \le k \le n} \mathrm{prob}(I_k = 1) = \sum_{1 \le k \le n} \frac{1}{k}.$$

So, $E[M_4] = 1 + 1/2 + 1/3 + 1/4 = (12 + 6 + 4 + 3)/12 = 25/12$. The sum $\sum_{1 \le k \le n} 1/k$ will appear several times in this book. It is known under the name "nth harmonic number" and is denoted by H_n. It is known that $\ln n \le H_n \le 1 + \ln n$, i.e., $H_n \approx \ln n$; see (A.13). We conclude that the average number of left-to-right maxima is much smaller than their maximum number.

Exercise 2.11. Show that $\sum_{k=1}^{n} \frac{1}{k} \le \ln n + 1$. Hint: Show first that $\sum_{k=2}^{n} \frac{1}{k} \le \int_{1}^{n} \frac{1}{x} \, dx$.

We now describe an alternative analysis. We introduce A_n as a shorthand for $E[M_n]$ and set $A_0 = 0$. The first element is always a left-to-right maximum, and each number is equally likely as the first element. If the first element is equal to i, then only the numbers $i + 1$ to n can be further left-to-right maxima. They appear in random order in the remaining sequence, and hence we shall see an expected number of A_{n-i} further maxima. Thus

$$A_n = 1 + \left(\sum_{1 \le i \le n} A_{n-i} \right) / n \quad \text{or} \quad nA_n = n + \sum_{0 \le i \le n-1} A_i.$$

A simple trick simplifies this recurrence. The corresponding equation for $n - 1$ instead of n is $(n - 1)A_{n-1} = n - 1 + \sum_{1 \le i \le n-2} A_i$. Subtracting the equation for $n - 1$ from the equation for n yields

$$nA_n - (n - 1)A_{n-1} = 1 + A_{n-1} \quad \text{or} \quad A_n = 1/n + A_{n-1},$$

and hence $A_n = H_n$.

2.9.3 Linear Search

We come now to our third example; this example is even more demanding. Consider the following search problem. We have items 1 to n, which we are required to arrange linearly in some order; say, we put item i in position ℓ_i. Once we have arranged the items, we perform searches. In order to search for an item x, we go through the sequence from left to right until we encounter x. In this way, it will take ℓ_i steps to access item i.

Suppose now that we also know that we shall access the items with fixed probabilities; say, we shall search for item i with probability p_i, where $p_i \ge 0$ for all i, $1 \le i \le n$, and $\sum_i p_i = 1$. In this situation, the *expected* or *average* cost of a search is equal to $\sum_i p_i \ell_i$, since we search for item i with probability p_i and the cost of the search is ℓ_i.

What is the best way of arranging the items? Intuition tells us that we should arrange the items in order of decreasing probability. Let us prove this.

Lemma 2.7. *An arrangement is optimal with respect to the expected search cost if it has the property that $p_i > p_j$ implies $\ell_i < \ell_j$. If $p_1 \geq p_2 \geq \ldots \geq p_n$, the placement $\ell_i = i$ results in the optimal expected search cost $Opt = \sum_i p_i i$.*

Proof. Consider an arrangement in which, for some i and j, we have $p_i > p_j$ and $\ell_i > \ell_j$, i.e., item i is more probable than item j and yet placed after it. Interchanging items i and j changes the search cost by

$$-(p_i\ell_i + p_j\ell_j) + (p_i\ell_j + p_j\ell_i) = (p_j - p_i)(\ell_i - \ell_j) < 0,$$

i.e., the new arrangement is better and hence the old arrangement is not optimal.

Let us now consider the case $p_1 > p_2 > \ldots > p_n$. Since there are only $n!$ possible arrangements, there is an optimal arrangement. Also, if $i < j$ and item i is placed after item j, the arrangement is not optimal by the argument in the preceding paragraph. Thus the optimal arrangement puts item i in position $\ell_i = i$ and its expected search cost is $\sum_i p_i i$.

If $p_1 \geq p_2 \geq \ldots \geq p_n$, the arrangement $\ell_i = i$ for all i is still optimal. However, if some probabilities are equal, we have more than one optimal arrangement. Within blocks of equal probabilities, the order is irrelevant.   □

Can we still do something intelligent if the probabilities p_i are not known to us? The answer is yes, and a very simple heuristic does the job. It is called the *move-to-front heuristic*. Suppose we access item i and find it in position ℓ_i. If $\ell_i = 1$, we are happy and do nothing. Otherwise, we place the item in position 1 and move the items in positions 1 to $\ell_i - 1$ by one position to the rear. The hope is that, in this way, frequently accessed items tend to stay near the front of the arrangement and infrequently accessed items move to the rear. We shall now analyze the expected behavior of the move-to-front heuristic.

We assume for the analysis that we start with an arbitrary, but fixed, initial arrangement of the n items and then perform search rounds. In each round, we access item i with probability p_i *independently of what happened in the preceding rounds*. Since the cost of the first access to any item is essentially determined by the initial configuration, we shall ignore it and assign a cost of 1 to it.[7] We now compute the expected cost in round t. We use C_{MTF} to denote this expected cost. Let ℓ_i be the position of item i at the beginning of round t. The quantities $\ell_1, \ldots, \ell_n$ are random variables that depend only on the accesses in the first $t - 1$ rounds; recall that we assume a fixed initial arrangement. If we access item i in round t, we incur a cost of $1 + Z_i$, where[8]

---

[7] The cost ignored in this way is at most $n(n-1)$. One can show that the expected cost in round t ignored in this way is no more than n^2/t.

[8] We define the cost as $1 + Z_i$, so that $Z_i = 0$ is the second case.

$$Z_i = \begin{cases} \ell_i - 1 & \text{if } i \text{ was accessed before round } t, \\ 0 & \text{otherwise.} \end{cases}$$

Of course, the random variables $Z_1, \ldots, Z_n$ also depend only on the sequence of accesses in the first $t - 1$ rounds. Thus

$$C_{\text{MTF}} = \sum_i p_i (1 + \text{E}[Z_i]) = 1 + \sum_i p_i \text{E}[Z_i].$$

We next estimate the expectation $\text{E}[Z_i]$. For this purpose, we define for each $j \neq i$ an indicator variable

$$I_{ij} = \begin{cases} 1 & \text{if } j \text{ is located before } i \text{ at the beginning of round } t \\ & \text{and at least one of the two items was accessed before round } t, \\ 0 & \text{otherwise.} \end{cases}$$

Then $Z_i \leq \sum_{j;\ j \neq i} I_{ij}$. Indeed, if i is accessed for the first time in round t, $Z_i = 0$. If i was accessed before round t, then $I_{ij} = 1$ for every j that precedes i in the list, and hence $Z_i = \sum_{j;\ j \neq i} I_{ij}$. Thus $\text{E}[Z_i] \leq \sum_{j;\ j \neq i} \text{E}[I_{ij}]$. We are now left with the task of estimating the expectations $\text{E}[I_{ij}]$.

If there was no access to either i or j before round t, $I_{ij} = 0$. Otherwise, consider the last round before round t in which either i or j was accessed. The (conditional) probability that this access was to item j and not to item i is $p_j/(p_i + p_j)$. Therefore, $\text{E}[I_{ij}] = \text{prob}(I_{ij} = 1) \leq p_j/(p_i + p_j)$, and hence $\text{E}[Z_i] \leq \sum_{j;\ j \neq i} p_j/(p_i + p_j)$. Summation over i yields

$$C_{\text{MTF}} = 1 + \sum_i p_i \text{E}[Z_i] \leq 1 + \sum_{i,j;\ i \neq j} \frac{p_i p_j}{p_i + p_j}.$$

Observe that for each i and j with $i \neq j$, the term $p_i p_j/(p_i + p_j) = p_j p_i/(p_j + p_i)$ appears twice in the sum above. In order to proceed with the analysis, we assume $p_1 \geq p_2 \geq \cdots \geq p_n$. We use this assumption in the analysis, but the algorithm has no knowledge of this. With $\sum_i p_i = 1$, we obtain

$$C_{\text{MTF}} \leq 1 + 2 \sum_{i,j;\ j < i} \frac{p_i p_j}{p_i + p_j} = \sum_i p_i \left(1 + 2 \sum_{j;\ j < i} \frac{p_j}{p_i + p_j} \right)$$

$$\leq \sum_i p_i \left(1 + 2 \sum_{j;\ j < i} 1 \right) < \sum_i p_i 2i = 2 \sum_i p_i i = 2Opt.$$

Theorem 2.8. *If the cost of the first access to each item is ignored, the expected search cost of the move-to-front-heuristic is at most twice the cost of the optimal fixed arrangement.*

2.10 Parallel-Algorithm Analysis

Analyzing a sequential algorithm amounts to estimating the execution time $T_{seq}(I)$ of a program for a given input instance I on a RAM. Now, we want to find the execution time $T_{par}(I, p)$ as a function of both the input instance and the number of available processors p of some parallel machine. As we are now studying a function of two variables, we should expect some complications. However, the basic tools introduced above – evaluating sums and recurrences, using asymptotics, … – will be equally useful for the analysis of parallel programs. We begin with some quantities derived from T_{seq} and T_{par} that will help us to understand the results of the analysis.

The (absolute) *speedup*

$$S(I, p) := \frac{T_{seq}(I)}{T_{par}(I, p)} \tag{2.3}$$

gives the factor of speed improvement compared with the best known sequential program for the same problem. Sometimes the relative speedup $T_{par}(I, 1)/T_{par}(I, p)$ is considered, but this is problematic because it does not tell us anything about how useful parallelization was. Of course, we would like to have a large speedup. But how large is good? Ideally, we would like to have $S = p$ – perfect speedup[9]. Even $S = \Theta(p)$ – *linear speedup* – is good. Since speedup $\Theta(p)$ means "good", it makes sense to normalize speedup to the *efficiency*

$$E(I, p) := \frac{S(I, p)}{p}, \tag{2.4}$$

so that we are now looking for constant efficiency. When do we call a parallel algorithm good or efficient? A common definition is to say that a parallel algorithm is efficient if it achieves constant efficiency for all sufficiently large inputs. The input size for which it achieves constant efficiency may grow with the number of processors. The *isoefficiency function* $I(p)$ measures this growth [192]. Let c be a constant. For any number p of processors, let $I(p)$ be the smallest n such that $E(I, p) \geq c$ for all instances I of size at least n. The isoefficiency function measures the scalability of an algorithm – the more slowly it grows as a function of p, the more scalable the algorithm is.

Brent's Principle. Brent's principle is a general method for converting inefficient parallel programs into efficient parallel programs for a smaller number of processors. It is best illustrated by an example. Assume we want to sum n numbers that are given in a global array. Clearly, $T_{seq}(n) = \Theta(n)$. With $n = p$ and the fast parallel sum algorithm presented in Sect. 2.4.2, we obtain a parallel execution time $O(\log p)$ and efficiency $O(1/\log p)$ – this algorithm is inefficient. However, when $n \gg p$ we can use a simple trick to get an efficient parallelization. We first use p copies of the sequential algorithm on subproblems of size n/p, and then

---

[9] There are situations where we can do even better. $S > p$ can happen if parallel execution mobilizes more resources. For example, on the parallel machine the input might fit into the aggregate cache of all PEs while a single PE needs to access the main memory a lot.

use the parallel algorithm with p processors to sum the p partial sums. Each PE adds n/p numbers sequentially and requires time $\Theta(n/p)$. The summation of the p partial results takes time $\Theta(\log p)$. Thus the overall parallel execution time is $T_{\text{par}} = \Theta(n/p + \log p)$. We obtain speedup $S = \Theta(n/(n/p + \log p))$ and efficiency $E = \Theta(n/(p(n/p + \log p))) = \Theta(1/(1 + (p\log p)/n))$. For E to be constant, we need $p\log(p)/n = O(1)$, i.e., $n = \Omega(p\log p)$. Hence, the isoefficiency of the algorithm is $I(p) = \Theta(p\log p)$.

Work and Span. The work, work(I), of a parallel algorithm performed on an instance I is the number of operations executed by the algorithm. Its span, span(I), is the execution time $T(I, \infty)$ if an unlimited number of processors is available. These two quantities allow us to obtain a rather abstract view of the efficiency and scalability of a parallel algorithm. Clearly, $T(I, p) \geq \text{span}(I)$ and $T(I, p) \geq \text{work}(I)/p$ are obvious lower bounds on the parallel execution time. So even with an infinite number of processors, the speed-up cannot be better than span$(I)/T_{\text{seq}}(I)$ (Amdahl's law). On the other hand, we can achieve $T(I, p) = O(\text{work}(I)/p + \text{span}(I))$ if we manage to schedule the computations in such a way that no PE is unnecessarily idle. In Sect. 14.5 we shall see that this is often possible. In this case, the algorithm is efficient provided that work$(I) = O(T_{\text{seq}}(I))$ and span$(I) = O(\text{work}(I)/p)$. Indeed,

$$
\begin{aligned}
T(I, p) &= O(\text{work}(I)/p + \text{span}(I)) \\
&= O(\text{work}(I)/p) && \text{since span}(I) = O(\text{work}(I)/p) \\
&= O(T_{\text{seq}}(I)) && \text{since work}(I) = O(T_{\text{seq}}(I)).
\end{aligned}
$$

For the array-sum example discussed above, we have work$(n) = \Theta(n)$ and span$(n) = \Theta(\log n)$, i.e., the algorithm is efficient if $\log n = O(n/p)$, i.e., when $p = O(n/\log n)$ or $n = \Omega(p\log p)$. Analyzing work and span allows us to consider scalability in a way similar to what we can with the isoefficiency function.

2.11 Randomized Algorithms

Suppose you are offered to participate in a TV game show. There are 100 boxes that you can open in an order of your choice. Box i contains an amount m_i of money. This amount is unknown to you but becomes known once the box is opened. No two boxes contain the same amount of money. The rules of the game are very simple:

- At the beginning of the game, the presenter gives you 10 tokens.
- When you open a box and the amount in the box is larger than the amount in all previously opened boxes, you have to hand back a token.[10]
- When you have to hand back a token but have no tokens, the game ends and you lose.
- When you manage to open all of the boxes, you win and can keep all the money.

---

[10] The amount in the first box opened is larger than the amount in all previously opened boxes, and hence the first token goes back to the presenter in the first round.

There are strange pictures on the boxes, and the presenter gives hints by suggesting the box to be opened next. Your aunt, who is addicted to this show, tells you that only a few candidates win. Now, you ask yourself whether it is worth participating in this game. Is there a strategy that gives you a good chance of winning? Are the presenters's hints useful?

Let us first analyze the obvious algorithm – you always follow the presenter. The worst case is that he makes you open the boxes in order of increasing value. Whenever you open a box, you have to hand back a token, and when you open the 11th box you are dead. The candidates and viewers would hate the presenter and he would soon be fired. Worst-case analysis does not give us the right information in this situation. The best case is that the presenter immediately tells you the best box. You would be happy, but there would be no time to place advertisements, so the presenterr would again be fired. Best-case analysis also does not give us the right information in this situation.

We next observe that the game is really the left-to-right maxima question of the preceding section in disguise. You have to hand back a token whenever a new maximum shows up. We saw in the preceding section that the expected number of left-to-right maxima in a random permutation is H_n, the nth harmonic number. For $n = 100$, $H_n < 6$. So if the presenter were to point to the boxes in random order, you would have to hand back only 6 tokens on average. But why should the presenter offer you the boxes in random order? He has no incentive to have too many winners.

The solution is to take your fate into your own hands: *Open the boxes in random order*. You select one of the boxes at random, open it, then choose a random box from the remaining ones, and so on. How do you choose a random box? When there are k boxes left, you choose a random box by tossing a die with k sides or by choosing a random number in the range 1 to k. In this way, you generate a random permutation of the boxes and hence the analysis in the previous section still applies. On average you will have to return fewer than 6 tokens and hence your 10 tokens will suffice. You have just seen a *randomized algorithm*. We want to stress that, although the mathematical analysis is the same, the conclusions are very different. In the average-case scenario, you are at the mercy of the presenter. If he opens the boxes in random order, the analysis applies; if he does not, it does not. You have no way to tell, except after many shows and with hindsight. In other words, the presenter controls the dice and it is up to him whether he uses fair dice. The situation is completely different in the randomized-algorithms scenario. You control the dice, and you generate the random permutation. The analysis is valid no matter what the presenter does.

We give a second example. Suppose that you are given an urn with white and red balls and your goal is to get a white ball. You know that at least half of the balls in the urn are white. Any deterministic strategy may be unlucky and look at all the red balls before it finds a white ball. It is much better to consider the balls in random order. Then the probability of picking a white ball is at least $1/2$ and hence an expected number of two draws suffices to get a white ball. Note that as long as you draw a red ball, the percentage of white balls in the urn is at least 50% and hence the probability of drawing a white ball stays at least $1/2$. The second example is an instance of the following scenario. You are given a large set of candidates and want to find a candidate that is good in some sense. You know that half of the candidates are good, but you have no idea which ones. Then you should examine the candidates in random order.

We come to a third example. Suppose, Alice and Bob are connected over a slow telephone line. Alice has an integer x and Bob has an integer y, each with six decimal digits. They want to determine whether they have the same number. As communication is slow, their goal is to minimize the amount of information exchanged. Local computation is not an issue.

In the obvious solution, Alice sends her number to Bob, and Bob checks whether the numbers are equal and announces the result. This requires them to transmit 6 digits. Alternatively, Alice could send the number digit by digit, and Bob would check for equality as the digits arrive and announce the result as soon as he knew it, i.e., as soon as corresponding digits differ or all digits had been transmitted. In the worst case, all 6 digits have to be transmitted. We shall now show how to use randomization for this task.

There are 21 two digit prime numbers, namely

$$11, 13, 17, 19, 23, 29, 31, 37, 41, 43, 47, 53, 59, 61, 67, 71, 73, 79, 83, 89, 97.$$

The protocol is now as follows. Alice chooses a random prime p from the list and sends p and $x \bmod p$ to Bob. Note that $x \bmod p$ is a 2-digit number, and hence Alice sends four digits. Bob computes $y \bmod p$ and compares it to $x \bmod p$. If the two remainders are different, he declares that x and y are different, otherwise he declares that x and y are the same. What is the probability that Bob declares an incorrect result? If x and y are equal, his answer is always correct. If x and y are different, their difference is a non-zero number which is less than 1 million in absolute value and hence is divided by at most six numbers on the list, since the product of any six numbers on the list exeeds 1 million. Thus the probability that Bob declares an incorrect result is at most $6/21$. Alice and Bob can reduce the probability of an incorrect result by computing the remainders with respect to more primes. Of course, this would also increase the number of bits sent. Can one live with incorrect answers? Yes, if the probability of an incorrect answer is sufficiently small. We continue this example in Section 2.11.2.

Exercise 2.12. Find out how many 3-digit primes there are? You should be able to find the answer in Wikipedia. Assume that x and y are less than 10^{20}. How many 3-digit primes can divide $x - y$, if $x \neq y$? What is the probability that Bob declares an incorrect result in the protocol above?

2.11.1 The Formal Model

Formally, we equip our RAM with an additional instruction: $R_i := randInt(C)$ which assigns a *random* integer between 0 and $C - 1$ to R_i. All C of these values are equally likely, and the value assigned is independent of the outcome of all previous random choices. In pseudocode, we write $v := randInt(C)$, where v is an integer variable. The cost of making a random choice is one time unit. Algorithms *not* using randomization are called *deterministic*.

The running time of a randomized algorithm will generally depend on the random choices made by the algorithm. So the running time on an instance i is no longer a number, but a random variable depending on the random choices. We may eliminate the dependency of the running time on random choices by equipping our machine with a timer. At the beginning of the execution, we set the timer to a value $T(n)$, which may depend on the size n of the problem instance, and stop the machine once the timer goes off. In this way, we can guarantee that the running time is bounded by $T(n)$. However, if the algorithm runs out of time, it does not deliver an answer.

The output of a randomized algorithm may also depend on the random choices made. How can an algorithm be useful if the answer on an instance i may depend on the random choices made by the algorithm – if the answer may be "Yes" today and "No" tomorrow? If the two cases are equally probable, the answer given by the algorithm has no value. However, if the correct answer is much more likely than the incorrect answer, the answer does have value.

2.11.2 *An Advanced Example

We continue with the third example from the introduction. Suppose, Alice and Bob are connected over a slow telephone line. Alice has an integer x and Bob has an integer y, each with n bits. They want to determine whether they have the same number. As communication is slow, their goal is to minimize the amount of information exchanged. Local computation is not an issue.

In the obvious solution, Alice sends her number to Bob, and Bob checks whether the numbers are equal and announces the result. This requires them to transmit n digits. Alternatively, Alice could send the number digit by digit, and Bob would check for equality as the digits arrived and announce the result as soon as he knew it, i.e., as soon as corresponding digits differed or all digits had been transmitted. In the worst case, all n digits have to be transmitted. We shall now show that randomization leads to a dramatic improvement. After transmission of only $O(\log n)$ bits, equality and inequality can be decided with high probability. Alice and Bob follow the following protocol. Each of them prepares an ordered list of prime numbers. The list consists of the L smallest primes $p_1, p_2, \ldots, p_L$ with a value of at least 2^k. We shall say more about the choice of L and k below. Clearly, Alice and Bob generate the same list. Then Alice chooses an index i, $1 \le i \le L$, at random and sends i and $x \bmod p_i$ to Bob. Bob computes $y \bmod p_i$. If $x \bmod p_i \ne y \bmod p_i$, he declares that the numbers are different. Otherwise, he declares the numbers the same. Clearly, if the numbers are the same, Bob will say so. If the numbers are different and $x \bmod p_i \ne y \bmod p_i$, he will declare them different. However, if $x \ne y$ and yet $x \bmod p_i = y \bmod p_i$, he will erroneously declare the numbers equal. What is the probability of an error?

An error occurs if $x \ne y$ but $x \equiv y \bmod p_i$. The latter condition is equivalent to p_i dividing the difference $D = x - y$. This difference is at most 2^n in absolute value. Since each prime p_i has a value of at least 2^k, our list contains at most n/k primes that divide[11] the difference, and hence the probability of error is at most $(n/k)/L$. We can make this probability arbitrarily small by choosing L large enough. If, say, we want to make the probability less than $0.000001 = 10^{-6}$, we choose $L = 10^6(n/k)$.

What is the appropriate choice of k? For sufficiently large k, about $2^k/\ln(2^k) = 1.4477 \cdot 2^k/k$ primes[12] are contained in the interval $[2^k..2^{k+1} - 1]$. Hence, if $2^k/k \geq 10^6 n/k$, the list will contain only $k + 1$-bit integers. The condition $2^k \geq 10^6 n$ is equivalent to $k \geq \log n + 6\log 10$. With this choice of k, the protocol transmits $\log L + k = \log n + 12\log 10$ bits. *This is exponentially better than the naive protocol.*

What can we do if we want an error probability less than 10^{-12}? We could redo the calculations above with $L = 10^{12}(n/k)$. Alternatively, we could run the protocol twice and declare the numbers different if at least one run declares them different. This two-stage protocol errs only if both runs err, and hence the probability of error is at most $10^{-6} \cdot 10^{-6} = 10^{-12}$.

Exercise 2.13. Compare the efficiency of the two approaches for obtaining an error probability of 10^{-12}.

Exercise 2.14. In the protocol described above, Alice and Bob have to prepare ridiculously long lists of prime numbers. Discuss the following modified protocol. Alice chooses a random $k + 1$-bit integer p (with leading bit 1) and tests it for primality. If p is not prime, she repeats the process. If p is prime, she sends p and $x \bmod p$ to Bob.

Exercise 2.15. Assume you have an algorithm which errs with a probability of at most $1/4$ and that you run the algorithm k times and output the majority output. Derive a bound on the error probability as a function of k. Do a precise calculation for $k = 2$ and $k = 3$, and give a bound for large k. Finally, determine k such that the error probability is less than a given ε.

2.11.3 Las Vegas and Monte Carlo Algorithms

Randomized algorithms come in two main varieties, the Las Vegas and the Monte Carlo variety. A *Las Vegas algorithm* always computes the correct answer but its running time is a random variable. Our solution for the game show is a Las Vegas algorithm (if the player is provided with enough tokens); it always finds the box containing the maximum; however, the number of tokens to be returned (the number of left-to-right maxima) is a random variable. A *Monte Carlo* algorithm always has the same running time, but there is a nonzero probability that it will give an incorrect answer. The probability that the answer is incorrect is at most $1/4$. Our algorithm

---

[11] Let d be the number of primes in our list that divide D. Then $2^n \geq |D| \geq (2^k)^d = 2^{kd}$ and hence $d \leq n/k$.

[12] For any integer x, let $\pi(x)$ be the number of primes less than or equal to x. For example, $\pi(10) = 4$ because there are four prime numbers (2, 3, 5, and 7) less than or equal to 10. Then $x/(\ln x + 2) < \pi(x) < x/(\ln x - 4)$ for $x \geq 55$. See the Wikipedia entry "Prime numbers" for more information.

for comparing two numbers over a telephone line is a Monte Carlo algorithm. In Exercise 2.15, it is shown that the error probability of a Monte Carlo algorithm can be made arbitrarily small by repeated execution.

Exercise 2.16. Suppose you have a Las Vegas algorithm with an expected execution time $t(n)$, and that you run it for $4t(n)$ steps. If it returns an answer within the allotted time, this answer is returned, otherwise an arbitrary answer is returned. Show that the resulting algorithm is a Monte Carlo algorithm.

Exercise 2.17. Suppose you have a Monte Carlo algorithm with an execution time $m(n)$ that gives a correct answer with probability p and a deterministic algorithm that verifies in time $v(n)$ whether the Monte Carlo algorithm has given the correct answer. Explain how to use these two algorithms to obtain a Las Vegas algorithm with expected execution time $(m(n) + v(n))/(1 - p)$.

We now come back to our game show example. You have 10 tokens available to you. The expected number of tokens required is less than 6. How sure should you be that you will go home a winner? We need to bound the probability that M_n is larger than 11, because you lose exactly if the sequence in which you order the boxes has 11 or more left-to-right maxima. *Markov's inequality* allows you to bound this probability. It states that, for a nonnegative random variable X and any constant $c \geq 1$, $\text{prob}(X \geq c \cdot \text{E}[X]) \leq 1/c$; see (A.5) for additional information. We apply the inequality with $X = M_n$ and $c = 11/6$. We obtain

$$\text{prob}(M_n \geq 11) \leq \text{prob}\left(M_n \geq \frac{11}{6}\text{E}[M_n]\right) \leq \frac{6}{11},$$

and hence the probability of winning is more than 5/11.

2.12 Graphs

Graphs are an extremely useful concept in algorithmics. We use them whenever we want to model objects and relations between them; in graph terminology, the objects are called *nodes*, and the relations between nodes are called *edges* or *arcs*. Some obvious applications are road maps and communication networks, but there are also more abstract applications. For example, nodes could be tasks to be completed when building a house, such as "build the walls" or "put in the windows", and edges could model precedence relations such as "the walls have to be built before the windows can be put in". We shall also see many examples of data structures where it is natural to view objects as nodes and pointers as edges between the object storing the pointer and the object pointed to.

When humans think about graphs, they usually find it convenient to work with pictures showing nodes as small disks and edges as lines and arrows. To treat graphs algorithmically, a more mathematical notation is needed: A *directed graph* $G = (V, E)$ is a pair consisting of a *node set* (or *vertex set*) V and an *edge set* (or *arc set*) $E \subseteq V \times V$. We sometimes abbreviate "directed graph" to *digraph*. For example, Fig. 2.11 shows a graph G with node set $\{s,t,u,v,w,x,y,z\}$ and edges (s,t), (t,u), (u,v), (v,w), (w,x), (x,y), (y,z), (z,s), (s,v), (z,w), (y,t), and (x,u). Throughout this book, we use the convention $n = |V|$ and $m = |E|$ if no other definitions for n or m are given. An edge $e = (u,v) \in E$ represents a connection from u to v. We call u and v the *source* and *target*, respectively, of e. We say that e is *incident* to u and v and that v and u are *adjacent*. The special case of a *self-loop* (v,v) is disallowed unless specifically mentioned otherwise. Modeling E as a set of edges also excludes multiple parallel edges between the same two nodes. However, sometimes it is useful to allow parallel edge, i.e., in a multigraph, E is a multiset where elements can appear multiple times.

The *outdegree* of a node v is the number of edges leaving it, and its *indegree* is the number of edges entering it. Formally, $outdegree(v) = |\{(v,u) \in E\}|$ and $indegree(v) = |\{(u,v) \in E\}|$. For example, node w in graph G in Fig. 2.11 has indegree two and outdegree one.

A *bidirected graph* is a digraph where, for any edge (u,v), the reverse edge (v,u) is also present. An *undirected graph* can be viewed as a streamlined representation of a bidirected graph, where we write a pair of edges (u,v), (v,u) as the two-element set $\{u,v\}$. Figure 2.11 includes a three-node undirected graph and its bidirected counterpart. Most graph-theoretic terms for undirected graphs have the same definition as for their bidirected counterparts, and so this section will concentrate on directed graphs and only mention undirected graphs when there is something special about them. For example, the number of edges of an undirected graph is only half the number of edges of its bidirected counterpart. Nodes of an undirected graph have identical indegree and outdegree, and so we simply talk about their *degree*. Undirected graphs are important because directions often do not matter and because many problems are easier to solve (or even to define) for undirected graphs than for general digraphs.

A graph $G' = (V',E')$ is a *subgraph* of G if $V' \subseteq V$ and $E' \subseteq E$. Given $G = (V,E)$ and a subset $V' \subseteq V$, the subgraph *induced* by V' is defined as $G' = (V',E \cap (V' \times V'))$. In Fig. 2.11, the node set $\{v,w\}$ in G induces the subgraph $H = (\{v,w\}, \{(v,w)\})$. A subset $E' \subseteq E$ of edges induces the subgraph (V,E').

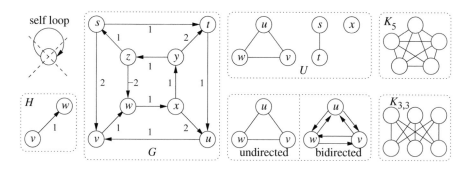

Fig. 2.11. Some graphs.

Often, additional information is associated with nodes or edges. In particular, we shall often need *edge weights* or *costs* $c : E \to \mathbb{R}$ that map edges to some numeric value. For example, the edge (z,w) in graph G in Fig. 2.11 has a weight $c((z,w)) = -2$. Note that an edge $\{u,v\}$ of an undirected graph has a unique edge weight, whereas, in a bidirected graph, we can have $c((u,v)) \neq c((v,u))$.

We have now seen rather many definitions on one page of text. If you want to see them at work, you may jump to Chap. 8 to see algorithms operating on graphs. But things are also becoming more interesting here.

An important higher-level graph-theoretic concept is the notion of a path. A *path* $p = \langle v_0, \ldots, v_k \rangle$ is a sequence of nodes in which consecutive nodes are connected by edges in E, i.e., $(v_0, v_1) \in E$, $(v_1, v_2) \in E$, $\ldots$, $(v_{k-1}, v_k) \in E$; p has length k and runs from v_0 to v_k. Sometimes a path is also represented by its sequence of edges. For example, $\langle u,v,w \rangle = \langle (u,v), (v,w) \rangle$ is a path of length 2 in Fig. 2.11. A sequence $p = \langle v_0, v_1, \ldots, v_k \rangle$ is a path in an undirected graph if it is a path in the corresponding bidirected graph and $v_{i-1} \neq v_{i+1}$ for $1 \leq i < k$, i.e., it is not allowed to use an edge and then immediately go back along the same edge. The sequence $\langle u,w,v,u,w,v \rangle$ is a path of length 5 in the graph U in Fig. 2.11. A path is *simple* if its nodes, except maybe for v_0 and v_k, are pairwise distinct. In Fig. 2.11, $\langle z,w,x,u,v,w,x,y \rangle$ is a nonsimple path in graph G. Clearly, if there is a path from u to v in some graph, there is also a simple path from u to v.

Cycles are paths of length at least 1 with a common first and last node. Cycles in undirected graphs have a length of at least three since consecutive edges must be distinct in a path in an undirected graph. In Fig. 2.11, the sequences $\langle u,v,w,x,y,z,w,x,u \rangle$ and $\langle u,w,v,u,w,v,u \rangle$ are cycles in G and U respectively. A simple cycle visiting all nodes of a graph is called a *Hamiltonian cycle*. For example, the cycle $\langle s,t,u,v,w,x,y,z,s \rangle$ in graph G in Fig. 2.11 is Hamiltonian. The cycle $\langle w,u,v,w \rangle$ in U is also Hamiltonian.

The concepts of paths and cycles allow us to define even higher-level concepts. A digraph is *strongly connected* if, for any two nodes u and v, there is a path from u to v. Graph G in Fig. 2.11 is strongly connected. A strongly connected component of a digraph is a maximal node-induced strongly connected subgraph. If we remove edge (w,x) from G in Fig. 2.11, we obtain a digraph without any cycles. A digraph without any cycles is called a *directed acyclic graph* (DAG). In a DAG, every strongly connected component consists of a single node. An undirected graph is *connected* if the corresponding bidirected graph is strongly connected. The connected components are the strongly connected components of the corresponding bidirected graph. Any two nodes in the same connected component are connected by a path, and there are no edges connecting nodes in distinct connected components. For example, graph U in Fig. 2.11 has connected components $\{u,v,w\}$, $\{s,t\}$, and $\{x\}$. The node set $\{u,w\}$ induces a connected subgraph, but it is not maximal and hence is not a component.

Exercise 2.18. Describe 10 substantially different applications that can be modeled using graphs; car and bicycle networks are not considered substantially different. At least five should be applications not mentioned in this book.

Exercise 2.19. A *planar graph* is a graph that can be drawn on a sheet of paper such that no two edges cross each other. Argue that street networks are *not* necessarily planar. Show that the graphs K_5 and $K_{3,3}$ in Fig. 2.11 are not planar.

2.12.1 A First Graph Algorithm

It is time for an example algorithm. We shall describe an algorithm for testing whether a directed graph is acyclic. We use the simple observation that a node v with outdegree 0 cannot lie on any cycle. Hence, by deleting v (and its incoming edges) from the graph, we obtain a new graph G' that is acyclic if and only if G is acyclic. By iterating this transformation, we either arrive at the empty graph, which is certainly acyclic, or obtain a graph G^* in which every node has an outdegree of at least 1. In the latter case, it is easy to find a cycle: Start at any node v and construct a path by repeatedly choosing an arbitrary outgoing edge until you reach a node v' that you have seen before. The constructed path will have the form $(v, \ldots, v', \ldots, v')$, i.e., the part $(v', \ldots, v')$ forms a cycle. For example, in Fig. 2.11, graph G has no node with outdegree 0. To find a cycle, we might start at node z and follow the path $\langle z, w, x, u, v, w \rangle$ until we encounter w a second time. Hence, we have identified the cycle $\langle w, x, u, v, w \rangle$. In contrast, if the edge (w, x) is removed, there is no cycle. Indeed, our algorithm will remove all nodes in the order w, v, u, t, s, z, y, x. In Chap. 8, we shall see how to represent graphs such that this algorithm can be implemented to run in linear time $O(|V| + |E|)$; see also Exercise 8.3. We can easily make our algorithm certifying. If the algorithm finds a cycle, the graph is certainly cyclic. Also it is easily checked whether the returned sequence of nodes is indeed a cycle. If the algorithm reduces the graph to the empty graph, we number the nodes in the order in which they are removed from G. Since we always remove a node v of outdegree 0 from the current graph, any edge out of v in the original graph must go to a node that was removed previously and hence has received a smaller number. Thus the ordering proves acyclicity: Along any edge, the node numbers decrease. Again this property is easily checked.

Exercise 2.20. Exhibit a DAG with n nodes and $n(n-1)/2$ edges for every n.

2.12.2 Trees

An undirected graph is a *tree* if there is *exactly* one path between any pair of nodes; see Fig. 2.12 for an example. An undirected graph is a *forest* if there is *at most* one path between any pair of nodes. Note that each connected component of a forest is a tree.

Lemma 2.9. *The following properties of an undirected graph G are equivalent:*

(a) G is a tree;
(b) G is connected and has exactly $n - 1$ edges;
(c) G is connected and contains no cycles.

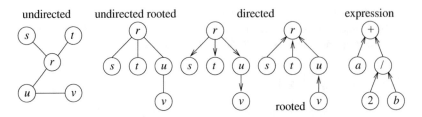

undirected   undirected rooted   directed   expression

Fig. 2.12. Different kinds of trees. From *left* to *right*, we see an undirected tree, an undirected rooted tree, a directed out-tree, a directed in-tree, and an arithmetic expression.

Proof. In a tree, there is a unique path between any two nodes. Hence the graph is connected and contains no cycles. Conversely, if there are two nodes that are connected by more than one path, the graph contains a cycle. Consider distinct paths p and q connecting the same pair of nodes. If the first edge of p is equal to the first edge of q, delete these edges from p and q to obtain p' and q'. The paths p' and q' are distinct and connect the same pair of nodes. Continuing in this way and also applying the argument to last edges, we end up with two paths connecting the same pair of nodes and having distinct first and last edges. Thus the concatenation of the first path with the reversal of the second forms a cycle. We have now shown that (a) and (c) are equivalent.

We next show the equivalence of (b) and (c). Assume that $G = (V, E)$ is connected, and let $m = |E|$. We perform the following experiment: We start with the empty graph and add the edges in E one by one. Addition of an edge can reduce the number of connected components by at most one. We start with n components and must end up with one component. Thus $m \geq n - 1$. Assume now that there is an edge $e = \{u, v\}$ whose addition does not reduce the number of connected components. Then u and v are already connected by a path, and hence addition of e creates a cycle. If G is cycle-free, this case cannot occur, and hence $m = n - 1$. Thus (c) implies (b). Assume next that G is connected and has exactly $n - 1$ edges. Again, add the edges one by one and observe that every addition must reduce the number of connected components by one, as otherwise we would not end up with a single component after $n - 1$ additions. Thus no addition can close a cycle, as such an addition would not reduce the number of connected components. Thus (b) implies (c).   □

Lemma 2.9 does not carry over to digraphs. For example, a DAG may have many more than $n - 1$ edges. A directed graph is an *out-tree* with a *root* node r if there is exactly one path from r to any other node. It is an *in-tree* with a root node r if there is exactly one path from any other node to r. Figure 2.12 shows examples. The *depth* of a node in a rooted tree is the length of the path to the root. The *height* of a rooted tree is the maximum of the depths of its nodes.

We can make an undirected tree rooted by declaring one of its nodes to be the root. Computer scientists have the peculiar habit of drawing rooted trees with the root at the top and all edges going downwards. For rooted trees, it is customary to denote relations between nodes by terms borrowed from family relations. Edges go between a unique *parent* and its *children*. Nodes with the same parent are *siblings*. Nodes without children are *leaves*. Nonroot, nonleaf nodes are *interior* nodes. Consider a path such that u is between the root and another node v. Then u is an *ancestor* of v, and v is a *descendant* of u. A node u and its descendants form the *subtree* rooted at u. For example, in Fig. 2.12, r is the root; s, t, and v are leaves; s, t, and u are siblings because they are children of the same parent r; u is an interior node; r and u are ancestors of v; s, t, u, and v are descendants of r; and v and u form a subtree rooted at u.

2.12.3 Ordered Trees

Trees are ideally suited for representing hierarchies. For example, consider the expression $a + 2/b$. We know that this expression means that a and $2/b$ are to be added. But deriving this from the sequence of characters $\langle a, +, 2, /, b \rangle$ is difficult. For example, the rule that division binds more tightly than addition has to be applied. Therefore compilers isolate this syntactical knowledge in *parsers* that produce a more structured representation based on trees. Our example would be transformed into the expression tree given in Fig. 2.12. Such trees are directed and, in contrast to graph-theoretic trees, they are *ordered*, i.e., the children of each node are ordered. In our example, a is the first, or left, child of the root, and $/$ is the right, or second, child of the root.

Expression trees are easy to evaluate by a simple recursive algorithm. Figure 2.13 shows an algorithm for evaluating expression trees whose leaves are numbers and whose interior nodes are binary operators (say $+$, $-$, $\cdot$, $/$).

We shall see many more examples of ordered trees in this book. Chapters 6 and 7 use them to represent fundamental data structures, and Chap. 12 uses them to systematically explore solution spaces.

Function *eval*(r) : $\mathbb{R}$
    if r *is a leaf* **then return** *the number stored in* r
    else   // r is an operator node
        $v_1 :=$ *eval*(*first child of* r)
        $v_2 :=$ *eval*(*second child of* r)
        return v_1 *operator*(r) v_2 // apply the operator stored in r

Fig. 2.13. Recursive evaluation of an expression tree rooted at r

2.13 P and NP

When should we call an algorithm efficient? Are there problems for which there is no efficient algorithm? Of course, drawing the line between "efficient" and "inefficient" is a somewhat arbitrary business. The following distinction has proved useful: An algorithm $\mathscr{A}$ runs in *polynomial time*, or is a *polynomial-time algorithm*, if there is a polynomial $p(n)$ such that its execution time on inputs of size n is $O(p(n))$. If not otherwise mentioned, the size of the input will be measured in bits. A problem can be solved in *polynomial time* if there is a polynomial-time algorithm that solves it. We equate "efficiently solvable" with "polynomial-time solvable". A big advantage of this definition is that implementation details are usually not important. For example, it does not matter whether a clever data structure can accelerate an $O(n^3)$ algorithm by a factor of n. All chapters of this book, except for Chap. 12, are about efficient algorithms. We use **P** to denote the class of problems solvable in polynomial time.

There are many problems for which no efficient algorithm is known. Here, we mention only six examples:

- *The Hamiltonian cycle problem*: Given an undirected graph, decide whether it contains a Hamiltonian cycle.
- *The Boolean satisfiability problem*: Given a Boolean expression in conjunctive form, decide whether it has a satisfying assignment. A Boolean expression in conjunctive form is a conjunction $C_1 \wedge C_2 \wedge \ldots \wedge C_k$ of clauses. A clause is a disjunction $\ell_1 \vee \ell_2 \vee \ldots \vee \ell_h$ of literals, and a literal is a variable or a negated variable. For example, $v_1 \vee \neg v_3 \vee \neg v_9$ is a clause.
- *The clique problem*: Given an undirected graph and an integer k, decide whether the graph contains a complete subgraph (= a clique) on k nodes. A graph is *complete* if every pair of nodes is connected by an edge. The graph K_5 in Fig. 2.11 is an example.
- *The knapsack problem*: Given n pairs of integers (w_i, p_i) and integers M and P, decide whether there is a subset $I \subseteq [1..n]$ such that $\sum_{i \in I} w_i \leq M$ and $\sum_{i \in I} p_i \geq P$. Informally, item i has volume w_i and value p_i and we want to know whether we can pack a knapsack of volume M such that its value is at least P. This problem will be heavily used as an example in Chap. 12.
- *The traveling salesman problem*: Given an edge-weighted undirected graph and an integer C, decide whether the graph contains a Hamiltonian cycle of cost at most C. See Sect. 11.7.2 for more details.
- *The graph-coloring problem*: Given an undirected graph and an integer k, decide whether there is a coloring of the nodes with k colors such that any two adjacent nodes are colored differently. This problem will also be used as an example in Chap. 12.

The fact that we know no efficient algorithms for these problems does not imply that none exist. It is simply not known whether efficient algorithms exist or not. In particular, we have no proof that such algorithms do not exist. In general, it is very hard to prove that a problem cannot be solved in a given time bound. We shall see some simple lower bounds in Sect. 5.5. Most algorithmicists believe that the six problems above have no efficient solution.

Complexity theory has found an interesting surrogate for the absence of lower-bound proofs. It clusters algorithmic problems into large groups that are equivalent with respect to some complexity measure. In particular, there is a large class of equivalent problems known as **NP**-*complete* problems. Here, **NP** is an abbreviation for "nondeterministic polynomial time". If the term "nondeterministic polynomial time" does not mean anything to you, ignore it and carry on. The six problems mentioned above are **NP**-complete, and so are many other natural problems.

*More on NP-Completeness

We shall now give formal definitions of the class **NP** and the class of **NP**-complete problems. We refer the reader to books about the theory of computation and complexity theory [22, 121, 299, 329] for a thorough treatment.

We assume, as is customary in complexity theory, that inputs are encoded in some fixed finite alphabet Σ. Think of the ASCII or Unicode alphabet or their binary encodings. In the latter case, $\Sigma = \{0,1\}$. We use Σ^* to denote all words (sequences of characters) over the alphabet Σ. The size of a word $x = a_1 \ldots a_n \in \Sigma^*$ is its length n. A *decision problem* is a subset $L \subseteq \Sigma^*$. We use χ_L (read "chi") to denote the characteristic function of L, i.e., $\chi_L(x) = 1$ if $x \in L$ and $\chi_L(x) = 0$ if $x \notin L$. A decision problem is polynomial-time solvable if and only if its characteristic function is polynomial-time computable. We use **P** to denote the class of polynomial-time-solvable decision problems.

A decision problem L is in **NP** if and only if there is a predicate $Q(x,y)$ (a subset $Q \subseteq (\Sigma^*)^2$) and a polynomial p such that

(a) for any $x \in \Sigma^*$, $x \in L$ if and only if there is a $y \in \Sigma^*$ with $|y| \leq p(|x|)$ and $Q(x,y)$, and

(b) Q is computable in polynomial time.

We call y satisfying (a) a *witness* for x or a *proof* of membership for x. For our example problems, it is easy to show that they belong to **NP**. In the case of the Hamiltonian cycle problem, the witness is a Hamiltonian cycle in the input graph. A witness for a Boolean formula is an assignment of truth values to variables that make the formula true. The solvability of an instance of the knapsack problem is witnessed by a subset of elements that fit into the knapsack and achieve the profit bound P.

Exercise 2.21. Prove that the clique problem, the traveling salesman problem, and the graph-coloring problem are in **NP**.

It is widely believed that **P** is a proper subset of **NP**. There are good arguments for this belief, as we shall see in a moment; however, there is no proof. In fact, the problem of whether **P** is equal to **NP** or properly contained in it is considered one of the major open problems in computer science and mathematics. A proof that the two classes are equal would have dramatic consequences: Thousands of problems which are currently believed to have no efficient algorithm would suddenly have one. A proof that the two classes are not equal would probably have no dramatic effect on computing, as most algorithmicists work under the assumption that these classes are distinct, but it would probably have a dramatic effect on theoretical computer science, logic, and mathematics, as the proof would probably introduce a new kind of argument. If **P** is properly contained in **NP**, **NP**-complete problems have no efficient algorithm.

A decision problem L is *polynomial-time reducible* (or simply *reducible*) to a decision problem L' if there is a polynomial-time-computable function g such that for all $x \in \Sigma^*$, we have $x \in L$ if and only if $g(x) \in L'$. If L is reducible to L' and $L' \in \mathbf{P}$, then $L \in \mathbf{P}$. Assume we have an algorithm for the reduction g with a polynomial time bound $p(n)$ and an algorithm for $\chi_{L'}$ with a polynomial time bound $q(n)$. An algorithm for χ_L operates as follows. On input of x, it first computes $g(x)$ using the first algorithm and then tests $g(x) \in L'$ using the second algorithm. The running time is at most $p(|x|) + q(|g(x)|)$. Since Turing machines can write at most one symbol in each step, we have $|g(x)| \leq |x| + p(|x|)$. Thus the running time is bounded by $p(|x|) + q(|x| + p(|x|))$; this is polynomial in $|x|$. A similar argument shows that reducibility is transitive.

A decision problem L is $\mathbf{NP}$-*hard* if every problem in $\mathbf{NP}$ is polynomial-time reducible to it. A problem is $\mathbf{NP}$-*complete* if it is $\mathbf{NP}$-hard and in $\mathbf{NP}$. At first glance, it might seem prohibitively difficult to prove any problem $\mathbf{NP}$-complete – one would have to show that *every* problem in $\mathbf{NP}$ was polynomial-time reducible to it. However, in 1971, Cook and Levin independently managed to do this for the Boolean satisfiability problem [79, 203]. From that time on, it was "easy". Assume you want to show that a problem L is $\mathbf{NP}$-complete. You need to show two things: (1) $L \in \mathbf{NP}$, and (2) there is *some* known $\mathbf{NP}$-complete problem L' that can be reduced to it. Transitivity of the reducibility relation then implies that all problems in $\mathbf{NP}$ are reducible to L. With every new $\mathbf{NP}$-complete problem, it becomes easier to show that other problems are $\mathbf{NP}$-complete. There is a Wikipedia page for the list of $\mathbf{NP}$-complete problems. We next give one example of a reduction.

Lemma 2.10. *The Boolean satisfiability problem is polynomial-time reducible to the clique problem.*

Proof. Let $F = C_1 \wedge \ldots \wedge C_m$, where $C_i = \ell_{i1} \vee \ldots \vee \ell_{ih_i}$ and $\ell_{ik} = x_{ik}^{\beta_{ik}}$, be a formula in conjunctive form. Here, x_{ik} is a variable and $\beta_{ik} \in \{0, 1\}$. A superscript 0 indicates a negated variable. Consider the following graph G. Its nodes V represent the literals in our formula, i.e., $V = \{(i, k) : 1 \leq i \leq m \text{ and } 1 \leq k \leq h_i\}$. Two nodes (i, k) and (j, k') are connected by an edge if and only if $i \neq j$ and either $x_{ik} \neq x_{jk'}$ or $\beta_{ik} = \beta_{jk'}$. In words, the representatives of two literals are connected by an edge if they belong to different clauses and an assignment can satisfy them simultaneously. We claim that F is satisfiable if and only if G has a clique of size m.

Assume first that there is a satisfying assignment α. The assignment must satisfy at least one literal in every clause, say literal ℓ_{ik_i} in clause C_i. Consider the subgraph of G induced by the node set $\{(i, k_i) : 1 \leq i \leq m\}$. This is a clique of size m. Assume otherwise; say, (i, k_i) and (j, k_j) are not connected by an edge. Then, $x_{ik_i} = x_{jk_j}$ and $\beta_{ik_i} \neq \beta_{jk_j}$. But then the literals ℓ_{ik_i} and ℓ_{jk_j} are complements of each other, and α cannot satisfy them both.

Conversely, assume that there is a clique M of size m in G. We can construct a satisfying assignment α. For each i, $1 \leq i \leq m$, M contains exactly one node (i, k_i). We construct a satisfying assignment α by setting $\alpha(x_{ik_i}) = \beta_{ik_i}$. Note that α is well defined because $x_{ik_i} = x_{jk_j}$ implies $\beta_{ik_i} = \beta_{jk_j}$; otherwise, (i, k_i) and (j, k_j) would not be connected by an edge. α clearly satisfies F. $\square$

Exercise 2.22. Show that the Hamiltonian cycle problem is polynomial-time reducible to the traveling salesman problem.

Exercise 2.23. Show that the clique problem is polynomial-time reducible to the graph-coloring problem.

All **NP**-complete problems have a common destiny. If anybody should find a polynomial-time algorithm for *one* of them, then **NP** = **P**. Since so many people have tried to find such solutions, it is becoming less and less likely that this will ever happen: The **NP**-complete problems are mutual witnesses of their hardness.

Does the theory of **NP**-completeness also apply to optimization problems? Optimization problems are easily turned into decision problems. Instead of asking for an optimal solution, we ask whether there is a solution with an objective value better than or equal to k, where k is an additional input. Here, better means greater in a maximization problem and smaller in a minimization problem. Conversely, if we have an algorithm to decide whether there is a solution with a value better than or equal to k, we can use a combination of exponential and binary search (see Sect. 2.7) to find the optimal objective value.

An algorithm for a decision problem returns yes or no, depending on whether the instance belongs to the problem or not. It does not return a witness. Frequently, witnesses can be constructed by applying the decision algorithm repeatedly to instances derived from the original instance. Assume we want to find a clique of size k, but have only an algorithm that decides whether a clique of size k exists. We first test whether G has a clique of size k. If not, there is no clique of size k. Otherwise, we select an arbitrary node v and ask whether $G' = G \setminus v$ has a clique of size k. If so, we search recursively for a clique of size k in G'. If not, we know that v must be part of the clique. Let V' be the set of neighbors of v. We search recursively for a clique C_{k-1} of size $k-1$ in the subgraph spanned by V'. Then $v \cup C_{k-1}$ is a clique of size k in G.

2.14 Implementation Notes

Our pseudocode is easily converted into actual programs in any imperative programming language. We shall give more detailed comments for C++ and Java below. The Eiffel programming language [226] has extensive support for assertions, invariants, preconditions, and postconditions.

Our special values $\bot$, $-\infty$, and ∞ are available for floating-point numbers. For other data types, we have to emulate these values. For example, we could use the smallest and largest representable integers for $-\infty$ and ∞, respectively. Undefined pointers are often represented by a null pointer **null**. Sometimes we use special values for convenience only, and a robust implementation should avoid using them. You will find examples in later chapters.

Randomized algorithms need access to a random source. You have a choice between a hardware generator that generates true random numbers and an algorithmic generator that generates pseudorandom numbers. We refer the reader to the Wikipedia page "Random number" for more information.

There has been a lot of research on parallel programming languages and software libraries for sequential languages. However, most users are conservative and use only a small number of tools that are firmly established and have wide industrial support in order to achieve high performance in a portable way. We shall take the same attitude in the implementation notes in this book. Moreover, some advanced, recently introduced, or rarely used features of firmly established tools may not deliver the performance you might expect, and should be used with care. However, we would like to point out that higher-level tools or features may be worth considering if you can validate your expectation that they will achieve sufficient performance and portability for your application.

2.14.1 C++

Our pseudocode can be viewed as a concise notation for a subset of C++. The memory management operations **allocate** and **dispose** are similar to the C++ operations *new* and *delete*. C++ calls the default constructor for each element of an array, i.e., allocating an array of n objects takes time $\Omega(n)$, whereas allocating an array n of *int*s takes constant time. In contrast, we assume that *all* arrays which are not explicitly initialized contain arbitrary values (garbage). In C++, you can obtain this effect using the C functions *malloc* and *free*. However, this is a deprecated practice and should only be used when array initialization would be a severe performance bottleneck. If memory management of many small objects is performance-critical, you can customize it using the *allocator* class of the C++ standard library.

Our parameterizations of classes using **of** is a special case of the C++ template mechanism. The parameters added in brackets after a class name correspond to the parameters of a C++ constructor.

Assertions are implemented as C macros in the include file `assert.h`. By default, violated assertions trigger a runtime error and print their position in the program text. If the macro *NDEBUG* is defined, assertion checking is disabled.

For many of the data structures and algorithms discussed in this book, excellent implementations are available in software libraries. Good sources are the standard template library STL [256], the Boost [50] C++ libraries, and the LEDA [195, 218] library of efficient algorithms and data structures.

C++ (together with C) is perhaps the most widely used programming language for nonnumerical[13] parallel computing because it has good, widely portable compilers and allows low-level tuning. However, only the recent C++11 standard begins to define some support for parallel programming. In Appendix C, we give a short introduction to the parallel aspects of C++11. We also say a few words about shared-memory parallel-programming tools used together with C++ such as OpenMP, Intel TBB, and Cilk.

In Appendix D we introduce MPI, a widely used software library for message-passing-based programming. It supports a wide variety of message-passing routines, including collective communication operations (see also Chap. 13).

---

[13] For *numerical* parallel computing, Fortran was traditionally the most widely used language. But even that is changing.

2.14.2 Graphics Processing Units (GPUs)

GPUs are often used for general-purpose parallel processing (general-purpose computing on graphics processing units, GPGPU). GPUs can be an order of magnitude more efficient than classical multicore processors with a comparable number of transistors. This is achieved using massive parallelism – the number of (very lightweight) threads used can be three orders of magnitude larger than for a comparable multicore processor. GPU programs therefore need highly scalable parallel algorithms. Further complications are coordination with the host CPU (heterogeneity), explicit management of several types of memory, and threads working in SIMD mode. In this book, we focus on simpler hardware but many of the algorithms discussed are also relevant for GPUs. For NVIDIA GPUs there is a C++ extension (part of the Compute Unified Device Architecture, CUDA) that allows rather high-level programming. A more portable but lower-level system is the C extension OpenCL (Open Computing Language).

2.14.3 Java

Java has no explicit memory management. Rather, a *garbage collector* periodically recycles pieces of memory that are no longer referenced. While this simplifies programming enormously, it can be a performance problem. Remedies are beyond the scope of this book. Generic types provide parameterization of classes. Assertions are implemented with the *assert* statement.

Implementations for many data structures and algorithms are available in the package *java.util.*

Java supports multithreaded programming of a shared-memory machine, including locks, support for atomic instructions, and data structures supporting concurrent access; see the documentation of the libraries beginning with `java.util.concurrent`. However, some high-level concepts such as parallel loops, collective operations, and efficient task-oriented programming are missing. There is also no direct support for message-passing programming. There are several software libraries and compilers addressing these deficits; see [305] for an overview. However, it is perhaps too early to say whether any of these techniques will gain a wide user base with efficient, widely portable implementations. More fundamentally, when an application is sufficiently performance-sensitive for one to consider parallelization, it is worth remembering that Java often incurs a significant performance penalty compared with using C++. This overhead can be worse for a multithreaded code than for a sequential code, since additional cache faults may expose a bottleneck in the memory subsystem and garbage collection incurs additional overheads and complications in a parallel setting.

2.15 Historical Notes and Further Findings

Shepherdson and Sturgis [294] defined the RAM model for use in algorithmic analysis. The RAM model restricts cells to holding a logarithmic number of bits. Dropping this assumption has undesirable consequences; for example, the complexity classes **P** and **PSPACE** collapse [145]. Knuth [186] has described a more detailed abstract machine model.

A huge number of algorithms have been developed for the PRAM model. Jájá's textbook [164] is a good introduction.

There are many variants of distributed-memory models. One can take a more detailed look at concrete network topologies and differentiate between nearby and far-away PEs. For example, Leighton's textbook [198] describes many such algorithms. We avoid these complications here because many networks can actually support our simple model to a reasonable approximation and because we shy away from the complications and portability problems of detailed network models.

One can also take a more abstract view. The *bulk synchronous parallel* (BSP) model [319] divides the computation into globally synchronized phases of local computation on the one hand and of global message exchange on the other hand. During a local computation phase, each PE can post send requests. During a message exchange phase, all these messages are delivered. In the terminology of this book, BSP programs are message-passing programs that use only the nonuniform all-to-all method described in Sect. 13.6.3 for communication. Let h denote the maximum number of machine words sent or received by any PE during a message exchange. Then the BSP model assumes that this message exchange takes time $\ell + gh$, where ℓ and g are machine parameters. This assumption simplifies the analysis of BSP algorithms. We do not adopt the BSP model here, since we want to be able to describe asynchronous algorithms and because, with very little additional effort for the analysis, we get more precise results, for example, when other collective communication operations presented in Chap. 13 are used.

A further abstraction looks only at the *communication volume* of a parallel algorithm, for example, by summing the h-values occurring in the communication steps in the BSP model [275]. This makes sense on large parallel systems, where global communication becomes the bottleneck for processing large data sets.

For modeling cache effects in shared-memory systems, the *parallel external-memory* (PEM) model [19] is useful. The PEM is a combination of the PRAM model and the external-memory model. PEs have local caches of size M each and access the shared main memory in cache lines of size B.

Floyd [106] introduced the method of invariants to assign meaning to programs and Hoare [152, 153] systematized their use. The book [137] is a compendium of sums and recurrences and, more generally, discrete mathematics.

Books on compiler construction (e.g., [233, 331]) will tell you more about the compilation of high-level programming languages into machine code.

3

Representing Sequences by Arrays and Linked Lists

Perhaps the world's oldest data structures were the tablets in cuneiform script used more than 5000 years ago by custodians in Sumerian temples. These custodians kept lists of goods, and their quantities, owners, and buyers. The picture on the left shows an example.[1] This was possibly the first application of written language. The operations performed on such lists have remained the same – adding entries, storing them for later, searching entries and changing them, going through a list to compile summaries, etc. The Peruvian quipu [225] that you see in the picture on the right served a similar purpose in the Inca empire, using knots in colored strings arranged sequentially on a master string. It is probably easier to maintain and use data on tablets than to use knotted string, but one would not want to haul stone tablets over Andean mountain trails. It is apparent that different representations make sense for the same kind of data.

The abstract notion of a sequence, list, or table is very simple and independent of its representation in a computer. Mathematically, the only important property is that the elements of a sequence $s = \langle e_0, \ldots, e_{n-1} \rangle$ are arranged in a linear order – in contrast to the trees and graphs discussed in Chaps. 7 and 8, or the unordered hash tables discussed in Chap. 4. There are two basic ways of referring to the elements of a sequence.

One is to specify the index of an element. This is the way we usually think about arrays, where $s[i]$ returns the ith element of a sequence s. Arrays are the basis of many parallel algorithms. In Sect. 3.1 we explain some of the basic approaches. Our pseudocode directly supports *static* arrays. In a *static* data structure, the size is known in advance, and the data structure is not modifiable by insertions and deletions. In a *bounded* data structure, the maximum size is known in advance. In Sect. 3.4, we

---

[1] This 4600-year-old tablet contains a list of gifts to the high priestess of Adab (see `commons.wikimedia.org/wiki/Image:Sumerian_26th_c_Adab.jpg`).

© Springer Nature Switzerland AG 2019
P. Sanders et al., *Sequential and Parallel Algorithms and Data Structures*,
https://doi.org/10.1007/978-3-030-25209-0_3

introduce *dynamic* or *unbounded arrays*, which can grow and shrink as elements are inserted and removed. The analysis of unbounded arrays introduces the concept of *amortized analysis*.

The second way of referring to the elements of a sequence is relative to other elements. For example, one could ask for the successor of an element e, the predecessor of an element e', or the subsequence $\langle e, \ldots, e' \rangle$ of elements between e and e'. Although relative access can be simulated using array indexing, we shall see in Sect. 3.2 that a list-based representation of sequences is more flexible. In particular, it becomes easier to insert or remove arbitrary pieces of a sequence. On the other hand, parallel processing of linked lists is difficult. In Sect. 3.3 we get a glimpse how to do it anyway.

Many algorithms use sequences in a quite limited way. Often only the front and/or the rear of the sequence is read and modified. Sequences that are used in this restricted way are called *stacks*, *queues*, and *deques*. We discuss them in Sects. 3.6 and 3.7. In Sect. 3.8, we summarize the findings of the chapter.

3.1 Processing Arrays in Parallel

Arrays are an important data structure for parallel processing, since we can easily assign operations on different array elements to different PEs. Suppose we want to assign the elements of an array $a[0..n-1]$ to p PEs numbered $0..p-1$. There are many ways for distributing the elements over the PEs. Figure 3.1 gives examples. Perhaps the most natural one – *blocked assignment* – assigns up to $\lceil n/p \rceil$ consecutive array elements to each PE, for example by mapping element $a[i]$ to PE $\lfloor i/\lceil n/p \rceil \rfloor$. This works well if the amount of work required for the different array elements is about the same. Moreover, this assignment is also cache-efficient. To simplify the notation, let us now assume that p divides n.

Another natural assignment is *round robin* (also called *cyclic*) where PE i works on array elements $a[j]$ with $j \bmod p = i$. This is less cache-efficient than blocked assignment but may sometimes assign the work more uniformly.

Exercise 3.1. Since cyclic assignment is less cache-efficient than blocked assignment, one also uses *block cyclic* assignment, where blocks of size B are cyclically assigned to PEs. Work out formulae that specify which elements are assigned to each PE (see Fig. 3.1 for an example).

Fig. 3.1. Simple assignments of 16 array elements $a[0..15]$ to PEs $0..3$

In Chapter 14 we shall discuss more advanced ways to assign work (e.g., array elements) to PEs in a load balanced way.

On a distributed-memory machine, we usually want to store an array element on the same PE that processes it. This important principle is known as *owner computes*. Hence, our logical array a will be stored in a distributed way – one piece of a is stored on each PE. This distribution is also known as *sharding* or *(horizontal) partitioning*. The distribution principle naturally transfers to more complex data structures such as multidimensional arrays or graphs (see also Sect. 8.6). Note that the "owner computes" principle can be applied to any approach to array partitioning – blocked, cyclic, block cyclic, explicitly load-balanced, ...

Let us look at a simple example. Suppose we want to double all elements of a using blocked assignment. Suppose PE i of a distributed-memory machine stores elements $i \cdot n/p..(i+1) \cdot n/p - 1$ in a local array $a[0..n/p - 1]$. Then then SPMD pseudocode for this doubling task is

for $i := 0$ **to** $n/p - 1$ **do** $a[i] *= 2$

and this takes time $O(n/p)$, requiring no communication.

Of course, parallel programming is less simple most of the time. In particular, computations usually involve several array elements at once. In Sect. 2.10, we already discussed the task of summing all elements of an array. As another example, suppose we have an array $a[0..n+1]$ and want to compute the average of $a[i-1]$, $a[i]$, and $a[i+1]$ for $i \in 1..n$; the boundary values $a[0]$ and $a[n+1]$ are fixed and are not changed.[2] Such a computation is frequent in the approximate numerical solution of partial differential equations. On a shared-memory machine, the task is quite simple. This time we use explicit loop parallelism. We allocate a second array b and say

for $i := 1$ **to** n **do**‖   // use blocked assignment of loop iterations to PEs
    $b[i] := (a[i-1] + a[i] + a[i+1])/3.$

On a distributed-memory machine, we need explicit communication to make sure that all the required data is available. We want PE i to compute components $i \cdot n/p + 1$ to $(i+1) \cdot n/p$ of the result vector, as shown below for $n = 8$ and $p = 2$:

global view	$a[0]$ $a[1]$ $a[2]$ $a[3]$ $a[4]$ $a[5]$ $a[6]$ $a[7]$ $a[8]$ $a[9]$
local view, PE 0	$a[0]^*$ $a[1]^*$ $a[2]^*$ $a[3]^*$ $a[4]^*$ $a[5]$
results computed by PE 0	$b[1]$ $b[2]$ $b[3]$ $b[4]$
local view, PE 1	$a[0]$ $a[1]^*$ $a[2]^*$ $a[3]^*$ $a[4]^*$ $a[5]^*$
results computed by PE 1	$b[1]$ $b[2]$ $b[3]$ $b[4]$

We therefore allocate to each PE a local array a with $n/p + 2$ cells, where $a[0]$ and $a[n/p+1]$ play a special role. PE i can compute its part of the output if it has access to elements $i \cdot n/p$ to $(i+1)n/p + 1$ of the input array, i.e., some elements of the input array must be stored in two local arrays as shown above. However, initially each element of the input array is available on only one of the PEs. We assume

---

[2] We are computing n values $b[1]$ to $b[n]$ and, for simplicity, want to stick to our convention that p divides n. Therefore, we use an input array of size $n+2$.

that initially the array is distributed over the PEs as follows. Elements 0 to n/p are stored in PE 0, elements $n/p+1$ to $2n/p$ are stored in PE 1, ..., elements $i \cdot n/p+1$ to $(i+1) \cdot n/p$ are stored in PE i, ..., and elements $(p-1) \cdot n/p+1$ to $n+1$ are stored in PE $p-1$. In the figure above, the initial distribution of the array elements is indicated by the symbol $*$.

As said above, PE i can compute its part of the output if it has access to elements $i \cdot n/p$ to $(i+1)n/p+1$ of the input array. It does not have the first element, which is stored only in PE $i-1$, and it does not have the last element, which is stored only in PE $i+1$. We are now ready for the SPMD pseudocode. In the first line, each PE i sends its local element $a[1]$ to the PE $i-1$ and receives its local element $a[n/p+1]$ from PE $i+1$; in the second line each PE i sends its local element $a[n/p]$ to PE $i+1$ and receives its local element $a[0]$ from PE $i-1$. The code exploits the convention that communication with nonexistent PEs does nothing (see Sect. 2.4.2):[3]

$$send(i_{\mathrm{proc}} - 1, a[1 \quad]) \;\|\; receive(i_{\mathrm{proc}} + 1, a[n/p+1]) \qquad\qquad // *$$
$$send(i_{\mathrm{proc}} + 1, a[n/p]) \;\|\; receive(i_{\mathrm{proc}} - 1, a[0]) \qquad\qquad\quad // **$$
$$\textbf{for } i := 1 \textbf{ to } n/p \textbf{ do } b[i] := (a[i-1] + a[i] + a[i+1])/3$$

The program takes $3n/p$ arithmetic operations and total communication effort $2(\alpha + \beta)$ (exploiting the fact that our distributed-memory machine can send and receive in parallel; see also Sect. 2.4.2). This is efficient in practice if the local computation is large compared with two startup overheads. Below, we shall see that we can sometimes do better.

For the above pseudocode, it is important that the send and receive operations in each line are actually executed in parallel. Suppose we were to replace the " $\|$ " by a ";" – sequential execution. Then all PEs would first send data to the left. However, no PEs are ready to receive that data. Only PE 0 suceeds, as its send operation to PE "-1" is interpreted as doing nothing. Subsequently, PE 0 is ready to receive from PE 1. Only when this first message transfer is finished will PE 1 be ready to receive from PE 2, and so on. Thus it would take time at least $(p-1)\alpha$ to complete line "*". A similar effect would happen in line "**".

Exercise 3.2. Suppose a parallel send and receive operation is not available. How can you change the above pseudocode fragment so that it executes in time $O(\alpha)$, independent of p? Hint: Distinguish between PEs with odd and even number.

3.1.1 *Reducing Latencies by Tiling

Suppose we want to perform the above averaging operation T times (this is typical of iterative numerical computations). The simple algorithm above would need $2T$ message startups and hence time at least $2T\alpha$. For $n/p \ll \alpha/\beta$, startup overheads would thus dominate the running time. We can overcome this bottleneck by partitioning the

---

[3] There has been considerable work on building parallel programming languages automating the reasoning in this paragraph. For the example presented here, this is possible. However, no such language has been very successful, since the situation is frequently more complex.

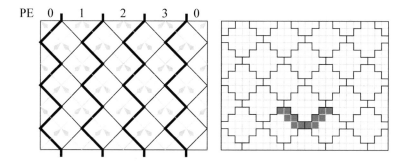

Fig. 3.2. Tiling of a rectangle of computations into diamond-shaped tiles for four PEs. One PE handles the triangular tiles at the left and right borders. *Left*: Continuous drawing for $n \to \infty$. Arrows indicate data dependencies between tiles. Thick lines separate tiles assigned to different PEs. *Right*: The case $n = 24$ leads to tiles of width $w = 24/4 = 6$. The 12 gray cells have to be communicated in order to perform the averaging operations in the tile above them.

averaging operations into more coarse-grained tasks. For this simple, regular computation, a geometrical interpretation is in order. The computations can be arranged as an $n \times T$ matrix, where entry (i,t) stands for computing the tth value of element $a[i]$. This array is partitioned into *tiles* that depend only on the results of a small number of other tiles in such a way that the dependencies form an acyclic graph – a *task DAG*, as described in Sect. 14.6. In the given case, we can minimize the number of startups by using *diamond-shaped* tiles. Figure 3.2 illustrates this situation (at the border of the computation domain, the tiles have a triangular shape); in this figure, time grows from the bottom to the top. The tiles have width $w = n/p$. The computation within a tile depends on $2w$ cells from two tiles below, as indicated in the figure. Since one of these tiles can be assigned to the same PE, only a single message containing the state of w cells must be received by each tile. This happens $2T/w$ times. Overall, we have $2T/w$ message startups and a communication volume of $w \cdot 2T/w = 2T$. Comparing this with the naive algorithm with two startups in each averaging step and two cells received, this reduces the startup overheads by a factor of n/p while keeping the number of exchanged cell states constant.

An alternative way of understanding the scheme is as follows. The area of a tile is essentially w^2. Hence a PE can do work w^2 after receiving a *single* message containing w cells. In the naive scheme, it could do w work w after receiving *two* messages containing two cells. Thus the ratio of work to message has improved by a factor of w.

Exercise 3.3. Explain how to implement diamond tiling using only two local arrays a_0 and a_1 of size $2n/p$ on each PE.

Real applications are more complex than the above simple example. Typically, they iterate on one or several two-dimensional or three-dimensional arrays, and the pattern of cells used for updating an array entry (the *stencil*) and the boundary conditions may vary. Still, the tiling techniques described above can be generalized.

3.2 Linked Lists

In this section, we study the representation of sequences by linked lists. In a doubly linked list, each item points to its successor and to its predecessor. In a singly linked list, each item points to its successor. We shall see that linked lists are easily modified in many ways: We may insert or delete items or sublists, and we may concatenate lists. The drawback is that random access (the operator $[\cdot]$) is not supported. We study doubly linked lists in Sect. 3.2.1, and singly linked lists in Sect. 3.2.3. Singly linked lists are more space-efficient and somewhat faster, and should therefore be preferred whenever their functionality suffices. A good way to think of a linked list is to imagine a chain, where one element is written on each link. Once we get hold of one link of the chain, we can retrieve all elements.

3.2.1 Doubly Linked Lists

Figure 3.3 shows the basic building blocks of a linked list. A list *item* stores an element, and pointers to its successor and predecessor. We call a pointer to a list item a *handle*. This sounds simple enough, but pointers are so powerful that we can make a big mess if we are not careful. What makes a consistent list data structure? We require that for each item *it*, the successor of its predecessor is equal to *it* and the predecessor of its successor is also equal to *it*.

A sequence of n elements is represented by a ring of $n+1$ items. There is a special dummy item h, which stores no element. The successor h_1 of h stores the first element of the sequence, the successor of h_1 stores the second element of the sequence, and so on. The predecessor of h stores the last element of the sequence; see Fig. 3.4. The empty sequence is represented by a ring consisting only of h. Since there are no elements in that sequence, h is its own successor and predecessor. Figure 3.5 defines a representation of sequences by lists. An object of class *List* contains a single list item h. The constructor of the class initializes the header h to an item containing $\perp$ and having itself as successor and predecessor. In this way, the list is initialized to the empty sequence.

We implement all basic list operations in terms of the single operation *splice* shown in Fig. 3.6. This operation cuts out a sublist from one list and inserts it after some target item. The sublist is specified by handles a and b to its first and its last

Class *Handle* = **Pointer to** *Item*

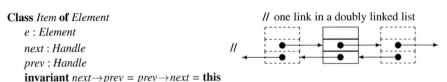

Class *Item* **of** *Element*   // one link in a doubly linked list
    e : *Element*
    next : *Handle*   //
    prev : *Handle*
    invariant *next*→*prev* = *prev*→*next* = **this**

Fig. 3.3. The items of a doubly linked list

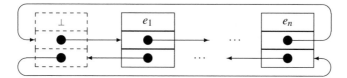

Fig. 3.4. The representation of a sequence $\langle e_1, \ldots, e_n \rangle$ by a doubly linked list. There are $n+1$ items arranged in a ring, a special dummy item h containing no element, and one item for each element of the sequence. The item containing e_i is the successor of the item containing e_{i-1} and the predecessor of the item containing e_{i+1}. The dummy item is between the item containing e_n and the item containing e_1.

Class *List* **of** *Element*
    // Item h is the predecessor of the first element and the successor of the last element.

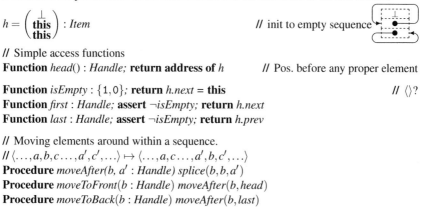

$h = \left(\begin{matrix} \perp \\ \textbf{this} \\ \textbf{this} \end{matrix} \right)$: *Item*   // init to empty sequence

    // Simple access functions
    Function *head*() : *Handle;* **return address of** h // Pos. before any proper element

    Function *isEmpty* : $\{1,0\}$; **return** *h.next* = **this**   // $\langle \rangle$?
    Function *first* : *Handle;* **assert** $\neg$*isEmpty;* **return** *h.next*
    Function *last* : *Handle;* **assert** $\neg$*isEmpty;* **return** *h.prev*

    // Moving elements around within a sequence.
    // $\langle \ldots, a, b, c \ldots, a', c', \ldots \rangle \mapsto \langle \ldots, a, c \ldots, a', b, c', \ldots \rangle$
    Procedure *moveAfter(b, a'* : *Handle) splice(b,b,a')*
    Procedure *moveToFront(b* : *Handle) moveAfter(b,head)*
    Procedure *moveToBack(b* : *Handle) moveAfter(b,last)*

Fig. 3.5. Some constant-time operations on doubly linked lists

element, respectively. In other words, b must be reachable from a by following zero or more next-pointers but without going through the dummy item. The target item t can be either in the same list or in a different list; in the former case, it must not be inside the sublist starting at a and ending at b.

splice does not change the number of items in the system. We assume that there is one special list, *freeList*, that keeps a supply of unused items. When inserting new elements into a list, we take the necessary items from *freeList*, and when removing elements, we return the corresponding items to *freeList*. The function *checkFreeList* allocates memory for new items when necessary. We defer its implementation to Exercise 3.6 and a short discussion in Sect. 3.9.

With these conventions in place, a large number of useful operations can be implemented as one-line functions that all run in constant time. Thanks to the power of *splice*, we can even manipulate arbitrarily long sublists in constant time. Figures 3.5 and 3.7 show many examples. In order to test whether a list is empty, we simply check whether h is its own successor. If a sequence is nonempty, its first and its last

// Remove $\langle a, \ldots, b \rangle$ from its current list and insert it after t

// $\ldots, a', a, \ldots, b, b', \ldots + \ldots, t, t', \ldots \mapsto \ldots, a', b', \ldots + \ldots, t, a, \ldots, b, t', \ldots$

Procedure $splice(a,b,t : Handle)$

    assert a and b belong to the same list, b is not before a, and $t \notin \langle a, \ldots, b \rangle$

    // cut out $\langle a, \ldots, b \rangle$

    $a' := a \rightarrow prev$

    $b' := b \rightarrow next$

    $a' \rightarrow next := b'$

    $b' \rightarrow prev := a'$

    // insert $\langle a, \ldots, b \rangle$ after t

    $t' := t \rightarrow next$

    $b \rightarrow next := t'$

    $a \rightarrow prev := t$

    $t \rightarrow next := a$

    $t' \rightarrow prev := b$

Fig. 3.6. Splicing lists

element are the successor and predecessor, respectively, of h. In order to move an item b to the position after an item a', we simply cut out the sublist starting and ending at b and insert it after a'. This is exactly what $splice(b,b,a')$ does. We move an element to the first or last position of a sequence by moving it after the head or after the last element, respectively. In order to delete an element b, we move it to *freeList*. To insert a new element e, we take the first item of *freeList*, store the element in it, and move it to the place of insertion.

Exercise 3.4 (alternative list implementation). Discuss an alternative implementation of *List* that does not need the dummy item h. Instead, this representation stores a pointer to the first list item in the list object. The position before the first list element is encoded as a null pointer. The interface and the asymptotic execution times of all operations should remain the same. Give at least one advantage and one disadvantage of this implementation compared with the one given in the text.

The dummy item is also useful for other operations. For example, consider the problem of finding the next occurrence of an element x starting at an item *from*. If x is not present, *head* should be returned. We use the dummy item as a *sentinel*. A sentinel is an item in a data structure that makes sure that some loop will terminate. In the case of a list, we store the key we are looking for in the dummy item. This ensures that x is present in the list structure and hence a search for it will always terminate. The search will terminate at a proper list item or the dummy item, depending on whether x was present in the list originally. It is no longer necessary to test whether the end of the list has been reached. In this way, the trick of using the dummy item h as a sentinel saves one test in each iteration improves the efficiency of the search:

Function *findNext*(*x* : *Element; from* : *Handle*) : *Handle*
    h.e = *x*   // Sentinel
    while *from*→*e* ≠ *x* **do**
        from := *from*→*next*
    return *from*

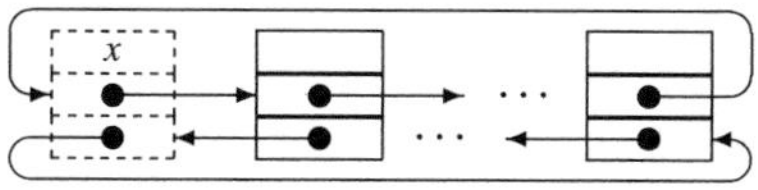

Exercise 3.5. Implement a procedure *swap* that swaps two sublists in constant time, i.e., sequences $(\langle \ldots, a', a, \ldots, b, b', \ldots \rangle, \langle \ldots, c', c, \ldots, d, d', \ldots \rangle)$ are transformed into $(\langle \ldots, a', c, \ldots, d, b', \ldots \rangle, \langle \ldots, c', a, \ldots, b, d', \ldots \rangle)$. Is *splice* a special case of *swap*?

Exercise 3.6 (memory management). Implement the function *checkFreeList* called by *insertAfter* in Fig. 3.7. Since an individual call of the programming-language primitive **allocate** for every single item might be slow, your function should allocate space for items in large batches. The worst-case execution time of *checkFreeList* should be independent of the batch size. Hint: In addition to *freeList*, use a small array of free items.

Exercise 3.7. Give a constant-time implementation of an algorithm for rotating a list to the right: $\langle a, \ldots, b, c \rangle \mapsto \langle c, a, \ldots, b \rangle$. Generalize your algorithm to rotate $\langle a, \ldots, b, c, \ldots, d \rangle$ to $\langle c, \ldots, d, a, \ldots, b \rangle$ in constant time.

// Deleting and inserting elements.
// $\langle \ldots, a, b, c, \ldots \rangle \mapsto \langle \ldots, a, c, \ldots \rangle$
Procedure *remove*(*b* : *Handle*) *moveAfter*(*b*, *freeList.head*)
Procedure *popFront remove*(*first*)
Procedure *popBack remove*(*last*)

// $\langle \ldots, a, b, \ldots \rangle \mapsto \langle \ldots, a, e, b, \ldots \rangle$
Function *insertAfter*(*x* : *Element; a* : *Handle*) : *Handle*
    checkFreeList   // make sure *freeList* is nonempty. See also Exercise 3.6
    a' := *freeList.first*   // Obtain an item *a'* to hold *x*,
    moveAfter(*a'*, *a*)   // put it at the right place,
    a'→*e* := *x*   // and fill it with the right content.
    return *a'*

Function *insertBefore*(*x* : *Element; b* : *Handle*) : *Handle* **return** *insertAfter*(*e*, *pred*(*b*))
Procedure *pushFront*(*x* : *Element*) *insertAfter*(*x*, *head*)
Procedure *pushBack*(*x* : *Element*) *insertAfter*(*x*, *last*)

// Manipulations of entire lists
// $(\langle a, \ldots, b \rangle, \langle c, \ldots, d \rangle) \mapsto (\langle a, \ldots, b, c, \ldots, d \rangle, \langle \rangle)$
Procedure *concat*(*L'* : *List*)
    splice(*L'.first*, *L'.last*, *last*)

// $\langle a, \ldots, b \rangle \mapsto \langle \rangle$
Procedure *makeEmpty*
    freeList.concat(**this**)   //

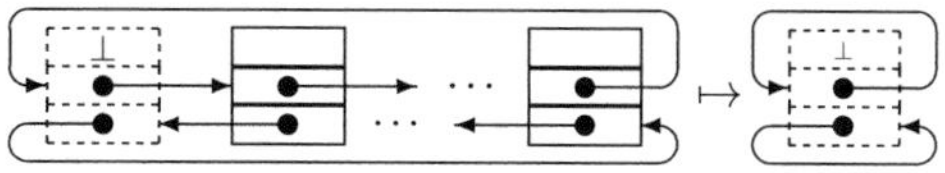

Fig. 3.7. More constant-time operations on doubly linked lists

Exercise 3.8. *findNext* using sentinels is faster than an implementation that checks for the end of the list in each iteration. But how much faster? What speed difference do you predict for many searches in a short list with 100 elements, and in a long list with 10 000 000 elements? Why is the relative speed difference dependent on the size of the list?

3.2.2 Maintaining the Size of a List

In our simple list data type, it is not possible to determine the length of a list in constant time. This can be fixed by introducing a member variable *size* that is updated whenever the number of elements changes. Operations that affect several lists now need to know about the lists involved, even if low-level functions such as *splice* only need handles to the items involved. For example, consider the following code for moving an element a from a list L to the position after a' in a list L':

> **Procedure** *moveAfter*(a, a' : Handle; L, L' : List)
>   *splice*(a,a,a'); L.size$--$; L'.size$++$

Maintaining the size of lists interferes with other list operations. When we move elements as above, we need to know the sequences containing them and, more seriously, operations that move sublists between lists cannot be implemented in constant time anymore. The next exercise offers a compromise.

Exercise 3.9. Design a list data type that allows sublists to be moved between lists in constant time and allows constant-time access to *size* whenever sublist operations have not been used since the last access to the list size. When sublist operations have been used, *size* is recomputed only when needed.

Exercise 3.10. Explain how the operations *remove*, *insertAfter*, and *concat* have to be modified to keep track of the length of a *List*.

3.2.3 Singly Linked Lists

The two pointers per item of a doubly linked list make programming quite easy. Singly linked lists are the lean sisters of doubly linked lists. We use *SItem* to refer to an item in a singly linked list. *SItems* scrap the predecessor pointer and store only a pointer to the successor. This makes singly linked lists more space-efficient and often faster than their doubly linked brothers. The downside is that some operations can no longer be performed in constant time or can no longer be supported in full generality. For example, we can remove an *SItem* only if we know its predecessor.

We adopt the implementation approach used with doubly linked lists. *SItems* form collections of cycles, and an *SList* has a dummy *SItem* h that precedes the first proper element and is the successor of the last proper element. Many operations on *Lists* can still be performed if we change the interface slightly. For example, the following implementation of *splice* needs the *predecessor* of the first element of the sublist to be moved:

$$// (\langle \ldots, a', a, \ldots, b, b' \ldots \rangle, \langle \ldots, t, t', \ldots \rangle) \mapsto (\langle \ldots, a', b' \ldots \rangle, \langle \ldots, t, a, \ldots, b, t', \ldots \rangle)$$

Procedure $splice(a', b, t : SHandle)$

$$\begin{pmatrix} a' \rightarrow next \\ t \rightarrow next \\ b \rightarrow next \end{pmatrix} := \begin{pmatrix} b \rightarrow next \\ a' \rightarrow next \\ t \rightarrow next \end{pmatrix}$$

Similarly, *findNext* should return not the handle of the next *SItem* containing the search key but its *predecessor*, so that it remains possible to remove the element found. Consequently, *findNext* can only start searching at the item *after* the item given to it. A useful addition to *SList* is a pointer to the last element because it allows us to support *pushBack* in constant time.

Exercise 3.11. Implement classes *SHandle*, *SItem*, and *SList* for singly linked lists in analogy to *Handle*, *Item*, and *List*. Show that the following functions can be implemented to run in constant time. The operations *head*, *first*, *last*, *isEmpty*, *popFront*, *pushFront*, *pushBack*, *insertAfter*, *concat*, and *makeEmpty* should have the same interface as before. The operations *moveAfter*, *moveToFront*, *moveToBack*, *remove*, *popFront*, and *findNext* need different interfaces.

We shall see several applications of singly linked lists in later chapters, for example in hash tables in Sect. 4.1 and in mergesort in Sect. 5.3. We may also use singly linked lists to implement free lists of memory managers – even for items in doubly linked lists.

3.3 Processing Linked Lists in Parallel

Linked lists are harder than arrays to process in parallel. In particular, we cannot easily split lists into equal-sized pieces of consecutive elements. The reason is that in arrays, proximity in memory corresponds to proximity in the logical structure. Moreover, arrays support access by index. For linked lists, proximity in memory does not correspond to proximity in the logical structure. The next-pointer of a list item may point to an arbitrary position in memory. Moreover, lists do not support access by index, but only sequential access. The only list element accessible for a given element is the next element, and hence linked lists seem to enfore sequential accessing. As a global rule, it is wise to avoid lists in parallel computing. However, the situation is not completely bleak. We shall see in this section how to convert a linked list into an array. Somewhat surprisingly, this process, which is also known as *list ranking*, is parallelizable. In the algorithms for list ranking, we shall exploit the fact that a parallel algorithm can start traversing the list simultaneously from many positions. The challenge lies in coordinating the different traversals. List ranking plays an important role in many theoretical PRAM algorithms, and parallel list-ranking algorithms are a good showcase of important parallelization ideas such as doubling, contraction, multilevel algorithms, and using inefficient subroutines in an overall efficient algorithm.

3.3.1 *List Ranking by Doubling

We shall work with a singly linked list whose items are stored in an array $L[0..n]$ in any order except that a dummy item is stored in $L[n]$. The dummy item is the last element of the list, and its next-pointer points to itself. We shall compute the distance of each item to the dummy item. Having the distance, reordering is easy. Actually, we shall solve a slightly more general problem. Items have an additional field *rank*. The dummy item has initial rank 0. The initial rank values for the other items are arbitrary; in the list-ranking task, the initial rank of all other items is 1. *The task is to compute for each item the sum of the rank values from the item to the dummy item following the next-pointers*; see Fig. 3.8(a) for an example. *Our algorithm will manipulate the next-pointers and rank values in such a way that this sum remains invariant.* At the end, the *next*-pointer of each item will point to the dummy item directly, and *rank* will contain the desired value.

Exercise 3.12. Give a sequential algorithm that computes the ranks in time $O(n)$.

We begin with an elegant and simple PRAM algorithm that repeatedly replaces the *next*-pointers with the result of following two *next*-pointers and compensates by adding the corresponding *rank* values:

> **Procedure** *doublingListRanking*(L : *Array* $[0..n]$ **of** *Item*)
>   **for** $j := 1$ **to** $\lceil \log n \rceil$ **do**
>   **for** $i := 0$ **to** $n - 1$ **do**$\|$ // Synchronize after each instruction!
> $L[i].rank \mathrel{+}= L[i].next \rightarrow rank$
> $L[i].next \; := L[i].next \rightarrow next$

The algorithm maintains the following invariant: Consider any list item $L[i]$ and consider the sublist of the original list starting at that item and ending just before the item $L[i].next$. Then $L[i].rank$ is the sum of the initial rank fields of the items in this sublist. Moreover, after j iterations, each *next*-pointer either points to the dummy item or jumps 2^j positions in the input sequence to the right. Se Fig. 3.8(a) for an example.

Exercise 3.13. Prove this by induction.

Hence, after $\lceil \log n \rceil$ iterations the *rank* values contain the final result.

The doubling algorithm has span $O(\log n)$ and work $O(n \log n)$. Since a sequential algorithm for list ranking needs only time $O(n)$, the efficiency of the doubling algorithm is only $O(1/\log n)$ even if we simulate several logical PEs on one physical PE. Still, the doubling algorithm is very interesting, since it demonstrates how sequentially following n pointers can be emulated using only $\log n$ iterations of a parallel algorithm. The parallel algorithm traverses the list simultaneously from all locations and uses pointer doubling to halve the distance of each item to the last item in each round. In the next section, we shall see how doubling can be used to accelerate an efficient algorithm.

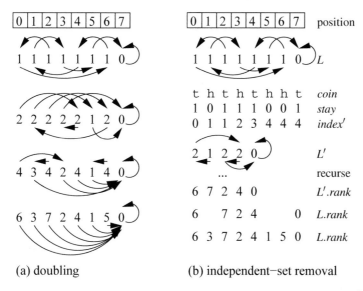

(a) doubling   (b) independent–set removal

Fig. 3.8. List-ranking algorithms. We have a list of eight items stored in an array $L[0..7]$. The last item is stored in $L[7]$ and points to itself. The next-pointers are shown as arrows. The initial ranks are 1 for all items but the dummy item. The initial rank of the dummy item is 0. The final ranks are shown in the last row. Note that the list element in position $L[6]$ has a rank of 5 because of the linking $L[6] \to L[4] \to L[1] \to L[3] \to L[5] \to L[7]$.

Exercise 3.14. Give a PRAM algorithm with work $O(n)$ and span $O(1)$ that converts an array of n singly-linked-list items into an array of doubly-linked-list items representing the same sequence.

3.3.2 *A Multilevel Algorithm for List Ranking

The pointer-doubling algorithm works on all list items in each round. Since the number of rounds is logarithmic, its work is $\Theta(n\log n)$. If we want an algorithm with linear work, we cannot work on all list items in every iteration of the algorithm. The *independent-set removal* algorithm is based on this idea. It first removes a constant fraction of the list items and builds a list of the remaining items. It then recurses on this *contracted* instance. Finally, it reinserts the removed items. In order to make the first and last steps efficient, an independent set of items is removed in the first step.

The algorithm is a good example of a *multilevel algorithm*: Build a new, smaller instance somehow representing the entire problem, solve the smaller problem recursively, and build the overall solution from the solution of the smaller problem. Multilevel algorithms are related to divide-and-conquer algorithms. One can view multilevel algorithms as divide-and-conquer algorithms which make only a single recursive call (e.g., see the quickselect algorithm in Sect. 5.8). The difference is a matter of interpretation – in a multilevel algorithm, the contracted instance represents the entire original input in some uniform way, whereas a divide-and-conquer algorithms splits off a subproblem by removing irrelevant information.

Procedure *isrListRanking*(*L* : *Array* [0..*n*] of *Item*, *B* : ℕ)
    if *n* ≤ *B* **then** *doublingListRanking*(*L*);   **return**   *// base case*
    findIndependentSet(*L*)   *// which items stay?*
    exclusivePrefixSum(*L*, *stay*, *index'*)   *// enumerate staying elements*
    L' : *Array* [0..*L*[*n*].*index'*] of *Item*   *// contracted instance*
    for *i* := 0 **to** *n* **do**∥   *// build L'*
        if *L*[*i*].*stay* **then**   *// move L*[*i*] *to L'*
            i' := *L*[*i*].*index'*   *// position in L'*
            r := *L*[*i*].*next*   *// right neighbor*
            L'[*i'*] := *L*[*i*]   *// copy*
            L'[*i'*].*next* := **addressof** *L'*[**if** *r* → *stay* **then** *r* → *index'* **else** *r* → *next* → *index'*]
            if ¬*r* → *stay* **then** *L'*[*i'*].*rank* += *r* → *rank*   *// establish invariant for L'*
    isrListRanking(*L'*, *B*)   *// recurse*
    for *i* := 0 **to** *n* − 1 **do**∥ **if** *L*[*i*].*stay* **then** *L*[*i*].*rank* := *L'*[*L*[*i*].*index'*].*rank*   *// assemble*
    for *i* := 0 **to** *n* − 1 **do**∥ **if** ¬*L*[*i*].*stay* **then** *L*[*i*].*rank* += *L*[*i*].*next* → *rank*   *// solution*

Fig. 3.9. List ranking by independent-set removal (explicitly parallel)

Figure 3.9 gives pseudocode for the algorithm. The items now have additional fields *stay* and *index'*. The Boolean *stay* indicates whether the item stays in the list (*stay* = 1) or is to be removed (*stay* = 0). For the dummy item, we always have *stay* = 1. For an element that stays, *index'* is the position of the item in the contracted instance. We shall explain in a moment how it is computed. The subroutine *findIndependentSet* is responsible for finding a (large) independent set of list items that can be removed (*stay* = 0). We never remove two adjacent elements. By requiring that the predecessors and successors of removed elements must stay, we make it easy to build the contracted instance and to reconstruct the overall solution. We defer the description of *findIndependentSet* for now. The position of a staying element in *L'* can be computed by computing a prefix sum; we have $L[j].index' = \sum_{i=1}^{j-1} L[i].stay$. In Sect. 13.3 it is explained how prefix sums can be computed using linear work and logarithmic span.

But how can *L'* represent the entire input? The idea is to define the input *rank* values of *L'* in a way such that for all staying elements the output *rank* value is the same as for *L*. To this end, we first copy rank values from *L* to *L'* and then add the rank of the successor item in *L* whenever the successor item is removed. After solving *L'* recursively, it is easy to build the overall solutions. The items of *L* which stayed in *L'* can simply take their *rank* value from the corresponding item in *L'*. The others add their input *rank* value to the *rank*value of their successor. Figure 3.8(b) gives an example.

Let us now analyze *isrListRanking* when the independent set encompasses a constant fraction of the elements. Assume αn elements stay, for some constant $\alpha < 1$.

Then we get the following recurrences for the work and span:

$$\text{work}(n) = \begin{cases} O(B\log B) & \text{if } n \leq B, \\ O(n+\text{work}(\alpha n)) & \text{else,} \end{cases} \qquad \text{span}(n) = \begin{cases} O(\log B) & \text{if } n \leq B, \\ O(\log n+\text{span}(\alpha n)) & \text{else.} \end{cases}$$

Roughly, the work shrinks geometrically, so that we get linear work overall plus $O(B\log B)$ for the base of the recursion. The span is basically the number of levels of recursion times $\log n$; the number of levels of recursion is $\lceil \log_{1/\alpha} n/B \rceil = O(\log n/B)$. More precisely, we can prove by induction that

$$\text{work}(n) = O(n+B\log B) \quad \text{and} \quad \text{span}(n) = O\left(\log B + \log n \log \frac{n}{B}\right).$$

How should we choose the tuning parameter B? Choosing B as a constant yields $\text{span}(n) = \Theta(\log^2 n)$. The span gets smaller for larger values of B. However, if we want linear work, $B\log B$ should be $O(n)$. This suggests that we should use $B = \Theta(n/\log n)$. We obtain

$$\text{work}(n) = O(n) \quad \text{and} \quad \text{span}(n) = O(\log n \log \log n). \tag{3.1}$$

Note that we are using an inefficient algorithm – *doublingListRanking* – to speed up an efficient algorithm – *isrListRanking*. The overall algorithm remains efficient, since we use the inefficient algorithm only when the problem size has been reduced sufficiently to make the inefficiency irrelevant. This is a general principle that works for many parallel algorithms. For example, in Sect. 5.2 we shall design fast, inefficient sorting algorithms that can be used to accelerate the sample sort algorithm described in Sect. 5.13 or the parallel selection algorithm in Sect. 5.9.

3.3.3 Computing an Independent Set

There is a very simple randomized algorithm for computing an independent set. For each item I, we throw a coin. If the coin shows a head and the coin for the successor $I.next$ shows a tail, we put I into the independent set. We always throw a tail for the dummy element. This algorithm has linear work and constant span on a PRAM. Note that there is no need to explicitly check the predecessor $I.prev$: Either we throw a tail for $I.prev$ and we are fine, or $I.prev$ sees that we threw a head for I and therefore stays out of the independent set. Figure 3.8(b) gives an example. Note that $L[3]$ does not enter the independent set even though we threw a head (h) for it, because we also threw a head for its successor $L[5]$.

The probability that I goes to the independent set is $\frac{1}{2} \cdot (1 - \frac{1}{2}) = \frac{1}{4}$ except for the predecessor of the dummy element, for which the probability is $\frac{1}{2}$.

Exercise 3.15. Show that using a biased coin does not help: The above probability is never more than $\frac{1}{4}$.

Lemma 3.1. *The expected size of the independent set is* $(n+1)/4$.

Proof. We define the indicator random variable X_i, $i \in 0..n-1$, to be 1 if and only if $L[i]$ goes to the independent set. The size of the independent set is $X := \sum_{i=0}^{n-1} X_i$. We have $\text{prob}(X_i = 1) = \frac{1}{4}$ except for the predecessor of the dummy element, where $\text{prob}(X_i = 1) = \frac{1}{2}$. Using the linearity of expectations we get

$$E[X] = E\left[\sum_{i=0}^{n-1} X_i\right] = \sum_{i=0}^{n-1} E[X_i] = (n-1)\frac{1}{4} + \frac{1}{2} = \frac{n+1}{4}.$$ □

Unfortunately, the analysis of *isrListRanking* assumes a deterministic algorithm for computing independent sets. There are various ways to fix this problem – none of them very elegant. One view is to ignore technicalities and simply hope that (3.1) is a good approximation of the expected work and span of the algorithm. We shall outline several ways to substantiate this hope. One way is to convert our Monte Carlo algorithm for finding an independent set into a Las Vegas algorithm by actually computing the size of the independent set and repeating until it is "large enough". If we choose the acceptance threshold right, we can prove that a constant number of iterations is enough. Since counting needs logarithmic span, we do not have constant span anymore. But this is no problem in the analysis of *isrListRanking*. However, it is bad style to make an algorithm more expensive just because one is too lazy to perform a tight analysis. We can use a similar argument to that in the analysis of quickselect in Sect. 5.8 to also allow for recursion levels that do not shrink enough. The only argument needed is a constant probability bound such as the following one.

*Exercise 3.16. Use Markov's inequality (A.5) to prove that with probability at most $\frac{4}{5}$, more than $\frac{15}{16}n$ items stay.

Unfortunately, the constant factor we get out of such an analysis is ridiculously pessimistic. We can do better by proving that the size of the independent set will be very close to its expectation with high probability.

**Exercise 3.17. Use the bounded difference inequality given in [210] to show that with probability $1 - O(1/n)$, the independent set will have size $n/4 - o(n)$. Hint: First show that the size of the independent set changes by at most two when we change the outcome of a single coin throw.

*Exercise 3.18. A *maximal independent set* is one that cannot be enlarged by including additional elements.

(a) Show that a maximal independent set of a list contains at least $n/3$ elements.
(b) Design a randomized parallel algorithm that computes a maximal independent set. Hint: Repeatedly throw coins. Fix the result for the staying items and their successors. Can you make this algorithm work efficiently?

3.4 Unbounded Arrays

Consider an array data structure that, besides the indexing operation $[\cdot]$, supports the following operations *pushBack*, *popBack*, and *size*:

$$\langle e_0, \ldots, e_{n-1}\rangle.pushBack(e) = \langle e_0, \ldots, e_{n-1}, e\rangle,$$
$$\langle e_0, \ldots, e_{n-1}\rangle.popBack = \langle e_0, \ldots, e_{n-2}\rangle \qquad \text{(for } n \geq 1\text{)},$$
$$size(\langle e_0, \ldots, e_{n-1}\rangle) = n.$$

Why are unbounded arrays important? Because in many situations we do not know in advance how large an array should be. Here is a typical example: Suppose you want to implement the Unix command `sort` for sorting the lines of a file. You decide to read the file into an array of lines, sort the array internally, and finally output the sorted array. With unbounded arrays, this is easy. With bounded arrays, you would have to read the file twice: once to find the number of lines it contains, and once again to actually load it into the array. The solution with unbounded arrays is clearly more elegant, in particular if you can use an implementation provided by a library. Also, there are situations where the input can be read only once.

We come now to the implementation of unbounded arrays. We emulate an unbounded array u with n elements by use of a dynamically allocated bounded array b with w entries, where $w \geq n$. The first n entries of b are used to store the elements of u. The last $w - n$ entries of b are unused. As long as $w > n$, *pushBack* simply increments n and uses the first unused entry of b for the new element. When $w = n$, the next *pushBack* allocates a new bounded array b' that is larger by a constant factor (say a factor of two). To reestablish the invariant that u is stored in b, the contents of b are copied to the new array so that the old b can be deallocated. Finally, the pointer defining b is redirected to the new array. Deleting the last element with *popBack* is even easier, since there is no danger that b may become too small. However, we might waste a lot of space if we allow b to be much larger than needed. The wasted space can be kept small by shrinking b when n becomes too small. Figure 3.10 gives the complete pseudocode for an unbounded-array class. Growing and shrinking are performed using the same utility procedure *reallocate*. Our implementation uses constants α and β, with $\beta = 2$ and $\alpha = 4$. Whenever the current bounded array becomes too small, we replace it by an array of β times the old size. Whenever the size of the current array becomes α times as large as its used part, we replace it by an array of size βn. The reasons for the choice of α and β will become clear later.

3.4.1 Amortized Analysis of Unbounded Arrays: The Global Argument

Our implementation of unbounded arrays follows the algorithm design principle "make the common case fast". Array access with the operator $[\cdot]$ is as fast as for bounded arrays. Intuitively, *pushBack* and *popBack* should "usually" be fast – we just have to update n. However, some insertions and deletions incur a cost of $\Theta(n)$. We shall show that such expensive operations are rare and that any sequence of m operations starting with an empty array can be executed in time $O(m)$.

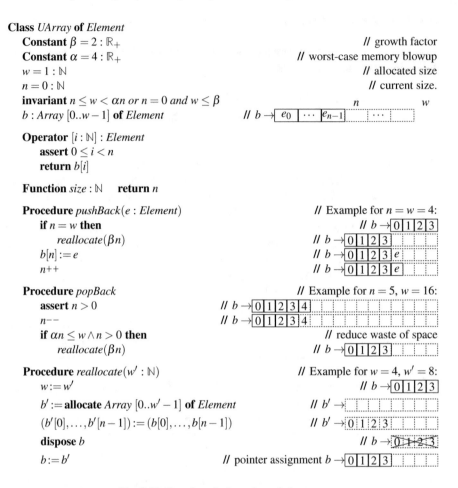

Fig. 3.10. Pseudocode for unbounded arrays

Lemma 3.2. *Consider an unbounded array u that is initially empty. Any sequence* $\sigma = \langle \sigma_1, \dots, \sigma_m \rangle$ *of pushBack or popBack operations on u is executed in time* $O(m)$.

Lemma 3.2 is a nontrivial statement. A small and innocent-looking change to the program invalidates it.

Exercise 3.19. Your manager asks you to change the initialization of α to $\alpha = 2$. He argues that it is wasteful to shrink an array only when three-fourths of it are unused. He proposes to shrink it when $n \leq w/2$. Convince him that this is a bad idea by giving a sequence of m *pushBack* and *popBack* operations that would need time $\Theta(m^2)$ if his proposal was implemented.

Lemma 3.2 makes a statement about the amortized cost of *pushBack* and *popBack* operations. Although single operations may be costly, the cost of a sequence of m

operations is $O(m)$. If we divide the total cost of the operations in σ by the number of operations, we get a constant. We say that the *amortized cost* of each operation is constant. Our usage of the term "amortized" is similar to its usage in everyday language, but it avoids a common pitfall. "I am going to cycle to work every day from now on, and hence it is justified to buy a luxury bike. The cost per ride will be very small – the investment will be amortized." Does this kind of reasoning sound familiar to you? The bike is bought, it rains, and all good intentions are gone. The bike has not been amortized. We shall, instead, insist that a large expenditure is justified by savings in the past and not by expected savings in the future. Suppose your ultimate goal is to go to work in a luxury car. However, you are not going to buy it on your first day of work. Instead, you walk and put a certain amount of money per day into a savings account. At some point, you will be able to buy a bicycle. You continue to put money away. At some point later, you will be able to buy a small car, and even later you can finally buy a luxury car. In this way, every expenditure can be paid for by past savings, and all expenditures are amortized. Using the notion of amortized costs, we can reformulate Lemma 3.2 more elegantly. The increased elegance also allows better comparisons between data structures.

Corollary 3.3. *Unbounded arrays implement the operation $[\cdot]$ in worst-case constant time and the operations pushBack and popBack in amortized constant time.*

To prove Lemma 3.2, we use the *bank account* or *potential* method. We associate an *account* or *potential* with our data structure and force every *pushBack* and *popBack* to put a certain amount into this account. Usually, we call our unit of currency a *token*. The idea is that whenever a call of *reallocate* occurs, the balance in the account is sufficiently high to pay for it. The details are as follows. A token can pay for moving one element from b to b'. Note that element copying in the procedure *reallocate* is the only operation that incurs a nonconstant cost in Fig. 3.10. More concretely, *reallocate* is always called with $w' = 2n$ and thus has to copy n elements. Hence, for each call of *reallocate*, we withdraw n tokens from the account. We charge two tokens for each call of *pushBack* and one token for each call of *popBack*. We now show that these charges guarantee that the balance of the account stays nonnegative and hence suffice to cover the withdrawals made by *reallocate*.

The first call of *reallocate* occurs when there is already one element in the array and a new element is to be inserted. The element already in the array has deposited two tokens in the account, and this more than covers the one token withdrawn by *reallocate*. The new element provides its tokens for the next call of *reallocate*.

After a call of *reallocate*, we have an array of w elements: $n = w/2$ slots are occupied and $w/2$ are free. The next call of *reallocate* occurs when either $n = w$ or $4n \leq w$. In the first case, at least $w/2$ elements have been added to the array since the last call of *reallocate*, and each one of them has deposited two tokens. So we have at least w tokens available and can cover the withdrawal made by the next call of *reallocate*. In the second case, at least $w/2 - w/4 = w/4$ elements have been removed from the array since the last call of *reallocate*, and each one of them has deposited one token. So we have at least $w/4$ tokens available. The call of *reallocate* needs at most $w/4$ tokens, and hence the cost of the call is covered. This completes the proof of Lemma 3.2. □

Exercise 3.20. Redo the argument above for general values of α and β, and charge $\beta/(\beta-1)$ tokens for each call of *pushBack* and $\beta/(\alpha-\beta)$ tokens for each call of *popBack*. Let n' be such that $w = \beta n'$. Then, after a *reallocate*, n' elements are occupied and $(\beta-1)n' = ((\beta-1)/\beta)w$ are free. The next call of *reallocate* occurs when either $n = w$ or $\alpha n \le w$. Argue that in both cases there are enough tokens.

Amortized analysis is an extremely versatile tool, and so we think that it is worthwhile to learn alternative proof methods[4] for amortized analysis. We shall now give two variants of the proof above.

Above, we charged two tokens for each *pushBack* and one token for each *popBack*. Alternatively, we could charge three tokens for each *pushBack* and not charge for *popBack* at all. The accounting is simple. The first two tokens pay for the insertion as above, and the third token is used when the element is deleted.

Exercise 3.21 (continuation of Exercise 3.20). Show that a charge of $\beta/(\beta-1) + \beta/(\alpha-\beta)$ tokens for each *pushBack* is enough. Determine values of α such that $\beta/(\alpha-\beta) \le 1/(\beta-1)$ and such that $\beta/(\alpha-\beta) \le \beta/(\beta-1)$.

3.4.2 Amortized Analysis of Unbounded Arrays: The Local Argument

We now describe our second modification of the proof. Above, we used a global argument in order to show that there are enough tokens in the account before each call of *reallocate*. We now show how to replace the global argument by a local argument. Recall that, immediately after a call of *reallocate*, we have an array of w elements, out of which $w/2$ are filled and $w/2$ are free. We argue that at any time after the first call of *reallocate*, the following token invariant holds:

the account contains at least $\max(2(n-w/2), w/2-n)$ tokens.

Observe that this number is always nonnegative. We use induction on the number of operations executed. Immediately after the first *reallocate*, there is one token in the account and the invariant requires none ($n = w/2 = 1$). A *pushBack* (ignoring the potential call of *reallocate*) increases n by one and adds two tokens. So the invariant is maintained. A *popBack* (again ignoring the potential call of *reallocate*) removes one element and adds one token. So the invariant is again maintained. We next turn to calls of *reallocate*. When a call of *reallocate* occurs, we have either $n = w$ or $4n \le w$. In the former case, the account contains at least n tokens, and n tokens are required for the reallocation. In the latter case, the account contains at least $w/4$ tokens, and n are required. So, in either case, the number of tokens suffices. Also, after the reallocation, $n = w/2$ and hence no tokens are required.

---
[4] Some induction proofs become easier if they are formulated in terms of a smallest counterexample. It is useful to know both methods. The situation is similar here.

Exercise 3.22. Charge three tokens for a *pushBack* and no tokens for a *popBack*. Argue that the account always contains at least $n + \max(2(n - w/2), w/2 - n) = \max(3n - w, w/2)$ tokens.

Exercise 3.23 (popping many elements). Implement an operation *popBack(k)* that removes the last k elements in amortized constant time. Of course, $0 < k \le n$, but k is arbitrary otherwise.

Exercise 3.24 (worst-case constant access time). Suppose, for a real-time application, you need an unbounded array data structure with a *worst-case* constant execution time for all operations. Design such a data structure. Hint: In an initial solution, support only $[.]$ and *pushBack*. Store the elements in up to two arrays. Start moving elements to a larger array well before the small array is completely exhausted. How do you generalize this approach if *popBack* must also be supported?

Exercise 3.25 (implicitly growing arrays). Implement unbounded arrays where the operation $[\cdot]$ accepts any positive index i as its argument. When $i \ge n$, the array is implicitly grown to size $n = i + 1$. When $n \ge w$, the array is reallocated as for *UArray*. Initialize entries that have never been written with some default value $\perp$.

Exercise 3.26 (sparse arrays). Implement bounded arrays with constant time for allocating arrays and constant time for the operation $[\cdot]$. All array elements should be (implicitly) initialized to $\perp$. You are not allowed to make any assumptions about the contents of a freshly allocated array. Hint: Use an extra array of the same size, and store the number t of array elements to which a value has already been assigned. Therefore $t = 0$ initially. An array entry i to which a value has been assigned stores that value and an index j, $1 \le j \le t$, of the extra array, and i is stored in that index of the extra array.

3.4.3 Amortized Analysis of Binary Counters

Amortized analysis is so important that it deserves a second introductory example. We consider the amortized cost of incrementing a binary counter. The value n of the counter is represented by a sequence $\ldots \beta_i \ldots \beta_1 \beta_0$ of binary digits, i.e., $\beta_i \in \{0, 1\}$ and $n = \sum_{i \ge 0} \beta_i 2^i$. The initial value is 0. Its representation is a string of 0's. We define the cost of incrementing the counter as 1 plus the number of trailing 1's in the binary representation, i.e., the transition

$$\ldots 01^k \to \ldots 10^k$$

has a cost $k + 1$. What is the total cost of m increments? We shall show that the cost is $O(m)$. Again, we give a global argument first and then a local argument.

If the counter is incremented m times, its final value is m. The representation of the number m requires $L = 1 + \lceil \log m \rceil$ bits. Among the numbers from 0 to $m - 1$, there are at most 2^{L-k-1} numbers whose binary representation ends with a 0 followed by k many 1's. For each one of them, an increment costs $1 + k$. Thus the total cost of the m increments is bounded by

$$\sum_{0\leq k<L}(k+1)2^{L-k-1}=2^L\sum_{1\leq k\leq L}k/2^k\leq 2^L\sum_{k\geq 1}k/2^k=2\cdot 2^L\leq 4m,$$

where the last equality uses (A.15). Hence, the amortized cost of an increment is $O(1)$.

The argument above is global, in the sense that it requires an estimate of the number of representations ending in a 0 followed by k many 1's. We now give a local argument which does not need such a bound. We associate a bank account with the counter. Its balance is the number of 1's in the binary representation of the counter. So, the balance is initially 0. Consider an increment of cost $k+1$. Before the increment, the representation ends in a zero followed by k many 1's, and after the increment, the representation ends in a 1 followed by k many 0's. So, the number of 1's in the representation decreases by $k-1$, i.e., the operation releases $k-1$ tokens from the account. The cost of the increment is $k+1$. We cover a cost of $k-1$ with the tokens released from the account, and charge a cost of two to the operation. Thus the total cost of m operations is at most $2m$.

3.5 *Amortized Analysis

We give here a general definition of amortized time bounds and amortized analysis. We recommend that you should read this section quickly and come back to it when needed. We consider an arbitrary data structure. The values of all program variables comprise the state of the data structure; we use S to denote the set of states. In the first example in the previous section, the state of our data structure is formed by the values of n, w, and b. Let s_0 be the initial state. In our example, we have $n=0$, $w=1$, and b is an array of size 1 in the initial state. We have operations to transform the data structure. In our example, we had the operations *pushBack*, *popBack*, and *reallocate*. The application of an operation X in a state s transforms the data structure to a new state s' and has a cost $T_X(s)$. In our example, the cost of a *pushBack* or *popBack* is 1, excluding the cost of the possible call to *reallocate*. The cost of a call *reallocate*(βn) is $\Theta(n)$.

Let F be a sequence of operations $\langle Op_1,\ Op_2,\ Op_3,\ \ldots,\ Op_m\rangle$. Starting at the initial state s_0, F takes us through a sequence of states to a final state s_m:

$$s_0\xrightarrow{Op_1}s_1\xrightarrow{Op_2}s_2\xrightarrow{Op_3}\cdots\xrightarrow{Op_m}s_m.$$

The cost $T(F)$ of F is given by

$$T(F)=\sum_{1\leq i\leq m}T_{Op_i}(s_{i-1}).$$

A family of functions $A_X(s)$, one for each operation X, is called a *family of amortized time bounds* if, for every sequence F of operations,

$$T(F)\leq A(F):=c+\sum_{1\leq i\leq m}A_{Op_i}(s_{i-1})$$

for some constant c not depending on F, i.e., up to an additive constant, the total actual execution time is bounded by the total amortized execution time.

This definition is a very general formulation of the bank account method. We start with a balance of c tokens and then execute the sequence of operations. If an operation X is executed in state s, we deposit $A_X(s)$ tokens into the account and also withdraw $T_X(s)$ tokens to pay for the execution of the operation. The functions A_X form a family of amortized time bounds if the balance of the account can never become negative. In order to use the bank account method, one has to define the functions A_X and the constant c and then *prove* that the balance can never become negative. The balance after the execution of a sequence F of operations is $c + A(F) - T(F)$.

There is always a trivial way to define a family of amortized time bounds, namely $A_X(s) := T_X(s)$ for all s. The challenge is to find a family of simple functions $A_X(s)$ (with small function values) that form a family of amortized time bounds. In our example, the functions $A_{pushBack}(s) = 2$, $A_{popBack}(s) = 1$, $A_{[\cdot]}(s) = 1$, and $A_{reallocate}(s) = 0$ for all s form a family of amortized time bounds (with $c = 0$). In order to prove that a set of functions is indeed a family of amortized time bounds, one uses induction with a suitable invariant which bounds from below the balance of the account after the execution of a sequence F of operations. For our example, the invariant states that after the execution of a sequence F of operations leading to the state (n, w), the balance is at least $\max(2(n - w/2), w/2 - n)$.

Some readers may find it counterintuitive that the amortized cost of *reallocate* is stated as 0. After a call of *reallocate* we have an array of size w, in which exactly half of the slots are occupied. The other half is free, i.e., $n = w/2$. According to the invariant, the balance of the account may be as low as 0 after the operation. The cost of the operation is $w/2$, as this is the number of elements that have to be moved. Before the call, we had either an array of $w/2$ slots, all of which were full ($n_{before} = w_{before} = w/2$), or an array of $2w$ slots, a quarter of which were full ($n_{before} = w/2$ and $w_{before} = 2w$). In the former case, the balance before the operation is at least $2(n_{before} - w_{before}/2) = w/2$. In the latter case, the balance before the operation is at least $w_{before}/2 - n_{before} = w - w/2 = w/2$. Thus, in either case, the cost of the operation is $w/2$ and the balance of the account is at least $w/2$. We can therefore completely pay for the cost of the operation out of the account and there is no need to charge any amortized cost.

3.5.1 The Potential Method for Amortized Analysis

Here, we introduce a powerful general technique for obtaining amortized time bounds: the potential method for amortized analysis. In Sect. 3.4.3, we analyzed the binary counter by associating with each bit string (state of the data structure) the number of 1-bits in the bit string (the potential of the state) and then using this potential to compute the charges required for the counter operations. We now formalize and generalize this method. The essence of the method is a function *pot* that associates a nonnegative potential with every state of the data structure, i.e., $pot \colon S \longrightarrow \mathbb{R}_{\geq 0}$. We call $pot(s)$ the potential of the state s. It requires ingenuity to come up with an appropriate function *pot*. For an operation X that transforms a state s into a state s' and has cost $T_X(s)$, we define the amortized cost $A_X(s)$ as the sum of the potential change and the actual cost, i.e., $A_X(s) = pot(s') - pot(s) + T_X(s)$. The functions obtained in this way form a family of amortized time bounds.

Theorem 3.4 (potential method). *Let S be the set of states of a data structure, let s_0 be the initial state, and let $pot: S \longrightarrow \mathbb{R}_{\geq 0}$ be a nonnegative function. For an operation X and a state s with $s \xrightarrow{X} s'$, we define*

$$A_X(s) = pot(s') - pot(s) + T_X(s).$$

The functions $A_X(s)$ are then a family of amortized time bounds with $c = pot(s_0)$.

Proof. A short computation suffices. Consider a sequence $F = \langle Op_1, \ldots, Op_m \rangle$ of operations. We have

$$\sum_{1 \leq i \leq m} A_{Op_i}(s_{i-1}) = \sum_{1 \leq i \leq m} (pot(s_i) - pot(s_{i-1}) + T_{Op_i}(s_{i-1}))$$

$$= pot(s_m) - pot(s_0) + \sum_{1 \leq i \leq m} T_{Op_i}(s_{i-1})$$

$$\geq \sum_{1 \leq i \leq m} T_{Op_i}(s_{i-1}) - pot(s_0),$$

since $pot(s_m) \geq 0$. Thus $T(F) \leq A(F) + pot(s_0)$ and the definition of amortized time bounds is satisfied with $c = pot(s_0)$. Note that c is a constant independent of F. $\square$

Let us formulate the analysis of unbounded arrays in the language above. The state of an unbounded array is characterized by the values of n and w. Following Exercise 3.22, the potential in state (n,w) is $\max(3n - w, w/2)$. The actual costs T of *pushBack* and *popBack* are 1 and the actual cost of *reallocate*(βn) is n. The potential of the initial state $(n,w) = (0,1)$ is $1/2$. A *pushBack* increases n by 1 and hence increases the potential by at most 3. Thus its amortized cost is bounded by 4. A *popBack* decreases n by 1 and hence does not increase the potential. Its amortized cost is therefore at most 1. The first *reallocate* occurs when the data structure is in the state $(n,w) = (1,1)$. The potential of this state is $\max(3-1, 1/2) = 2$, and the actual cost of the *reallocate* is 1. After the *reallocate*, the data structure is in the state $(n,w) = (1,2)$ and has a potential of $\max(3-2, 1) = 1$. Therefore the amortized cost of the first *reallocate* is $1 - 2 + 1 = 0$. Consider any other call of *reallocate*. We have either $n = w$ or $4n \leq w$. In the former case, the potential before the *reallocate* is $2n$, the actual cost is n, and the new state is $(n, 2n)$ and has a potential of n. Thus the amortized cost is $n - 2n + n = 0$. In the latter case, the potential before the operation is $w/2$, the actual cost is n, which is at most $w/4$, and the new state is $(n, w/2)$ and has a potential of $w/4$. Thus the amortized cost is at most $w/4 - w/2 + w/4 = 0$. We conclude that the amortized costs of *pushBack* and *popBack* are $O(1)$ and the amortized cost of *reallocate* is 0 or less. Thus a sequence of m operations on an unbounded array has cost $O(m)$.

Exercise 3.27 (amortized analysis of binary counters). Consider a nonnegative integer c represented by an array of binary digits, and a sequence of m increment and decrement operations. Initially, $c = 0$. This exercise continues the discussion at the end of Sect. 3.4.

(a) What is the worst-case execution time of an increment or a decrement as a function of m? Assume that you can work with only one bit per step.
(b) Prove that the amortized cost of the increments is constant if there are no decrements. Hint: Define the potential of c as the number of 1's in the binary representation of c.
(c) Give a sequence of m increment and decrement operations with cost $\Theta(m \log m)$.
(d) Give a representation of counters such that you can achieve worst-case constant time for increments and decrements.
(e) Allow each digit d_i to take values from $\{-1, 0, 1\}$. The value of the counter is $c = \sum_i d_i 2^i$. Show that in this *redundant ternary* number system, increments and decrements have constant amortized cost. Is there an easy way to tell whether the value of the counter is 0?

3.5.2 Universality of the Potential Method

We argue here that the potential-function technique is strong enough to obtain any family of amortized time bounds.

Theorem 3.5. *Let $B_X(s)$ be a family of amortized time bounds. Then there is a potential function pot such that $A_X(s) \leq B_X(s)$ for all states s and all operations X, where $A_X(s)$ is defined according to Theorem 3.4.*

Proof. For a sequence $F = \langle Op_1, \dots, Op_m \rangle$ of operations which generates the sequence $s_0, s_1, \dots, s_m$ from the start state s_0, define $B(F) = \sum_{1 \leq i \leq m} B_{Op_i}(s_{i-1})$. Let c be a constant such that $T(F) \leq B(F) + c$ for any such sequence F.

For any state s, we define its potential $pot(s)$ by

$$pot(s) = \inf\{B(F) + c - T(F) : F \text{ is a sequence of operations with final state } s\}.$$

We need to write inf instead of min here, since there might be infinitely many sequences leading to s. We have $pot(s) \geq 0$ for any s, since $T(F) \leq B(F) + c$ for any sequence F. Thus pot is a potential function, and the functions $A_X(s)$ defined according to Theorem 3.4 form a family of amortized time bounds.

We need to show that $A_X(s) \leq B_X(s)$ for all X and s. Let $\varepsilon > 0$ be arbitrary. We shall show that $A_X(s) \leq B_X(s) + \varepsilon$. Since ε is arbitrary, this proves that $A_X(s) \leq B_X(s)$.

Fix $\varepsilon > 0$, let s be an arbitrary state, and let X be an operation. Let F be a sequence with final state s and $B(F) + c - T(F) \leq pot(s) + \varepsilon$. The operation X transforms s into some state s'. Let F' be F followed by X, i.e.,

$$s_0 \xrightarrow{\;F\;} s \xrightarrow{\;X\;} s'.$$

Then $pot(s') \leq B(F') + c - T(F')$ by the definition of $pot(s')$, $pot(s) \geq B(F) + c - T(F) - \varepsilon$ by the choice of F, $B(F') = B(F) + B_X(s)$ and $T(F') = T(F) + T_X(s)$ since $F' = F \circ X$, and $A_X(s) = pot(s') - pot(s) + T_X(s)$ by the definition of $A_X(s)$. Combining these inequalities, we obtain

$$
\begin{aligned}
A_X(s) &\leq (B(F') + c - T(F')) - (B(F) + c - T(F) - \varepsilon) + T_X(s) \\
&= (B(F') - B(F)) - (T(F') - T(F) - T_X(s)) + \varepsilon \\
&= B_X(s) + \varepsilon.
\end{aligned}
$$

$\square$

3.5.3 Amortization in Parallel Processing

Amortized solutions are also useful in parallel processing, but there is a potential pitfall that forces us to use them with more care. This is because local amortization may have global consequences. For example, assume that each PE performs a *pushBack* operation on a local unbounded array and that the PEs have to synchronize after the *pushBack*. Then, if one of the PEs has to copy its array, all other PEs have to wait for it, and the amortized analysis breaks down.

Exercise 3.28. Show that the following SPMD pseudocode takes time $\Omega(pn)$:

> $a : UArray$ **of** $\mathbb{N}$ // one on each PE!
> **for** $i := 1$ **to** n **do**   // synchronize after each iteration
>   **if** $i > i_{\text{proc}}$ **then** $a.pushBack(i)$

Hint: Investigate when different PEs perform expensive *pushBacks*.

In order to avoid this pitfall, either we should avoid synchronization to such an extent that delays due to expensive events provably cannot lead to excessive waiting times, or we should synchronize the PEs such that all expensive events happen on all PEs at the same time. When amortization is used globally in the first place, we often want a global trigger for rare expensive events and we then synchronize all PEs to work collectively on them. Examples are epoch FIFO queues, described in Sect. 3.7, where we use a probabilistic asynchronous trigger in order to avoid contention due to the triggering mechanism, and the unbounded distributed-memory hash table described in Sect. 4.6.1 where we use explicit synchronization and collective communication in order to find the exact global size of the data structure.

3.6 Stacks and Queues

Sequences are often used in a rather limited way. Let us start with some examples from precomputer days. Sometimes a clerk will work in the following way: The clerk keeps a *stack* of unprocessed files on their desk. New files are placed on the top of the stack. When the clerk processes the next file, she also takes it from the top of the stack. The easy handling of this "data structure" justifies its use; of course, files may stay in the stack for a long time. In the terminology of the preceding sections, a stack is a sequence that supports only the operations *pushBack*, *popBack*, *last*, and *isEmpty*. We shall use the simplified names *push*, *pop*, and *top* for the three main stack operations.

The behavior is different when people are standing in line waiting for service at a post office: Customers join the line at one end and leave it at the other end. Such sequences are called *FIFO* (first in, first out) *queues* or simply *queues*. In the terminology of the *List* class, FIFO queues use only the operations *first*, *pushBack*, *popFront*, and *isEmpty*.

The more general *deque* (pronounced "deck"), or *double-ended queue*, allows the operations *first*, *last*, *pushFront*, *pushBack*, *popFront*, *popBack*, and *isEmpty*; it can also be observed at a post office when some not so nice individual jumps the line, or when the clerk at the counter gives priority to a pregnant woman at the end of the line. Figure 3.11 illustrates the access patterns of stacks, queues, and deques.

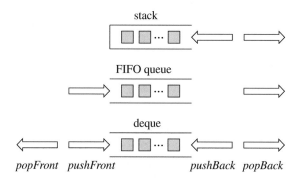

Fig. 3.11. Operations on stacks, queues, and double-ended queues (deques).

Exercise 3.29 (the Tower of Hanoi). *In the great temple of Brahma in Benares, on a brass plate under the dome that marks the center of the world, there are 64 disks of pure gold that the priests carry one at a time between three diamond needles according to Brahma's immutable law: No disk may be placed on a smaller disk. At the beginning of the world, all 64 disks formed the Tower of Brahma on one needle. Now, however, the process of transfer of the tower from one needle to another is in mid-course. When the last disk is finally in place, once again forming the Tower of Brahma but on a different needle, then the end of the world will come and all will turn to dust* [154].[5]

Describe the problem formally for any number k of disks. Write a program that uses three stacks for the piles and produces a sequence of stack operations that transforms the state $(\langle k, \dots, 1 \rangle, \langle \rangle, \langle \rangle)$ into the state $(\langle \rangle, \langle \rangle, \langle k, \dots, 1 \rangle)$.

Exercise 3.30. Explain how to implement a FIFO queue using two stacks so that each FIFO operation takes amortized constant time.

---

[5] In fact, this mathematical puzzle was invented by the French mathematician Édouard Lucas in 1883.

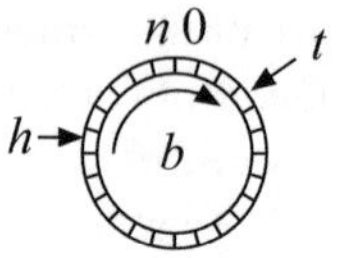

Class *BoundedFIFO(n : ℕ)* **of** *Element*

   b : *Array* [0..*n*] **of** *Element*

   $h = 0 : ℕ$ // index of first element

   $t = 0 : ℕ$ // index of first free entry

Function *isEmpty* : $\{1,0\}$; **return** $h = t$

Function *first* : *Element;* **assert** ¬*isEmpty;* **return** $b[h]$

Function *size* : ℕ; **return** $(t - h + n + 1) \bmod (n + 1)$

Procedure *pushBack(x : Element)*

   assert *size* $< n$

   $b[t] := x$

   $t := (t + 1) \bmod (n + 1)$

Procedure *popFront* **assert** ¬*isEmpty;* $h := (h + 1) \bmod (n + 1)$

Fig. 3.12. An array-based bounded FIFO queue implementation

Why should we care about these specialized types of sequence if we already know a list data structure which supports all of the operations above and more in constant time? There are at least three reasons. First, programs become more readable and are easier to debug if special usage patterns of data structures are made explicit. Second, simple interfaces also allow a wider range of implementations. In particular, the simplicity of stacks and queues allows specialized implementations that are more space-efficient than general *List*s. We shall elaborate on this algorithmic aspect in the remainder of this section. In particular, we shall strive for implementations based on arrays rather than lists. Third, lists are not suited for external-memory use because any access to a list item may cause an I/O operation. The sequential access patterns of stacks and queues translate into good reuse of cache blocks when stacks and queues are represented by arrays.

Bounded stacks, where we know the maximum size in advance, are readily implemented with bounded arrays. For unbounded stacks, we can use unbounded arrays. Stacks can also be represented by singly linked lists: The top of the stack corresponds to the front of the list. FIFO queues are easy to realize with singly linked lists with a pointer to the last element. However, deques cannot be represented efficiently by singly linked lists.

We discuss next an implementation of bounded FIFO queues by use of arrays; see Fig. 3.12. We view an array as a cyclic structure where entry 0 follows the last entry. In other words, we have array indices 0 to *n*, and view the indices modulo $n + 1$. We maintain two indices *h* and *t* that delimit the range of valid queue entries; the queue comprises the array elements indexed by $h..t - 1$. The indices travel around the cycle as elements are queued and dequeued. The cyclic semantics of the indices can be implemented using arithmetic modulo the array size.[6] We always leave at least one

---

[6] On some machines, one might obtain a significant speedup by choosing the array size to be a power of two and replacing **mod** by bit operations.

entry of the array empty, because otherwise it would be difficult to distinguish a full queue from an empty queue. The implementation is readily generalized to bounded deques. Circular arrays also support the random access operator $[\cdot]$:

Operator $[i : \mathbb{N}] : Element;$ **return** $b[i + h \bmod (n+1)]$

Bounded queues and deques can be made unbounded using techniques similar to those used for unbounded arrays in Sect. 3.4.

We have now seen the major techniques for implementing stacks, queues, and deques. These techniques may be combined to obtain solutions that are particularly suited for very large sequences or for external-memory computations.

Exercise 3.31 (lists of arrays). Here we aim to develop a simple data structure for stacks, FIFO queues, and deques that combines all the advantages of lists and un-bounded arrays and is more space-efficient than either lists or unbounded arrays. Use a list (doubly linked for deques) where each item stores an array of K elements for some large constant K. Implement such a data structure in your favorite program-ming language. Compare the space consumption and execution time with those for linked lists and unbounded arrays in the case of large stacks.

Exercise 3.32 (external-memory stacks and queues). Design a stack data struc-ture that needs $O(1/B)$ I/Os per operation in the I/O model described in Sect. 2.2. It suffices to keep two blocks in internal memory. What can happen in a naive imple-mentation with only one block in memory? Adapt your data structure to implement FIFO queues, again using two blocks of internal buffer memory. Implement deques using four buffer blocks.

3.7 Parallel Queue-Like Data Structures

All operations on stacks, queues, and deques concentrate on one or two logical po-sitions. These positions constitute potential bottlenecks and thus can be problematic for parallel processing. Hence, queue-like data structures should only be used with great care. However, there are situations where we need them and where they can even simplify parallelization. We shall discuss such situations in this section.

The straightforward implementation of queue-like data structures on a shared-memory machine protects the data structure with a lock, which has to be acquired before performing any operation. The lock serializes all operations on the queue. Although it seems hard to avoid serialization in the worst case, for example for a queue that alternates between being empty and being nonempty, it is undesirable for parallel computing. Therefore there has been considerable work on better implemen-tations [151] for special situations. What exactly can be done depends not only on the structure (stack, queue, deque) but also on which PEs are allowed to perform which operations. We shall now concentrate on FIFO queues as a concrete example. We will therefore use the abbreviations *push* for *pushFront* and *pop* for *popBack* here. Stacks are much less important for parallelization. At the end of this section, we shall discuss an important special case of a deque.

3.7.1 Single-Producer/Single-Consumer FIFO Queues

Let us call PEs allowed to push elements *producers* and PEs allowed to pop elements *consumers*. A very simple case of a shared-memory FIFO queue is a *single-producer/single-consumer FIFO queue* (1/1-FIFO queue). This case is important since 1/1-FIFOs queues can be used to decouple PEs dedicated to different functions. This way, we can build a *pipeline* that processes a stream of tasks in several stages, with one PE dedicated to each stage. For example, in the UNIX shell, the operator " | " *pipes* data from one command to the next. In 1/1-FIFO queues, locks can be completely avoided. We explain the FastForward queue [124], which is a variant of the bounded circular array FIFO queue shown in Fig. 3.12. *We assume that full and empty queue entries can be distinguished.* As a consequence, there is no need to access h (head) and t (tail) to find out whether the queue is full or empty. The code simplifies and the performance increases. Also, there is no need anymore for an extra location to distinguish an empty from a full queue. However, we now need to explicitly return error conditions, since there is no safe way to find out beforehand whether an operation will be successful. We describe the operations.

Function *push*(x : *Element*) : $\{$OK, FULL$\}$
   if $\neg b[t].isEmpty$ **then return** FULL
   $b[t] := x;$ $t := t + 1 \bmod n;$ **return** OK

The function *pop* looks very similar except that it has to explicitly empty the entry just popped:

Function *pop*(x : *Element*) : $\{$OK, EMPTY$\}$
   if $b[h].isEmpty$ **then return** EMPTY
   $x := b[h];$ $b[h] :=$ EMPTY; $h := h + 1 \bmod n;$ **return** OK

3.7.2 Relaxed FIFO Queues and Bulk Operations

More general FIFO queues allowing fully concurrent access are much more complicated, and it is astonishing how much can go wrong (see [151]). The trouble comes from making sure that the distributed execution is equivalent to a serial execution in which each operation is atomically executed at some point in time between the start and the end of the distributed execution. This behavior is not guaranteed by currently available memory consistency models (see also Sect. B.3). Moreover, even achieving a global ordering does not mean that the order in which elements are processed corresponds to the actual time when the operations are called – this would be almost impossible to ensure, for fundamental physical reasons. We view these difficulties as an indication that the semantics of a FIFO queue is unnecessarily strict. If the strict FIFO property is not required, significant simplications are possible. The FIFO semantics is violated when an element is pushed before another element but is popped after it. Then the second element *overtakes* the first. Figure 3.13 shows an example. We can quantify the degree of violation of the FIFO property by the number of elements that can overtake another element or how long the time lapses can be.

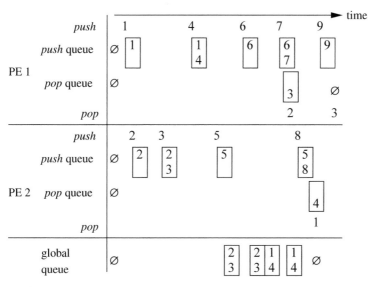

Fig. 3.13. Nine *push* operations (elements 1..9) and three *pop* operations on a batched FIFO queue with $B = 2$ and two PEs. First PE 1 pushes 1, then PE 2 pushes 2 and 3, then PE 1 pushes 4, and PE 2 pushes 5. The elements 2 and 3 fill a block. This block is moved to the global queue when PE 2 pushes 5. Next PE 1 pushes 6, and the block containing 1 and 4 is moved to the global queue. The global queue now contains two blocks. When a *pop* is executed and the PE has no block in its pop buffer, it fetches a buffer from the global queue. So when PE 1 performs a *pop*, it fetches the block containing 2 and 3, and when PE 2 performs a *pop*, it fetches the block containing 1 and 4. PE 1 then serves the *pop* operations from the block in its *pop* buffer.

Once we accept a relaxed semantics, we can use it to reduce contention. For example, we can process elements in batches, performing expensive access to shared variables only occasionally for an entire batch of elements. Assume each PE maintains separate local push and pop buffers taking up to B elements. When a push buffer is full, it is moved to a global FIFO queue F that takes batches of B elements. Similarly, rather than popping individual elements, a PE pops an entire batch of size B from F into its *pop* buffer and uses this to answer up to B single-element pops. The advantage of this approach is that the cost and contention of accessing the global queue are amortized over B elements. Assuming that elements are produced and consumed continuously, incurring only constant delays between operations, our simple batched FIFO with $B = \Theta(p)$ allows at most $O(p^2)$ elements to overtake another one. Figure 3.13 gives an example. Note that the first element pushed (1 on PE 2) is popped after element 2. If the above assumptions are not fulfilled, things can be much worse. For example, a PE might push an element x into an empty buffer, and then will be delayed for a long time while all other PEs rapidly *push* and *pop* an unbounded number of elements that all overtake x.

3.7.3 Distributed-Memory FIFO Queues

On a distributed-memory machine, the message startup overhead is so big that even for a 1/1-FIFO queue, bulk operations make sense. Indeed, 1/1-FIFO queues with bulk operations have a semantics very similar to point-to-point message passing. Most message-passing systems ensure that messages exchanged between a fixed pair of PEs are exchanged in FIFO order – there is no overtaking. A more general FIFO queue allowing arbitrary producers and consumers could be managed by a designated PE f, which receives all *push* and *pop* requests. The order in which these requests are received represents a natural global ordering. If the queue becomes large or if large batches of elements are involved, PE f can delegate the communication and storage of the actual queue content to further PEs. For example, it could just maintain a global counter c, which assigns the cth enqueued batch to PE $b := c$ mod p. A *push* by a PE a would then involve three messages: a request from a to f, a reply from f to a telling it the value of counter c, and a message from a to b delivering the actual data. PE b would then maintain a local queue for the batches delivered to it. In order to ensure the global ordering determined by PE f, PE b has to reorder incoming messages according to their c-values. Since b can receive batches from any other PE, there is no guarantee that the batches will arrive in the order in which they were sent. Hence the necessity for reordering.

3.7.4 *The Epoch FIFO Queue

We discuss here a scalable, fully distributed relaxed FIFO queue called the epoch FIFO queue. We achieve scalability by using efficient distributed mechanisms, such as prefix sums (Sect. 13.3) and work stealing (Sect. 14.5), for global control. Pushs are done in *epochs* and the guarantee is that all elements pushed in one epoch are popped before all elements pushed in later epochs. Each PE maintains a local push buffer taking all of the elements it pushes during the current epoch. When an epoch ends, all elements pushed during this epoch are moved to a single global array, which is pushed as one big batch into a global queue Q. We discuss below how the length of an epoch is determined. Long epochs reduce the amount of communication required, and short epochs guarantee a smaller violation of the FIFO property. Note that the length of an epoch bounds the number of elements that can overtake any element.

Exercise 3.33. Explain how the elements can be moved efficiently using prefix sums.

All elements from a batch are popped before switching to the next batch. Assigning the elements from a particular batch to the PEs is a special case of the loop-scheduling problem discussed in Sect. 14.5, in which a subinterval of the current batch is assigned to each PE. When this local interval is exhausted, it can steal pieces of intervals from other PEs. It can be shown that this can be done in expected time $O(\log p + B/p)$ for a batch of size B using randomized work stealing. This is work-efficient if we ensure that the average batch size is $\Omega(p \log p)$. Since shorter batches guarantee a smaller violation of the FIFO property, our goal is to have batch sizes $\Theta(p \log p)$.

The easiest way to control the batch size is to maintain a global counter c which counts the number of push operations performed in the current epoch. Keeping an exact count is a source of contention. Fortunately, it suffices to keep an approximate count. To avoid contention, c is only incremented with probability $\Theta(1/p)$ when a push is performed. When c exceeds a limit $\ell = \Omega(\log p)$, the next epoch is triggered. In this way, an epoch contains $\Theta(p\ell) = \Theta(p\log p)$ pushes with high probability. Since the work per epoch is $O(p\log p)$, the amortized cost per operation is constant.

The epoch length bounds the number of elements that can overtake any element. Thus only $O(p\log p)$ elements can overtake any element – much less than the $\Theta(p^2)$ overtaking elements for simple batched FIFO queues, and independent of additional assumptions about delays between operations. The disadvantage of epoch FIFO queues is that they require global synchronization of all PEs in order to collectively perform the operations needed for switching from one epoch to the next one.[7]

Epoch FIFO queues can also be implemented efficiently on distributed memory. The global counter c is maintained by PE 0, to which all increment requests are sent. Since PE 0 receives only a logarithmic number of requests per epoch, it does not constitute a bottleneck. PE 0 notifies the other PEs about the end of an epoch using an asynchronous broadcast (see Sect. 13.1). To avoid moving elements around, each entry of the global queue Q is split into p pieces – one for each PE holding the elements pushed by that PE. These elements also constitute the initial interval assigned to the PE. Possible imbalances will then be equalized during work stealing.

3.7.5 Deques for Work Stealing

A variant of deques which is important for the work-stealing load balancers described in Sect. 14.5 has a single PE that uses the deque like a stack (*pushBack*, *popBack*), and any number of PEs that are allowed to do *popFront*s. A lock-free implementation is given in [20] that exploits the special set of operations and additional properties of the application.

3.8 Lists versus Arrays

Table 3.1 summarizes the findings of this chapter. Arrays are better at indexed access, whereas linked lists have their strength in manipulations of sequences at arbitrary positions. Both of these approaches realize the operations needed for stacks and queues efficiently. However, arrays are more cache-efficient here, whereas lists provide worst-case performance guarantees.

Singly linked lists can compete with doubly linked lists in most but not all respects. The only advantage of cyclic arrays over unbounded arrays is that they can implement *pushFront* and *popFront* efficiently.

---

[7] In a more sophisticated variant of the epoch FIFO queue, this work could also be done in the background without working threads noticing, by using $\Theta(p)$ additional server threads which are triggered by the end of an epoch or the exhaustion of the elements in the batch.

Table 3.1. Running times of operations on sequences with n elements. The entries have an implicit $O(\cdot)$ around them. *List* stands for doubly linked lists, *SList* stands for singly linked lists, *UArray* stands for unbounded arrays, and *CArray* stands for circular arrays.

Operation	*List*	*SList*	*UArray*	*CArray*	Explanation of "*"
$[\cdot]$	n	n	1	1	
size	1^*	1^*	1	1	Not with interlist *splice*
first	1	1	1	1	
last	1	1	1	1	
insert	1	1^*	n	n	*insertAfter* only
remove	1	1^*	n	n	*removeAfter* only
pushBack	1	1	1^*	1^*	Amortized
pushFront	1	1	n	1^*	Amortized
popBack	1	n	1^*	1^*	Amortized
popFront	1	1	n	1^*	Amortized
concat	1	1	n	n	
splice	1	1	n	n	
findNext, ...	n	n	n^*	n^*	Cache-efficient

Space efficiency is also a nontrivial issue. Linked lists are very compact if the elements are much larger than the pointers. For small *Element* types, arrays are usually more compact because there is no overhead for pointers. This is certainly true if the sizes of the arrays are known in advance so that bounded arrays can be used. Unbounded arrays have a trade-off between space efficiency and copying overhead during reallocation.

3.9 Implementation Notes

Every decent programming language supports bounded arrays. In addition, unbounded arrays, lists, stacks, queues, and deques are provided in libraries that are available for the major imperative languages. Nevertheless, you will often have to implement list-like data structures yourself, for example when your objects are members of several linked lists. In such implementations, memory management is often a major challenge.

3.9.1 C++

The class *vector*⟨*Element*⟩ in the STL realizes unbounded arrays. However, most implementations never shrink the array. There is functionality for manually setting the allocated size. Usually, you will give some initial estimate of the sequence size n when the *vector* is constructed. This can save you many grow operations. Often, you also know when the array will stop changing size, and you can then force $w = n$. With these refinements, there is little reason to use the built-in C-style arrays. An added benefit of *vector*s is that they are automatically destroyed when the variable goes out of scope. Furthermore, during debugging, you may switch to implementations with bound checking.

There are some additional issues that you may want to address if you need very high performance for arrays that grow or shrink a lot. During reallocation, *vector* has to move array elements using the copy constructor of *Element*. In most cases, a call to the low-level byte copy operation *memcpy* would be much faster. Another low-level optimization is to implement *reallocate* using the standard C function *realloc*. The memory manager might be able to avoid copying the data entirely.

A stumbling block with unbounded arrays is that pointers to array elements become invalid when the array is reallocated. You should make sure that the array does not change size while such pointers are being used. If reallocations cannot be ruled out, you can use array indices rather than pointers.

The STL and LEDA [195] offer doubly linked lists in the class *list⟨Element⟩*, and singly linked lists in the class *slist⟨Element⟩*. Their memory management uses free lists for all objects of (roughly) the same size, rather than only for objects of the same class.

If you need to implement a list-like data structure, note that the operator *new* can be redefined for each class. The standard library class *allocator* offers an interface that allows you to use your own memory management while cooperating with the memory managers of other classes.

The STL provides the classes *stack⟨Element⟩* and *deque⟨Element⟩* for stacks and double-ended queues, respectively. *Deque*s also allow constant-time indexed access using the operator $[\cdot]$. LEDA offers the classes *stack⟨Element⟩* and *queue⟨Element⟩* for unbounded stacks, and FIFO queues implemented via linked lists. It also offers bounded variants that are implemented as arrays.

Iterators are a central concept of the STL; they implement our abstract view of sequences independent of the particular representation.

3.9.2 Java

Since version 6 of Java, the *util* package provides *ArrayList* for unbounded arrays and *LinkedList* for doubly linked lists. There is a *Deque* interface, with implementations by use of *ArrayDeque* and *LinkedList*. A *Stack* is implemented as an extension to *Vector*.

Many book on Java proudly announce that Java has no pointers, so you might wonder how to implement linked lists. The solution is that object references in Java are essentially pointers. In a sense, Java has *only* pointers, because members of non-simple type are always references, and are never stored in the parent object itself.

Explicit memory management is optional in Java, since it provides garbage collection of all objects that are not referenced anymore.

3.10 Historical Notes and Further Findings

All of the algorithms described in this chapter are "folklore", i.e., they have been around for a long time and nobody claims to be their inventor. Indeed, we have seen that many of the underlying concepts predate computers.

Amortization is as old as the analysis of algorithms. The *bank account* and *potential* methods were introduced at the beginning of the 1980s by Brown, Huddlestone, Mehlhorn, Sleator, and Tarjan [57, 159, 300, 301]. The overview article [309] popularized the term *amortized analysis*, and Theorem 3.5 first appeared in [214].

There is an array-like data structure that supports indexed access in constant time and arbitrary element insertion and deletion in amortized time $O(\sqrt{n})$. The trick is relatively simple. The array is split into subarrays of size $n' = \Theta(\sqrt{n})$. Only the last subarray may contain fewer elements. The subarrays are maintained as cyclic arrays, as described in Sect. 3.6. Element i can be found in entry $i \bmod n'$ of subarray $\lfloor i/n' \rfloor$. A new element is inserted into its subarray in time $O(\sqrt{n})$. To repair the invariant that subarrays have the same size, the last element of this subarray is inserted as the first element of the next subarray in constant time. This process of shifting the extra element is repeated $O(n/n') = O(\sqrt{n})$ times until the last subarray is reached. Deletion works similarly. Occasionally, one has to start a new last subarray or change n' and reallocate everything. The amortized cost of these additional operations can be kept small. With some additional modifications, all deque operations can be performed in constant time. We refer the reader to [178] for more sophisticated implementations of deques and an implementation study.

4

Hash Tables and Associative Arrays

If you want to get a book from the central library of the Karlsruhe Institute of Technology (KIT), you have to order the book in advance. The library personnel fetch the book from the stacks and deliver it to a room with 100 shelves. You find your book on a shelf numbered with the last *two digits of your library card. Why the last digits and not the leading digits? Probably because this distributes the books more evenly among the shelves. The library cards are numbered consecutively as students register. The University of Karlsruhe, the predecessor of KIT, was founded in 1825. Therefore, the students who enrolled at the same time are likely to have the same leading digits in their card number, and only a few shelves would be in use if the leading digits were used.*

The subject of this chapter is the robust and efficient implementation of the above "delivery shelf data structure". In computer science, this data structure is known as a *hash*[1] *table*. Hash tables are one implementation of *associative arrays*, or *dictionaries*. The other implementation is the tree data structures which we shall study in Chap. 7. An associative array is an array with a potentially infinite or at least very large index set, out of which only a small number of indices are actually in use. For example, the potential indices may be all strings, and the indices in use may be all identifiers used in a particular C++ program. Or the potential indices may be all ways of placing chess pieces on a chess board, and the indices in use may be the placements required in the analysis of a particular game. Associative arrays are versatile data structures. Compilers use them for their *symbol table*, which associates identifiers with information about them. Combinatorial search programs often use them for detecting whether a situation has already been looked at. For example, chess programs have to deal with the fact that board positions can be reached by different sequences of moves. However, each position needs to be evaluated only once. The solution is to store positions in an associative array. One of the most widely used implementations of the *join* operation in relational databases temporarily stores one

---

[1] Photograph of the mincer above by Kku, Rainer Zenz (Wikipedia), License CC-by-SA 2.5.

of the participating relations in an associative array. Scripting languages such as AWK
[8] and Perl [326] use associative arrays as their *main* data structure. In all of the
examples above, the associative array is usually implemented as a hash table. The
exercises in this section ask you to develop some further uses of associative arrays.

Formally, an associative array S stores a set of elements. Each element e has an
associated key $key(e) \in Key$. We assume keys to be unique, i.e., distinct elements
in S have distinct keys. Frequently, elements are key-value pairs, i.e., *Element* =
Key × *Value*. Associative arrays support the following operations:

- $S.build(\{e_1,\dots,e_n\})$: $S := \{e_1,\dots,e_n\}$.
- $S.insert(e : Element)$: If S contains no element with key $key(e)$, $S := S \cup \{e\}$.
  Otherwise, nothing is done.[2]
- $S.remove(x : Key)$: if there is an $e \in S$ with $key(e) = x$: $S := S \setminus \{e\}$.
- $S.find(x : Key)$: if there is an $e \in S$ with $key(e) = x$, return e,
  otherwise return $\bot$.

If only operations *build* and *find* are used, the data structure is called *static*. Otherwise, it is called a *dynamic* data structure. The operation $build(\{e_1,\dots,e_n\})$ requires
that the keys of the elements e_1 to e_n are pairwise distinct. If operation *find*(x) returns
a reference to an element it can also be *updated* subsequently by replacing the value
associated with key x. The operations *find*, *insert*, and *remove* essentially correspond
to reading from or writing to an array at an arbitrary position (random access). This
explains the name "associative array".

In addition, we assume a mechanism that allows us to retrieve all elements in S.
Since this *forall* operation is usually easy to implement, we defer its discussion to
the exercises.

The set *Key* is the set of potential array indices, whereas the set $\{key(s) : e \in S\}$
comprises the indices in use at any particular time. Throughout this chapter, we use
n to denote the size of S, and N to denote the size of *Key*. In a typical application of
associative arrays, N is humongous and hence the use of an array of size N is out of
the question. We are aiming for solutions which use space $O(n)$.

On a parallel machine, we also need atomic operations for finding and updating
a hash table element. Equally useful can be an operation *insertOrUpdate* that inserts
an element if it is not yet present and updates it otherwise. See Sect. 4.6 for more
details. We can also consider bulk operations – inserting, removing and updating
many elements in a batched fashion.

In the library example, *Key* is the set of all library card numbers, and the elements
are book orders. Another precomputer example is provided by an English-German
dictionary. The keys are English words, and an element is an English word together
with its German translations.

The basic idea behind the hash table implementation of associative arrays is simple. We use a *hash function* h to map the set *Key* of potential array indices to a small
range $0..m-1$ of integers. We also have an array t with index set $0..m-1$, the *hash*

---

[2] An alternative implementation replaces the old element with key $key(e)$ by e.

table. In order to keep the space requirement low, we want m to be about the number of elements in S. The hash function associates with each element e a *hash value* $h(key(e))$. In order to simplify the notation, we write $h(e)$ instead of $h(key(e))$ for the hash value of e. In the library example, h maps each library card number to its last two digits. Ideally, we would like to store element e in the table entry $t[h(e)]$. If this works, we obtain constant execution time[3] for our three operations *insert*, *remove*, and *find*.

Unfortunately, storing e in $t[h(e)]$ will not always work, as several elements might *collide*, i.e., map to the same table entry. The library example suggests a fix: Allow several book orders to go to the same shelf. The entire shelf then has to be searched to find a particular order. A generalization of this fix leads to *hashing with chaining*. In each table entry, we store a *set* of elements and implement the set using singly linked lists. Section 4.1 analyzes hashing with chaining using some rather optimistic (and hence unrealistic) assumptions about the properties of the hash function. In this model, we achieve constant expected time for all three dictionary operations.

In Sect. 4.2, we drop the unrealistic assumptions and construct hash functions that come with good (probabilistic) performance guarantees. Our simple examples already show that finding good hash functions is nontrivial. For example, if we were to apply the least-significant-digit idea from the library example to an English–German dictionary, we might come up with a hash function based on the last four letters of a word. But then we would have many collisions for words ending in "tion", "able", etc.

We can simplify hash tables (but not their analysis) by returning to the original idea of storing all elements in the table itself. When a newly inserted element e finds the entry $t[h(e)]$ occupied, it scans the table until a free entry is found. In the library example, assume that shelves can hold exactly one book. The librarians would then use adjacent shelves to store books that map to the same delivery shelf. Section 4.3 elaborates on this idea, which is known as *hashing with open addressing and linear probing*. After comparing the two approaches in Sect. 4.4, we turn to parallel hashing in Sect. 4.6. The main issue here is to avoid or mitigate the effects of multiple PEs trying to access the same entry of the hash table.

Why are hash tables called hash tables? The dictionary defines "to hash" as "to chop up, as of potatoes". This is exactly what hash functions usually do. For example, if keys are strings, the hash function may chop up the string into pieces of fixed size, interpret each fixed-size piece as a number, and then compute a single number from the sequence of numbers. A good hash function creates disorder and, in this way, avoids collisions. A good hash function should distribute every subset of the key space about evenly over the hash table. Hash tables are frequently used in time-critical parts of computer programs.

---

[3] Strictly speaking, we have to add additional terms for evaluating the hash function and for moving elements around. To simplify the notation, we assume in this chapter that all of this takes constant time.

Exercise 4.1. Assume you are given a set M of pairs of integers. M defines a binary relation R_M. Use an associative array to check whether R_M is symmetric. A relation is symmetric if $\forall (a,b) \in M : (b,a) \in M$.

Exercise 4.2. Write a program that reads a text file and outputs the 100 most frequent words in the text.

Exercise 4.3 (a billing system). Assume you have a large file consisting of triples (transaction, price, customer ID). Explain how to compute the total payment due for each customer. Your algorithm should run in linear time.

Exercise 4.4 (scanning a hash table). Show how to realize the *forall* operation for hashing with chaining and for hashing with open addressing and linear probing. What is the running time of your solution?

Exercise 4.5 ((database) hash join). Consider two relations $R \subseteq A \times B$ and $Q \subseteq B \times C$ with $A \neq C$. The (natural) *join* of R and Q is

$$R \bowtie Q := \{(a,b,c) \subseteq A \times B \times C : (a,b) \in R \wedge (b,c) \in Q\}.$$

Give an algorithm for computing $R \bowtie Q$ in expected time $O(|R| + |Q| + |R \bowtie Q|)$ assuming that elements of B can be hashed in constant time. Hint: The hash table entries may have to store sets of elements.

4.1 Hashing with Chaining

Hashing with chaining maintains an array t of linear lists (see Fig. 4.1); the linear list $t[k]$ contains all elements $e \in S$ with $key(e) = k$. The associative-array operations are easy to implement. To find an element with key k, we scan through $t[h(k)]$. If an element e with $key(e) = k$ is encountered, we return it. Otherwise, we return $\bot$. To remove an element with key k, we scan through $t[h(k)]$. If an element e with $key(e) = k$ is encountered, we remove it and return. To insert an element e, we also scan through the sequence $t[h(k)]$. If an element e' with $key(e') = k$ is encountered, we do nothing, otherwise, we add e to the sequence. The operation $build(\{e_1, \ldots, e_n\})$ is realized by n insert operations. Since the precondition of the operation guarantees that the elements have distinct keys, there is no need to check whether there is already an element with the same key and hence every element e can be inserted at the beginning of the list $t[h(e)]$. Therefore, the running time is $O(n)$.

The space consumption of the data structure is $O(n+m)$. To remove, find or insert an element with key k, we have to scan the sequence $t[h(k)]$. In the worst case, for example, if *find* looks for an element that is not there, the entire list has to be scanned. If we are unlucky, all elements are mapped to the same table entry and the execution time is $\Theta(n)$. So, in the worst case, hashing with chaining is no better than linear lists.

Are there hash functions that guarantee that all sequences are short? The answer is clearly no. A hash function maps the set of keys to the range $0..m-1$, and hence

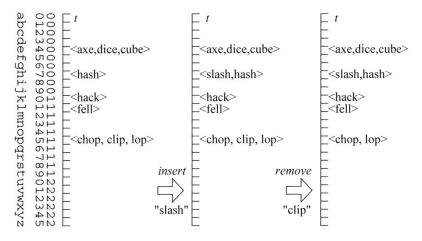

Fig. 4.1. Hashing with chaining. We have a table t of sequences. This figure shows an example where a set of words (short synonyms of "hash") is stored using a hash function that maps the last character to the integers $0..25$. We see that this hash function is not very good.

for every hash function there is always a set of N/m keys that all map to the same table entry. In most applications, $n < N/m$ and hence hashing can always deteriorate to linear search. We shall study three approaches to dealing with the worst-case. The first approach is average-case analysis, where we average either over all possible hash functions (Theorem 4.1) or over the possible inputs (Exercise 4.8). The second approach is to use randomization, and to choose the hash function at random from a collection of hash functions. This is equivalent to average case analysis where we average over the possible hash functions. We shall study this approach in this section and the next. The third approach is to change the algorithm. For example, we could make the hash function depend on the set of keys in actual use. We shall investigate this approach in Sect. 4.5 and shall show that it leads to good worst-case behavior.

Let H be the set of all functions from *Key* to $0..m-1$. We assume that the hash function h is chosen randomly[4] from H and shall show that for any fixed set S of n keys, the expected execution time of *insert*, *remove*, and *find* is $O(1+n/m)$. Why do we prove a theorem based on unrealistic assumptions? It shows us what might be possible. We shall later obtain the same time bounds with realistic assumptions.

Theorem 4.1. *If n elements are stored in a hash table with m entries and a random hash function is used, the expected execution time of insert, remove, and find is* $O(1+n/m)$.

Proof. The proof requires the probabilistic concepts of random variables, their expectation, and the linearity of expectations as described in Sect. A.3. Consider the ex-

---

[4] This assumption is completely unrealistic. There are m^N functions in H, and hence it requires $N\log m$ bits to specify a function in H. This defeats the goal of reducing the space requirement from N to n.

ecution time of *insert*, *remove*, and *find* for a fixed key k. These operations need constant time for evaluating the hash function and accessing the list $t[h(k)]$ plus the time for scanning the sequence $t[h(k)]$. Hence the expected execution time is $O(1 + E[X])$, where the random variable X stands for the length of the sequence $t[h(k)]$. Let S be the set of n elements stored in the hash table. For each $e \in S$, let X_e be the *indicator* variable which tells us whether e hashes to the same location as k, i.e., $X_e = 1$ if $h(e) = h(k)$ and $X_e = 0$ otherwise. In shorthand, $X_e = [h(e) = h(k)]$. There are two cases. If there is no entry in S with key k, then $X = \sum_{e \in S} X_e$. If there is an entry e_0 in S with $key(e_0) = k$, then $X = 1 + \sum_{e \in S \setminus \{e_0\}} X_e$. Using the linearity of expectations, we obtain in the first case

$$E[X] = E\left[\sum_{e \in S} X_e\right] = \sum_{e \in S} E[X_e] = \sum_{e \in S} \mathrm{prob}(X_e = 1).$$

A random hash function maps e to all m table entries with the same probability, independent of $h(k)$. Hence, $\mathrm{prob}(X_e = 1) = 1/m$ and therefore $E[X] = n/m$. In the second case (key k occurs in S), we obtain analogously $X \le 1 + (n-1)/m \le 1 + n/m$. Thus, the expected execution time of *insert*, *find*, and *remove* is $O(1 + n/m)$.   □

We can achieve a linear space requirement and a constant expected execution time for all three operations by guaranteeing that $m = \Theta(n)$ at all times. Adaptive reallocation, as described for unbounded arrays in Sect. 3.4, is the appropriate technique.

Exercise 4.6 (unbounded hash tables). Explain how to guarantee $m = \Theta(n)$ in hashing with chaining. You may assume the existence of a hash function $h' : Key \to \mathbb{N}$. Set $h(k) = h'(k) \bmod m$ and use adaptive reallocation.

Exercise 4.7 (waste of space). In part, the waste of space in hashing with chaining is due to empty table entries. Assuming a random hash function, compute the expected number of empty table entries as a function of m and n. Hint: Define indicator random variables $Y_0, \dots, Y_{m-1}$, where $Y_i = 1$ if $t[i]$ is empty.

Exercise 4.8 (average-case behavior). Assume that the hash function distributes the set of potential keys evenly over the table, i.e., for each i, $0 \le i \le m - 1$, we have $|\{k \in Key : h(k) = i\}| \le \lceil N/m \rceil$. Assume also that a random set S of n keys is stored in the table, i.e., S is a random subset of *Key* of size n. Show that for any table position i, the expected number of elements in S that hash to i is at most $\lceil N/m \rceil \cdot n/N \approx n/m$.

4.2 Universal Hashing

Theorem 4.1 is unsatisfactory, as it presupposes that the hash function is chosen randomly from the set of all functions[5] from keys to table positions. The class of all such functions is much too big to be useful. We shall show in this section that

---

[5] In the context of hashing, one usually talks about a *class* or a *family* of functions and reserves the word "set" for the set of keys or elements stored in the table.

the same performance can be obtained with much smaller classes of hash functions. The families presented in this section are so small that a member can be specified in constant space. Moreover, the functions are easy to evaluate.

Definition 4.2. *Let c be a positive constant. A family H of functions from Key to* $0..m-1$ *is called c-universal if any two distinct keys collide with a probability of at most* c/m, *i.e., for all x, y in Key with* $x \neq y$,

$$|\{h \in H : h(x) = h(y)\}| \leq \frac{c}{m}|H|.$$

In other words, for random $h \in H$,

$$\mathrm{prob}(h(x) = h(y)) \leq \frac{c}{m}.$$

This definition has been formulated so as to guarantee that the proof of Theorem 4.1 continues to work.

Theorem 4.3. *If n elements are stored in a hash table with m entries using hashing with chaining and a random hash function from a c-universal family is used, the expected execution time of insert, remove and find is* $O(1 + cn/m)$.

Proof. We can reuse the proof of Theorem 4.1 almost literally. Observe that we have changed the probability space. We are now choosing the hash function h from a c-universal class. Nevertheless, the wording of the argument does not basically change. Consider the execution time of *insert*, *remove*, or *find* for a fixed key k. They need constant time plus the time for scanning the sequence $t[h(k)]$. Hence the expected execution time is $O(1 + E[X])$, where the random variable X stands for the length of the sequence $t[h(k)]$. Let S be the set of n elements stored in the hash table. For each $e \in S$, let X_e be the indicator variable which tells us whether e hashes to the same location as k, i.e., $X_e = 1$ if $h(e) = h(k)$ and $X_e = 0$ otherwise. In shorthand, $X_e = [h(e) = h(k)]$. There are two cases. If there is no entry in S with key k, then $X = \sum_{e \in S} X_e$. If there is an entry e_0 in S with $key(e_0) = k$, then $X = 1 + \sum_{e \in S \setminus \{e_0\}} X_e$. Using the linearity of expectations, we obtain in the first case

$$E[X] = E\left[\sum_{e \in S} X_e\right] = \sum_{e \in S} E[X_e] = \sum_{e \in S} \mathrm{prob}(X_e = 1).$$

Since h is chosen uniformly from a c-universal class, we have $\mathrm{prob}(X_e = 1) \leq c/m$, and hence $E[X] = cn/m$. In the second case (key k occurs in S), we obtain analogously $X \leq 1 + c(n-1)/m \leq 1 + cn/m$. Thus, the expected execution time of *insert*, *find*, and *remove* is $O(1 + cn/m)$. $\square$

It now remains to find c-universal families of hash functions that are easy to construct and easy to evaluate. We shall describe a simple and quite practical 1-universal family in detail and give further examples in the exercises. We assume that our keys are bit strings of a certain fixed length; in the exercises, we discuss how the fixed-length assumption can be overcome. We also assume that the table size m is a prime number. Why a prime number? Because arithmetic modulo a prime is particularly nice; in particular, the set $\mathbb{Z}_m = \{0, \ldots, m-1\}$ of numbers modulo m forms a field.[6] Let $w = \lfloor \log m \rfloor$. We subdivide the keys into pieces of w bits each, say k pieces. We interpret each piece as an integer in the range $0..2^w - 1$ and keys as k-tuples of such integers. For a key $\mathbf{x}$, we write $\mathbf{x} = (x_1, \ldots, x_k)$ to denote its partition into pieces. Each x_i lies in $0..2^w - 1$. We can now define our class of hash functions. For each $\mathbf{a} = (a_1, \ldots, a_k) \in \{0..m-1\}^k$, we define a function $h_\mathbf{a}$ from Key to $0..m-1$ as follows. Let $\mathbf{x} = (x_1, \ldots, x_k)$ be a key and let $\mathbf{a} \cdot \mathbf{x} = \sum_{i=1}^k a_i x_i$ denote the scalar product (over $\mathbb{Z}$) of $\mathbf{a}$ and $\mathbf{x}$. Then

$$h_\mathbf{a}(\mathbf{x}) = \mathbf{a} \cdot \mathbf{x} \bmod m.$$

This is the scalar product of a and $\mathbf{x}$ over $\mathbb{Z}_m$. It is time for an example. Let $m = 17$ and $k = 4$. Then $w = 4$ and we view keys as 4-tuples of integers in the range $0..15$, for example $\mathbf{x} = (11, 7, 4, 3)$. A hash function is specified by a 4-tuple of integers in the range $0..16$, for example $\mathbf{a} = (2, 4, 7, 16)$. Then $h_\mathbf{a}(\mathbf{x}) = (2 \cdot 11 + 4 \cdot 7 + 7 \cdot 4 + 16 \cdot 3) \bmod 17 = 7$.

Theorem 4.4. *Let m be a prime. Then*

$$H^{\cdot} = \left\{ h_\mathbf{a} : \mathbf{a} \in \{0..m-1\}^k \right\}$$

is a 1-universal family of hash functions.

In other words, the scalar product of the representation of a key as a tuple of numbers in $\{0..m-1\}$ and a random vector modulo m defines a good hash function if the product is computed modulo a prime number.

Proof. Consider two distinct keys $\mathbf{x} = (x_1, \ldots, x_k)$ and $\mathbf{y} = (y_1, \ldots, y_k)$. To determine $\text{prob}(h_\mathbf{a}(\mathbf{x}) = h_\mathbf{a}(\mathbf{y}))$, we count the number of choices for $\mathbf{a}$ such that $h_\mathbf{a}(\mathbf{x}) = h_\mathbf{a}(\mathbf{y})$. Choose an index j such that $x_j \neq y_j$. Then $(x_j - y_j) \not\equiv 0 \pmod{m}$, and hence any equation of the form $a_j(x_j - y_j) \equiv b \pmod{m}$, where $b \in \mathbb{Z}_m$, has a unique solution for a_j, namely $a_j \equiv (x_j - y_j)^{-1} b \pmod{m}$. Here $(x_j - y_j)^{-1}$ denotes the *multiplicative inverse*[7] of $(x_j - y_j)$.

We claim that for each choice of the a_i's with $i \neq j$, there is exactly one choice of a_j such that $h_\mathbf{a}(\mathbf{x}) = h_\mathbf{a}(\mathbf{y})$. Indeed,

---

[6] A field is a set with special elements 0 and 1 and with addition and multiplication operations. Addition and multiplication satisfy the usual laws known for the field of rational numbers.

[7] In a field, any element $z \neq 0$ has a unique multiplicative inverse, i.e., there is a unique element z^{-1} such that $z^{-1} \cdot z = 1$. For example, in $\mathbb{Z}_7$, we have $1^{-1} = 1$, $2^{-1} = 4$, $3^{-1} = 5$, $4^{-1} = 2$, and $5^{-1} = 3$. Multiplicative inverses allow one to solve linear equations of the form $zx = b$, where $z \neq 0$. The solution is $x = z^{-1}b$.

$$h_{\mathbf{a}}(\mathbf{x}) = h_{\mathbf{a}}(\mathbf{y}) \Leftrightarrow \sum_{1\le i\le k} a_i x_i \equiv \sum_{1\le i\le k} a_i y_i \qquad (\bmod\ m)$$

$$\Leftrightarrow a_j(x_j - y_j) \equiv \sum_{i\ne j} a_i(y_i - x_i) \qquad (\bmod\ m)$$

$$\Leftrightarrow a_j \equiv (x_j - y_j)^{-1} \sum_{i\ne j} a_i(y_i - x_i) \quad (\bmod\ m).$$

There are m^{k-1} ways to choose the a_i with $i \ne j$, and for each such choice there is a unique choice for a_j. Since the total number of choices for $\mathbf{a}$ is m^k, we obtain

$$\text{prob}(h_{\mathbf{a}}(\mathbf{x}) = h_{\mathbf{a}}(\mathbf{y})) = \frac{m^{k-1}}{m^k} = \frac{1}{m}. \qquad \square$$

Is it a serious restriction that table sizes need to be prime? At first glance, yes. We certainly cannot burden users with the task of providing appropriate primes. Also, when we grow or shrink an array adaptively, it is not clear how to find a prime in the vicinity of the desired new table size. A closer look, however, shows that the problem is easy to resolve.

Number theory tells us that primes are abundant. More precisely, it is an easy consequence of the prime number theorem [138, p. 264] that for every fixed $\alpha > 1$ and every sufficiently large m, the interval $[m, \alpha m]$ contains about $(\alpha - 1)m/\ln m$ prime numbers. The easiest solution is then to precompute a table which contains, for example, for each interval $2^\ell..2^{\ell+1} - 1$, a prime number in this interval. Such tables are also available on the internet.

If one does not want to use a table, the required prime numbers can also be computed on the fly. We make use of the following statement (A_k), where $k \ge 1$ is an integer:

The interval $k^3..(k+1)^3$ contains at least one prime. $\qquad (A_k)$

It is known that (A_k) holds for $k \le 8\cdot 10^7$ and for $k > e^{e^{15}}$. For "small" k, the statement was established by computation; we shall tell you more about this computation in Sect. 4.8. For "large" k, the statement was established by mathematical proof [69]. For "intermediate" k, the question of whether (A_k) holds is open. Fortunately, the "small k" result suffices for our purposes. If we want to use a hash table of size approximately m, we determine a k with $k^3 \le m \le (k+1)^3$ and then search for a prime in the interval $k^3..(k+1)^3$. The search is guaranteed to succeed for $m \le 64\cdot 10^{21}$; it may succeed also for larger m.

How does this search work? We use a variant of the "sieve of Eratosthenes" (cf. Exercise 2.5). Any nonprime number in the interval must have a divisor which is at most $\sqrt{(k+1)^3} = (k+1)^{3/2}$. We therefore iterate over the numbers from 2 to $\lfloor (k+1)^{3/2} \rfloor$ and, for each such j, remove its multiples in $k^3..(k+1)^3$. For each fixed j, this takes time $((k+1)^3 - k^3)/j = O(k^2/j)$. The total time required is

$$\sum_{j \leq (k+1)^{3/2}} O\left(\frac{k^2}{j}\right) = k^2 \sum_{j \leq (k+1)^{3/2}} O\left(\frac{1}{j}\right)$$

$$= O\left(k^2 \ln\left((k+1)^{3/2}\right)\right) = O\left(k^2 \ln k\right) = o(m)$$

and hence is negligible compared with the cost of initializing a table of size m. The second equality in the equation above uses the harmonic sum (A.13).

Exercise 4.9 (strings as keys). Adapt the class $H^{\cdot}$ to strings of arbitrary length. Assume that each character requires eight bits (= a byte). You may assume that the table size is at least $m = 257$. The time for evaluating a hash function should be proportional to the length of the string being processed. Input strings may have arbitrary lengths not known in advance. Hint: Use "lazy evaluation" for choosing the random vector **a**, i.e., fix only the components that have already been in use and extend if necessary. You may assume at first that strings do not start with the character 0 (whose byte representation consists of eight 0's); note that the strings x and $0x$ are different but have the same hash value for every function $h_{\mathbf{a}}$. Once you have solved this restricted case, show how to remove the restriction.

Exercise 4.10 (hashing using bit matrix multiplication). For this exercise, keys are bit strings of length k, i.e., $Key = \{0,1\}^k$, and the table size m is a power of two, say $m = 2^w$. Each $w \times k$ matrix M with entries in $\{0,1\}$ defines a hash function h_M. For $x \in \{0,1\}^k$, let $h_M(x) = Mx \bmod 2$, i.e., $h_M(x)$ is a matrix–vector product computed modulo 2. The resulting w-bit vector is interpreted as a number in $0..m-1$. Let

$$H^{\text{lin}} = \left\{ h_M : M \in \{0,1\}^{w \times k} \right\}.$$

For $M = \begin{pmatrix} 1 & 0 & 1 & 1 \\ 0 & 1 & 1 & 1 \end{pmatrix}$ and $x = (1,0,0,1)^T$, we have $Mx \bmod 2 = (0,1)^T$. This represents the number $0 \cdot 2^1 + 1 \cdot 2^0 = 1$. Note that multiplication modulo two is the logical AND operation, and that addition modulo two is the XOR operation $\oplus$.

(a) Explain how $h_M(x)$ can be evaluated using k bit-parallel exclusive OR operations. Hint: The ones in x select columns of M. Add the selected columns.
(b) Explain how $h_M(x)$ can be evaluated using w bit-parallel AND operations and w parity operations. Many machines provide an instruction $parity(y)$ that returns 1 if the number of ones in y is odd, and 0 otherwise. Hint: Multiply each row of M by x.
(c) We now want to show that H^{lin} is 1-universal. (1) Show that for any two keys $x \neq y$, any bit position j where x and y differ, and any choice of the columns M_i of the matrix with $i \neq j$, there is exactly one choice of a column M_j such that $h_M(x) = h_M(y)$. (2) Count the number of ways to choose $k-1$ columns of M. (3) Count the total number of ways to choose M. (4) Compute the probability $prob(h_M(x) = h_M(y))$ for $x \neq y$ if M is chosen randomly.

*Exercise 4.11 (more on hashing using matrix multiplication). Let p be a prime number and assume that *Key* is the set of k-tuples with elements in $0..p-1$. Let $w \geq 1$ be an integer. Generalize the class H^{lin} to a class of hash functions

$$H^{\times} = \left\{ h_M : M \in \{0..p-1\}^{w \times k} \right\}$$

that map keys to $(0..p-1)^w$. The matrix multiplication is now performed modulo p. Show that $H^{\times}$ is 1-universal. Explain how $H^{\cdot}$ is a special case of $H^{\times}$.

Exercise 4.12 (simple linear hash functions). Assume that $Key \subseteq 0..p-1 = \mathbb{Z}_p$ for some prime number p. Assume also that $m \leq p$, where m is the table size. For $a, b \in \mathbb{Z}_p$, let $h_{(a,b)}(x) = ((ax+b) \bmod p) \bmod m$. For example, if $p = 97$ and $m = 8$, we have $h_{(23,73)}(2) = ((23 \cdot 2 + 73) \bmod 97) \bmod 8 = 22 \bmod 8 = 6$. Let

$$H^* = \left\{ h_{(a,b)} : a, b \in 0..p-1 \right\}.$$

Show that this family is $(\lceil p/m \rceil / (p/m))^2$-universal.

Exercise 4.13 (continuation of Exercise 4.12). Show that the following holds for the class H^* defined in the previous exercise. If x and y are distinct keys, i and j in $0..m-1$ are arbitrary, and $h_{(a,b)}$ is chosen randomly in H^* then

$$\text{prob}(h_{(a,b)}(x) = i \text{ and } h_{(a,b)}(y) = j) \leq c/m^2$$

for some constant c.

Exercise 4.14 (a counterexample). Let $Key = 0..p-1$, and consider the set of hash functions

$$H^{\text{fool}} = \left\{ h_{(a,b)} : a, b \in 0..p-1 \right\}$$

with $h_{(a,b)}(x) = (ax+b) \bmod m$. Show that there is a set S of $\lceil p/m \rceil$ keys such that for any two keys x and y in S, all functions in H^{fool} map x and y to the same value. Hint: Let $S = \{0, m, 2m, \ldots, \lfloor p/m \rfloor m\}$.

Exercise 4.15 (table size 2^ℓ). Let $Key = 0..2^k - 1$. Show that the family of hash functions

$$H^{\gg} = \left\{ h_a : 0 < a < 2^k \wedge a \text{ is odd} \right\}$$

with $h_a(x) = (ax \bmod 2^k) \operatorname{div} 2^{k-\ell}$ is 2-universal. Note that the binary representation of ax consists of at most $2k$ bits. The hash function select the first ℓ bits of the last k bits. Hint: See [93].

Exercise 4.16 (tabulation hashing, [335]). Let $m = 2^w$, and view keys as $k+1$-tuples, where the zeroth element is a w-bit number and the remaining elements are

a-bit numbers for some small constant a. A hash function is defined by tables t_1 to t_k, each having a size $s = 2^a$ and storing bit strings of length w. We then have

$$h_{\oplus(t_1,\ldots,t_k)}((x_0,x_1,\ldots,x_k)) = x_0 \oplus \bigoplus_{i=1}^{k} t_i[x_i],$$

i.e., x_i selects an element in table t_i, and then the bitwise exclusive OR of x_0 and the $t_i[x_i]$ is formed. Show that

$$H^{\oplus[]} = \left\{ h_{(t_1,\ldots,t_k)} : t_i \in \{0..m-1\}^s \right\}$$

is 1-universal.

4.3 Hashing with Linear Probing

Hashing methods are categorized as being either open or closed. Hashing with chaining is categorized as an *open* hashing approach as it uses space outside the hash table to store elements. In contrast, *closed* hashing schemes store all elements in the table, but not necessarily at the table position given by the hash value. Closed schemes have no need for secondary data structures such as linked lists; this comes at the expense of more complex insertion and deletion algorithms. Closed hashing schemes are also known under the name *open addressing*, the adjective "open" indicating that elements are not necessarily stored at their hash value. Similarly, hashing with chaining is also referred to as *closed addressing*, the adjective "closed" indicating that elements are stored at their hash value. This terminology is confusing, but standard. Many ways of organizing closed hashing have been investigated [252]; see also [131, Ch. 3.3]. We shall explore only the simplest scheme. It goes under the name of *hashing with linear probing* and is based on the following principles. Unused entries are filled with a special element $\perp$. An element e is stored in the entry $t[h(e)]$ or further to the right. But we only go away from the index $h(e)$ with good reason: If e is stored in $t[i]$ with $i > h(e)$, then the positions $h(e)$ to $i-1$ are occupied by other elements. This invariant is maintained by the implementations of the dictionary operations.

The implementations of *insert* and *find* are trivial. To insert an element e, we scan the table linearly starting at $t[h(e)]$, until either an entry storing an element e' with $key(e') = key(e)$ or a free entry is found. In the former case, we do nothing, in the latter case, we store e in the free entry. Figure 4.2 gives an example. Similarly, to find an element e, we scan the table, starting at $t[h(e)]$, until that element is found. The search is aborted when an empty table entry is encountered. So far, this sounds easy enough, but we have to deal with one complication. What happens if we reach the end of the table during an insertion? We discuss two solutions. A simple fix is to allocate m' additional table entries to the right of the largest index produced by the hash function h. For "benign" hash functions, it should be sufficient to choose m' much smaller than m in order to avoid table overflows. Alternatively, one may treat the table as a cyclic array; see Exercise 4.17 and Sect. 3.6. This alternative is more robust but slightly slower.

insert : axe, chop, clip, cube, dice, fell, hack, hash, lop, slash

an	bo	cp	dq	er	fs	gt	hu	iv	jw	kx	ly	mz
t 0	1	2	3	4	5	6	7	8	9	10	11	12
⊥	⊥	⊥	⊥	axe	⊥	⊥	⊥	⊥	⊥	⊥	⊥	⊥
⊥	⊥	chop	⊥	axe	⊥	⊥	⊥	⊥	⊥	⊥	⊥	⊥
⊥	⊥	chop	clip	axe	⊥	⊥	⊥	⊥	⊥	⊥	⊥	⊥
⊥	⊥	chop	clip	axe	cube	⊥	⊥	⊥	⊥	⊥	⊥	⊥
⊥	⊥	chop	clip	axe	cube	dice	⊥	⊥	⊥	⊥	⊥	⊥
⊥	⊥	chop	clip	axe	cube	dice	⊥	⊥	⊥	⊥	fell	⊥
⊥	⊥	chop	clip	axe	cube	dice	⊥	⊥	⊥	hack	fell	⊥
⊥	⊥	chop	clip	axe	cube	dice	hash	⊥	⊥	⊥	fell	⊥
⊥	⊥	chop	clip	axe	cube	dice	hash	lop	⊥	hack	fell	⊥
⊥	⊥	chop	clip	axe	cube	dice	hash	lop	slash	hack	fell	⊥

remove ⇩ clip

⊥	⊥	chop	~~clip~~	axe	cube	dice	hash	lop	slash	hack	fell	⊥
⊥	⊥	chop	lop	axe	cube	dice	hash	~~lop~~	slash	hack	fell	⊥
⊥	⊥	chop	lop	axe	cube	dice	hash	slash	~~slash~~	hack	fell	⊥
⊥	⊥	chop	lop	axe	cube	dice	hash	slash	⊥	hack	fell	⊥

Fig. 4.2. Hashing with linear probing. We have a table *t* with 13 entries storing synonyms of "(to) hash". The hash function maps the last character of the word to the integers 0..12 as indicated above the table: a and n are mapped to 0, b and o are mapped to 1, and so on. First, the words are inserted in alphabetical order. Then "clip" is removed. The figure shows the state changes of the table. Gray areas show the range that is scanned between the state changes.

The implementation of *remove* is nontrivial. Simply overwriting the element with ⊥ does not suffice, as it may destroy the invariant. The following example illustrates this point. Assume that $h(x) = h(z)$ and $h(y) = h(x) + 1$. Now, x, y, and z are inserted in that order. Then z is stored at position $h(x) + 2$. Assume that we next want to delete y. Simply overwriting y with ⊥ is not a solution as it will make z inaccessible. There are three solutions. First, we can disallow removals. Second, we can mark y but not actually remove it. Searches are allowed to stop at ⊥, but not at marked elements. The problem with this approach is that the number of nonempty cells (occupied or marked) keeps increasing, so that searches eventually become slow. This can be mitigated only by introducing the additional complication of periodic cleanup of the table. A third and much better approach is to actively restore the invariant. Assume that we want to remove the element at i. We overwrite it with ⊥ leaving a "hole". We then scan the entries to the right of i to check for violations of the invariant. We set j to $i + 1$. If $t[j] = \bot$, we are finished. Otherwise, let f be the element stored in $t[j]$. If $h(f) > i$, there is no hole between $h(f)$ and j and we increment j. If $h(f) \leq i$, leaving the hole would violate the invariant for f, and f would not be found anymore. We therefore move f to $t[i]$ and write $\perp$ into $t[j]$. In other words, we swap f and the hole. We set the hole position i to its new position j and continue with $j := j + 1$. Figure 4.2 gives an example.

The analysis of linear probing is beyond the scope of this book. We only mention that we need stronger properties of the hash function than guaranteed by universal hash functions. See also Sect. 4.8.

Exercise 4.17 (cyclic linear probing). Implement a variant of linear probing where the table size is m rather than $m + m'$. To avoid overflow at the right-hand end of the array, make probing wrap around. (1) Adapt *insert* and *remove* by replacing increments with $i := i + 1 \mod m$. (2) Specify a predicate *between*(i, j, k), where $i, j, k \in 1..m - 1$, that is true if and only if i is cyclically strictly between j and k. (3) Reformulate the invariant using *between*. (4) Adapt *remove*. (5) Can one allow the table to become completely full, i.e., store m elements? Consider a search for an element that is not in the table.

Exercise 4.18 (unbounded linear probing). Implement unbounded hash tables using linear probing and universal hash functions. Pick a new hash function whenever the table is reallocated. Let α, β, and γ denote constants with $1 < \gamma < \beta < \alpha$ that we are free to choose. Keep track of the number of stored elements n. Expand the table to $m = \beta n$ if $n > m/\gamma$. Shrink the table to $m = \beta n$ if $n < m/\alpha$. If you do not use cyclic probing as in Exercise 4.17, set $m' = \delta m$ for some $\delta < 1$ and choose a new hash function (without changing m and m') whenever the right-hand end of the table overflows.

4.4 Chaining versus Linear Probing

We have seen two different approaches to hash tables, chaining and linear probing. Which one is better? This question is beyond theoretical analysis, as the answer depends on the intended use and many technical parameters. We shall therefore discuss some qualitative issues and report on some experiments performed by us.

An advantage of chaining is referential integrity. Subsequent find operations for the same element will return the same location in memory, and hence references to the results of find operations can be established. In contrast, linear probing moves elements during element removal and hence invalidates references to them.

An advantage of linear probing is that each table access touches a contiguous piece of memory. The memory subsystems of modern processors are optimized for this kind of access pattern, whereas they are quite slow at chasing pointers when the data does not fit into cache memory. A disadvantage of linear probing is that search times become high when the number of elements approaches the table size. For chaining, the expected access time remains small. On the other hand, chaining wastes space on pointers that linear probing could use for a larger table. A fair comparison must be based on space consumption and not just on table size.

We have implemented both approaches and performed extensive experiments. The outcome was that both techniques performed almost equally well when they were given the same amount of memory. The differences were so small that details of the implementation, compiler, operating system, and machine used could reverse the picture. Hence we do not report exact figures.

However, we found chaining harder to implement. Only the optimizations discussed in Sect. 4.7 made it competitive with linear probing. Chaining is much slower if the implementation is sloppy or memory management is not implemented well.

In Theorem 4.3, we showed that the combination of hashing with chaining and c-universal classes of hash functions guarantees good expected behavior. A similar result does *not* hold for the combination of hashing with linear probing and c-universal hash functions. For a guarantee of expected constant search time, linear probing needs hash families with stronger randomness properties or the assumption of full randomness. We come back to this point in Sect. 4.8.

4.5 *Perfect Hashing

The hashing schemes discussed so far guarantee only *expected* constant time for the operations *find*, *insert*, and *remove*. This makes them unsuitable for real-time applications that require a worst-case guarantee. In this section, we shall study *perfect hashing*, which guarantees constant worst-case time for *find*. To keep things simple, we shall restrict ourselves to the *static* case, where we consider a fixed set S of n elements. For simplicity, we identify elements and their keys, i.e., $S = \{x_1, \ldots, x_n\}$ is the set of keys occuring.

In this section, we use H_m to denote a family of c-universal hash functions with range $0..m-1$. In Exercise 4.12, it was shown that 2-universal classes exist for every m. For $h \in H_m$, we use $C(h)$ to denote the number of collisions produced by h, i.e., the number of (ordered) pairs of distinct keys in S which are mapped to the same position:

$$C(h) = |\{(x,y) : x, y \in S, \; x \neq y \text{ and } h(x) = h(y)\}|.$$

If h is chosen randomly in H_m, C is a random variable. As a first step, we derive a bound on the expectation of C.

Lemma 4.5. $E[C] \leq cn(n-1)/m$. Also, for at least half of the functions $h \in H_m$, we have $C(h) \leq 2cn(n-1)/m$.

Proof. We define $n(n-1)$ indicator random variables $X_{ij}(h)$. For $i \neq j$, let $X_{ij}(h) = 1$ if $h(x_i) = h(x_j)$ and $X_{ij} = 0$ otherwise. Then $C(h) = \sum_{ij} X_{ij}(h)$, and hence

$$E[C] = E\left[\sum_{ij} X_{ij}\right] = \sum_{ij} E[X_{ij}] = \sum_{ij} \text{prob}(X_{ij} = 1) \leq \sum_{ij} c/m = n(n-1) \cdot c/m,$$

where the second equality follows from the linearity of expectations (see (A.3)) and the inequality follows from the universality of H_m. The second claim follows from Markov's inequality (A.5). $\qquad \square$

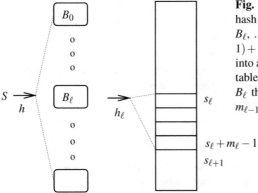

Fig. 4.3. Perfect hashing. The top-level hash function h splits S into subsets $B_0, \ldots,$ $B_\ell, \ldots$. Let $b_\ell = |B_\ell|$ and $m_\ell = cb_\ell(b_\ell - 1) + 1$. The function h_ℓ maps B_ℓ injectively into a table of size m_ℓ. We arrange the subtables into a single table. The subtable for B_ℓ then starts at position $s_\ell = m_0 + \ldots + m_{\ell-1}$ and ends at position $s_\ell + m_\ell - 1$.

If we are willing to work with a quadratic-size table, our problem is solved.

Lemma 4.6. *If $m \geq cn(n-1) + 1$, at least half of the functions $h \in H_m$ operate injectively on S.*

Proof. By Lemma 4.5, we have $C(h) < 2$ for at least half of the functions in H_m. Since $C(h)$ is even (recall that it counts ordered pairs), $C(h) < 2$ implies $C(h) = 0$, and so h operates injectively on S. $\square$

So we fix a m with $m \geq cn(n-1) + 1$, choose a random $h \in H_m$, and check whether or not it is injective on S. If not, we iterate until we have found an injective h. After an average of two trials, we shall be successful.

In the remainder of this section, we show how to bring the table size down to linear. The idea is to use a two-stage mapping of keys (see Fig. 4.3). The first stage maps keys to buckets such that the sum of the squared bucket sizes is linear in n. The second stage uses an amount of space for each bucket that is quadratic in the number of elements contained in the bucket. For $\ell \in 0..m-1$ and $h \in H_m$, let B_ℓ^h be the elements in S that are mapped to ℓ by h and let b_ℓ^h be the cardinality of B_ℓ^h.

Lemma 4.7. *For every $h \in H_m$, $C(h) = \sum_\ell b_\ell^h(b_\ell^h - 1)$.*

Proof. For any ℓ, the keys in B_ℓ^h give rise to $b_\ell^h(b_\ell^h - 1)$ ordered pairs of distinct keys mapping to the same location. Summation over ℓ completes the proof. $\square$

We are now ready for the construction of a perfect hash function. Let α be a constant, which we shall fix later. We choose a hash function $h \in H_{\lceil \alpha n \rceil}$ to split S into subsets B_ℓ. Of course, we choose h to be in the good half of $H_{\lceil \alpha n \rceil}$, i.e., we choose $h \in H_{\lceil \alpha n \rceil}$ with $C(h) \leq 2cn(n-1)/\lceil \alpha n \rceil \leq 2cn/\alpha$. For each ℓ, let B_ℓ be the elements in S mapped to ℓ and let $b_\ell = |B_\ell|$.

Now consider any B_ℓ. Let $m_\ell = cb_\ell(b_\ell - 1) + 1$. We choose a function $h_\ell \in H_{m_\ell}$ which maps B_ℓ injectively into $0..m_\ell - 1$. At least half of the functions in H_{m_ℓ} have this property, by Lemma 4.6 applied to B_ℓ. In other words, h_ℓ maps B_ℓ injectively into a table of size m_ℓ. We stack the various tables on top of each other to obtain one large table of size $\sum_\ell m_\ell$. In this large table, the subtable for B_ℓ starts at position $s_\ell = m_0 + m_1 + \ldots + m_{\ell-1}$. Then

$$\ell := h(x); \quad \textbf{return } s_\ell + h_\ell(x)$$

computes an injective function on S. The values of this function are bounded by

$$\sum_\ell m_\ell - 1 \leq \lceil \alpha n \rceil + c \cdot \sum_\ell b_\ell(b_\ell - 1) - 1$$
$$\leq 1 + \alpha n + c \cdot C(h) - 1$$
$$\leq \alpha n + c \cdot 2cn/\alpha$$
$$\leq (\alpha + 2c^2/\alpha)n,$$

and hence we have constructed a perfect hash function that maps S into a linearly sized range, namely $0.. \lfloor (\alpha + 2c^2/\alpha)n \rfloor$. In the derivation above, the first inequality uses the definition of the m_ℓ's, the second inequality uses Lemma 4.7, and the third inequality uses $C(h) \leq 2cn/\alpha$. The choice $\alpha = \sqrt{2}c$ minimizes the size of the range. For $c = 1$, the size of the range is $2\sqrt{2}n$. Besides the table, we need space for storing the representation of the hash function. This space is essentially determined by the space needed for storing the parameters of the functions h_ℓ and the starting value s_ℓ of the ℓth subtable, $\ell \in 0.. \lceil \alpha n \rceil - 1$. We need to store three numbers for each ℓ and hence the space needed for the representation of the function is linear. The expected time for finding the function h is $O(n)$ and the expected time for finding h_ℓ is $O(b_\ell)$. Thus the total construction time is linear.

Theorem 4.8. *For any set of n keys, a perfect hash function with range $0.. \lfloor 2\sqrt{2}n \rfloor$ can be constructed in linear expected time. The space needed to store the function is linear.*

Constructions with smaller ranges are known. Also, it is possible to support insertions and deletions.

Exercise 4.19 (dynamization). We outline a scheme for "dynamization" of perfect hashing, i.e., a method that supports insertions and deletions and guarantees constant access time. Consider a fixed S of size n and choose $h \in H_m$, where $m = 2 \lceil \alpha n \rceil$. For each ℓ, let $m_\ell = \lceil 2cb_\ell(b_\ell - 1) \rceil$, i.e., all m_ℓ's are chosen to be twice as large as in the static scheme. Construct a perfect hash function as above. Insertion of a new x is handled as follows. Assume that h maps x onto ℓ. Increment b_ℓ. If h_ℓ is not injective on $B_\ell \cup \{x\}$ and $m_\ell \geq \lceil cb_\ell(b_\ell - 1) \rceil$, we choose a new h_ℓ. Repeat until the hash function is injective. Once $m_\ell < \lceil cb_\ell(b_\ell - 1) \rceil$, we allocate a new table for B_ℓ of size $m_\ell = \lceil 2cb_\ell(b_\ell - 1) \rceil$. We also keep track of $n = |S|$ and $C(h)$. Once n exceeds m/α, we set $m = 2 \lceil \alpha n \rceil$, choose a new function h for the first level, and move S to a new table of size m. If $n \leq m/\alpha$ but $C(h)$ exceeds $2cn/\alpha$, we keep m fixed and choose a new first-level function h. We move S to a new table of size m. Work out the details of the algorithm and of the analysis. Hint: See [94].

4.6 Parallel Hashing

A shared hash table is a powerful way for multiple PEs to share information in a fine-grained way. However, a shared hash table raises a number of nontrivial issues with respect to correctness, debugging, and performance. What happens when several PEs want to access the same element or the same position in the table? Let us first define what we want to happen, i.e., the programming interface of a shared hash table.

The operation *build* has the same meaning as before, except that each PE might contribute to the initial table content.

Insertions are as before. When several PEs attempt to insert an element with the same key concurrently, only one of those elements will be inserted. The application program should not make any assumptions about which of these concurrent operations succeeds. Similarly, when several PEs attempt to *remove* an element, it will be removed only once.

The operation *find*(x) is almost as before. It returns $\perp$ if x is not part of the table. Otherwise, rather than a reference, it should return a *copy* of the value currently stored with key x.[8] On a shared-memory machine, many concurrent read accesses to the same element should be possible without performance penalty.

Concurrent updates have to be performed atomically in order to avoid chaos. In many situations, the value written depends on the previously stored value v'. We therefore encapsulate this behavior in an atomic update operation *update*(x, v, f) that is passed not only a value v but also a function $f : Value \times Value \rightarrow Value$. This update function stores the value $f(v', v)$. For example, in order to increment a counter associated with key x, one could call *update*$(x, 1, +)$. Sometimes we also need a combined operation *insertOrUpdate*. If an element with the given key x is not in the table yet, then the key value pair (x, v) is inserted. Otherwise, if (x, v') is already in the table, then v' is replaced by $f(v', v)$.

An important alternative to a shared hash table is the use of local hash tables. Additional code will then be needed to coordinate the PEs. This is often more efficient than a shared hash table, and may be easier to debug because the PEs work independently most of the time.

Let us consider both approaches for a concrete example. Suppose we have a multiset M of objects and want to count how often each element of M occurs. The following sequential pseudocode builds a hash table T that contains the counts of all elements of M:

Class *Entry*(*key* : *Element, val* : $\mathbb{N}$)
T : *HashTable* **of** *Entry*
forall $m \in M$ **do** $e := T.find(m)$;   **if** $e = \perp$ **then** $T.insert((m, 1))$ **else** $e.val$++

A simple parallelization in shared memory is to make T a shared hash table and to make the loop a parallel loop. The only complication is that the sequential code

---

[8] Returning a reference – with the implication that the element could be updated using this reference – would put the responsibility of ensuring consistent updates on the user. This would be a source of hard to trace errors.

contains a *find* operation, a conditional insertion, and a possible write access. Our concurrent operation *insertOrUpdate* nicely covers what we actually want. We can simply write

forall $m \in M$ **do**$\|$ $T.insertOrUpdate(m, 1, +)$

When several PEs call this operation concurrently with the same key, this will be translated into several atomic increment operations (see also Sect. 2.4.1). If key m was not present before, one of the concurrent calls will initialize the counter to 1 and the others will perform atomic increment operations.

Having avoided possible correctness pitfalls, there is still a potential performance problem. Suppose most elements of M in our example have the same key. Then the above parallel loop would perform a huge number of atomic increment operations on a single table entry. As already discussed in Sect. 2.4.1, this leads to high contention and consequently low performance.

We next discuss the use of local hash tables. In our example, each element of M will then be counted in some local table. At the end, we need additional code to merge the local tables into a global table. If the number of elements in M is much larger than p times the number of different keys occuring, then this merging operation will have negligible cost and we shall get good parallel performance. Note that this covers, in particular, the high-contention case, where the shared table was bad. A potential performance problem arises when the number of different keys is large. Not only will merging then be expensive, but the total memory footprint of the parallel program may also be up to p times larger than that of the sequential program. Thus, it might happen that the sequential program runs in cache while the parallel program makes many random accesses to main memory. In such a situation, just filling the local tables may take longer than solving the overall problem sequentially. There are more sophisticated algorithms for such *aggregation problems* that interpolate between hashing-based and sorting-based algorithms (e.g., [234]). We refer our readers to the literature for a detailed discussion.

4.6.1 Distributed-Memory Hashing

We start with the obvious way of realizing a distributed hash table. Each processor is made responsible for a fraction $1/p$ of the table and handles all requests to this part of the table. More precisely, we distribute the hash table $t[0..m-1]$ to p PEs by assigning the part $t[i \cdot m/p..(i+1) \cdot m/p]$ to PE $i \in 0..p-1$. The operation $find(x)$ is translated into a find request message to PE $h(x) \operatorname{div} m/p$. Each PE provides a hash table server thread which processes requests affecting its part. The operations *insert*, *remove*, and *update* are handled analogously. This approach works for any representation of the local part, for example hashing with chaining or hashing with linear probing. In the latter case, wraparound for collision resolution should be local, i.e., within the part itself, to avoid unnecessary communication. Each PE knows the hash function and hence can evaluate it efficiently.

The distributed hash table just outlined is simple and elegant but leads to fine-grained messages. On most machines, sending messages is much more expensive than hash table accesses to local memory. Furthermore, in the worst case, there could be considerable contention due to many accesses to the same entry of the hash table. We shall next describe a scheme that overcomes these shortcomings by using globally synchronized processing of batches of operations. It provides high performance for large batches even in the presence of contention, and also simplifies unbounded hash tables.

Each PE has a local batch O containing up to o operations. The batches are combined as follows. Each PE first sorts its O by the ID of the PE responsible for handling an operation. This can be done in time $O(o + p)$ using bucket sort (see Sect. 5.10). On average, we expect a PE to access at most o/p *distinct* locations in any other PE. In parallel computing, the maximum bucket size is important. It will be part of the analysis to show that the probability that some PE accesses more than $2o/p$ distinct locations in some other PE is small.

There is another problem that we have to deal with. If a PE accesses the same location many times, this may lead to buckets with many more than $2o/p$ operations. Buckets whose size "significantly" exceeds the expected bucket size o/p might contain many operations addressing the same key.[9] Therefore, after bucket sorting, the operations in each bucket are inserted into a temporary local hash table in order to condense them: An element inserted/removed/searched several times needs only to be inserted/removed/searched once. We assume here that update operations to the same location can be combined into a single update operation. For example, several counter increment operations can be combined into a single add operation.

Using an all-to-all data exchange (see Sect. 13.6), the condensed operations are delivered to the responsible PEs. Each PE then performs the received operations on its local part. The results of the *find* operations are returned to their initiators using a further all-to-all operation. Figure 4.4 gives an example.

Theorem 4.9. *Assuming $o = \Omega(p \log p)$ and that the hash function behaves like a truly random hash function, a batch of hash table operations can be performed in expected time $O(T_{\text{all}\to\text{all}}(o/p))$, where $T_{\text{all}\to\text{all}}(x)$ is the execution time for a regular all-to-all data exchange with messages of size at most x (see Sect. 13.6).*

Proof. Recall our assumption that we resolve *all* locally duplicated keys by first building local hash tables. Preparing the messages to be delivered takes time $O(o)$, as explained above. Since $T_{\text{all}\to\text{all}}(o/p) \geq p \cdot (o/p) \cdot \beta = \Omega(o)$, the $O(o)$ term is covered by the bound stated in the theorem.

Each PE sends a message to every other PE. More precisely, the message sent by PE i to PE j contains all operations originating from PE i to elements stored in PE j. We have to show that the expected maximum message size is $O(o/p)$. Since duplicate keys have been resolved, at most o keys originate from any processor. Since we assume a random hash function, these keys behave like balls thrown uniformly

---

[9] There are several ways here to define what "significantly" means. For the analysis, we assume that all common keys will be condensed. In practice, this step is often completely skipped. The truth for a robust but practically efficient solution lies somewhere in the middle. A simple threshold such as e2 · o/p should be a good compromise.

at random into p bins. Hence, we can employ the Chernoff tail bound (A.7): The probability that a message is larger than twice its expectation o/p is bounded by

$$\left(\frac{e^1}{2^2}\right)^{o/p} = \left(\frac{e}{4}\right)^{o/p}.$$

This probability is bounded by $1/p^3$ if $o \geq 3\log(4/e) \cdot p\log p = \Theta(p\log p)$. In this situation, the probability that any of the p^2 buckets (p buckets in each PE) is larger then $2o/p$ is at most $p^2/p^3 = 1/p$. Furthermore, no bucket can be larger then o. Combining both bounds we obtain an upper bound of

$$\left(1 - \frac{1}{p}\right) \cdot 2\frac{o}{p} + \frac{1}{p}o \leq 2\frac{o}{p} + \frac{o}{p} = 3\frac{o}{p}$$

on the expectation of the maximum bucket size.

A similar argument can be used to bound the expected number of operations any PE has to execute locally – only with probability at most $1/p$ will any message contain more than $2o/p$ operations and, even in the worst case, the total number of received operations cannot exceed $p \cdot o$. Hence, the expected work on any PE is bounded by

$$\left(1 - \frac{1}{p}\right) \cdot 2\frac{o}{p} \cdot p + \frac{1}{p}o \cdot p \leq 2o + o = 3o.$$

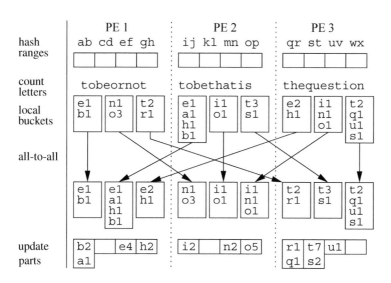

Fig. 4.4. Counting letters on three PEs using a distributed hash table. PE 1 is responsible for the key range a..h, PE 2 for the range i..p, and PE 3 for the range q..r. Each PE first processes its part of the text and prepares appropriate messages. For example, the text of PE 1 contains three occurrences of the character o and hence PE 1 prepares the message "o3" to be sent to PE 2. After the messages are delivered, each PE updates its part. The local hash tables have buckets of size 2.

To summarize, we have expected time $O(o) = O(T_{\text{all}\rightarrow\text{all}}(o/p))$ for local prepro-cessing and final processing and time $O(T_{\text{all}\rightarrow\text{all}}(o/p))$ for message exchange when $o \geq 3\log(4/e) \cdot p \log p$. For smaller values of $o \in \Theta(p \log p)$, the running time cannot be larger, and remains in $O(T_{\text{all}\rightarrow\text{all}}(\log p)) = O(T_{\text{all}\rightarrow\text{all}}(o/p))$.   □

*Exercise 4.20. Redo the proof of Theorem 4.9 for the case where an all-to-all data exchange with variable message sizes is used (see Sect. 13.6.3). Show that in this case time $O(T^*_{\text{all}\rightarrow\text{all}}(o))$ is already achieved when $o = \Omega(p)$.

In sequential hashing, theorems that have been proved for random hash func-tions can frequently be transferred to universal hash functions. This is *not* true for parallel hashing. Theorem 4.9 does not transfer if we replace random hash functions by universal hash functions. We need stronger properties of the hash function em-ployed in order to prove that it is unlikely that any message will become too long. The tabulation hash function in Exercise 4.16 can be shown to have the required properties [259].

To make the hash table unbounded, we determine the *maximum* number of el-ements $n_{\max}$ in any local table using an all-reduce operation (see Sect. 13.2). We choose the local table size based on $n_{\max}$. This way, decisions about resizing the table are made identically on all PEs.

Exercise 4.21 (MapReduce programming model). [86] A widely used pattern for processing large data sets is as follows: Consider the (key) sets K and K' and the value sets V and V'. The input is a set of key-value pairs $P \subseteq K \times V$. There are two-user defined functions, $map : K \times V \rightarrow 2^{K' \times V'}$ and $reduce : K' \times V'^* \rightarrow K' \times V'$, where V'^* denotes the set of sequences consisting of elements of V'. The resulting MapReduce computation first *maps* each pair $(k, v) \in P$ to a new set of key-value pairs. Consider the union $P' \subseteq K' \times V'$ of all these sets. Now, pairs in P' with the same key are collected together, i.e.,

$$P'' := \left\{ (k', s) \in K' \times V'^* : s = \langle v' : (k', v') \in P' \rangle \wedge |s| > 0 \right\}.$$

Finally, the *reduce* function is applied to each element of P''.

Assuming that the elements of P and P'' are distributed over the PEs of a distributed-memory parallel computer, mapping and reducing are local computa-tions. However, the collection step requires communication. Explain how you can implement it using batched distributed hashing. Analyze its performance. Under what circumstances is a speedup $\Omega(p)$ possible?

Explain how the billing system discussed in Exercise 4.3 can be viewed as an instantiation of the MapReduce pattern.

4.6.2 Shared-Memory Hashing

The distributed-memory hash table from Sect. 4.6.1 is readily transformed into a shared-memory setting: The hash table is split into p pieces of equal size, with one

PE responsible for all operations on one of the pieces. This approach works efficiently only for large batches of operations. Moreover, the need for synchronization reduces flexibility.

We shall next work out how a single hash table can be accessed concurrently and asynchronously. We start with fast, simple, and more specialized solutions, going to more general solutions in several steps. This approach reflects the fact that concurrent hash tables exhibit a trade-off between simplicity and efficiency on the one side and generality on the other side.[10]

Only finds. The most simple case is that only (read-only) *find* operations are processed asynchronously. In this case, we can simply adopt the sequential implementation of the operation *find* – parallel executions of several calls do not interfere with each other. However, before processing a batch of update/insert/remove operations, all PEs need to synchronize. In particular, while a batch is being processed, no thread can execute *find* operations.

The above data structure also allows concurrent asynchronous update operations under the following circumstances: Updates have to be performed in an atomic way, for example using a CAS instruction, and *find* operations also have to read the updateable part of an element atomically.

Exercise 4.22. Work out an example where a *find* operation returns an inconsistent value when an *update* operation modifies a table entry in several steps.

Insertions. Similarly, we can support concurrent asynchronous insertions of an element e using atomic operations on table entries. We shall discuss this for hashing with linear probing (see also [304, 206]). The operation looks for a free table entry i as in sequential linear probing and attempts to write e. This write operation has to be done atomically, for example using a CAS instruction. Without atomic writing, concurrent *find* operations could return inconsistent, partially constructed table entries. If the CAS instruction fails because another concurrent *insert* got there first, the insertion operation continues to look for a free entry starting at position i. Entry i has to be reinspected, since another PE might have inserted an element with the same key as e, in which case the operation is terminated immediately. Note that reinspecting the same position cannot lead to an infinite loop, since the succeeding CAS will fill the position and our implementation never reverts a filled entry to free.

Unbounded Hash Tables. How about adaptive growing? In the sequential case, we simply remarked that this can be handled as with unbounded arrays – reallocation and copying of the table content. In the concurrent case the same approach works, but is quite difficult to implement correctly – we have to make sure that all PEs switch from one version of the table to the next in a consistent fashion. We are only aware of a single paper showing that this can be done efficiently [206].

---

[10] An interesting related paper by Shun and Blelloch [295] achieves a good compromise between simplicity and generality by requiring global synchronizations between phases with different operation mixes.

Removals. We can support removals by marking the corresponding table entry as deleted. This requires that *find* operations read the entire table entry atomically. In Sect. 4.3 we have already pointed out that removal by marking has considerable disadvantages over properly removing the element, since we can never free any space. Hence, for most uses of deletion, we should also have a mechanism for growing the table or cleaning it up (see [206] for details).

Performance Analysis. Let us now analyze the performance of the above asynchronous data structure and compare it with batched hashing (see Sect. 4.6.1). The asynchronous implementation works well as long as the application does not access particular keys very frequently. In this situation, we have low contention for memory accesses and we get constant expected execution time for all supported operations. For such workloads, the asynchronous implementation is more flexible than the distributed one and will often be faster. The situation is different when particular keys are used very often. The most problematic operations here are updates, since they actually modify the elements. This leads to massive write contention. Here, the distributed implementation has big advantages, since it resolves the contention locally. For example, when all PEs update the same element $\Theta(p \log p)$ times, this takes expected time $O(p \log p)$ in the distributed implementation by Theorem 4.9, whereas the asynchronous implementation needs time $\Omega(p^2 \log p)$. The situation is less severe for *find* operations since modern shared-memory machines accelerate concurrent reading via caching. The aCRQW-PRAM model introduced in Sect. 2.4.1 reflects this difference by predicting constant expected running time for *find* even in the case of high contention. The situation is somewhere in between for insertions and removals. Concurrent execution of p operations affecting the same key is guaranteed to make only a single modification to the memory. However, if all these operations are executed at the same time, all PEs might initiate a write instruction. In the aCRQW-PRAM model, this would take time $\Theta(p)$. However, concurrent attempts to *insert* the same element multiple times will be fast if the element was originally inserted sufficiently long ago.

Shared-Memory Hashing with Chaining. In hashing with chaining, we can lock individual buckets to allow for full support of *insert*, *remove* and *update*, including support for complex objects that cannot be written atomically. However, even *find* operations then have to lock the bucket they want to access. What sounds like a triviality at first glance can be a severe performance bottleneck. In particular, when many *find* operations address the same table entry, we could have a lot of contention for writing the lock variable, even though we actually only want to read the value (which, by itself, is fast in the aCRQW-PRAM-model).

Exercise 4.23 (parallel join). Consider two relations $R \subseteq A \times B$ and $Q \subseteq (B \times C)$. Refine your algorithm obtained in Exercise 4.5 for computing

$$R \bowtie Q = \{(a,b,c) \subseteq A \times B \times C : (a,b) \in R \land (b,c) \in Q\}$$

to work on an aCREW PRAM. Each PE should perform only $O(|Q|/p)$ *find* operations and $O(|R|/p)$ *insertOrUpdate* operations. Now assume that each value of B appears only once in R. Show how to achieve running time $O((|R| + |Q| + |R \bowtie Q|)/p)$. Discuss what can go wrong if this assumption is lifted.

4.6.3 Implementation of Shared-Memory Hashing

We now discuss an implementation of shared-memory linear probing. We aim for simplicity and efficiency rather than generality and portability. Since linear probing requires atomic operations, there is a strong dependency on the processor instruction set and also on its level of support by the compiler (see also Sects. B.4 and C.4). We give an implementation of a hash table supporting atomic *insert*, *update*, and *find* for the Linux *gcc* compiler on the x86 architecture, which supports 16-byte atomic CAS and 8-byte atomic loads. If the key and the data fit together into 8 bytes, a similar implementation would work for a wider range of architectures. We could support longer key or data fields by introducing additional indirections – the table entries would then contain pointers to the actual data. Transactional synchronization instructions (i.e., the Intel Transactional Synchronization Extensions (Intel TSX) described in Sect. B.5) can provide larger atomic transactions but often require a fallback implementation based on ordinary atomic instructions. We come back to this point at the end of this section. We assume that the key and the data of an element require 8 bytes each, that an empty table entry is indicated by a special key value (in our case the largest representable key), and that a find returns an element. If the find is unsuccessful, the key of the element returned is the special key value. Under these assumptions, we can work with 16-byte reads implemented as two atomic 8-byte reads. Consider the possible cases for the execution of a *find* operation:

- Table entry $t[i]$ is empty: The first 8-byte read copies the special key into the element to be returned. It is irrelevant what the second 8-byte read copies into the element returned; it may copy data that was written by an insert that started after the find, but completed before it. In any case, the returned element has a key indicating an empty table entry. The *find* operation returns an outdated but consistent result.
- Table entry $t[i]$ contains a nonempty element: The first 8-byte read copies a valid *key* and the second 8-byte read copies the latest value written by an *update* before the second read. A valid element is returned, because updates do not change the *key*. Recall that there are no deletions.

With this reasoning, we can use a single `movups` x86 instruction on 16-byte data that issues two 8-byte loads which are guaranteed to be atomic if the data is 8-byte aligned. On most compilers, this instruction can be generated by calling the `_mm_loadu_ps` intrinsic.

The class `MyElement` below encapsulates an element data type, including most architecture-specific issues. Here, we use 64-bit integers for both key and data. Empty table entries are encoded as the largest representable key. Other representations are possible as long as the public methods are implemented in an atomic way.

```
class MyElement                                                                   1
{                                                                                  2
public:                                                                            3
  typedef long long int Key; //64-bit key                                          4
  typedef long long int Data; //64-bit data or a pointer                           5
private:                                                                           6
  Key key;                                                                         7
  Data data;                                                                       8
  template <class T> T & asWord() { // a helper cast                               9
    return *reinterpret_cast<T *> (this);                                         10
  }                                                                               11
  template <class T> const T & asWord() const {                                   12
    return *reinterpret_cast<const T *> (this);                                   13
  }                                                                               14
public:                                                                           15
  MyElement() {}                                                                  16
  MyElement(const Key & k, const Data & d):key(k),data(d){}                       17
  Key getKey() const { return key; }                                             18
  static MyElement getEmptyValue() {                                              19
    return MyElement(numeric_limits<Key>::max(), 0);                             20
  }                                                                               21
  bool isEmpty() const {                                                          22
    return key == numeric_limits<Key>::max();                                    23
  }                                                                               24
  bool CAS(MyElement & expected, const MyElement & desired) {                     25
    return __sync_bool_compare_and_swap_16(&asWord<__int128>(),                  26
      expected.asWord<__int128>(), desired.asWord<__int128>());                  27
  }                                                                               28
  MyElement(const MyElement & e) {                                                29
    asWord<__m128i>() = _mm_loadu_si128((__m128i*)&e);                           30
  }                                                                               31
  MyElement & operator = (const MyElement & e) {                                  32
    asWord<__m128i>() = _mm_loadu_si128((__m128i*)&e);                           33
    return *this;                                                                 34
  }                                                                               35
  void update(const MyElement & e) { data = e.data; }                            36
};//SPDX-License-Identifier: BSD-3-Clause; Copyright(c) 2018 Intel Corporation   37
```

Given the class MyElement above, the implementation of the hash table is mostly straightforward. The constructor allocates the table array using the instruction _aligned_malloc available in Linux[11] in order to have table entries start at multiples of 16. This is required in order to use the 16-byte x86 CAS instructions. The hash function is taken from the C++standard library. To upper-bound the lookup time for densely populated tables, we limit the maximum length of the scan to a large enough *maxDist*. If a scan reaches this limit, the table needs to be enlarged to avoid bad performance. We implement cycling probing (Sect. 4.3) and avoid modulo operations by using only powers of two for the table capacity. If the table capacity m is

---

[11] In Windows, one would use memalign.

equal to 2^ℓ, $x \bmod 2^\ell = x \wedge (2^l - 1)$, where $\wedge$ is a bitwise AND operation, which is much faster than a general modulo operation.

```
template <class Element>                                              1
class HashTable                                                       2
{                                                                     3
  typedef typename Element::Key Key;                                  4
  size_t h(const Key & k) const { return hash(k) & mask; }            5
  enum { maxDist = 100 };                                             6
public:                                                               7
  HashTable(size_t logSize = 24) : mask((1ULL << logSize) -1){        8
    t = (Element *)_aligned_malloc((mask + 1)*sizeof(Element), 16);   9
    if (t == NULL) std::bad_alloc();                                  10
    std::fill(t, t + mask + 1, Element::getEmptyValue());             11
  }                                                                   12
  virtual ~HashTable() { if (t) _aligned_free(t); }                   13
  bool insert(const Element & e) {                                    14
    const Key k = e.getKey();                                         15
    const size_t H = h(k), end = H + maxDist;                         16
    for (size_t i = H; i < end; ++i) {                                17
      /* copy the element guaranteeing that a concurrent update       18
         of the source will not result in an inconsistent state */    19
      Element current(t[i&mask]);                                     20
      if (current.getKey() == k) return false; // key already exists  21
      if (current.isEmpty()) { // found free space                    22
        if (t[i&mask].CAS(current, e)) return true;                   23
        // potentially collided with another insert                   24
        --i; // need to reinspect position i;                         25
      }                                                               26
    }                                                                 27
    throw bad_alloc(); // no space found for the element              28
    return false;                                                     29
  }                                                                   30
  Element find(const Key & k) {                                       31
    const size_t H = h(k), end = H + maxDist;                         32
    for (size_t i = H; i < end; ++i) {                                33
      const Element e(t[i&mask]);                                     34
      if (e.isEmpty() || (e.getKey() == k)) return e;                 35
    }                                                                 36
    return Element::getEmptyValue();                                  37
  }                                                                   38
private:                                                              39
  Element * t;                                                        40
  std::hash<Key> hash;                                                41
  const size_t mask;                                                  42
};//SPDX-License-Identifier: BSD-3-Clause; Copyright(c) 2018 Intel Corporation   43
```

We implement a powerful update function *insertOrUpdate* that inserts an element *e* if the key of *e* is not already present in the table. Otherwise, it updates the existing table entry using a function $f(c,e)$ of the old element c and the new element e. This function guarantees an atomic update of the element.

```
template <class F>                                                          1
bool insertOrUpdate(const Element & e, F f = F()) {                          2
  const Key k = e.getKey();                                                  3
  const size_t H = h(k), end = H + maxDist;                                  4
  for (size_t i = H; i < end; ++i) {                                         5
    Element current(t[i&mask]);                                              6
    if (current.getKey() == k) { // key already exists                       7
      while (!t[i&mask].atomicUpdate(current, e, f)) {                       8
        // potentially collided with another update                         9
        current = t[i&mask]; // need to reinspect position i                10
      }                                                                      11
      return false;                                                          12
    }                                                                        13
    if (current.isEmpty()) { // found free space                            14
      if (t[i&mask].CAS(current, e)) return true;                           15
      // potentially collided with another insert                           16
      --i; // need to reinspect position i                                  17
    }                                                                        18
  }                                                                          19
  }                                                                          20
  throw bad_alloc(); // no space found for the element                      21
  return false;                                                              22
}                                                                            23
template <class F>                                                          24
bool MyElement::atomicUpdate(MyElement & expected,                          25
          const MyElement & desired, F f) {                                 26
  return __sync_bool_compare_and_swap(&data,                                27
      expected.data, f(expected, desired).data);                            28
}//SPDX-License-Identifier: BSD-3-Clause; Copyright(c) 2018 Intel Corporation 29
```

The update function object f can be also specified using the C++11 lambda notation. The software design allows us to take advantage of specialized update function objects for insert-or-increment and insert-or-decrement that use more efficient fetch-and-increment/fetch-and-decrement processor instructions instead of general CAS instructions. Additionally, overwriting the data for existing keys can be specialized without using the heavy CAS instruction.

```
struct Overwrite {};                                                        1
struct Increment {};                                                        2
struct Decrement {};                                                        3
// atomic update by overwriting                                             4
bool atomicUpdate(MyElement & expected,                                     5
    const MyElement & desired, Overwrite f) {                               6
  update(desired);                                                          7
  return true;                                                              8
}                                                                           9
```

```
// atomic update by increment                                      10
bool atomicUpdate(MyElement & expected,                            11
    const MyElement & desired, Increment f) {                      12
  __sync_fetch_and_add(&data, 1);                                  13
  return true;                                                     14
}                                                                  15
// atomic update by decrement                                      16
bool atomicUpdate(MyElement & expected,                            17
    const MyElement & desired, Decrement f) {                      18
  __sync_fetch_and_sub(&data, 1);                                  19
  return true;                                                     20
}//SPDX-License-Identifier: BSD-3-Clause; Copyright(c) 2018 Intel Corporation   21
```

We finally give an implementation of *insertOrUpdate* that wraps a simpler sequential code into a memory (Intel TSX) transaction (see Sect. B.5). If the transaction fails, the version using an atomic CAS operation is used as a fallback.

```
bool insertOrUpdateTSX(const Element & e, F f = F()) {             1
  if(_xbegin() == _XBEGIN_STARTED) // successful transaction start  2
  {                                                                3
    const Key k = e.getKey();                                      4
    const size_t H = h(k), end = H + maxDist;                      5
    for (size_t i = H; i < end; ++i) {                             6
      Element & current = t[i&mask];                               7
      if (current.getKey() == k) { //key already exists            8
        current.update(e,f);                                       9
        _xend();                                                   10
        return true;                                               11
      }                                                            12
      if (current.isEmpty()) { //found free space                  13
        current = e;                                               14
        _xend();                                                   15
        return true;                                               16
      }                                                            17
    }_xend();                                                      18
    //no space found for the element, use a table with a larger capacity  19
    throw bad_alloc();                                             20
  }                                                                21
  // transaction fall-back using CAS                               22
  return insertOrUpdate(e,f);                                      23
}//SPDX-License-Identifier: BSD-3-Clause; Copyright(c) 2018 Intel Corporation   24
```

4.6.4 Experiments

We conducted a series of experiments to evaluate the scalability of the implementation using the machine described in Appendix B. We compared our implementation with two hash table implementations from the Intel Threading Building Blocks (TBB) Library, both based on chaining with buckets. We used a table with a capacity of 2^{28} elements which was initially populated with 2^{26} elements.

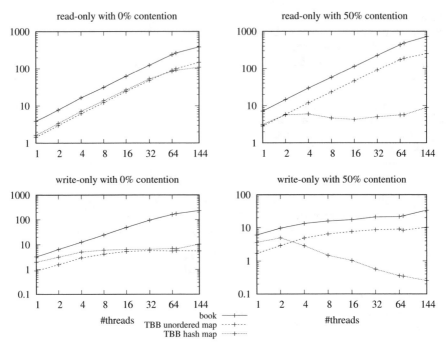

Fig. 4.5. Throughput (million operations/second) of concurrent hash table implementations. The horizontal axis indicates the number of threads, and the vertical axis indicates the throughput. Higher is better.

To demonstrate the effects of contention, we conducted the following simple experiment. The measurement performed 2^{28} queries with $c\%$ operations using a particular "hot" key value and $(100 - c)\%$ operations using a random key. The operations were distributed over p threads. Figure 4.5 shows the resulting performance for a read-only workload (*find* operations) and write-only workload (increment operations). Our implementation used *insertOrUpdate* specialized to a fetch-and-add operation. The Intel TBB hash map returns an accessor object as a result of the insert operation, which locks the existing element. We performed the increment operation under this lock. For the *find* operation on the TBB hash map, we used the *const* accessor, which implies a read lock. The TBB unordered map returns iterators (pointers) to the elements in the hash table without any implicit locks. To implement the update function, we had to use an atomic fetch-and-add instruction on the data referenced by the returned iterator.

Without contention, the read-only workload is easy for all three implementations. However, the book implementation is more than twice as fast as the TBB implementations. Since hyper-threading hides memory latencies, we observed a significant speedup when using 144 threads on the 72-core machine.

When there is read contention, the TBB hash maps do not scale well, because of the contention on bucket and element read locks (the locks have to perform atomic writes to change their internal state). The book implementation, in contrast, profits from read contention, since it does not modify any locks or data and the hot element can be held in the L1-cache.

For write-only workloads without contention, the book implementation demonstrates very good scaling, whereas the TBB implementations show only a relative speedup of 6 using all cores. Both TBB implementations suffer from contended atomic operations in their internal structures.

With write contention, the scaling reduces, being dependent on the speed of the contended atomic increment. We also tried to use generic 8-byte CAS on the data part of the element instead of an atomic fetch-and-add instruction. This resulted in worse scaling. This experiment confirms that the specialized hardware implementation is preferable when applicable.

Our work on this book motivated us to write a scientific paper on hash tables [206]. This paper describes several generalizations of the data structure described above, in particular, how automatic adaptive growing and shrinking of the table can be implemented. There is also an extensive comparison with other concurrent hash tables.

4.7 Implementation Notes

Although hashing is an algorithmically simple concept, a clean, efficient, and robust implementation can be surprisingly difficult. Less surprisingly, the hash functions used are an important issue. Most applications seem to use simple, very fast hash functions based on exclusive OR, shifting, and table lookup rather than universal hash functions; see, for example, `www.burtleburtle.net/bob/hash/doobs.html`, `github.com/aappleby/smhasher/wiki/SMHasher`, `cyan4973.github.io/xxHash/`, `github.com/minio/highwayhash`. Although these functions seem to work well in practice, we believe that the universal families of hash functions described in Sect. 4.2 are competitive. The Wikipedia pages on "*Universal hashing*" and "*List of hash functions*" mention universal hash functions also.

Unfortunately, there is no implementation study covering all of the fastest families. Thorup [312] implemented a fast 1-universal family with additional properties. We suggest using the family $H^{\oplus[]}$ considered in Exercise 4.16 for integer keys, and the functions in Exercise 4.9 for strings. It might be possible to implement the latter function to run particularly fast using the SIMD instructions of modern processors that allow the parallel execution of several operations.

Hashing with chaining uses only very specialized operations on sequences, for which singly linked lists are ideally suited. Since we are dealing with many short lists, some deviations from the implementation scheme described in Sect. 3.2 are in order. In particular, it would be wasteful to store a dummy item with each list. Instead, one should use a single, shared dummy item to mark the ends of all lists. This item can then be used as a sentinel element for *find* and *remove*, as in the function *findNext* in Sect. 3.2.1. This trick not only saves space, but also makes it likely that the dummy item will reside in the cache memory.

With respect to the first element of the lists, there are two alternatives. One can either use a table of pointers and store the first element outside the table, or store the first element of each list directly in the table. We refer to these alternatives as *slim tables* and *fat tables*, respectively. Fat tables are usually faster and more space-efficient. Slim tables are superior when the elements are very large. In comparison, a slim table wastes the space occupied by m pointers and a fat table wastes the space of the unoccupied table positions (see Exercise 4.7). Slim tables also have the advantage of referential integrity even when tables are reallocated. We have already observed this complication for unbounded arrays in Sect. 3.9.

Comparing the space consumption of hashing with chaining and hashing with linear probing is even more subtle than what is outlined in Sect. 4.4. On the one hand, linked lists burden the memory management with many small pieces of allocated memory; see Sect. 3.2.1 for a discussion of memory management for linked lists. On the other hand, the slim table implementations of unbounded hash tables based on chaining can avoid occupying two tables during reallocation by use of the following method. First, concatenate all lists into a single list L. Deallocate the old table. Only then, allocate the new table. Finally, scan L, moving the elements to the new table. For fat tables and for hashing with linear probing, the use of two tables during reallocation seems necessary at first sight. However, even for them reallocation can be avoided. The results are hash tables that *never* consume significantly more space than what is needed anyway just to store the elements [205].

Exercise 4.24. Implement hashing with chaining and hashing with linear probing on your own machine using your favorite programming language. Compare their performance experimentally. Also, compare your implementations with hash tables available in software libraries. Use elements of size 8 bytes.

Exercise 4.25 (large elements). Repeat the above measurements with element sizes of 32 and 128. Also, add an implementation of *slim chaining*, where table entries store only pointers to the first list element.

Exercise 4.26 (large keys). Discuss the impact of large keys on the relative merits of chaining versus linear probing. Which variant will profit? Why?

Exercise 4.27. Implement a hash table data type for very large tables stored on disk. Should you use chaining or linear probing? Why?

4.7.1 C++

The C++ standard library did not define a hash table data type until 2011. The new standard, C++11 introduced, such a data type. It offers several variants that are all realized by hashing with chaining: *unordered_set*, *unordered_map*, *unordered_multiset*, and *unordered_multimap*. Here "set" stands for the kind of interface used in this chapter, whereas a "map" is an associative array indexed by keys. The prefix "multi" indicates that multiple elements with the same key are allowed. Hash functions are implemented as *function objects*, i.e., the class *hash<T>* overloads the operator "( )"

so that an object can be used like a function. This approach allows the hash function to store internal state such as random coefficients.

Unfortunately, the current implementations of these data structures are in many situations considerably slower than what is possible. For example, consider the case of bounded tables with small elements and suppose that space is not too much at a premium. Here, a specialized implementation of linear probing can be several times faster than library implementations that pay a high price for generality.

LEDA [195] offers several hashing-based implementations of associative arrays. The class $h\_array\langle Key, T \rangle$ offers associative arrays for storing objects of type T with keys of type Key. This class requires a user-defined hash function $int\ Hash(Key\&)$ that returns an integer value which is then mapped to a table index by LEDA. The implementation uses hashing with chaining and adapts the table size to the number of elements stored. The class *map* is similar but uses a built-in hash function.

Exercise 4.28 (associative arrays). Implement a C++ class for associative arrays. The implementation should offer the *operator*[ ] for any index type that supports a hash function. Overload the assignment operator such that the assignment $H[x] = \cdots$ works as expected if x is the key of a new element.

Concurrent hash tables. Our implementations [206] are available at github. com/TooBiased and provide good scalability in many situations. More general functionality is provided by the concurrent hash table in the Intel TBB library. However, the price paid for this generality is quite high, for this and other implementations. Currently, one should use such libraries only if the hash table accesses do not consume a significant fraction of the overall work [206].

4.7.2 Java

The class *java.util.HashMap* implements unbounded hash tables using the function *hashCode* as a hash function. The function *hashCode* must be defined for the objects stored in the hash table. A concurrent hash table is available as *java.util.concurrent.ConcurrentHashMap*. The caveat about performance mentioned in the preceding paragraph also applies to this implementation [206].

4.8 Historical Notes and Further Findings

Hashing with chaining and hashing with linear probing were used as early as the 1950s [252]. The analysis of hashing began soon after. In the 1960s and 1970s, average-case analysis in the spirit of Theorem 4.1 and Exercise 4.8 prevailed. Various schemes for random sets of keys or random hash functions were analyzed. An early survey paper was written by Morris [232]. The book [185] contains a wealth of material. For example, it analyzes linear probing assuming random hash functions. Let n denote the number of elements stored, let m denote the size of the table and set $\alpha = n/m$. The expected number T_{fail} of table accesses for an unsuccessful search and the number T_{success} for a successful search are about

$$T_{\text{fail}} \approx \frac{1}{2}\left(1+\left(\frac{1}{1-\alpha}\right)^2\right) \text{ and } T_{\text{success}} \approx \frac{1}{2}\left(1+\frac{1}{1-\alpha}\right),$$

respectively. Note that these numbers become very large when n approaches m, i.e., it is not a good idea to fill a linear-probing table almost completely.

Universal hash functions were introduced by Carter and Wegman [61]. The original paper proved Theorem 4.3 and introduced the universal classes discussed in Exercise 4.12. More on universal hashing can be found in [15].

In Sect. 4.2, we described a method for finding a prime number in an interval $k^3..(k+1)^3$. Of course, the method will only be successful if the interval actually contains a prime, i.e., the statement (A_k) holds. For $k > e^{e^k}$ [69] the statement has been proven. If the *Riemann Hypothesis*, one of the most famous, yet unproven conjectures in number theory, is true, (A_k) holds for all $k \geq 1$ [60]. For the application to hashing, the range $k \leq 8 \cdot 10^7$ is more than sufficient: (A_k) holds for all such k. On the internet, one can find tables of primes and also tables of gaps between primes. The web page `primes.utm.edu/notes/GapsTable.html` lists *largest gaps* between primes for the range up to $4 \cdot 10^{17}$. One can conclude from this table that (A_k) holds for $k \leq 3000$. It is also known [262] that, for all primes $p > 11 \cdot 10^9$, the distance from p to the next larger prime is at most $p/(2.8 \cdot 10^7)$. A simple computation shows that this implies (A_k) for all k between 3000 and $8.4 \cdot 10^7$.

The sieve of Eratosthenes is, by far, not the most efficient method of finding primes, and should only be used for finding primes that are less than a few billion. A better method of finding larger primes is based on the fact that primes are quite abundant. Say we want to find a prime in the vicinity of a given number m. We repeatedly choose a random number from the interval $[m,2m]$ (in this interval, there will be $\Omega(m/\ln m)$ primes) and test whether the number is a prime using an efficient randomized primality test [138, p. 254]. Such an algorithm will require storage space $O(\log p)$ and even a naive implementation will run in expected time $O((\log p)^3)$. With such an algorithm, one can find primes with thousands of decimal digits.

Perfect hashing was a black art until Fredman, Komlós, and Szemeredi [113] introduced the construction shown in Theorem 4.8. Dynamization is due to Dietzfelbinger et al. [94]. Cuckoo hashing [248] is an alternative approach to dynamic perfect hashing, where each element can be stored in two [248] or more [109] places of a table. If these places consist of buckets with several slots, we obtain a highly space efficient data structure [95, 205].

A *minimal perfect hash function* bijectively maps a set $S \subseteq 0..U-1$ to the range $0..n-1$, where $n = |S|$. The goal is to find a function that can be evaluated in constant time and requires little space for its representation – $\Omega(n)$ bits is a lower bound. For a few years now, there have been practicable schemes that achieve this bound [37, 52, 235]. One variant assumes three truly random hash functions[12] $h_i : 0..U-1 \to im/3..(i+1)m/3-1$ for $i \in 0..2$ and $m = \alpha n$, where $\alpha \approx 1.23n$. In a first step, called

---

[12] Actually implementing such hash functions would require $\Omega(n\log n)$ bits. However, this problem can be circumvented by first splitting S into many small *buckets*. We can then use the same set of fully random hash functions for all the buckets [95].

the *mapping step*, one searches for an injective function $p\colon S \to 0..m-1$ such that $p(x) \in \{h_0(x), h_1(x), h_2(x)\}$ for all $x \in S$. It can be shown that such a p can be found in linear time with high probability. Next one determines a function $g\colon 0..m-1 \to \{0,1,2\}$ such that

$$p(x) = h_i(x), \text{ where } i = g(h_0(x)) \oplus g(h_1(x)) \oplus g(h_2(x)) \bmod 3 \quad \text{for all } x \in S.$$

The function g is not hard to find by a greedy algorithm. It is stored as a table of $O(n)$ bits. In a second step, called the *ranking step*, the set $0..m$ is mapped to $0..n-1$ via a function $rank(i) = |\{k \in S : p(k) \leq i\}|$. Then $h(x) = rank(p(x))$ is an injective function from S to $0..n-1$. The task of computing a representation of *rank* that uses only $O(n)$ bits and can be evaluated in constant time is a standard problem in the field of *succinct data structures* [238].

Universal hashing bounds the probability of any two keys colliding. A more general notion is *k*-way independence, where *k* is a positive integer. A family H of hash functions is *k-way c-independent*, where $c \geq 1$ is a constant, if for any k distinct keys x_1 to x_k, and any k hash values a_1 to a_k, $\mathrm{prob}(h(x_1) = a_1 \wedge \cdots \wedge h(x_k) = a_k) \leq c/m^k$. The polynomials of degree at most $k-1$ with random coefficients in a prime field $\mathbb{Z}_p$ and evaluated in $\mathbb{Z}_p$ are a simple *k*-way 1-independent family of hash functions [61] (see Exercise 4.13 for the case $k=2$).

The combination of linear probing with simple universal classes such as H^* and $H^{\gg}$ may lead to nonconstant insertion and search time [247, 249]. Only 5-independent classes guarantee constant insertion and search time [247]. Also, tabulation hashing [335] (Exercise 4.16 with $w=0$) makes linear probing provably efficient [259].

Cryptographic hash functions need stronger properties than what we need for hash tables. Roughly, for a value x, it should be difficult to come up with a value x' such that $h(x') = h(x)$.

5

Sorting and Selection

Telephone directories are sorted alphabetically by last name. Why? Because a sorted index can be searched quickly. Even in the telephone directory of a huge city, one can find a name in a few seconds. In an unsorted index, nobody would even try to find a name. This chapter teaches you how to turn an unordered collection of elements into an ordered collection, i.e., how to sort *the collection. The sorted collection can then be searched fast. We will get to know several algorithms for sorting; the different algorithms are suited for different situations, for example sorting in main memory or sorting in external memory, and illustrate different algorithmic paradigms. Sorting has many other uses as well. An early example of a massive data-processing task was the statistical evaluation of census data; 1500 people needed seven years to manually process data from the US census in 1880. The engineer Herman Hollerith,[1] who participated in this evaluation as a statistician, spent much of the 10 years to the next census developing counting and sorting machines for mechanizing this gigantic endeavor. Although the 1890 census had to evaluate more people and more questions, the basic evaluation was finished in 1891. Hollerith's company continued to play an important role in the development of the information-processing industry; since 1924, it has been known as International Business Machines (IBM). Sorting is important for census statistics because one often wants to form subcollections, for example, all persons between age 20 and 30 and living on a farm. Two applications of sorting solve the problem. First, we sort all persons by age and form the subcollection of persons between 20 and 30 years of age. Then we sort the subcollection by the type of the home (house, apartment complex, farm, ...) and extract the subcollection of persons living on a farm.*

Although we probably all have an intuitive concept of what *sorting* is about, let us give a formal definition. The input is a sequence $s = \langle e_1, \ldots, e_n \rangle$ of n elements. Each element e_i has an associated *key* $k_i = key(e_i)$. The keys come from an ordered

---

[1] The photograph was taken by C. M. Bell; US Library of Congress' Prints and Photographs Division, ID cph.3c15982.

© Springer Nature Switzerland AG 2019
P. Sanders et al., *Sequential and Parallel Algorithms and Data Structures*,
https://doi.org/10.1007/978-3-030-25209-0_5

universe, i.e., there is a *linear order* (also called a *total order*) $\leq$ defined on the keys.[2] For ease of notation, we extend the comparison relation to elements so that $e \leq e'$ if and only if $key(e) \leq key(e')$. Since different elements may have equal keys, the relation $\leq$ on elements is only a linear preorder. The task is to produce a sequence $s' = \langle e'_1, \ldots, e'_n \rangle$ such that s' is a permutation of s and such that $e'_1 \leq e'_2 \leq \cdots \leq e'_n$. Observe that the ordering of elements with equal key is arbitrary.

Although different comparison relations for the same data type may make sense, the most frequent relations are the obvious order for numbers and the *lexicographic order* (see Appendix A) for tuples, strings, and sequences. The lexicographic order for strings comes in different flavors. We may treat corresponding lower-case and upper-case characters as being equivalent, and different rules for treating accented characters are used in different contexts.

Exercise 5.1. Given linear orders $\leq_A$ for A and $\leq_B$ for B, define a linear order on $A \times B$.

Exercise 5.2. Consider the relation R over the complex numbers defined by $x\,R\,y$ if and only if $|x| \leq |y|$. Is it total? Is it transitive? Is it antisymmetric? Is it reflexive? Is it a linear order? Is it a linear preorder?

Exercise 5.3. Define a total order for complex numbers with the property that $x \leq y$ implies $|x| \leq |y|$.

Sorting is a ubiquitous algorithmic tool; it is frequently used as a preprocessing step in more complex algorithms. We shall give some examples.

- *Preprocessing for fast search.* In Sect. 2.7 on binary search, we have already seen that a sorted directory is easier to search, both for humans and for computers. Moreover, a sorted directory supports additional operations, such as finding all elements in a certain range. We shall discuss searching in more detail in Chap. 7. Hashing is a method for searching unordered sets.
- *Grouping.* Often, we want to bring equal elements together to count them, eliminate duplicates, or otherwise process them. Again, hashing is an alternative. But sorting has advantages, since we shall see rather fast, space-efficient, deterministic sorting algorithm that scale to huge data sets.
- *Processing in a sorted order.* Certain algorithms become very simple if the inputs are processed in sorted order. Exercise 5.4 gives an example. Other examples are Kruskal's algorithm presented in Sect. 11.3, and several of the algorithms for the knapsack problem presented in Chap. 12. You may also want to remember sorting when you solve Exercise 8.6 on interval graphs.

In Sect. 5.1, we shall introduce several simple sorting algorithms. They have quadratic complexity, but are still useful for small input sizes. Moreover, we shall

---

[2] A linear or total order is a reflexive, transitive, total, and antisymmetric relation such as the relation $\leq$ on the real numbers. A reflexive, transitive, and total relation is called a linear preorder or linear quasiorder. An example is the relation $R \subseteq \mathbb{R} \times \mathbb{R}$ defined by $x\,R\,y$ if and only if $|x| \leq |y|$; see Appendix A for details.

learn some low-level optimizations. Section 5.3 introduces *mergesort*, a simple divide-and-conquer sorting algorithm that runs in time $O(n \log n)$. Section 5.5 establishes that this bound is optimal for all *comparison-based* algorithms, i.e., algorithms that treat elements as black boxes that can only be compared and moved around. The *quicksort* algorithm described in Sect. 5.6 is again based on the divide-and-conquer principle and is perhaps the most frequently used sorting algorithm. Quicksort is also a good example of a randomized algorithm. The idea behind quicksort leads to a simple algorithm for a problem related to sorting. Section 5.8 explains how the kth smallest of n elements can be *selected* in time $O(n)$. Sorting can be made even faster than the lower bound obtained in Sect. 5.5 by exploiting the internal structure of the keys, for example by exploiting the fact that numbers are sequences of digits. This is the content of Sect. 5.10. Section 5.12 generalizes quicksort and mergesort to very good algorithms for sorting inputs that do not fit into internal memory.

Most parallel algorithms in this chapter build on the sequential algorithms. We begin in Sect. 5.2 with an inefficient yet fast and simple algorithm that can be used as a subroutine for sorting very small inputs very quickly. Parallel mergesort (Sect. 5.4) is efficient for inputs of size $\Omega(p \log p)$ and a good candidate for sorting relatively small inputs on a shared-memory machine. Parallel quicksort (Sect. 5.7) can be used in similar circumstances and might be a good choice on distributed-memory machines. There is also an almost in-place variant for shared memory. Selection (Sect. 5.9) can be parallelized even better than sorting. In particular, there is a communication-efficient algorithm that does not need to move the data. The noncomparison-based algorithms in Sect. 5.10 are rather straightforward to parallelize (Sect. 5.11). The external-memory algorithms in Sect. 5.12 are the basis of very efficient parallel algorithms for large inputs. Parallel sample sort (Sect. 5.13) and parallel multiway mergesort (Sect. 5.14) are only efficient for rather large inputs of size $\omega(p^2)$, but the elements need to be moved only once. Since sample sort is a good compromise between simplicity and efficiency, we give two implementations – one for shared memory and the other for distributed memory. Finally, in Sect. 5.15 we outline a sophisticated algorithm that is asymptotically efficient even for inputs of size p. This algorithm is a recursive generalization of sample sort that uses the fast, inefficient algorithm in Sect. 5.2 for sorting the sample.

Exercise 5.4 (a simple scheduling problem). A hotel manager has to process n advance bookings of rooms for the next season. His hotel has k identical rooms. Bookings contain an arrival date and a departure date. He wants to find out whether there are enough rooms in the hotel to satisfy the demand. Design an algorithm that solves this problem in time $O(n \log n)$. Hint: Consider the multiset of all arrivals and departures. Sort the set and process it in sorted order.

Exercise 5.5 ((database) sort join). As in Exercise 4.5, consider two relations $R \subseteq A \times B$ and $Q \subseteq B \times C$ with $A \neq C$ and design an algorithm for computing the natural *join* of R and Q

$$R \bowtie Q := \{(a,b,c) \subseteq A \times B \times C : (a,b) \in R \wedge (b,c) \in Q\}.$$

Show how to obtain running time $O((|R|+|Q|)\log(|R|+|Q|)+|R\bowtie Q|)$ with a deterministic algorithm.

Exercise 5.6 (sorting with a small set of keys). Design an algorithm that sorts n elements in $O(k\log k+n)$ expected time if there are only k different keys appearing in the input. Hint: Combine hashing and sorting, and use the fact that k keys can be sorted in time $O(k\log k)$.

Exercise 5.7 (checking). It is easy to check whether a sorting routine produces a sorted output. It is less easy to check whether the output is also a permutation of the input. But here is a fast and simple Monte Carlo algorithm for integers: (a) Show that $\langle e_1,\dots,e_n\rangle$ is a permutation of $\langle e'_1,\dots,e'_n\rangle$ if and only if the polynomial $q(z) :=$ $\prod_{i=1}^n(z-e_i)-\prod_{i=1}^n(z-e'_i)$ is identically 0. Here, z is a variable. (b) For any $\varepsilon>0$, let p be a prime with $p>\max\{n/\varepsilon,e_1,\dots,e_n,e'_1,\dots,e'_n\}$. Now the idea is to evaluate the above polynomial mod p for a random value $z\in 0..p-1$. Show that if $\langle e_1,\dots,e_n\rangle$ is *not* a permutation of $\langle e'_1,\dots,e'_n\rangle$, then the result of the evaluation is 0 with probability at most ε. Hint: A polynomial of degree n that is not identically 0 modulo p has at most n 0's in $0..p-1$ when evaluated modulo p.

Exercise 5.8 (permutation checking by hashing). Consider sequences A and B where A is not a permutation of B. Suppose $h : Element \to 0..U-1$ is a random hash function. Show that $\text{prob}(\sum_{e\in A}h(e)=\sum_{e\in B}h(e))\le 1/U$. Hint: Focus on one element that occurs a different number of times in A and B.

5.1 Simple Sorters

We shall introduce two simple sorting techniques: *selection sort* and *insertion sort*.

Selection sort repeatedly selects the smallest element from the input sequence, deletes it, and adds it to the end of the output sequence. The output sequence is initially empty. The process continues until the input sequence is exhausted. For example,

$$\langle\rangle,\langle 4,7,1,1\rangle \rightsquigarrow \langle 1\rangle,\langle 4,7,1\rangle \rightsquigarrow \langle 1,1\rangle,\langle 4,7\rangle \rightsquigarrow \langle 1,1,4\rangle,\langle 7\rangle \rightsquigarrow \langle 1,1,4,7\rangle,\langle\rangle.$$

The algorithm can be implemented such that it uses a single array of n elements and works *in-place*, i.e., it needs no additional storage beyond the input array and a constant amount of space for loop counters, etc. The running time is quadratic.

Exercise 5.9 (simple selection sort). Implement selection sort so that it sorts an array with n elements in time $O(n^2)$ by repeatedly scanning the input sequence. The algorithm should be in-place, i.e., the input sequence and the output sequence should share the same array. Hint: The implementation operates in n phases numbered 1 to n. At the beginning of the ith phase, the first $i-1$ locations of the array contain the $i-1$ smallest elements in sorted order and the remaining $n-i+1$ locations contain the remaining elements in arbitrary order.

Procedure *insertionSort*(*a* : *Array* [1..*n*] **of** *Element*)
    for $i := 2$ **to** n **do**
        invariant $a[1] \leq \cdots \leq a[i-1]$
        // move $a[i]$ to the right place
        $e := a[i]$
        if $e < a[1]$ **then**   // new minimum
            for $j := i$ **downto** 2 **do** $a[j] := a[j-1]$
            $a[1] := e$
        else   // use $a[1]$ as a sentinel
            for $j := i$ **downto** $-\infty$ **while** $a[j-1] > e$ **do** $a[j] := a[j-1]$
            $a[j] := e$

Fig. 5.1. Insertion sort

In Sect. 6.6, we shall learn about a more sophisticated implementation where the input sequence is maintained as a *priority queue*. Priority queues support efficient repeated selection of the minimum element. The resulting algorithm runs in time $O(n \log n)$ and is frequently used. It is efficient, it is deterministic, it works in-place, and the input sequence can be dynamically extended by elements that are larger than all previously selected elements. The last feature is important in discrete-event simulations, where events have to be processed in increasing order of time and processing an event may generate further events in the future.

Selection sort maintains the invariant that the output sequence is sorted by carefully choosing the element to be deleted from the input sequence. *Insertion sort* maintains the same invariant by choosing an arbitrary element of the input sequence but taking care to insert this element in the right place in the output sequence. For example,

$$\langle\rangle, \langle 4,7,1,1\rangle \rightsquigarrow \langle 4\rangle, \langle 7,1,1\rangle \rightsquigarrow \langle 4,7\rangle, \langle 1,1\rangle \rightsquigarrow \langle 1,4,7\rangle, \langle 1\rangle \rightsquigarrow \langle 1,1,4,7\rangle, \langle\rangle.$$

Figure 5.1 gives an in-place array implementation of insertion sort. The implementation is straightforward except for a small trick that allows the inner loop to use only a single comparison. When the element e to be inserted is smaller than all previously inserted elements, it can be inserted at the beginning without further tests. Otherwise, it suffices to scan the sorted part of a from right to left while e is smaller than the current element. This process has to stop, because $a[1] \leq e$.

In the worst case, insertion sort is quite slow. For example, if the input is sorted in decreasing order, each input element is moved all the way to $a[1]$, i.e., in iteration i of the outer loop, i elements have to be moved. Overall, we obtain

$$\sum_{i=2}^{n}(i-1) = -n + \sum_{i=1}^{n} i = \frac{n(n+1)}{2} - n = \frac{n(n-1)}{2} = \Omega(n^2)$$

movements of elements; see also (A.12).

Nevertheless, insertion sort is useful. It is fast for small inputs (say, $n \leq 10$) and hence can be used as the base case in divide-and-conquer algorithms for sorting.

Furthermore, in some applications the input is already "almost" sorted, and in this situation insertion sort will be fast.

Exercise 5.10 (almost sorted inputs). Prove that insertion sort runs in time $O(n+D)$, where $D = \sum_i |r(e_i) - i|$ and $r(e_i)$ is the *rank* (position) of e_i in the sorted output.

Exercise 5.11 (average-case analysis). Assume that the input to an insertion sort is a permutation of the numbers 1 to n. Show that the average execution time over all possible permutations is $\Omega(n^2)$. Hint: Argue formally that about one-third of the input elements in the right third of the array have to be moved to the left third of the array. Can you improve the argument to show that, on average, $n^2/4 - O(n)$ iterations of the inner loop are needed?

Exercise 5.12 (insertion sort with few comparisons). Modify the inner loops of the array-based insertion sort algorithm in Fig. 5.1 so that it needs only $O(n\log n)$ comparisons between elements. Hint: Use binary search as discussed in Chap. 7. What is the running time of this modification of insertion sort?

Exercise 5.13 (efficient insertion sort?). Use the data structure for sorted sequences described in Chap. 7 to derive a variant of insertion sort that runs in time $O(n\log n)$.

*Exercise 5.14 (formal verification). Use your favorite verification formalism, for example Hoare calculus, to prove that insertion sort produces a permutation of the input.

5.2 Simple, Fast, and Inefficient Parallel Sorting

In parallel processing, there are also cases where spending a quadratic number of comparisons to sort a small input makes sense. Assume that the PEs are arranged as a quadratic matrix with PE indices written as pairs. Assume furthermore that we have input elements e_i at the diagonal PEs with index (i,i). For simplicity, assume also that all elements are different. In this situation, there is a simple and fast algorithm that sorts in logarithmic time: PE (i,i) first broadcasts its element along row i and column i. This can be done in logarithmic time; see Sect. 13.1. Now, for every pair (i,j) of input elements, there is a dedicated processor that can compare them in constant time. The rank of element i can then be determined by adding the 0–1 value $[e_i \geq e_j]$ along each row. We can already view this mapping of elements to ranks as the output of the sorting algorithm. If desired, we can also use this information to permute the elements. For example, we could send the elements with rank i to PE (i,i).

At the first glance, this sounds like a rather useless algorithm, since its efficiency is $o(1)$. However, there are situations where speed is more important than efficiency, for example for the fast parallel selection algorithm discussed in Sect. 5.9, where we use sorting a sample to obtain a high-quality pivot. Also, note that in that situation, even finding a single random pivot requires a prefix sum and a broadcast, i.e., taking

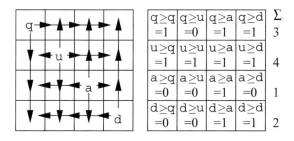

	q≥q	q≥u	q≥a	q≥d	Σ
q	=1	=0	=1	=1	3
u	u≥q =1	u≥u =1	u≥a =1	u≥d =1	4
a	a≥q =0	a≥u =0	a≥a =1	a≥d =0	1
d	d≥q =0	d≥u =0	d≥a =1	d≥d =1	2

Fig. 5.2. Brute force ranking of four elements on 4×4 PEs

a random pivot is only a constant factor faster than choosing the median of a sample of size $\sqrt{p}$. Figure 5.2 gives an example.

We can obtain wider applicability by generalizing the algorithm to handle larger inputs. Here, we outline an algorithm described in more detail in [23] and care only about computing the rank of each element. Now, the PEs are arranged into an $a \times b$ matrix. Each PE has a (possibly empty) set of input elements. Each PE sorts its elements locally. Then we redistribute the elements such that PE (i, j) has two sequences I and J, where I contains all elements from row i and J contains all elements from column J. This can be done using all-gather operations along the rows and columns; see Sect. 13.5. Additionally, we ensure that the sequences are sorted by replacing the local concatenation operations in the all-gather algorithm by a merge operation. Subsequently, elements from I are ranked with respect to the elements in J, i.e., for each element $x \in I$, we count how many elements $y \in J$ have $y \le x$. This can be done in linear time by merging I and J. The resulting local rank vectors are then added along the rows. Figure 5.3 gives an example.

Overall, if all rows and columns contain a balanced number of elements, we get a total execution time

$$O\left(\alpha \log p + \beta \frac{n}{\sqrt{p}} + \frac{n}{p} \log \frac{n}{p} \right). \tag{5.1}$$

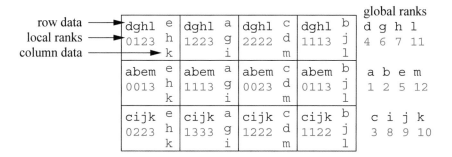

Fig. 5.3. Fast, inefficient ranking of $\langle d,g,h,l,a,b,e,m,c,i,j,k \rangle$ on 3×4 PEs.

5.3 Mergesort – an $O(n \log n)$ Sorting Algorithm

Mergesort is a straightforward application of the divide-and-conquer principle. The unsorted sequence is split into two parts of about equal size. The parts are sorted recursively, and the sorted parts are merged into a single sorted sequence. This approach is efficient because merging two sorted sequences a and b is quite simple. The globally smallest element is either the first element of a or the first element of b. So, we move the smaller element to the output, find the second smallest element using the same approach, and iterate until all elements have been moved to the output. Figure 5.4 gives pseudocode, and Fig. 5.5 illustrates a sample execution. If the sequences are represented as linked lists (see Sect. 3.2), no allocation and deallocation of list items is needed. Each iteration of the inner loop of *merge* performs one element comparison and moves one element to the output. Each iteration takes constant time. Hence, merging runs in linear time.

Function *mergeSort*($\langle e_1, \ldots, e_n \rangle$) : *Sequence* **of** *Element*
    if $n = 1$ **then return** $\langle e_1 \rangle$
    else return *merge*(*mergeSort*($\langle e_1, \ldots, e_{\lfloor n/2 \rfloor} \rangle$),
                    mergeSort($\langle e_{\lfloor n/2 \rfloor + 1}, \ldots, e_n \rangle$)))

// merging two sequences represented as lists
Function *merge*(a, b : *Sequence* **of** *Element*) : *Sequence* **of** *Element*
    $c := \langle \rangle$
    loop
        invariant a, b, and c are sorted and $\forall e \in c, e' \in a \cup b : e \leq e'$
        if *a.isEmpty*   **then**   *c.concat*(b); **return** c
        if *b.isEmpty*   **then**   *c.concat*(a); **return** c
        if *a.first* $\leq$ *b.first* **then**   *c.moveToBack*(*a.PopFront*)
        else   *c.moveToBack*(*b.PopFront*)

Fig. 5.4. Mergesort

a	b	c	operation
$\langle 1,2,7 \rangle$	$\langle 1,2,8,8 \rangle$	$\langle \rangle$	move a
$\langle 2,7 \rangle$	$\langle 1,2,8,8 \rangle$	$\langle 1 \rangle$	move b
$\langle 2,7 \rangle$	$\langle 2,8,8 \rangle$	$\langle 1,1 \rangle$	move a
$\langle 7 \rangle$	$\langle 2,8,8 \rangle$	$\langle 1,1,2 \rangle$	move b
$\langle 7 \rangle$	$\langle 8,8 \rangle$	$\langle 1,1,2,2 \rangle$	move a
$\langle \rangle$	$\langle 8,8 \rangle$	$\langle 1,1,2,2,7 \rangle$	concat b
$\langle \rangle$	$\langle \rangle$	$\langle 1,1,2,2,7,8,8 \rangle$	

Fig. 5.5. Execution of *mergeSort*($\langle 2,7,1,8,2,8,1 \rangle$). The *left* part illustrates the recursion in *mergeSort* and the *right* part illustrates the *merge* in the outermost call.

Theorem 5.1. *The function merge, applied to sequences of total length n, executes in time* $O(n)$ *and performs at most* $n - 1$ *element comparisons.*

For the running time of mergesort, we obtain the following result.

Theorem 5.2. *Mergesort runs in time* $O(n \log n)$ *and performs no more than* $\lceil n \log n \rceil$ *element comparisons.*

Proof. Let $C(n)$ denote the worst-case number of element comparisons performed. We have $C(1) = 0$ and $C(n) \le C(\lfloor n/2 \rfloor) + C(\lceil n/2 \rceil) + n - 1$, using Theorem 5.1. The master theorem for recurrence relations (2.5) suggests $C(n) = O(n \log n)$. We next give two proofs that show explicit constants. The first proof shows $C(n) \le 2n \lceil \log n \rceil$, and the second proof shows $C(n) \le n \lceil \log n \rceil$.

For n a power of two, we define $D(1) = 0$ and $D(n) = 2D(n/2) + n$. Then $D(n) = n \log n$ for n a power of two by either the master theorem for recurrence relations or by a simple induction argument.[3] We claim that $C(n) \le D(2^k)$, where k is such that $2^{k-1} < n \le 2^k$. Then $C(n) \le D(2^k) = 2^k k \le 2n \lceil \log n \rceil$. It remains to argue the inequality $C(n) \le D(2^k)$. We use induction on k. For $k = 0$, we have $n = 1$ and $C(1) = 0 = D(1)$, and the claim certainly holds. For $k > 1$, we observe that $\lfloor n/2 \rfloor \le \lceil n/2 \rceil \le 2^{k-1}$, and hence

$$C(n) \le C(\lfloor n/2 \rfloor) + C(\lceil n/2 \rceil) + n - 1 \le 2D(2^{k-1}) + 2^k - 1 \le D(2^k).$$

This completes the first proof.

We turn now to the second, refined proof. We prove

$$C(n) \le n \lceil \log n \rceil - 2^{\lceil \log n \rceil} + 1 \le n \log n$$

by induction over n. For $n = 1$, the claim is certainly true. So, assume $n > 1$. Let k be such that $2^{k-1} < \lceil n/2 \rceil \le 2^k$, i.e. $k = \lceil \log \lceil n/2 \rceil \rceil$. Then $C(\lceil n/2 \rceil) \le \lceil n/2 \rceil k - 2^k + 1$ by the induction hypothesis. If $\lfloor n/2 \rfloor > 2^{k-1}$, then k is also equal to $\lceil \log \lfloor n/2 \rfloor \rceil$ and hence $C(\lfloor n/2 \rfloor) \le \lfloor n/2 \rfloor k - 2^k + 1$ by the induction hypothesis. If $\lfloor n/2 \rfloor = 2^{k-1}$ and hence $k - 1 = \lceil \log \lfloor n/2 \rfloor \rceil$, the induction hypothesis yields $C(\lfloor n/2 \rfloor) = \lfloor n/2 \rfloor (k - 1) - 2^{k-1} + 1 = 2^{k-1}(k - 1) - 2^{k-1} + 1 = \lfloor n/2 \rfloor k - 2^k + 1$. Thus we have the same bound for $C(\lfloor n/2 \rfloor)$ in both cases, and hence

$$
\begin{aligned}
C(n) &\le C(\lfloor n/2 \rfloor) + C(\lceil n/2 \rceil) + n - 1 \\
&\le \left(\lfloor n/2 \rfloor k - 2^k + 1 \right) + \left(\lceil n/2 \rceil k - 2^k + 1 \right) + n - 1 \\
&= nk + n - 2^{k+1} + 1 = n(k + 1) - 2^{k+1} + 1 = n \lceil \log n \rceil - 2^{\lceil \log n \rceil} + 1.
\end{aligned}
$$

It remains to argue that $nk - 2^k + 1 \le n \log n$ for $k = \lceil \log n \rceil$. If $n = 2^k$, the inequality clearly holds. If $n < 2^k$, we have $nk - 2^k + 1 \le n(k - 1) + (n - 2^k + 1) \le n(k - 1) \le n \log n$.

The bound for the execution time can be verified using a similar recurrence relation. □

---

[3] For $n = 1 = 2^0$, we have $D(1) = 0 = n \log n$, and for $n = 2^k$ and $k \ge 1$, we have $D(n) = 2D(n/2) + n = 2(n/2) \log(n/2) + n = n(\log n - 1) + n = n \log n$.

Mergesort is the method of choice for sorting linked lists and is therefore frequently used in functional and logical programming languages that have lists as their primary data structure. In Sect. 5.5, we shall see that mergesort is basically optimal as far as the number of comparisons is concerned; so it is also a good choice if comparisons are expensive. When implemented using arrays, mergesort has the additional advantage that it streams through memory in a sequential way. This makes it efficient in memory hierarchies. Section 5.12 has more on that issue. However, mergesort is not the usual method of choice for an efficient array-based implementation, since it does not work in-place, but needs additional storage space; but see Exercise 5.20.

Exercise 5.15. Explain how to insert k new elements into a sorted list of size n in time $O(k \log k + n)$.

Exercise 5.16. We have discussed *merge* for lists but used abstract sequences for the description of *mergeSort*. Give the details of *mergeSort* for linked lists.

Exercise 5.17. Implement mergesort in a functional programming language.

Exercise 5.18. Give an efficient array-based implementation of mergesort in your favorite imperative programming language. Besides the input array, allocate one auxiliary array of size n at the beginning and then use these two arrays to store all intermediate results. Can you improve the running time by switching to insertion sort for small inputs? If so, what is the optimal switching point in your implementation?

Exercise 5.19. The way we describe *merge*, there are three comparisons for each loop iteration – one element comparison and two termination tests. Develop a variant using sentinels that needs only one termination test. Can you do this task without appending dummy elements to the sequences?

Exercise 5.20. Exercise 3.31 introduced a list-of-blocks representation for sequences. Implement merging and mergesort for this data structure. During merging, reuse emptied input blocks for the output sequence. Compare the space and time efficiency of mergesort for this data structure, for plain linked lists, and for arrays. Pay attention to constant factors.

5.4 Parallel Mergesort

The recursive mergesort from Fig. 5.4 contains obvious task-based parallelism – one simply performs the recursive calls in parallel. However, this algorithm needs time $\Omega(n)$ regardless of the number of processors available, since the final, sequential merge takes that time. In other words, the maximum obtainable speedup is $O(\log n)$ and the corresponding isoefficiency function is exponential in p. This is about as far away from a scalable parallel algorithm as it gets.

In order to obtain a scalable parallel mergesort, we need to parallelize merging. Our approach to merging two sorted sequences a and b in parallel is to split both sequences into p pieces $a_1, \ldots, a_p$ and $b_1, \ldots, b_p$ such that $merge(a,b)$ is the concatenation of $merge(a_1, b_1), \ldots, merge(a_p, b_p)$. The p merges are performed in parallel by assigning one PE each. For this to be correct, the elements in a_i and b_i must be no larger than the elements in a_{i+1} and b_{i+1}. Additionally, to achieve good load balance, we want to ensure that $|a_i| + |b_i| \approx (|a| + |b|)/p$ for $i \in 1..p$. All these properties can be achieved by defining the elements in a_i and b_i to be the elements with positions in $(i-1)\lceil(|a|+|b|)/p\rceil + 1..i\lceil(|a|+|b|)/p\rceil$ in the merged sequence. The strategy is now clear. PE i first determines where a_i ends in a and b_i ends in b. It then merges a_i and b_i.

Let $k = i\lceil(|a|+|b|)/p\rceil$. In order to find where a_i and b_i end in a and b, we need to find the smallest k elements in the two sorted arrays. This is a special case of the selection problem discussed in Sect. 5.8, where we can exploit the sortedness of the arrays a and b to accelerate the computation. We now develop a sequential deterministic algorithm $twoSequenceSelect(a, b, k)$ that locates the k smallest elements in two sorted arrays a and b in time $O(\log|a| + \log|b|)$. The idea is to maintain subranges $a[\ell_a..r_a]$ and $b[\ell_b..r_b]$ with the following properties:

(a) The elements $a[1..\ell_a - 1]$ and $b[1..\ell_b - 1]$ belong to the k smallest elements.
(b) The k smallest elements are contained in $a[1..r_b]$ and $b[1..r_b]$.

We shall next describe a strategy which allows us to halve one of the ranges $[\ell_a..r_a]$ or $[\ell_b..r_b]$. For simplicity, we assume that the elements are pairwise distinct. Let $m_a = \lfloor(\ell_a + r_a)/2\rfloor$, $\bar a = a[m_a]$, $m_b = \lfloor(\ell_b + r_b)/2\rfloor$, and $\bar b = b[m_b]$. Assume that $\bar a < \bar b$, the other case being symmetric. If $k < m_a + m_b$, then the elements in $b[m_b..r_b]$ cannot belong to the k smallest elements and we may set r_b to $m_b - 1$. If $k \geq m_a + m_b$, then all elements in $a[\ell_a..m_a]$ belong to the k smallest elements and we may set ℓ_a to $m_a + 1$. In either case, we have reduced one of the ranges to half its size. This is akin to binary search. We continue until one of the ranges becomes empty, i.e., $r_a = \ell_a - 1$ or $r_b = \ell_b - 1$. We complete the search by setting $r_b = k - r_a$ in the former case and $r_a = k - r_b$ in the latter case.

Since one of the ranges is halved in each iteration, the number of iterations is bounded by $\log|a| + \log|b|$. Table 5.1 gives an example.

Table 5.1. Example calculation for selecting the $k = 4$ smallest elements from the sequences $a = \langle 4, 5, 6, 8 \rangle$ and $b = \langle 1, 2, 3, 7 \rangle$. In the first line, we have $\bar a > \bar b$ and $k \geq m_a + m_b$. Therefore, the first two elements of b belong to the k smallest elements and we may increase ℓ_b to 3. Similarly, in the second line, we have $m_a = 2$ and $m_b = 3$, $\bar a > \bar b$, and $k < m_a + m_b$. Therefore all elements of a except maybe the first do not belong to the k smallest. We may therefore set r_a to 1.

a	ℓ_a	m_a	r_a	$\bar a$	b	ℓ_b	m_b	r_b	$\bar b$	$k < m_a + m_b$?	$\bar a < \bar b$?	action
[4$\bar5$|68]	1	2	4	5	[1$\bar2$|37]	1	2	4	2	no	no	$\ell_b := 3$
[4$\bar5$|68]	1	2	4	5	12[$\bar3$|7]	3	3	4	3	yes	no	$r_a := 1$
[$\bar4$|]568	1	1	1	4	12[$\bar3$|7]	3	3	4	3	no	no	$\ell_b := 4$
[$\bar4$|]568	1	1	1	4	123[$\bar7$|]	4	4	4	7	yes	yes	$r_b := 3$
[$\bar4$|]568	1	1	1	4	123[|]7	4	3	3		finish		$r_a := 1$
4|568		1			123|7		3			done		

*Exercise 5.21. Assume initially that all elements have different keys.

(a) Implement the algorithm *twoSequenceSelect* outlined above. Test it carefully and make sure that you avoid off-by-one errors.

(b) Prove that your algorithm terminates by giving a *loop variant*. Show that at least one range shrinks in every iteration of the loop. Argue as in the analysis of binary search.

(c) Now drop the assumption that all keys are different. Modify your function so that it outputs splitting positions m_a and m_b in a and b such that $m_a + m_b = k$, $a[m_a] \leq b[m_b + 1]$, and $b[m_b] \leq a[m_a + 1]$. Hint: Stop narrowing a range once all its elements are equal. At the end choose the splitters within the ranges such that the above conditions are met.

We can now define a shared-memory parallel binary mergesort algorithm. To keep things simple, we assume that n and p are powers of two. First we build p runs by letting each PE sort a subset of n/p elements. Then we enter the merge loop. In iteration i of the main loop ($i \in 0.. \log p - 1$), we merge pairs of sorted sequences of size $2^i \cdot n/p$ using 2^i PEs. The merging proceeds as described above, i.e., both input sequences are split into 2^i parts each and then each processor merges corresponding pieces.

Let us turn to the analysis. Run formation uses a sequential sorting algorithm and takes time $O((n/p) \log(n/p))$. Each iteration takes time $O(\log(2^i \cdot (n/p))) = O(\log n)$ for splitting (each PE in parallel finds one splitter) and time $O(n/p)$ for merging pieces of size n/p. Overall, we get a parallel execution time

$$
T_{par} = O\left(\frac{n}{p} \log \frac{n}{p} + \log p \left(\log n + \frac{n}{p} \right) \right)
$$
$$
= O\left(\log^2 n + \frac{n \log n}{p} \right).
$$

This algorithm is efficient[4] for $n = \Omega(p \log p)$. The algorithm is a good candidate for an implementation on real-world shared-memory machines since it does sequential merging and sorting in its inner loops and since it can effectively adapt to the memory hierarchy. However, its drawback is that it moves the data logarithmically often. In Sects. 5.13 and 5.14, we shall see algorithms that move the data less frequently.

On the theoretical side, it is worth noting that there is an ingenious but complicated variant of parallel mergesort by Cole [75] which works in time $O(\log p + (n \log n)/p)$, i.e., it is even more scalable. We shall present a randomized algorithm in Sect. 5.15 that is simpler and also allows logarithmic time.

*Exercise 5.22. Design a task-based parallel mergesort with work $O(n \log n)$ and span $O(\log^3 n)$. Hint: You may want to use parallel recursion both for independent subproblems and for merging. For the latter, you may want to use the function *twoSequenceSelect* from Exercise 5.21. Be careful with the size of base case inputs. Compare the scalability of this recursive algorithm with the bottom-up parallel mergesort described above.

---

[4] Note that $\log^2 n \leq (n \log n)/p$ if and only if $p \leq n/\log n$ if $n = \Omega(p \log p)$.

**Exercise 5.23. Develop a practical distributed-memory parallel mergesort. Can you achieve running time $O(\frac{n}{p}\log n + \log^2 p)$? A major obstacle may be that our shared-memory algorithm assumes that concurrent reading is fast. In particular, naive access to the midpoints of the current search ranges may result in considerable contention.

Exercise 5.24 (parallel sort join). As in Exercises 4.5, 4.23, and 5.5, consider two relations $R \subseteq A \times B$ and $Q \subseteq B \times C$ with $A \neq C$ and design an algorithm for computing the natural join of R and Q

$$R \bowtie Q := \{(a,b,c) \subseteq A \times B \times C : (a,b) \in R \wedge (b,c) \in Q\}.$$

Give a parallel algorithm with run time $O(((|R| + |Q|)\log(|R| + |Q|) + |R \bowtie Q|)/p)$ for sufficiently large inputs. How large must the input be? How can the limitation in Exercise 4.23 be lifted? Hint: You have to ensure that the work of outputting the result is well balanced over the PEs.

5.5 A Lower Bound

Algorithms give upper bounds on the complexity of a problem. By the preceding discussion, we know that we can sort n items in time $O(n\log n)$. Can we do better, and maybe even achieve linear time? A "yes" answer requires a better algorithm and its analysis. How could we potentially argue a "no" answer? We would have to argue that no algorithm, however ingenious, can run in time $o(n\log n)$. Such an argument is called a *lower bound*. So what is the answer? The answer is both "no" and "yes". The answer is "no" if we restrict ourselves to comparison-based algorithms, and the answer is "yes" if we go beyond comparison-based algorithms. We shall discuss noncomparison-based sorting in Sect. 5.10.

What is a comparison-based sorting algorithm? The input is a set $\{e_1, \ldots, e_n\}$ of n elements, and the only way the algorithm can learn about its input is by comparing elements. In particular, it is not allowed to exploit the representation of keys, for example as bit strings. When the algorithm stops, it must return a sorted permutation of the input, i.e., a permutation $\langle e'_1, \ldots, e'_n \rangle$ of the input such that $e'_1 \leq e'_2 \leq \ldots \leq e'_n$. Deterministic comparison-based algorithms can be viewed as trees. They make an initial comparison; for instance, the algorithm asks "$e_i \leq e_j$?", with outcomes yes and no. Since the algorithm cannot learn anything about the input except through comparisons, this first comparison must be the same for all inputs. On the basis of the outcome, the algorithm proceeds to the next comparison. There are only two choices for the second comparison: one is chosen if $e_i \leq e_j$, and the other is chosen if $e_i > e_j$. Proceeding in this way, the possible executions of the sorting algorithm define a tree. The key point is that the comparison made next depends only on the outcome of all preceding comparisons and nothing else. Figure 5.6 shows a sorting tree for three elements.

Formally, a comparison tree for inputs e_1 to e_n is a binary tree whose nodes have labels of the form "$e_i \leq e_j$?". The two outgoing edges correspond to the outcomes $\leq$

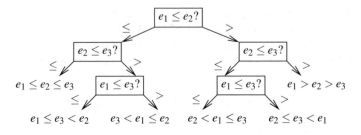

Fig. 5.6. A tree that sorts three elements. We first compare e_1 and e_2. If $e_1 \leq e_2$, we compare e_2 with e_3. If $e_2 \leq e_3$, we have $e_1 \leq e_2 \leq e_3$ and are finished. Otherwise, we compare e_1 with e_3. For either outcome, we are finished. If $e_1 > e_2$, we compare e_2 with e_3. If $e_2 > e_3$, we have $e_1 > e_2 > e_3$ and are finished. Otherwise, we compare e_1 with e_3. For either outcome, we are finished. The worst-case number of comparisons is three. The average number is $(2+3+3+2+3+3)/6 = 8/3$.

and $>$. The computation proceeds in the natural way. We start at the root. Suppose the computation has reached a node labeled $e_i : e_j$. If $e_i \leq e_j$, we follow the edge labeled $\leq$, and if $e_i > e_j$, we follow the edge labeled $>$. The leaves of the comparison tree correspond to the different outcomes of the algorithm.

We next formalize what it means that a comparison tree solves the sorting problem of size n. We restrict ourselves to inputs in which all keys are distinct. When the algorithm terminates, it must have collected sufficient information so that it can tell the ordering of the input. For a permutation π of the integers 1 to n, let ℓ_π be the leaf of the comparison tree reached on input sequences $\{e_1, \ldots, e_n\}$ with $e_{\pi(1)} < e_{\pi(2)} < \ldots < e_{\pi(n)}$. Note that this leaf is welldefined since π fixes the outcome of all comparisons. *A comparison tree solves the sorting problem of size n if, for any two distinct permuations π and σ of $\{1, \ldots, n\}$, the leaves ℓ_π and ℓ_σ are distinct.*

Any comparison tree for sorting n elements must have at least $n!$ leaves. Since a tree of depth T has at most 2^T leaves, we must have

$$2^T \geq n! \quad \text{or} \quad T \geq \log n!.$$

Via Stirling's approximation to the factorial (A.10), we obtain

$$T \geq \log n! \geq \log \left(\frac{n}{e}\right)^n = n \log n - n \log e.$$

Theorem 5.3. *Any comparison-based sorting algorithm needs $n \log n - O(n)$ comparisons in the worst case.*

We state without proof that this bound also applies to randomized sorting algorithms and to the average-case complexity of sorting, i.e., worst-case instances are not much more difficult than random instances.

Theorem 5.4. *Any comparison-based sorting algorithm for n elements needs $n \log n - O(n)$ comparisons on average, i.e.,*

$$\frac{\sum_\pi d_\pi}{n!} = n\log n - O(n),$$

where the sum extends over all n! permutations of the set $\{1,\dots,n\}$ and d_π is the depth of the leaf ℓ_π.

The *element uniqueness problem* is the task of deciding whether, in a set of n elements, all elements are pairwise distinct.

Theorem 5.5. *Any comparison-based algorithm for the element uniqueness problem of size n requires $\Omega(n\log n)$ comparisons.*

Proof. The algorithm has two outcomes "all elements are distinct" and "there are equal elements" and hence, at first sight, we know only that the corresponding comparison tree has at least two leaves. We shall argue that there are $n!$ leaves for the outcome "all elements are distinct". For a permutation π of $\{1,\dots,n\}$, let ℓ_π be the leaf reached on input sequences $\langle e_1,\dots,e_n\rangle$ with $e_{\pi(1)} < e_{\pi(2)} < \dots < e_{\pi(n)}$. This is one of the leaves for the outcome "all elements are distinct".

Let $i \in 1..n-1$ be arbitrary and consider the computation on an input with $e_{\pi(1)} < e_{\pi(2)} < \dots < e_{\pi(i)} = e_{\pi(i+1)} < \dots < e_{\pi(n)}$. This computation has outcome "equal elements" and hence cannot end in the leaf ℓ_π. Since only the outcome of the comparison $e_{\pi(i+1)} : e_{\pi(i)}$ differs for the two inputs (it is $>$ if the elements are distinct and $\leq$ if they are the same), this comparison must have been made on the path from the root to the leaf ℓ_π, and the comparison has established that $e_{\pi(i+1)}$ is larger than $e_{\pi(i)}$. Thus the path to ℓ_π establishes that $e_{\pi(1)} < e_{\pi(2)}, e_{\pi(2)} < e_{\pi(3)}, \dots, e_{\pi(n-1)} < e_{\pi(n)}$, and hence $\ell_\pi \neq \ell_\sigma$ whenever π and σ are distinct permutations of $\{1,\dots,n\}$. □

Exercise 5.25. Why does the lower bound for the element uniqueness problem not contradict the fact that we can solve the problem in linear expected time using hashing?

Exercise 5.26. Show that any comparison-based algorithm for determining the smallest of n elements requires $n-1$ comparisons. Show also that any comparison-based algorithm for determining the smallest and second smallest elements of n elements requires at least $n-1+\log n$ comparisons. Give an algorithm with this performance.

Exercise 5.27 (lower bound for average case). With the notation above, let d_π be the depth of the leaf ℓ_π. Argue that $A = (1/n!)\sum_\pi d_\pi$ is the average-case complexity of a comparison-based sorting algorithm. Try to show that $A \geq \log n!$. Hint: Prove first that $\sum_\pi 2^{-d_\pi} \leq 1$. Then consider the minimization problem "minimize $\sum_\pi d_\pi$ subject to $\sum_\pi 2^{-d_\pi} \leq 1$". Argue that the minimum is attained when all d_i's are equal.

Exercise 5.28 (sorting small inputs optimally). Give an algorithm for sorting k elements using at most $\lceil \log k! \rceil$ element comparisons. (a) For $k \in \{2,3,4\}$, use merge-sort. (b) For $k=5$, you are allowed to use seven comparisons. This is difficult. Merge-sort does not do the job, as it uses up to eight comparisons. (c) For $k \in \{6,7,8\}$, use the case $k=5$ as a subroutine.

5.6 Quicksort

Quicksort is a divide-and-conquer algorithm that is, in a certain sense, complementary to the mergesort algorithm of Sect. 5.3. Quicksort does all the difficult work *before* the recursive calls. The idea is to distribute the input elements into two or more sequences so that the corresponding key ranges do not overlap. Then, it suffices to sort the shorter sequences recursively and concatenate the results. To make the duality to mergesort complete, we would like to split the input into two sequences of equal size. Unfortunately, this is a nontrivial task. However, we can come close by picking a random splitter element. The splitter element is usually called the *pivot*. Let p denote the pivot element chosen. Elements are classified into three sequences of elements that are smaller than, equal to, and larger than the pivot. Figure 5.7 gives a high-level realization of this idea, and Fig. 5.8 depicts a sample execution. Quicksort has an expected execution time of $O(n \log n)$, as we shall show in Sect. 5.6.1. In Sect. 5.6.2, we discuss refinements that have made quicksort the most widely used sorting algorithm in practice.

Function *quickSort(s : Sequence* **of** *Element*) : *Sequence* **of** *Element*
    if $|s| \leq 1$ **then return** s *// base case*
    pick $p \in s$ uniformly at random   *// pivot key*
    $a := \langle e \in s : e < p \rangle$
    $b := \langle e \in s : e = p \rangle$
    $c := \langle e \in s : e > p \rangle$
    return *concatenation of quickSort(a), b, and quickSort(c)*

Fig. 5.7. High-level formulation of quicksort for lists

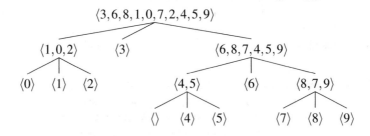

Fig. 5.8. Execution of *quickSort* (Fig. 5.7) on $\langle 3,6,8,1,0,7,2,4,5,9 \rangle$ using the first element of a subsequence as the pivot. The first call of quicksort uses 3 as the pivot and generates the subproblems $\langle 1,0,2 \rangle$, $\langle 3 \rangle$, and $\langle 6,8,7,4,5,9 \rangle$. The recursive call for the third subproblem uses 6 as a pivot and generates the subproblems $\langle 4,5 \rangle$, $\langle 6 \rangle$, and $\langle 8,7,9 \rangle$.

5.6.1 Analysis

To analyze the running time of quicksort for an input sequence $s = \langle e_1,\dots,e_n\rangle$, we focus on the number of element comparisons performed. We allow *three-way* comparisons here, with possible outcomes "smaller", "equal", and "larger". Other operations contribute only constant factors and small additive terms to the execution time.

Let $C(n)$ denote the worst-case number of comparisons needed for any input sequence of size n and any choice of pivots. The worst-case performance is easily determined. The subsequences a, b, and c in Fig. 5.7 are formed by comparing the pivot with all other elements. This requires $n-1$ comparisons. Let k denote the number of elements smaller than the pivot and let k' denote the number of elements larger than the pivot. We obtain the following recurrence relation: $C(0) = C(1) = 0$ and

$$C(n) \leq n - 1 + \max\left\{C(k) + C(k') : 0 \leq k \leq n-1, 0 \leq k' < n-k\right\}.$$

It is easy to verify by induction that

$$C(n) \leq \frac{n(n-1)}{2} = \Theta\left(n^2\right).$$

This worst case occurs if all elements are different and we always pick the largest or smallest element as the pivot.

The expected performance is much better. We first give a plausibility argument for an $O(n\log n)$ bound and then show a bound of $2n\ln n$. We concentrate on the case where all elements are different. Other cases are easier because a pivot that occurs several times results in a larger middle sequence b that need not be processed any further. Consider a fixed element e_i, and let X_i denote the total number of times e_i is compared with a pivot element. Then $\sum_i X_i$ is the total number of comparisons. Whenever e_i is compared with a pivot element, it ends up in a smaller subproblem. Therefore, $X_i \leq n-1$, and we have another proof of the quadratic upper bound. Let us call a comparison "good" for e_i if e_i moves to a subproblem of at most three-quarters the size. Any e_i can be involved in at most $\log_{4/3} n$ good comparisons. Also, the probability that a pivot which is good for e_i is chosen is at least $1/2$; this holds because a bad pivot must belong to either the smallest or the largest quarter of the elements. So $E[X_i] \leq 2\log_{4/3} n$, and hence $E[\sum_i X_i] = O(n\log n)$. We shall next prove a better bound by a completely different argument.

Theorem 5.6. *The expected number of comparisons performed by quicksort is*

$$\bar{C}(n) \leq 2n\ln n \leq 1.39n\log n.$$

Proof. Let $s' = \langle e_1',\dots,e_n'\rangle$ denote the elements of the input sequence in sorted order. Every comparison involves a pivot element. If an element is compared with a pivot, the pivot and the element end up in different subsequences. Hence any pair of elements is compared at most once, and we can therefore count comparisons by looking at the indicator random variables X_{ij}, $i < j$, where $X_{ij} = 1$ if e'_i and e'_j are compared and $X_{ij} = 0$ otherwise. We obtain

$$\bar{C}(n) = E\left[\sum_{i=1}^{n}\sum_{j=i+1}^{n} X_{ij}\right] = \sum_{i=1}^{n}\sum_{j=i+1}^{n} E[X_{ij}] = \sum_{i=1}^{n}\sum_{j=i+1}^{n} \mathrm{prob}(X_{ij} = 1).$$

The middle transformation follows from the linearity of expectations (A.3). The last equation uses the definition of the expectation of an indicator random variable $E[X_{ij}] = \mathrm{prob}(X_{ij} = 1)$. Before we can simplify further the expression for $\bar{C}(n)$, we need to determine the probability of X_{ij} being 1.

Lemma 5.7. *For any $i < j$,* $\mathrm{prob}(X_{ij} = 1) = \dfrac{2}{j-i+1}$.

Proof. Consider the $j-i+1$-element set $M = \{e'_i, \ldots, e'_j\}$. As long as no pivot from M is selected, e'_i and e'_j are not compared, but all elements from M are passed to the same recursive calls. Eventually, a pivot p from M is selected. Each element in M has the same chance $1/|M|$ of being selected. If $p = e'_i$ or $p = e'_j$, we have $X_{ij} = 1$. Otherwise, e'_i and e'_j are passed to different recursive calls, so that they will never be compared. Thus $\mathrm{prob}(X_{ij} = 1) = 2/|M| = 2/(j-i+1)$. □

We can now complete the proof of Theorem 5.6 by a relatively simple calculation:

$$\bar{C}(n) = \sum_{i=1}^{n}\sum_{j=i+1}^{n} \mathrm{prob}(X_{ij} = 1) = \sum_{i=1}^{n}\sum_{j=i+1}^{n} \frac{2}{j-i+1} = \sum_{i=1}^{n}\sum_{k=2}^{n-i+1} \frac{2}{k}$$

$$\leq \sum_{i=1}^{n}\sum_{k=2}^{n} \frac{2}{k} = 2n\sum_{k=2}^{n} \frac{1}{k} = 2n(H_n - 1) \leq 2n(1+\ln n - 1) = 2n\ln n.$$

For the last three steps, recall the properties of the nth harmonic number $H_n := \sum_{k=1}^{n} 1/k \leq 1+\ln n$ (A.13). □

Note that the calculations in Sect. 2.11 for left-to-right maxima were very similar, although we had quite a different problem at hand.

5.6.2 *Refinements

We shall now discuss refinements of the basic quicksort algorithm. The resulting algorithm, called *qSort*, works in-place, and is fast and space-efficient. Figure 5.9 shows the pseudocode, and Fig. 5.10 shows a sample execution. The refinements are nontrivial and we need to discuss them carefully.

The function *qSort* operates on an array a. The arguments ℓ and r specify the sub-array to be sorted. The outermost call is *qSort*$(a, 1, n)$. If the size of the subproblem is smaller than some constant n_0, we resort to a simple algorithm[5] such as the insertion

---

[5] Some authors propose leaving small pieces unsorted and cleaning up at the end using a single insertion sort that will be fast, according to Exercise 5.10. Although this nice trick reduces the number of instructions executed, the solution shown is faster on modern machines because the subarray to be sorted will already be in cache.

Procedure $qSort(a : Array$ **of** $Element; \ell, r : \mathbb{N})$ // Sort the subarray $a[\ell..r]$
    while $r - \ell + 1 > n_0$ **do**   // Use divide-and-conquer.
        $j := pickPivotPos(a, \ell, r)$ // Pick a pivot element and
        $swap(a[\ell], a[j])$ // bring it to the first position.
        $p := a[\ell]$ // p is the pivot now.
        $i := \ell; \ j := r$
        repeat   // $a:$ [$\ell \quad i \to \leftarrow j \quad r$]
            while $a[i] < p$ **do** $i{+}{+}$ // Skip over elements
            while $a[j] > p$ **do** $j{-}{-}$ // already in the correct subarray.
            if $i \leq j$ **then**   // If partitioning is not yet complete,
                $swap(a[i], a[j]); i{+}{+}; j{-}{-}$ // (*) swap misplaced elements and go on.
        until $i > j$ // Partitioning is complete.
        if $i < (\ell + r)/2$ **then** $qSort(a, \ell, j); \ \ell := i$ // Recurse on
        else $qSort(a, i, r); \ r := j$ // smaller subproblem.
    endwhile
    $insertionSort(a[\ell..r])$ // Faster for small $r - \ell$

Fig. 5.9. Refined quicksort for arrays

```
i →           ← j              3 6 8 1 0 7 2 4 5 9
3 6 8 1 0 7 2 4 5 9            2 0 1|8 6 7 3 4 5 9
2 6 8 1 0 7 3 4 5 9            1 0|2 5 6 7 3 4|8 9
2 0 8 1 6 7 3 4 5 9            0 1  |4 3|7 6 5|8 9
2 0 1 8 6 7 3 4 5 9                 3 4|5 6|7
     j i                              |5 6|
```

Fig. 5.10. Execution of $qSort$ (Fig. 5.9) on $\langle 3, 6, 8, 1, 0, 7, 2, 4, 5, 9 \rangle$ using the first element as the pivot and $n_0 = 1$. The *left-hand side* illustrates the first partitioning step, showing elements in **bold** that have just been swapped. The *right-hand side* shows the result of the recursive partitioning operations.

sort shown in Fig. 5.1. The best choice for n_0 depends on many details of the machine and compiler and needs to be determined experimentally; a value somewhere between 10 and 40 should work fine under a variety of conditions.

The pivot element is chosen by a function *pickPivotPos* that we shall not specify further. The correctness does not depend on the choice of the pivot, but the efficiency does. Possible choices are the first element; a random element; the median ("middle") element of the first, middle, and last elements; and the median of a random sample consisting of k elements, where k is either a small constant, say 3, or a number depending on the problem size, say $\lceil \sqrt{r - \ell + 1} \rceil$. The first choice requires the least amount of work, but gives little control over the size of the subproblems; the last choice requires a nontrivial but still sublinear amount of work, but yields balanced subproblems with high probability. After selecting the pivot p, we swap it into the first position of the subarray (= position ℓ of the full array).

The repeat–until loop partitions the subarray into two proper (smaller) subarrays. It maintains two indices i and j. Initially, i is at the left end of the subarray and j is at the right end; i scans to the right, and j scans to the left. After termination of the loop, we have $i = j + 1$ or $i = j + 2$, all elements in the subarray $a[\ell..j]$ are no larger than p, all elements in the subarray $a[i..r]$ are no smaller than p, each subarray is a proper subarray, and, if $i = j + 2$, $a[j + 1]$ is equal to p. So, recursive calls $qSort(a, \ell, j)$ and $qSort(a, i, r)$ will complete the sort. We make these recursive calls in a nonstandard fashion; this is discussed below.

Let us see in more detail how the partitioning loops work. In the first iteration of the repeat loop, i does not advance at all but remains at ℓ, and j moves left to the rightmost element no larger than p. So, j ends at ℓ or at a larger value; generally, the latter is the case. In either case, we have $i \le j$. We swap $a[i]$ and $a[j]$, increment i, and decrement j. In order to describe the total effect more generally, we distinguish cases.

If p is the unique smallest element of the subarray, j moves all the way to ℓ, the swap has no effect, and $j = \ell - 1$ and $i = \ell + 1$ after the increment and decrement. We have an empty subproblem $a[\ell..\ell - 1]$ and a subproblem $a[\ell + 1..r]$. Partitioning is complete, and both subproblems are proper subproblems.

If j moves down to $i + 1$, we swap, increment i to $\ell + 1$, and decrement j to ℓ. Partitioning is complete, and we have the subproblems $a[\ell..\ell]$ and $a[\ell + 1..r]$. Both subarrays are proper subarrays.

If j stops at an index larger than $i + 1$, we have $\ell < i \le j < r$ after executing the line marked (*) in Fig. 5.9. Also, all elements to the left of i are at most p *(and there is at least one such element)*, and all elements to the right of j are at least p *(and there is at least one such element)*. Since the scan loop for i skips only over elements smaller than p and the scan loop for j skips only over elements larger than p, further iterations of the repeat loop maintain this invariant. Also, all further scan loops are guaranteed to terminate by the italicized claims above and so there is no need for an index-out-of-bounds check in the scan loops. In other words, the scan loops are as concise as possible; they consist of a test and an increment or decrement.

Let us next study how the repeat loop terminates. If we have $i \le j + 2$ after the scan loops, we have $i \le j$ in the termination test. Hence, we continue the loop. If we have $i = j - 1$ after the scan loops, we swap, increment i, and decrement j. So $i = j + 1$, and the repeat loop terminates with the proper subproblems $a[\ell..j]$ and $a[i..r]$. The case $i = j$ after the scan loops can occur only if $a[i] = p$. In this case, the swap has no effect. After incrementing i and decrementing j, we have $i = j + 2$, resulting in the proper subproblems $a[\ell..j]$ and $a[j + 2..r]$, separated by one occurrence of p. Finally, when $i > j$ after the scan loops, then either i goes beyond j in the first scan loop or j goes below i in the second scan loop. By our invariant, i must stop at $j + 1$ in the first case, and then j does not move in its scan loop or j must stop at $i - 1$ in the second case. In either case, we have $i = j + 1$ after the scan loops. The line marked (*) is not executed, so we have subproblems $a[\ell..j]$ and $a[i..r]$, and both subproblems are proper.

We have now shown that the partitioning step is correct, terminates, and generates proper subproblems.

Exercise 5.29. Does the algorithm stay correct if the scan loops skip over elements equal to p? Does it stay correct if the algorithm is run only on inputs for which all elements are pairwise distinct?

The refined quicksort handles recursion in a seemingly strange way. Recall that we need to make the recursive calls $qSort(a,\ell,j)$ and $qSort(a,i,r)$. We may make these calls in either order. We exploit this flexibility by making the call for the smaller subproblem first. The call for the larger subproblem would then be the last thing done in $qSort$. This situation is known as *tail recursion* in the programming-language literature. Tail recursion can be eliminated by setting the parameters (ℓ and r) to the right values and jumping to the first line of the procedure. This is precisely what the while-loop does. Why is this manipulation useful? Because it guarantees that the size of the recursion stack stays logarithmically bounded; the precise bound is $\lceil \log(n/n_0) \rceil$. This follows from the fact that in a call for $a[\ell..r]$, we make a single recursive call for a subproblem which has size at most $(r-\ell+1)/2$.

Exercise 5.30. What is the maximal depth of the recursion stack without the "smaller subproblem first" strategy? Give a worst-case example.

*Exercise 5.31 (sorting strings using multikey quicksort [43]). Let s be a sequence of n strings. We assume that each string ends in a special character that is different from all "normal" characters. Show that the function $mkqSort(s,1)$ below sorts a sequence s consisting of *different* strings. What goes wrong if s contains equal strings? Solve this problem. Show that the expected execution time of $mkqSort$ is $O(N+n\log n)$ if $N=\sum_{e\in s}|e|$.

> **Function** $mkqSort(s : Sequence$ **of** $String, i : \mathbb{N}) : Sequence$ **of** $String$
>   **assert** $\forall e,e' \in s : e[1..i-1] = e'[1..i-1]$
>   **if** $|s| \leq 1$ **then return** s // base case
>   pick $p \in s$ uniformly at random   // pivot character
>   **return** concatenation of $mkqSort(\langle e \in s : e[i] < p[i] \rangle, i)$,
> $mkqSort(\langle e \in s : e[i] = p[i] \rangle, i+1)$, *and*
> $mkqSort(\langle e \in s : e[i] > p[i] \rangle, i)$

Exercise 5.32. Implement several different versions of $qSort$ in your favorite programming language. Use and do not use the refinements discussed in this section, and study the effect on running time and space consumption.

*Exercise 5.33 (Strictly inplace quicksort). Develop a version of quicksort that requires only constant additional memory. Hint: Develop a nonrecursive algorithm where the subproblems are marked by storing their largest element at their first array entry.

5.7 Parallel Quicksort

Analogously to parallel mergesort, there is a trivial parallelization of quicksort that performs only the recursive calls in parallel. We strive for a more scalable solution that also parallelizes partitioning. In principle, parallel partitioning is also easy: Each PE is assigned an equal share of the array to be partitioned and partitions it. The partitioned pieces have to be reassembled into sequences. Compared with mergesort, parallel partitioning is simpler than parallel merging. However, since the pivots we choose will not split the input perfectly into equal pieces, we face a load-balancing problem: Which processors should work on which recursive subproblem? Overall, we get an interesting kind of parallel algorithm that combines data parallelism with task parallelism. We first explain this in the distributed-memory setting and then outline a shared-memory solution that works almost in-place.

Exercise 5.34. Adapt Algorithm 5.7 to become a task-parallel algorithm with work $O(n \log n)$ and span $O(\log^2 n)$.

5.7.1 Distributed-Memory Quicksort

Figure 5.11 gives high-level pseudocode for distributed-memory parallel quicksort. Figure 5.12 gives an example. In the procedure *parQuickSort*, every PE has a local array s of elements. The PEs cooperate in groups and together sort the union of their arrays. Each group is an interval $i..j$ of PEs. Initially $i = 1$, $j = p$, and each processor has an about equal share of the input, say PEs $1..j$ have $\lceil n/p \rceil$ elements and PEs $j + 1..p$ have $\lfloor n/p \rfloor$ elements, where $j = p \cdot (n/p - \lfloor n/p \rfloor)$. The recursion bottoms out when there is a single processor in the group, i.e., $i = j$. The PE completes the sort by calling sequential quicksort for its piece of the input. When further partitioning is needed, the PEs have to agree on a common pivot. The choice of pivot has a significant influence on the load balance and is even more crucial than for sequential quicksort. For now, we shall only explain how to select a random pivot; we shall discuss alternatives at the end of the section. The group $i..j$ of PEs needs to select a random element from the union of their local arrays. This can be implemented

---

Function *parQuickSort*(s : *Sequence* **of** *Element*, i, j : $\mathbb{N}$) : *Sequence* **of** *Element*
    $p' := j - i + 1$ // # of PEs working together
    if $i = j$ **then** *quickSort*(s)   ; **return** s // sort locally
    $v := pickPivot(s, i, j)$
    $a := \langle e \in s : e \le v \rangle;$ $b := \langle e \in s : e > v \rangle$ // partition
    $n_a := \sum_{i \le k \le j} |a| @ k;$ $n_b := \sum_{i \le k \le j} |b| @ k$ // all-reduce in segment $i..j$ *1*
    $k' := \frac{n_a}{n_a + n_b} p'$ // fractional number of PEs responsible for a *2*
    choose $k \in \{\lfloor k' \rfloor, \lceil k' \rceil\}$ such that $\max \left\{ \lceil \frac{n_a}{k} \rceil, \lceil \frac{n_b}{p'-k} \rceil \right\}$ is minimized   *3*
    send the a's to PEs $i..i+k-1$ such that no PE receives more than $\lceil \frac{n_a}{k} \rceil$ of them
    send the b's to PEs $i+k..j$ such that no PE receives more than $\lceil \frac{n_b}{p'-k} \rceil$ of them
    receive data sent to PE i_{proc} into s
    if $i_{\text{proc}} < i + k$ **then** *parQuickSort*($s, i, i+k-1$) **else** *parQuickSort*($s, i+k, j$)

Fig. 5.11. SPMD pseudocode for parallel quicksort. Each PE has a local array s. The group $i..j$ of PEs work together to sort the union of their local arrays.

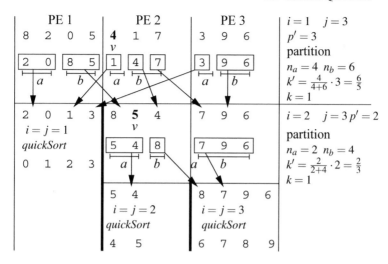

Fig. 5.12. Example of distributed-memory parallel quicksort.

efficiently using prefix sums as follows: We compute the prefix sum over the local values of $|s|$, i.e., PE $\ell \in i..j$ obtains the number $S@\ell := \sum_{i \le k \le \ell} |s|@k$ of elements stored in PEs $i..\ell$. Moreover all PEs in $i..j$ need the total size $S@j$; see Sect. 13.3 for the realization of prefix sums. Now we pick a random number $x \in 1..S@j$. This can be done without communication if we assume that we have a replicated pseudorandom number generator, i.e., a generator that computes the same number on all participating PEs. The PE where $x \in S - |s| + 1..S$ picks $s[x - (S - |s|)]$ as the pivot and broadcasts it to all PEs in the group.

In practice, an even simpler algorithm can be used that approximates random sampling if all PEs hold a similar number of elements. We fist pick a random PE index ℓ using a replicated random number generator. Then PE ℓ broadcasts a random element of $s@\ell$. Note that the only nonlocal operation here is a single broadcast; see Sect. 13.1.

Once each PE knows the pivot, local partitioning is easy. Each PE splits its local array into the sequence a of elements no larger than the pivot and the sequence b of elements larger than the pivot. We next need to set up the two subproblems. We split the range of PEs $i..j$ into subranges $i..i+k-1$ and $i+k..j$ such that the left subrange sorts all the a's and the right subrange sorts the b's. A crucial decision is how to choose the number k of PEs dedicated to the a's. We do this so as to minimize load imbalance. The load balance would be perfect if we could split PEs into fractional pieces. This calculation is done in lines 1 and 2. Then line 3 rounds this so that load imbalance is minimized.

Now the data has to be redistributed accordingly. We explain the redistribution for a. For b, this can be done in an analogous fashion. A similar redistribution procedure is explained as a general load-balancing principle in Sect. 14.2. Conceptually, we assign global numbers to the array elements – element $a[x]$ of PE i_{proc} gets a number

$\sum_{\ell < i_{\text{proc}}} |a| @\ell + x$. Note that this can be done using yet another prefix sum calculation. Let $L = \lceil n_a/k \rceil$ denote the maximum number of elements we want to send to one PE. We send the element with global number y to PE $i + \lfloor (y-1)/L \rfloor$. This way, each PE gets at most L elements and the receiving PEs range from $i + \lfloor (1-1)/L \rfloor = i$ to $i + \lfloor (n_a - 1)/L \rfloor \leq i + k - 1$. Since the elements of a have consecutive numbers, they are sent to at most $\lceil |a|/L \rceil + 1$ PEs with consecutive PE numbers. In other words, a is split into up to $\lceil |a|/L \rceil + 1$ pieces of consecutive elements. Each piece can be sent as a single message.

Exercise 5.35. Give detailed pseudocode for a procedure actually doing the message exchange.

We shall not give a detailed analysis of parallel quicksort but restrict ourselves to an outline. The first analysis of the expected performance of sequential quicksort given in Sect. 5.6.1 can be generalized to show that with high probability the depth of the parallel recursion is $O(\log p)$ – a longer recursion branch would require a sequence of bad pivots that is very unlikely.

Exercise 5.36. Give a formal proof similar to the one of Lemma 5.13.

A single level of recursion takes time

$$O\left(\max_{k \in 1..p} |s| @k + \log p \right),$$

where the logarithmic term stems from the collective broadcast, reduction, and prefix-sum operations needed to coordinate the PEs. Summing over all levels of recursion, we get a term $O(\log^2 p)$ for the collective communication operations. If all PEs always had the same number of elements $|s| = \frac{n}{p}$, the remaining work would be $O(\frac{n}{p} \log p)$ for the recursion and $O(\frac{n}{p} \log \frac{n}{p})$ for the base-case sequential sorting. Overall, we would get time

$$O(\frac{n}{p} \log n + \log^2 p).$$

This includes $O(\log^2 p)$ message startup overhead, $O(\frac{n}{p} \log p)$ communication volume, and $O(\frac{n}{p} \log n)$ element comparisons. Hence, we can hope for similar performance as for parallel mergesort if we can bound the load imbalance.

So, let us have a closer look at load balancing. Assuming perfect load balance for the input, the good news is that load imbalance stems only from rounding effects that "should" be small. However, the bad news is that these rounding errors have to be multiplied in each level of recursion. How bad can the rounding errors get? The worst that can happen is that one of the recursive subproblems gets (almost) one PE load's worth of elements more than the other one. Since we always round in an optimal way, we can assume that this additional load is allocated to the *larger* subproblem – of course, it can also happen that the smaller subproblem gets more elements per PE, but only if this results in a smaller imbalance. To make the analysis simple, we shall analyze a modified algorithm that only uses "good" pivots where the smaller subproblem has size at least $|s|/4$. The worst case is then that we have $k = \log_{4/3} p$ levels of recursion and an imbalance factor bounded by

$$\prod_{i=1}^{k}\left(1+\frac{1}{p\,(3/4)^i}\right) = e^{\sum_{i=1}^{k}\ln\left(1+\frac{1}{p(3/4)^i}\right)} \qquad \text{Estimate A.18}$$

$$\leq e^{\sum_{i=0}^{k}\frac{1}{p(3/4)^i}} = e^{\frac{1}{p}\sum_{i=0}^{k}(4/3)^i} \qquad \text{Equation A.14}$$

$$= e^{\frac{1}{p}\frac{(4/3)^{k+1}-1}{4/3-1}} \leq e^{\frac{1}{p}4\overbrace{(4/3)^k}^{=p}} = e^4 \approx 54.6.$$

The good news is that this is a constant, i.e., our algorithm achieves constant efficiency. The bad news is that e^4 is a rather large constant, and even a more detailed analysis will not get an imbalance factor close to one. However, we can refine the algorithm to get a better load balance. A key observation is that $\prod_{i=1}^{k'}\left(1+1/(p(3/4)^i)\right)$ is close to one if $(4/3)^{k'} = o(p)$. For example, once $j - i \leq \log p$, we could switch to another algorithm with better load balance. For example, we can choose the pivot carefully based on a large sample. Or, we could switch to the sample sort algorithm described in Sect. 5.13. This hybrid algorithm combines the high scalability of pure quicksort with the good load balance of pure sample sort. Another interesting approach is JanusSort [24] that actually splits the PEs fractionally and thus achieves perfect load balance. This is possible by spawning an additional thread on PEs that are fractionally assigned to two subproblems.

5.7.2 *In-Place Shared-Memory Quicksort

A major reason for the popularity of sequential quicksort is its small memory footprint. Besides the space for the input array, it only requires space for the recursion stack. The depth of the recusion stack can be kept logarithmic in the size of the input if the smaller subproblem is always solved first. Is there also a parallel quicksort which is basically in-place? Tsigas and Zhang [317] described such an algorithm whose innermost loop is similar to sequential quicksort. Suppose we want to use p processors to partition an array. We logically split the input array into blocks of size B and keep two global counters ℓ and r, with $\ell \leq r$. The blocks with indices $[\ell + 1..r - 1]$ are untouched. In the innermost loop, each PE works on two blocks L and R, where the index of L is at most ℓ and the index of R is at least r. As in sequential array-based partitioning (Sect. 5.6.2), the PE scans L from left to right and R from right to left, exchanging small elements of L with large elements of R. When the right end of block L is reached, L is "clean" – all its elements are small. Block L is set aside and the PE chooses the block with index $\ell + 1$ as its new block. To this end, the PE increments ℓ atomically and, at the same time, makes sure that $\ell \leq r$. A single CAS instruction suffices provided it can access both counters at the same time.[6] Similarly, a new block from the right is acquired by atomically decrementing

---

[6] On machines providing only CAS on a single machine word, this can be achieved by making the block size sufficiently large, so that two block counters fit into one machine word.

r. The initial values of ℓ and r are 1 and $\lceil |s|/p \rceil$. Once $\ell = r$, no further blocks remain and the *parallel* partitioning step terminates. It is followed by a cleanup phase. Note that for each PE, there are up to two blocks that are not yet clean. These are cleaned using a sequential algorithm.

It is instructive to analyze the scalability of this partitioning algorithm. First of all, we need $B = \Omega(p)$, since there would otherwise be too much contention for updating the counters ℓ and r. The sequential cleaning step looks at $\Theta(p)$ blocks and hence needs time $\Omega(p^2)$. Apparently, we pay a high price for the in-place property – our noninplace algorithm in Sect. 5.7.1 has a span of only $O(\log^2 p)$.

**Exercise 5.37. (Research problem) Design a practical in-place parallel sorting algorithm with polylogarithmic span. Hints: One possibility is to improve the Tsigas–Zhang algorithm by using a smaller block size, a relaxed data structure for assigning blocks (see also Sect. 3.7.2), and a parallel cleanup algorithm. Another possibility is to make the algorithm in Sect. 5.7.1 in-place – partition locally and then permute the data such that we obtain a global partition.

5.8 Selection

Selection refers to a class of problems that are easily reduced to sorting but do not require the full power of sorting. Let $s = \langle e_1, \ldots, e_n \rangle$ be a sequence and call its sorted version $s' = \langle e'_1, \ldots, e'_n \rangle$. Selection of the smallest element amounts to determining e'_1, selection of the largest amounts to determining e'_n, and selection of the kth smallest amounts to determining e'_k. Selection of the median[7] refers to determining $e'_{\lceil n/2 \rceil}$. Selection of the median and also of quartiles[8] is a basic problem in statistics. It is easy to determine the smallest element or the smallest and the largest element by a single scan of a sequence in linear time. We now show that the kth smallest element can also be determined in linear time. The simple recursive procedure shown in Fig. 5.13 solves the problem.

This procedure is akin to quicksort and is therefore called *quickselect*. The key insight is that it suffices to follow one of the recursive calls. As before, a pivot is chosen, and the input sequence s is partitioned into subsequences a, b, and c containing the elements smaller than the pivot, equal to the pivot, and larger than the pivot, respectively. If $|a| \geq k$, we recurse on a, and if $k > |a| + |b|$, we recurse on c with a suitably adjusted k. If $|a| < k \leq |a| + |b|$, the task is solved: The pivot has rank k and we return it. Observe that the latter case also covers the situation $|s| = k = 1$, and hence no special base case is needed. Figure 5.14 illustrates the execution of quickselect.

---

[7] The standard definition of the median of an even number of elements is the average of the two middle elements. Since we do not want to restrict ourselves to the situation where the inputs are numbers, we have chosen a slightly different definition. If the inputs are numbers, the algorithm discussed in this section is easily modified to compute the average of the two middle elements.

[8] The elements with ranks $\lceil \alpha n \rceil$, where $\alpha \in \{1/4, 1/2, 3/4\}$.

As with quicksort, the worst-case execution time of quickselect is quadratic. But the expected execution time is linear and hence a logarithmic factor faster than quicksort.

Theorem 5.8. *Algorithm quickselect runs in expected time* $O(n)$ *on an input of size n.*

Proof. We give an analysis that is simple and shows a linear expected execution time. It does not give the smallest constant possible. Let $T(n)$ denote the maximum expected execution time of quickselect on any input of size at most n. Then $T(n)$ is a nondecreasing function of n. We call a pivot *good* if neither $|a|$ nor $|c|$ is larger than $2n/3$. Let γ denote the probability that a pivot is good. Then $\gamma \geq 1/3$, since each element in the middle third of the sorted version $s' = \langle e'_1, \ldots, e'_n \rangle$ is good. We now make the conservative assumption that the problem size in the recursive call is reduced only for good pivots and that, even then, it is reduced only by a factor of $2/3$, i.e., reduced to $\lfloor 2n/3 \rfloor$. For bad pivots, the problem size stays at n. Since the work outside the recursive call is linear in n, there is an appropriate constant c such that

$$T(n) \leq cn + \gamma T\left(\left\lfloor \frac{2n}{3} \right\rfloor\right) + (1-\gamma)T(n).$$

Solving for $T(n)$ yields

$$T(n) \leq \frac{cn}{\gamma} + T\left(\left\lfloor \frac{2n}{3} \right\rfloor\right) \leq 3cn + T\left(\left\lfloor \frac{2n}{3} \right\rfloor\right) \leq 3c\left(n + \frac{2n}{3} + \frac{4n}{9} + \ldots\right)$$

$$\leq 3cn \sum_{i \geq 0} \left(\frac{2}{3}\right)^i \leq 3cn \frac{1}{1-2/3} = 9cn. \qquad \square$$

```
// Find an element with rank k
Function select(s : Sequence of Element; k : ℕ) : Element
    assert |s| ≥ k
    pick p ∈ s uniformly at random                          // pivot key
    a := ⟨e ∈ s : e < p⟩
    if |a| ≥ k then return select(a,k)                      //
    b := ⟨e ∈ s : e = p⟩
    if |a| + |b| ≥ k then return p                          //
    c := ⟨e ∈ s : e > p⟩
    return select(c, k − |a| − |b|)                         //
```

Fig. 5.13. Quickselect

s	k	p	a	b	c
$\langle 3,1,4,5,9,\mathbf{2},6,5,3,5,8 \rangle$	6	2	$\langle 1 \rangle$	$\langle 2 \rangle$	$\langle 3,4,5,9,6,5,3,5,8 \rangle$
$\langle 3,4,5,9,\mathbf{6},5,3,5,8 \rangle$	4	6	$\langle 3,4,5,5,3,4 \rangle$	$\langle 6 \rangle$	$\langle 9,8 \rangle$
$\langle 3,4,\mathbf{5},5,3,5 \rangle$	4	5	$\langle 3,4,3 \rangle$	$\langle 5,5,5 \rangle$	$\langle \rangle$

Fig. 5.14. The execution of $select(\langle 3,1,4,5,9,2,6,5,3,5,8,6 \rangle, 6)$. The middle element (**bold**) of the current sequence s is used as the pivot p.

Exercise 5.38. Modify quickselect so that it returns the k smallest elements.

Exercise 5.39. Give a selection algorithm that permutes an array in such a way that the k smallest elements are in entries $a[1], \dots, a[k]$. No further ordering is required except that $a[k]$ should have rank k. Adapt the implementation tricks used in the array-based quicksort to obtain a nonrecursive algorithm with fast inner loops.

Exercise 5.40 (streaming selection). A data stream is a sequence of elements presented one by one.

(a) Develop an algorithm that finds the kth smallest element of a sequence that is presented to you one element at a time in an order you cannot control. You have only space $O(k)$ available. This models a situation where voluminous data arrives over a network at a compute node with limited storage capacity.
(b) Refine your algorithm so that it achieves a running time $O(n \log k)$. You may want to read some of Chap. 6 first.
*(c) Refine the algorithm and its analysis further so that your algorithm runs in average-case time $O(n)$ if $k = O(n/\log n)$. Here, "average" means that all orders of the elements in the input sequence are equally likely.

5.9 Parallel Selection

Essentially, our selection algorithm in Fig. 5.13 is already a parallel algorithm. We can perform the partitioning into a, b, and c in parallel using time $O(n/p)$. Determining $|a|$, $|b|$, and $|c|$ can be done using a reduction operation in time $O(\log p)$. Note that all PEs recurse on the same subproblem so that we do not have the load-balancing issues we encountered with parallel quicksort. We get an overall expected parallel execution time of $O(\frac{n}{p} + \log p \log n) = O(\frac{n}{p} + \log^2 p)$. The simplification of the asymptotic complexity can be seen from a simple case distinction. If $n = O(p \log^2 p)$, then $\log n = O(\log p)$. Otherwise, the term n/p dominates the term $\log n \log p$.

For parallel selection on a distributed-memory machine, an interesting issue is the communication volume involved. One approach is to redistribute the data evenly before a recursive call, using an approach similar to distributed-memory parallel quicksort. We then get an overall communication volume $O(n/p)$ per PE, i.e., essentially all the data is moved.

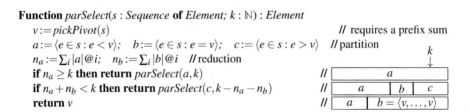

Function *parSelect*(s : *Sequence* **of** *Element*; k : $\mathbb{N}$) : *Element*
 $v := pickPivot(s)$ // requires a prefix sum
 $a := \langle e \in s : e < v \rangle$; $b := \langle e \in s : e = v \rangle$; $c := \langle e \in s : e > v \rangle$ // partition
 $n_a := \sum_i |a|@i$; $n_b := \sum_i |b|@i$ // reduction
 if $n_a \geq k$ **then return** *parSelect*(a, k)   //
 if $n_a + n_b < k$ **then return** *parSelect*($c, k - n_a - n_b$) //
 return v //

Fig. 5.15. SPMD pseudocode for communication-efficient parallel selection

From the point of view of optimizing communication volume, we can do much better by always keeping the data where it is. We get the simple algorithm outlined in Fig. 5.15. However, in the worst case, the elements with ranks near k are all at the same PE. On that PE, the size of s will only start to shrink after $\Omega(\log p)$ levels of recursion. Hence, we get a parallel execution time of $\Omega(\frac{n}{p}\log p + \log p \log n)$, which is not efficient.

5.9.1 *Using Two Pivots

The $\mathrm{O}(\log p \log n)$ term in the running time of the parallel selection algorithm stems from the fact that the recursion depth is $\mathrm{O}(\log n)$ since the expected problem size is reduced by a constant factor in each level of the recursion and that time $\mathrm{O}(\log p)$ time is needed in each level for the reduction operation. We shall now look at an algorithm that manages to shrink the problem size by a factor $f := \Theta(p^{1/3})$ in each level of the recursion and reduces the running time to $\mathrm{O}(n/p + \log p)$. Floyd and Rivest [107] (see also [158, 271]) had the idea of choosing two pivots ℓ and r where, with high probability, ℓ is a slight underestimate of the sought element with rank k and where r is a slight overestimate. Figure 5.16 outlines the algorithm and Fig. 5.17 gives an example.

Function *parSelect2*(s : *Sequence* **of** *Element*; k : $\mathbb{N}$) : *Element*
  if $\sum_i |s@i| < n/p$ **then** // small total remaining input size? (reduction)
    gather all data on a single PE
    solve the problem sequentially there
  else
    $(\ell, r) := pickPivots(s)$ // requires a prefix sum
    $a := \langle e \in s : e < \ell \rangle$; $b := \langle e \in s : \ell \le e \le r \rangle$; $c := \langle e \in s : e > r \rangle$ //partition
    $n_a := \sum_i |a|@i$; $n_b := \sum_i |b|@i$ //reduction
    if $n_a \ge k$ **then return** $parSelect2(a, k)$ //
    if $n_a + n_b < k$ **then return** $parSelect2(c, k - n_a - n_b)$ //
    return $parSelect2(b, k - n_a)$ //

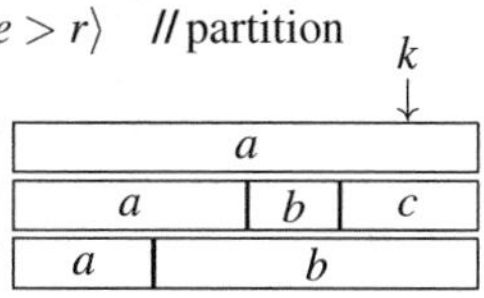

Fig. 5.16. Efficient parallel selection with two splitters

PE 1	PE 2	PE 3	
8 2 0 5	4 1 7 **6**	**3** 9 6	$k = 5$
	r	l	partition
2 0 5 8	1 4 6 7	3 6 9	$n_a = 3 < 5$
			$n_b = 5$
a b c	a b c	b c	$n_a + n_b = 9 \ge 5$
5	4 6	3 6	$k = 2$

Fig. 5.17. Selecting the median of $\langle 8, 2, 0, 5, 4, 1, 7, 6, 3, 9, 6 \rangle$ using distributed-memory parallel selection with two pivots. The figure shows the first level of recursion using three PEs.

The improved algorithm is similar to the single-pivot algorithm. The crucial difference lies in the selection of the pivots. The idea is to choose a random sample S of the input s and to sort S. Now, $v = S[\lfloor k|S|/|s| \rfloor]$ will be an element with rank close to k. However, we do not know whether v is an underestimate or an overestimate of the element with rank k. We therefore introduce a safety margin Δ and set $\ell = S[\lfloor k|S|/|s| \rfloor - \Delta]$ and $r = S[\lfloor k|S|/|s| \rfloor + \Delta]$. The tricky part is to choose $|S|$ and Δ such that sampling and sorting the sample are fast, and that with high probability $rank(\ell) \leq k \leq rank(r)$ and $rank(r) - rank(\ell)$ is small. With the right choice of the parameters $|S|$ and Δ, the resulting algorithm can be implemented to run in time $O(n/p + \log p)$.

The basic idea is to choose $|S| = \Theta(\sqrt{p})$ so that we can sort the sample in time $O(\log p)$ using the fast, inefficient algorithm in Sect. 5.2. Note that this algorithm assumes that the elements to be sorted are uniformly distributed over the PEs. This may not be true in all levels of the recursion. However, we can achieve this uniform distribution in time $O(n/p + \log p)$ by redistributing the sample.

*Exercise 5.41. Work out the details of the redistribution algorithm. Can you do it also in time $O(\beta n/p + \alpha \log p)$?

We choose $\Delta = \Theta(p^{1/6})$. Working only with expectations, each sample represents $\Theta(n/\sqrt{p})$ input elements, so that with $\Delta = \Theta(p^{1/6})$, the expected number of elements between ℓ and r is $\Theta(n/\sqrt{p} \cdot p^{1/6}) = \Theta(n/p^{1/3})$.

**Exercise 5.42. Prove using Chernoff bounds (see Sect. A.3) that for any constant c, with probability at least $1 - p^{-c}$, the following two propositions hold: The number of elements between ℓ and r is $\Theta(n/p^{1/3})$ and the element with rank k is between ℓ and r.

Hence, a constant number of recursion levels suffices to reduce the remaining input size to $O(n/p)$. The remaining small instance can be gathered onto a single PE in time $O(n/p)$ using the algorithm described in Section 13.5. Solving the problem sequentially on this PE then also takes time $O(n/p)$.

The communication volume of the algorithm above can be reduced [158]. Further improvements are possible if the elements on each PE are sorted and when k is not exactly specified. Bulk deletion from parallel priority queues (Sect. 6.4) is a natural generalization of the parallel selection problem.

5.10 Breaking the Lower Bound

The title of this section is, of course, nonsense. A lower bound is an absolute statement. It states that, in a certain model of computation, a certain task cannot be carried out faster than the bound. So a lower bound cannot be broken. But be careful. It cannot be broken within the model of computation used. The lower bound does not exclude the possibility that a faster solution exists in a richer model of computation. In fact, we may even interpret the lower bound as a guideline for getting faster. It tells us that we must enlarge our repertoire of basic operations in order to get faster.

Procedure *KSort*(*s* : *Sequence* **of** *Element*)
  $b = \langle\langle\rangle,\ldots,\langle\rangle\rangle$: *Array* $[0..K-1]$ **of** *Sequence* **of** *Element*
  foreach $e \in s$ **do** $b[key(e)].pushBack(e)$
  $s := $ *concatenation of* $b[0],\ldots,b[K-1]$

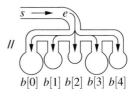

Fig. 5.18. Sorting with keys in the range $0..K-1$

Procedure *LSDRadixSort*(*s* : *Sequence* **of** *Element*)
  for $i := 0$ **to** $d-1$ **do**
    redefine $key(x)$ as $(x\operatorname{div} K^i) \bmod K$
    KSort(*s*)
    invariant *s* is sorted with respect to digits $i..0$

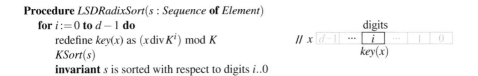

Fig. 5.19. Sorting with keys in $0..K^d - 1$ using **l**east **s**ignificant **d**igit (LSD) radix sort

What does this mean in the case of sorting? So far, we have restricted ourselves to comparison-based sorting. The only way to learn about the order of items was by comparing two of them. For structured keys, there are more effective ways to gain information, and this will allow us to break the $\Omega(n\log n)$ lower bound valid for comparison-based sorting. For example, numbers and strings have structure: they are sequences of digits and characters, respectively.

Let us start with a very simple algorithm, *Ksort* (or *bucket sort*), that is fast if the keys are small integers, say in the range $0..K-1$. The algorithm runs in time $O(n+K)$. We use an array $b[0..K-1]$ of *buckets* that are initially empty. We then scan the input and insert an element with key k into bucket $b[k]$. This can be done in constant time per element, for example by using linked lists for the buckets. Finally, we concatenate all the nonempty buckets to obtain a sorted output. Figure 5.18 gives the pseudocode. For example, if the elements are pairs whose first element is a key in the range $0..3$ and

$$s = \langle(3,a),(1,b),(2,c),(3,d),(0,e),(0,f),(3,g),(2,h),(1,i)\rangle,$$

we obtain $b = [\langle(0,e),(0,f)\rangle,\ \langle(1,b),(1,i)\rangle,\ \langle(2,c),(2,h)\rangle,\ \langle(3,a),(3,d),(3,g)\rangle]$ and output $\langle(0,e),(0,f),(1,b),(1,i),(2,c),(2,h),(3,a),(3,d),(3,g)\rangle$. This example illustrates an important property of *Ksort*. It is *stable*, i.e., elements with the same key inherit their relative order from the input sequence. Here, it is crucial that elements are *appended* to their respective bucket.

Comparison-based sorting uses two-way branching. We compare two elements and follow different branches of the program depending on the outcome. In *KSort*, we use K-way branching. We put an element into the bucket selected by its key and hence may proceed in K different ways. The K-way branch is realized by array access and is visualized in Figure 5.18.

KSort can be used as a building block for sorting larger keys. The idea behind *radix sort* is to view integer keys as numbers represented by digits in the range $0..K-1$. Then *KSort* is applied once for each digit. Figure 5.19 gives a radix-sorting algorithm for keys in the range $0..K^d - 1$ that runs in time $O(d(n+K))$. The elements are first sorted by their least significant digit (*LSD radix sort*), then by the second least significant digit, and so on until the most significant digit is used for sorting. It is not obvious why this works. The correctness rests on the stability of *Ksort*. Since *KSort* is stable, the elements with the same ith digit remain sorted with respect to digits $i - 1..0$ during the sorting process with respect to digit i. For example, if $K = 10$, $d = 3$, and

$$s = \langle 017,042,666,007,111,911,999 \rangle, \text{ we successively obtain}$$
$$s = \langle 111,911,042,666,017,007,999 \rangle,$$
$$s = \langle 007,111,911,017,042,666,999 \rangle, \text{ and}$$
$$s = \langle 007,017,042,111,666,911,999 \rangle.$$

*Exercise 5.43 (variable length keys). Assume that input element x is a number with d_x digits.

(a) Extend LSD radix sort to this situation and show how to achieve a running time of $O(d_{\max}(n+K))$, where $d_{\max}$ is the maximum d_x of any input.
(b) Modify the algorithm so that an element x takes part only in the first d_x rounds of radix sort, i.e., only in the rounds corresponding to the last d_x digits. Show that this improves the running time to $O(L + Kd_{\max})$, where $L = \sum_x d_x$ is the total number of digits in the input.
(c) Modify the algorithm further to achieve a running time of $O(L + K)$. Hint: From an input $x = \sum_{0 \le \ell < d_x} x_i K^i$ generate the d_x pairs (ℓ, x_ℓ), $0 \le \ell < d_x$, and sort them using radix sort. Use K buckets for the first round of radix sort and $d_{\max}$ buckets for the second round. Observe that the ℓth bucket, $0 \le \ell < d_{\max}$, will contain the multiset of digits that occur as the ℓth least significant digit. Now run LSD radix sort on the original inputs with the following modification: Whenever you concatenate buckets at the end of a call of Ksort, concatenate only the nonempty buckets and do not touch the empty buckets.

*Exercise 5.44 (string sorting). Modify the algorithm from Exercise 5.43 to sort strings of total length N over an alphabet of size K in time $O(N+K)$.

Radix sort starting with the most significant digit (*MSD radix sort*) is also possible. Here, we apply *KSort* to the most significant digit and then sort each bucket recursively. The only problem is that a bucket may contain much fewer than K elements and then it would be wasteful to sort it further using maybe several rounds of *KSort*. The solution is to switch to another sorting algorithm when the buckets become small. This works particularly well if we can assume that the keys are uniformly distributed. More specifically, let us now assume that the keys are real numbers with $0 \le key(e) < 1$. The algorithm *uniformSort* in Fig. 5.20 scales these keys to integers between 0 and $n - 1 = |s| - 1$ and groups them into n buckets, where bucket $b[i]$ is responsible for keys in the range $[i/n, (i+1)/n)$. For example, if $s = \langle 0.8, 0.4, 0.7, 0.6, 0.3 \rangle$, we obtain five buckets responsible for intervals of size 0.2, and $b = [\langle \rangle, \langle 0.3 \rangle, \langle 0.4 \rangle, \langle 0.7, 0.6 \rangle, \langle 0.8 \rangle]$. Only $b[3] = \langle 0.7, 0.6 \rangle$ is a nontrivial subproblem. *uniformSort* is very efficient for *random* keys.

Procedure *uniformSort*(*s* : *Sequence* **of** *Element*)
  $n := |s|$
  $b = \langle\langle\rangle,\dots,\langle\rangle\rangle$: *Array* $[0..n-1]$ **of** *Sequence* **of** *Element*
  foreach $e \in s$ **do** $b[\lfloor key(e) \cdot n \rfloor].pushBack(e)$
  for $i := 0$ **to** $n-1$ **do** sort $b[i]$ in time $O(|b[i]| \log |b[i]|)$
  $s := concatenation\ of\ b[0],\dots,b[n-1]$

Fig. 5.20. Sorting random keys in the range $[0,1)$

Theorem 5.9. *If the keys are independent uniformly distributed random values in* $[0,1)$, *uniformSort sorts n keys in expected time* $O(n)$ *and worst-case time* $O(n \log n)$. *The linear time bound for the average case holds even if an algorithm with quadratic running time is used for sorting the buckets.*

Proof. We leave the worst-case bound as an exercise and concentrate on the average case. The total execution time T is $O(n)$ for setting up the buckets and concatenating the sorted buckets, plus the time for sorting the buckets. Let T_i denote the time for sorting the ith bucket. We obtain

$$E[T] = O(n) + E\left[\sum_{i<n} T_i\right] = O(n) + \sum_{i<n} E[T_i] = O(n) + nE[T_0].$$

The second equality follows from the linearity of expectations (A.3), and the third equality uses the fact that all bucket sizes have the same distribution for uniformly distributed inputs. Hence, it remains to show that $E[T_0] = O(1)$. The analysis is similar to the arguments used to analyze the behavior of hashing in Chap. 4.

Let $B_0 = |b[0]|$. We have $E[T_0] = O(E[B_0^2])$. The random variable B_0 obeys a binomial distribution (A.8) with n trials and success probability $1/n$, and hence

$$\text{prob}(B_0 = i) = \binom{n}{i}\left(\frac{1}{n}\right)^i\left(1 - \frac{1}{n}\right)^{n-i} \leq \frac{n^i}{i!}\frac{1}{n^i} = \frac{1}{i!} \leq \left(\frac{e}{i}\right)^i,$$

where the last inequality follows from Stirling's approximation to the factorial (A.10). We obtain

$$E[B_0^2] = \sum_{i \leq n} i^2 \text{prob}(B_0 = i) \leq \sum_{i \leq n} i^2 \left(\frac{e}{i}\right)^i$$

$$\leq \sum_{i \leq 5} i^2 \left(\frac{e}{i}\right)^i + e^2 \sum_{i \geq 6} \left(\frac{e}{i}\right)^{i-2}$$

$$\leq O(1) + e^2 \sum_{i \geq 6} \left(\frac{1}{2}\right)^{i-2} = O(1),$$

and hence $E[T] = O(n)$ (note that $e/i \leq 1/2$ for $i \geq 6$). □

*Exercise 5.45 (inplace bucket sort). Develop an sorting algorithm for elements with keys in the range $0..K-1$ that uses the data structure of Exercise 3.31. The space consumption should be $n + O(n/B + KB)$ for n elements, and blocks of size B.

5.11 *Parallel Bucket Sort and Radix Sort

We shall first describe a stable, distributed-memory implementation of bucket sort. Each PE builds a local array of K buckets and distributes its locally present elements to these buckets. Now we need to concatenate local buckets to form global buckets. Note that the bucket sizes can be extremely skewed. For example, 90% of all elements could have key 42. We need to ensure good load balance even for highly skewed inputs – each PE gets at most $L := \lceil n/p \rceil$ elements. We use a similar strategy to that in parallel quicksort and use prefix sums to assign global numbers to all elements. To do this, we compute both the global bucket sizes and a vector-valued prefix sum over the bucket sizes. This can be done in time $O(K + \log p)$; see Sect. 13.3. Let $m_i := \sum_j |b[i]@j|$ denote the global size of bucket i. Then the kth element, $0 \le k < |b[i]@i_{\mathrm{proc}}|$, in bucket i on PE i_{proc} gets a global number $\sum_{j<i} m_j + \sum_{j<i_{\mathrm{proc}}} |b[i]@j| + k$. An element with global number y is assigned to PE $1 + \lfloor (y-1)/L \rfloor$. Assuming that no PE initially has more then L elements, no local bucket will spread over more than two PEs, and hence a PE sends at most $2K$ messages. The average number of received messages is the same. However, in the worst case, there might be a situation where many small buckets are assigned to one PE.

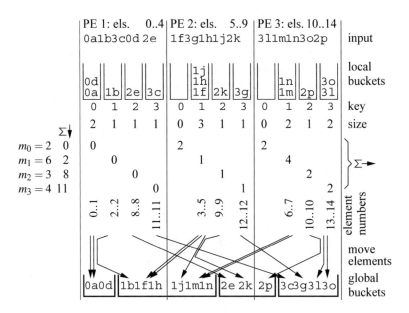

Fig. 5.21. Parallel bucket sort of 15 elements with $K = 4$. The elements have a letter as associated information. For example, $2k$ stands for an element with key 2 and information k. The middle part of the figure shows various sums and prefix sums. m_i is the number of elements in buckets i summed over all processors. The column $\Sigma\downarrow$ shows the prefix sums for the m_i's. The rows $\Sigma\rightarrow$ show the prefix sums of the $|b[i]@j|$. The elements in bucket 1 of processor 2 have global numbers $\sum_{j<1} m_i + \sum_{j<2} |b[1]@j| + k = 2 + 1 + k = 3 + k$, $0 \le k < 3$, i.e., their global numbers span the interval 3..5. PE 2 sends one message to itself; it contains $1j$ and $2k$.

This PE might have to receive $\Theta(pK)$ messages. Therefore, pieces of buckets moved to the same PE should be packed into a single message. This limits the number of sent and received messages to p. Figure 5.21 gives an example. The overall cost of the data exchange is $T_{\text{all}\rightarrow\text{all}}(L)$ using an all-to-all communication; see Sect. 13.6.

The above stable distributed bucket sort can be used to implement parallel LSD radix sort. A disadvantage of LSD radix sort is that all elements are moved d times in d-digit radix sort. An alternative is to start radix sort with the *Most Significant Digit* (*MSD radix sort*). Buckets of size $m_i < L$ can then be assigned to a single PE where they can be sorted locally looking at the remaining digits. Larger buckets still need to be split between several PEs but at least the number of PEs involved decreases. Also, if the PEs are numbered so as to respect locality of communication, then communication between PEs assigned to the same bucket may be faster than global communication.

5.12 *External Sorting

Sometimes the input is so large that it does not fit into internal memory. In this section, we learn how to sort such data sets in the external-memory model introduced in Sect. 2.2. This model distinguishes between a fast internal memory of size M and a large external memory. Data is moved in blocks of size B between the two levels of the memory hierarchy. Scanning data is fast in external memory, and mergesort is based on scanning. We therefore take mergesort as the starting point for external-memory sorting.

Assume that the input is given as an array in external memory. We shall describe a nonrecursive implementation of mergesort for the case where the number of elements n is divisible by B. We load subarrays of size M into internal memory, sort them using our favorite algorithm, for example *qSort*, and write the sorted subarrays back to external memory. We refer to the sorted subarrays as *runs*. The *run formation phase* takes n/B block reads and n/B block writes, i.e., a total of $2n/B$ I/Os. We then merge pairs of runs into larger runs in $\lceil \log(n/M) \rceil$ *merge phases*, ending up with a single sorted run. Figure 5.22 gives an example for $n = 48$ and runs of length 12.

How do we merge two runs? We keep one block from each of the two input runs and one from the output run in internal memory. We call these blocks *buffers*.

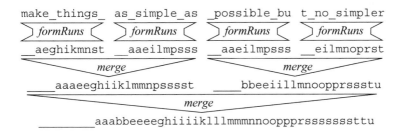

Fig. 5.22. An example of two-way mergesort with initial runs of length 12.

Initially, the input buffers are filled with the first B elements of the input runs, and the output buffer is empty. We compare the leading elements of the input buffers and move the smaller element to the output buffer. If an input buffer becomes empty, we fetch the next block of the corresponding input run; if the output buffer becomes full, we write it to external memory.

Each merge phase reads all current runs and writes new runs of twice the length. Therefore, each phase needs n/B block reads and n/B block writes. Summing over all phases, we obtain $(2n/B)(1 + \lceil \log n/M \rceil)$ I/Os. This technique works provided that $M \geq 3B$.

5.12.1 Multiway Mergesort

In general, internal memory can hold many blocks and not just three. We shall describe how to make full use of the available internal memory during merging. The idea is to merge more than just two runs; this will reduce the number of phases. In *k-way merging*, we merge k sorted sequences into a single output sequence. In each step, we find the input sequence with the smallest first element. This element is removed and appended to the output sequence. External-memory implementation is easy as long as we have enough internal memory for k input buffer blocks, one output buffer block, and a small amount of additional storage.

For each sequence, we need to remember which element we are currently considering. To find the smallest element out of all k sequences, we keep their current elements in a *priority queue*. A priority queue maintains a set of elements supporting the operations of insertion and deletion of the minimum. Chapter 6 explains how priority queues can be implemented so that insertion and deletion take time $O(\log k)$ for k elements. The priority queue tells us, at each step, which sequence contains the smallest element. We delete this element from the priority queue, move it to the output buffer, and insert the next element from the corresponding input buffer into the priority queue. If an input buffer runs dry, we fetch the next block of the corresponding sequence, and if the output buffer becomes full, we write it to the external memory.

How large can we choose k? We need to keep $k + 1$ blocks in internal memory and we need a priority queue for k keys. So, we need $(k + 1)B + O(k) \leq M$ or $k = O(M/B)$. The number of merging phases is reduced to $\lceil \log_k(n/M) \rceil$, and hence the total number of I/Os becomes

$$2\frac{n}{B}\left(1 + \left\lceil \log_{M/B} \frac{n}{M} \right\rceil\right). \tag{5.2}$$

The difference from binary merging is the much larger base of the logarithm. Interestingly, the above upper bound for the I/O complexity of sorting is also a lower bound [6], i.e., under fairly general assumptions, no external sorting algorithm with fewer I/O operations is possible.

In practice, the number of merge phases will be very small. Observe that a single merge phase suffices as long as $n \leq M^2/B$. We first form M/B runs of length M each and then merge these runs into a single sorted sequence. If internal memory stands

for DRAM and "external memory" stands for hard disks or solid state disks, this
bound on n is no real restriction, for all practical system configurations.

Exercise 5.46. Show that a multiway mergesort needs only $O(n \log n)$ element comparisons.

Exercise 5.47 (balanced systems). Study the current market prices of computers,
internal memory, and mass storage (currently hard disks and solid state disks). Also,
estimate the block size needed to achieve good bandwidth for I/O. Can you find
any configuration where multiway mergesort would require more than one merging
phase for sorting an input that fills all the disks in the system? If so, what fraction of
the cost of that system would you have to spend on additional internal memory to go
back to a single merging phase?

5.12.2 Sample Sort

The most popular internal-memory sorting algorithm is not mergesort but quicksort.
So it is natural to look for an external-memory sorting algorithm based on quicksort. We shall sketch *sample sort* [110]. It has the same performance guarantees as
multiway mergesort (5.2), but only in expectation not in the worst case. On the positive side, sample sort is easier to adapt to parallel disks and parallel processors than
merging-based algorithms. Furthermore, similar algorithms can be used for fast external sorting of integer keys along the lines of Sect. 5.10.

Instead of the single pivot element of quicksort, we now use $k - 1$ *splitter elements* $s_1, \ldots, s_{k-1}$ to split an input sequence into k output sequences, or *buckets*.
Bucket i gets the elements e for which $s_{i-1} \le e < s_i$. To simplify matters, we define
the artificial splitters $s_0 = -\infty$ and $s_k = \infty$ and assume that all elements have different keys. The splitters should be chosen in such a way that the buckets have a size
of roughly n/k. The buckets are then sorted recursively. In particular, buckets that fit
into the internal memory can subsequently be sorted internally. Note the similarity
to the MSD radix sort described in Sect. 5.10.

The main challenge is to find good splitters quickly. Sample sort uses a fast,
simple randomized strategy. For some integer a, we randomly choose $(a+1)k-1$
sample elements from the input. The sample S is then sorted internally, and we define the splitters as $s_i = S[(a+1)i]$ for $1 \le i \le k-1$, i.e., consecutive splitters are
separated by a samples, the first splitter is preceded by a samples, and the last splitter is followed by a samples. Taking $a = 0$ results in a small sample set, but the
splitting will not be very good. Moving all n elements to the sample will result in
perfect splitters, but the sample will be too big. The following analysis shows that
setting $a = \Theta(\log k)$ achieves roughly equal bucket sizes at low cost for sampling and
sorting the sample.

The most I/O-intensive part of sample sort is the k-way distribution of the input
sequence to the buckets. We keep one buffer block for the input sequence and one
buffer block for each bucket. These buffers are handled analogously to the buffer
blocks in k-way merging. If the splitters are kept in a sorted array, we can find the
right bucket for an input element e in time $O(\log k)$ using binary search.

Theorem 5.10. *Sample sort uses*

$$O\left(\frac{n}{B}\left(1+\left\lceil\log_{M/B}\frac{n}{M}\right\rceil\right)\right)$$

expected I/O steps for sorting n elements. The internal work is $O(n\log n)$.

We leave a detailed proof to the reader and describe only the key ingredient of the analysis here. We use $k = \Theta(\min(n/M, M/B))$ buckets and a sample of size $O(k\log k)$. The following lemma shows that with this sample size, it is unlikely that any bucket has a size much larger than the average. We hide the constant factors behind $O(\cdot)$ notation because our analysis is not very tight in this respect.

Lemma 5.11. *Let $k \geq 2$ and $a + 1 = 12\ln k$. A sample of size $(a+1)k - 1$ suffices to ensure with probability at least $1/2$ that no bucket receives more than $4n/k$ elements.*

Proof. As in our analysis of quicksort (Theorem 5.6), it is useful to study the sorted version $s' = \langle e'_1, \ldots, e'_n \rangle$ of the input. Assume that there is a bucket with at least $4n/k$ elements assigned to it. We estimate the probability of this event.

We split s' into $k/2$ segments of length $2n/k$. The jth segment t_j contains elements $e'_{2jn/k+1}$ to $e'_{2(j+1)n/k}$. If $4n/k$ elements end up in some bucket, there must be some segment t_j such that all its elements end up in the same bucket. This can only happen if fewer than $a + 1$ samples are taken from t_j, because otherwise at least one splitter would be chosen from t_j and its elements would not end up in a single bucket. Let us concentrate on a fixed j.

We use a random variable X to denote the number of samples taken from t_j. Recall that we take $(a+1)k - 1$ samples. For each sample i, $1 \leq i \leq (a+1)k - 1$, we define an indicator variable X_i with $X_i = 1$ if the ith sample is taken from t_j and $X_i = 0$ otherwise. Then $X = \sum_{1 \leq i \leq (a+1)k - 1} X_i$. Also, the X_i's are independent, and $\text{prob}(X_i = 1) = 2/k$. Independence allows us to use the Chernoff bound (A.6) to estimate the probability that $X < a + 1$. We have

$$\text{E}[X] = ((a+1)k - 1) \cdot \frac{2}{k} = 2(a+1) - \frac{2}{k} \geq \frac{3(a+1)}{2}.$$

Hence $X < a + 1$ implies $X < (1 - 1/3)\text{E}[X]$, and so we can use (A.6) with $\varepsilon = 1/3$. Thus

$$\text{prob}(X < a + 1) \leq e^{-(1/9)\text{E}[X]/2} \leq e^{-(a+1)/12} = e^{-\ln k} = \frac{1}{k}.$$

The probability that an insufficient number of samples is chosen from a fixed t_j is thus at most $1/k$, and hence the probability that an insufficient number is chosen from some t_j is at most $(k/2) \cdot (1/k) = 1/2$. Thus, with probability at least $1/2$, each bucket receives fewer than $4n/k$ elements.   □

Exercise 5.48. Work out the details of an external-memory implementation of sample sort. In particular, explain how to implement multiway distribution using $2n/B + k + 1$ I/O steps if the internal memory is large enough to store $k + 1$ blocks of data and $O(k)$ additional elements.

Exercise 5.49 (many equal keys). Explain how to generalize multiway distribution so that it still works if some keys occur very often. Hint: There are at least two different solutions. One uses the sample to find out which elements are frequent. Another solution makes all elements unique by interpreting an element e at an input position i as the pair (e, i).

*Exercise 5.50 (more accurate distribution). A larger sample size improves the quality of the distribution. Prove that a sample of size $O\big((k/\varepsilon^2)\log(km/\varepsilon)\big)$ guarantees, with probability at least $1 - 1/m$, that no bucket has more than $(1 + \varepsilon)n/k$ elements. Can you get rid of the ε in the logarithmic factor?

5.13 Parallel Sample Sort with Implementations

We learned about sample sort as an external-memory algorithm in Sect. 5.12.2. It is equally useful for parallel sorting [47]. We begin with a high-level description as a distributed-memory algorithm. Then we describe shared memory and MPI implementations in sections 5.13.1 and 5.13.2 respectively.

Figure 5.23 gives high-level pseudocode for distributed-memory sample sort. The number of splitters is now $p - 1$: We divide the input into p pieces of about equal size, one for each PE. The procedure *selectSplitters* encapsulates the task of selecting samples in parallel and extracting splitters from them. A pragmatic approximation to a uniform random sample is that each PE chooses a number of samples from its local data proportional to its share in the overall amount of data. A more sophisticated algorithm that actually chooses a uniformly distributed global random sample is described in [274]. This algorithm needs only $O(\log p)$ communication cost.

Perhaps the simplest way to extract splitters from the sample is to gather (see Sect. 13.5) the local samples at PE 0, sort them there and take the splitters as equidistant elements of the sorted sample as described in Sect. 5.12.2. The splitters are then broadcast to all PEs; see Sect. 13.1. A variant of this idea is to use an all-gather operation and to do the sorting and splitter selection redundantly. This saves us the broadcast but makes other operations slower. If p or n is large, it may be good to use

---

Procedure *parSampleSort*(s : *Sequence* **of** *Element*)
$\quad b = \langle \langle \rangle, \ldots, \langle \rangle \rangle$: *Array* $[1..p]$ **of** *Sequence* **of** *Element*
$\quad \langle s_0 = -\infty, s_1, \ldots, s_{p-1}, s_p = \infty \rangle := selectSplitters(s)$
$\quad$ **foreach** $e \in s$ **do**
$\quad\quad$ *determine i such that* $s_{i-1} \le e < s_i$
$\quad\quad$ $b[i].pushBack(e)$
$\quad$ *send $b[i]$ to PE i for $i \in 1..p$* $\hspace{2cm}$ // all-to-all
$\quad$ *receive buckets into s*
$\quad$ *sort*(s)

Fig. 5.23. SPMD pseudocode for distributed-memory sample sort

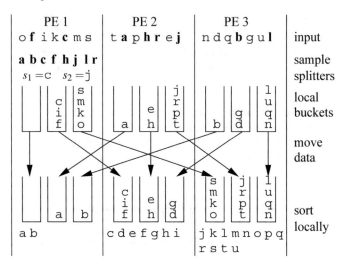

Fig. 5.24. Sample sort of 21 characters on three PEs using oversampling parameter $a = 2$.

a parallel algorithm to sort the sample – perhaps a fast one such as the fast, inefficient algorithm in Sect. 5.2 or the parallel quicksort in Sect. 5.7.1.

In the partitioning phase, we need to locate the right bucket for each element. This can be achieved in time $O(\log p)$ per element by binary search in the sorted array of splitters.[9]

Delivering the buckets to their destination is an all-to-all operation; see Sect. 13.6. The data can be received into the input sequence s, but a size update is needed since the splitters do not perfectly partition the data into equal-sized pieces. Finally, s is sorted locally. Figure 5.24 gives an example.

We now complete the analysis of parallel sample sort.

Theorem 5.12. *Parallel sample sort takes time* $O(\frac{n}{p}\log n + p\log^2 p)$ *assuming sequential sorting of the sample. Using a fast parallel algorithm for sorting the sample, the running time reduces to* $O(\frac{n}{p}\log n + p)$.

Proof. (Outline.) We leave the variant with centralized sorting of the sample as an exercise. By Lemma 5.11, a sample size of $O(p\log p)$ suffices to achieve pieces of size $O(n/p)$. Determining the sample in parallel takes time $O(\log p) = o(p)$ as described above. Sorting the sample using the fast, inefficient algorithm in Sect. 5.2 needs time

$$O\left(\log p + \frac{p\log p}{\sqrt{p}} + \frac{p\log p}{p}\log\frac{p\log p}{\log p}\right) = O(\sqrt{p}\log p) = o(p);$$

see (5.1). Distributing the elements to local buckets takes time $O(\frac{n}{p}\log p)$. Delivering the local buckets to the PEs responsible for them is the time for a nonuniform all-to-

---

[9] For a more efficient way to find buckets, see [280].

all communication with $h = O(n/p)$, i.e., $O(n/p+p)$; see Sect. 13.6.3. Local sorting takes time $O(\frac{n}{p}\log n)$. Summing all these terms yields the bound. ☐

We can see that sample sort with parallel sorting of the sample is efficient when $n \gg p^2/\log p$. In order to actually achieve efficiency close to 1, it is important to balance the work between the PEs very well. The following exercise works out how this affects the required sample size.

Exercise 5.51. For any constant $\varepsilon > 0$, show that a sample size $O(p\log(n)/\varepsilon^2)$ ensures with high probability that no piece is larger than $(1+\varepsilon)n/p$. Hint: Consider a potential piece A that is larger than $(1+\varepsilon)n/p$. Show using Chernoff bounds that it is unlikely that so few samples are taken from A that A is not split into several pieces. Can you show that $O(p\log(p)/\varepsilon^2)$ samples are also enough?

5.13.1 Shared-Memory Sample Sort Implementation

We now explain how to implement sample sort on a shared-memory machine using C++11 and the standard library. Compared with the distributed-memory version, matters simplify. In particular, the input and output are just a single global array s. With centralized splitter determination, a single PE can take random samples from s, sort this sample, and copy the splitters to a global splitter array. Hardware caching will make sure that each PE has a local copy of the splitter array in its cache. Instead of invoking an all-to-all collective communication, each PE will copy the local buckets to the appropriate places in s. However, we shall see that a few special measures are needed in order to adapt to NUMA effects; see Sect. 2.4.3.

Listing 5.1 shows our first attempt. Besides the random access iterator[10] s pointing to the beginning of the input array, $n = |s|$, and the number of threads (PEs) p, the (sequential) routine *pSampleSort* is passed a three-dimensional array *buckets*: *buckets*[i][j] is a vector into which thread i puts all elements from his batch that should go to thread j. Line 6 declares a global *barrier* object, which is later used to synchronize the worker threads. In lines 8–12, the calling thread takes a random sample S using a Mersenne Twister pseudorandom generator of 32-bit numbers with a state size of 19 937 bits and oversampling factor $a = 16\log p$; the declaration of SampleDistribution states that we want to generate integers in $0..n-1$. In lines 13–16, S is then sorted and condensed to contain only the splitters.

The parallel part of the program starts in line 19. Lines 19–45 create p threads running a worker function defined in-place. The worker function is called with the argument i, the second argument in the call *thread*(definition of worker function, i), i.e., the local value of *iPE* is i. In the while-loop of the worker function (lines 24–28), thread *iPE* scans its region of the input $s[iPE \cdot n/p..(iPE+1) \cdot n/p-1]$. Each element $e = *current$ is located in the array *splitters* using the function *upper_bound* from the standard library. This function uses binary search to locate the first splitter position

---

[10] An iterator is an object similar to a pointer that allows a programmer to traverse a data structure.

Listing 5.1. Sample sort n elements in s using p threads

```
template <class Iterator>                                                              1
void pSampleSort(const Iterator & s, const size_t n, const unsigned p,                 2
    vector<vector<vector<Element> > > & buckets)                                        3
{                                                                                      4
  mt19937 rndEngine;                                                                   5
  Barrier barrier(p); // for barrier synchronization                                   6
  // Choose random samples                                                             7
  vector<KeyType> S; // random sample of a elements from s                             8
  uniform_int_distribution<size_t> SampleDistribution(0, n−1);                         9
  const int a = (int)(16*log(p)/log(2.0)); // oversampling ratio                      10
  for(size_t i=0; i < (size_t)(a+1)*p − 1; ++i)                                       11
    S.push_back((s + SampleDistribution(rndEngine))−>key);                            12
  sort(S.begin(),S.end()); // sort samples sequentially                              13
  for(size_t i=0; i < p−1 ; ++i) // select splitters                                 14
    S[i] = S[(a+1)*(i+1)];                                                            15
  S.resize(p−1);                                                                      16
  vector<size_t> bucketSize(p, 0ULL);                                                17
  vector<thread> threads(p);                                                         18
  for (unsigned i = 0; i < p; ++i) { // go parallel                                  19
    threads[i] = thread( [&](const unsigned iPE) // the worker function              20
    { // distribute elements                                                         21
      auto current = s + iPE*n/p, end = s + (iPE+1)*n/p;                              22
      auto & myBuckets = buckets[iPE];                                               23
      while(current != end) {                                                        24
        const size_t i = upper_bound(S.begin(),S.end(),                              25
          current−>key) − S.begin(); // binary search                               26
        myBuckets[i].push_back(*current++);                                          27
      }                                                                              28
      barrier.wait(iPE, p);                                                          29
      // now each thread works on bucket "iPE". First compute the total size of bucket iPE:  30
      size_t myBuckSize = 0;//accumulate into local variable (prevent false sharing) 31
      for (const auto & b : buckets) myBuckSize += b[iPE].size();                    32
      bucketSize[iPE] = myBuckSize;                                                  33
      barrier.wait(iPE, p);                                                          34
      // find the bucket start in s by summing the sizes of the previous buckets (<iPE)  35
      auto bucketBegin = s;                                                          36
      for(size_t b = 0; b < iPE ; ++b) bucketBegin += bucketSize[b];                 37
      // copy the bucket 'iPE' from all PEs into s                                   38
      auto currOut = bucketBegin;                                                    39
      for(const auto & b: buckets)                                                   40
        currOut = copy(b[iPE].cbegin(), b[iPE].cend(), currOut);                     41
      sort(bucketBegin, currOut); // sort the bucket                                 42
    } // end of the worker function                                                 43
    , i);                                                                            44
  }                                                                                  45
  for (auto & t : threads) t.join();                                                 46
}//SPDX−License−Identifier: BSD−3−Clause; Copyright(c) 2018 Intel Corporation        47
```

larger than e. If no such splitter exists, *splitters.end*() is returned. Note that the specification of *upper_bound* elegantly avoids the need for explicitly storing a sentinel key ∞. Subtracting *splitters.begin*() yields the index of the bucket where e should be stored. Each thread uses its own local bucket array *myBuckets* = *buckets*[*iPE*].

Before sorting can continue, a barrier synchronization is necessary (see Sect. 13.4.2), i.e., all worker threads have to wait until all other worker threads have finished the while-loop.

After another barrier synchronization (line 34), each worker thread computes the beginning of the part of the output it is responsible for. This is done by adding the *bucketSizes* of the threads with smaller number (lines 36 and 37). Then, in lines 39–41, the *iPE*-th local bucket from each worker thread is copied to the output array s. Note that bucketBegin is an iterator into the array s and hence all the copying is done into s. Note also that the use of the standard library function *copy* allows the compiler to use highly tuned code here. Finally, in line 42, the *iPE*-th bucket, which now resides in a contiguous piece of the output array s, is sorted by calling the standard library function *sort*.

We measured the speedup of this simple implementation compared with sequential `std::sort` on the machine described in Appendix B. We performed a typical *weak scaling* experiment, i.e., the input size was scaled linearly with the number p of threads used – here, $n = p \cdot 8 \cdot 10^6$. We sort key-value pairs with 8-byte integers each. The lowest line in Fig. 5.25 shows the speedup obtained as a function of the number of threads. The outcome is disappointing. The speedup is never more than seven on a system with 72 physical cores, i.e., it is below 10% of what is suggested by the core count.

Using profiling tools (see Sect. B.8) one can see that for larger thread counts, the CPU cycles are spent mostly in the Linux kernel page fault handler and not on sorting itself. In the Linux kernel used, page fault handling is mostly serial, exposing a serious scalability bottleneck for applications with many page faults. In our implementation, the page faults stem from filling the buckets. The vectors used there are essentially the unbounded arrays in Sect. 3.4. Not only does this imply that data is copied as the arrays grow, but we also suffer from an "improvement" in the operating system that only lazily allocates physical memory in small pages (typically 4 KB) as the allocated virtual memory is actually written to; see `linux.die.net/man/2/set_mempolicy`. This is a clever strategy for sequential execution or a small number of threads but apparently becomes very slow for a large number of threads.

In other words, we face a frequent situation in parallel programming – we have to program around a performance bug of software that we cannot influence. We circumvented the Linux kernel page fault handler by using the Intel `tbb::memory_pool_allocator`, which redirects bucket allocation requests to a pre-assigned internal memory pool. This version performs significantly better; see the line marked "psamplesort-mpool" in Fig. 5.25. However, using all four sockets still does not give any significant speedup compared with two sockets.

An analysis of the profiler and processor performance data (see Sect. B.8) for the improved implementation shows that the number of CPU cycles required per instruction (CPI) significantly increases in the case of threads scattered over all sockets.

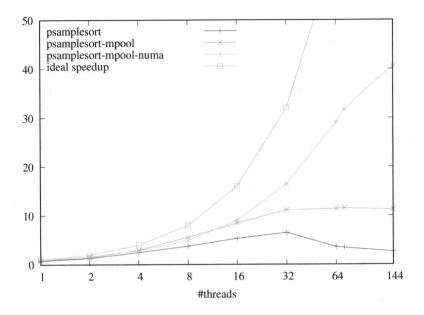

Fig. 5.25. Speedup of different parallel sorting algorithms over `std::sort` sorting $8 \cdot 10^6$ elements per thread.

Increased CPI is particularly prominent for memory access instructions. Since this only happens in the multiple-socket case, NUMA effects seem to be the cause, i.e., memory access latencies go up for accessing memory allocated on a remote socket. This is not surprising, given the first-touch NUMA memory assignment policy in the Linux kernel: If the capacity of the socket local memory permits, an allocation uses the physical memory pages on the socket of the running thread. Using memory from only a single socket can constitute a serious scalability bottleneck, since memory controllers on this socket will be responsible for servicing memory requests of all worker threads in the system. Even if the memory allocation is distributed uniformly over all sockets, the order-of-magnitude higher latency for remote accesses will degrade performance. Most operating systems expose NUMA allocation interfaces, allowing programmers to allocate memory from specific sockets. Unfortunately, these interfaces are not standardized yet and have many differences in semantics (Sect. C.6). An alternative way to control NUMA allocation in Linux is to exploit the first-touch policy. As a first measure, we *pin* worker thread i to physical core i (Sect. C.7); see the function *pinToCore* in the source code. The concrete measures to improve locality are twofold; see function *initialize* in the source code. First, each thread initializes the part of s it later distributes to the buckets. Note that this part is also a good approximation to the part it will finally sort sequentially. Second, each thread reserves a local memory pool for the buckets.

Figure 5.25 shows that the NUMA-optimized implementation exhibits good scaling. As a further experiment (Fig. 5.26), we compare different parallel sorting algorithms using up to 144 threads. An algorithm from the Intel TBB library achieves a speedup that is always below 8 even for large inputs. This algorithm lacks scalability because it uses a very simple parallelization of quicksort that uses sequential partitioning. However, TBB has the advantage to yield some speedup even for small inputs. Likely this is due to the use of efficient light-weight parallel tasks instead of threads. The curve labelled "std parallel mode" refers of an implementation of parallel multiway mergesort, that is available with the parallel version of the STL for g++ [298]. This algorithm as well as our best sample sort implementation (psamplesort-mpool-numa) achieve speedups around 30 for large inputs. This can be viewed as a success for our sample sort since it is much simpler. The MPI implementation of sample sort discussed in the next section performs even better for large but not too large inputs – achieving speedup of up to 47. We discuss this surprising effect below. Finally, the line labelled ipS^{4}o (inplace super scalar sample sort) refers to a recent inplace variant of sample sort [25]. Somewhat surprisingly, it significantly outperforms all the other algorithms besides saving on memory. With up to 87, the achieved speedup even exceeds the number of cores (72). There are two reasons for this good performance. First, ipS^{4}o performs element comparisons very efficiently and without incurring conditional branch instructions (indeed, ipS^{4}o also outperforms `std::sort` as a sequential algorithm). Second it avoids several sources of overhead involved with noninplace sorting.

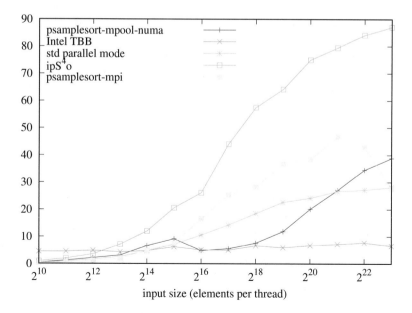

Fig. 5.26. Speedup over `std::sort` as a function of input size with 144 threads.

**Exercise 5.52. Design and implement a shared-memory parallel sorter that works well over a wide range of values of n and p. A few ideas: Avoid creating threads just for a single sorting call. Switch between different algorithms for different p and n, e.g., the algorithms from Sects. 5.2, 5.4, 5.7, and 5.13. Use profiling tools for tuning. Combine experiments and asymptotic complexity to find the right switching points.

5.13.2 MPI Implementation

Listing 5.2 gives an implementation of parallel sample sort using the message passing interface; see also App. D.[11] The routine *parallelSort* has a template parameter *Element* specifying the element data type from the perspective of C++. Unfortunately, MPI does not know about this and needs to get another specification of this data type, *mpiType*. Each PE samples $1 + 16 \lfloor \log p \rfloor$ of its local elements into the local sample vector *locS* (lines 5–10). These samples are collected in the global sample s using all-gather; see lines 11–13 and also Sect. 13.5). The samples are then sorted using the standard library (line 14). The vector s is reused as a splitter array – the splitter at $s[a \cdot i]$ is moved to $s[i]$ (lines 15–16).

Then the local data is distributed into a vector of bucket vectors *bucket* (lines 17–23). As in the shared-memory implementation, the function *upper_bound* from the standard library is used to find the right bucket. Note that we do not take any special measures with respect to memory management or NUMA effects. Our measurements indicate that MPI and the operating system take care of this quite well.

Now the buckets have to be delivered to the PEs responsible for sorting them. This is done in line 37 using the operation *MPI_Alltoallv*. Doing this requires some preparation (lines 25–36) though. MPI expects senders *and* receivers to specify the length and address of all messages to be delivered. An all-to-all operation with uniform message lengths is used to deliver this information (line 32). These preparations may be a bit cumbersome, but note that they are not a big performance issue. For sorting large data sets, the cost of the preparatory *MPI_Alltoall* is dwarfed by the cost of the subsequent *MPI_Alltoallv*.

Finally, actually sorting the local data is a simple library call (line 39).

Overall, MPI does not get a first prize for extreme elegance, but we end up with a code that has comparable length to the basic shared-memory code and outperforms it significantly.

Figure 5.26 shows the performance of our code when run on a shared-memory system.[12] The MPI code outperforms the shared memory code for large but not too large inputs. This is surprising since a direct shared-memory implementation should usually be faster than a message passing code especially if it goes into a number of complications to handle NUMA effects and to avoid operating system bottlenecks. The point is that MPI does these things implicitly. The kernel bottleneck discussed

---

[11] We would like to thank Michael Axtmann for providing this implementation and the measurements.

[12] We used GCC 4.8.5 with optimization -o2 and OpenMPI using the Byte Transfer Layer TCP.

Listing 5.2. MPI sample sort

```
template<class Element>                                                    1
void parallelSort(MPI_Comm comm, vector<Element>& data,                    2
                MPI_Datatype mpiType, int p, int myRank)                    3
{ random_device rd;                                                        4
  mt19937 rndEngine(rd());                                                 5
  uniform_int_distribution<size_t> dataGen(0, data.size() − 1);            6
  vector<Element> locS; // local sample of elements from input <data>      7
  const int a = (int)(16*log(p)/log(2.)); // oversampling ratio            8
  for (size_t i=0; i < (size_t)(a+1); ++i)                                 9
    locS.push_back(data[dataGen(rndEngine)]);                             10

  vector<Element> s(locS.size() * p); // global samples                   11
  MPI_Allgather(locS.data(), locS.size(), mpiType,                        12
      s.data(), locS.size(), mpiType, comm);                              13

  sort(s.begin(), s.end()); // sort global sample                         14
  for (size_t i=0; i < p−1; ++i) s[i] = s[(a+1) * (i+1)]; //select splitters 15
  s.resize(p−1);                                                          16

  vector<vector<Element>> buckets(p); // partition data                   17
  for(auto& bucket : buckets) bucket.reserve((data.size() / p) * 2);      18
  for( auto& el : data) {                                                 19
    const auto bound = upper_bound(s.begin(), s.end(), el);               20
    buckets[bound − s.begin()].push_back(el);                            21
  }                                                                       22
  data.clear();                                                           23

  // gather bucket sizes and calculate send/recv information              24
  vector<int> sCounts, sDispls, rCounts(p), rDispls(p + 1);               25
  sDispls.push_back(0);                                                   26
  for (auto& bucket : buckets) {                                          27
    data.insert(data.end(), bucket.begin(), bucket.end());               28
    sCounts.push_back(bucket.size());                                     29
    sDispls.push_back(bucket.size() + sDispls.back());                    30
  }                                                                       31
  MPI_Alltoall(sCounts.data(),1,MPI_INT,rCounts.data(),1,MPI_INT,comm);   32
  // exclusive prefix sum of recv displacements                           33
  rDispls[0] = 0;                                                         34
  for(int i = 1; i <= p; i++) rDispls[i] = rCounts[i−1]+rDispls[i−1];     35

  vector<Element> rData(rDispls.back()); // data exchange                 36
  MPI_Alltoallv(data.data(), sCounts.data(), sDispls.data(), mpiType,     37
    rData.data(), rCounts.data(), rDispls.data(), mpiType, comm);         38

  sort(rData.begin(), rData.end());                                       39
  rData.swap(data);                                                       40
}                                                                         41
```

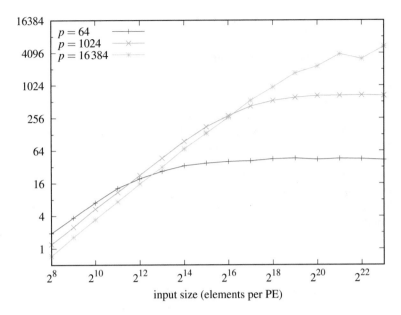

Fig. 5.27. Speedup over `std::sort` as a function of input size on BlueGene/Q. The running times are the median of five trials. The sorted elements were 64-bit random integers.

above does not apply when each PE has an operating system process of its own. Since each process of the MPI code generates its own input data, it automatically allocates the data on the right NUMA-node. Also, MPI pins its processes, i.e., it forces them to be executed on the same core all the time; see Sect. C.7. We have also run the MPI program on up to 16 384 cores of an IBM BlueGene/Q supercomputer.[13] Measuring speedup for large p and large inputs is difficult, since the biggest inputs do not fit into the internal memory of a single node and hence the sequential running time cannot be measured. We overcame this problem by extrapolating the running time of the sequential algorithm. This gives us an optimistic estimate of sequential running time on a hypothetical machine with sufficient memory. Note that using such numbers for computing speedups yields pessimistic estimates for the speedup. Figure 5.27 shows the achievable (extrapolated) speedup as a function of n/p. For $p = 1024$, we achieve a speedup of up to 663 – an efficiency of up to 65%. For such a simple algorithm, this is remarkably efficient. For larger p, the efficiency goes down because we get more and more contention in the interconnection network; see Sect. B.7. Moreover, even for the biggest inputs, with 2^{23} elements per PE, the individual messages in the all-to-all operation have a size of only 256 bytes, so that the startup overheads for

---

[13] We would like to thank the Gauss Centre for Supercomputing (GCS) for providing computing time through the John von Neumann Institute for Computing (NIC) on the GCS share of the supercomputer JUQUEEN [303] at the Jülich Supercomputing Centre (JSC).

message exchange dominate. Nevertheless, for $p = 16384$ we observe a speedup of up to 5391 which is still an efficiency of 33%.

5.14 *Parallel Multiway Mergesort

Multiway mergesort is another external-memory sorting algorithm that is a good candidate for an efficient parallel sorting algorithm. We first describe the shared-memory variant for its elegance and simplicity.

To implement parallel p-way mergesort, we first split the input array s into p equally sized pieces, possibly trying to allocate PEs on the same NUMA node as the RAM storing that piece of data. Each PE then locally sorts the data allocated to it. This takes time $O(\frac{n}{p} \log \frac{n}{p})$.

For parallel p-way merging, we generalize the splitting idea used in parallel binary (two-way) mergesort (Sect. 5.4). Rather than splitting two sequences into p pieces each, we now split p sequences into p pieces each such that all elements in the first pieces are smaller than all elements in the second pieces, which in turn are smaller than all elements in the third pieces, and so on. We can then obtain the sorted output by sorting the union of the first pieces, sorting the union of the second pieces, and so on. This description assumes that elements are pairwise distinct.

The function $smmSort$ in Fig. 5.28 realizes parallel multiway mergesort. After sorting locally, we run p multisequence selections in parallel to find the splitters. The ith processor is responsible for finding the split vector $x@i$ such that the total length of the sequences up to the split elements is equal to $i \cdot n/p$. We then run p incarnations of sequential multiway merging in parallel. The ith incarnation merges the subsequences delineated by $x@(i-1)$ and $x@i$.

We next discuss the search for the splitters. The function $multiSequenceSelect$ in Fig. 5.29 describes a sequential algorithm for finding one set of splitting positions. Its input is an array of p sorted sequences and an integer k. It determines, for each sequence, a split index ℓ_i, $1 \le i \le p$, such that $\sum_i \ell_i = k$ and all elements up to any split index are smaller than all elements following a split index. The function maintains two vectors ℓ and r and the invariants $\ell \le r$, $\sum_i \ell_i \le k \le \sum_i r_i$, and

$$\max_i \cup_i S_i[1..\ell_i] < \min_i \cup_i S_i[\ell_i + 1..r_i] \le \max_i \cup_i S_i[\ell_i + 1..r_i] < \min_i \cup_i S_i[r_i + 1..|S_i|],$$

Function $smmSort(s : Sequence$ **of** $Element) : Sequence$ **of** $Element$
   $sort(s);$ $barrier$ // sort locally then synchronize globally
   $x := multiSequenceSelect(\langle s@1, \ldots, s@p \rangle, \lceil i_{\mathrm{proc}} \frac{\sum_i |s@i|}{p} \rceil);$ $barrier$ // find splitters
   return $multiwayMerge(\langle\ s@1[x_1 @ (i_{\mathrm{proc}} - 1) + 1..x_1], \ldots,$ // assume
                                 $s@p[x_p @ (i_{\mathrm{proc}} - 1) + 1..x_p] \rangle)$ // $x@0 = \langle 0, \ldots, 0 \rangle$

Fig. 5.28. SPMD pseudocode for shared-memory multiway mergesort.

Function *multiSequenceSelect(S : Array of Sequence of Element; k : ℕ) : Array of ℕ*
   for $i := 1$ **to** $|S|$ **do** $(\ell_i, r_i) := (0, |S_i|)$
   invariant $\forall i : \ell_i..r_i$ *contains the splitting position of* S_i
   invariant $\forall i, j : \forall a \le \ell_i, b > r_j : S_i[a] \le S_j[b]$
   while $\exists i : \ell_i < r_i$ **do**
      $v := pickPivot(S, \ell, r)$
      for $i := 1$ **to** $|S|$ **do** $m_i := binarySearch(v, S_i[\ell_i..r_i])$ // $S_i[m_i] \le v < S_i[m_i + 1]$
      if $\sum_i m_i \le k$ **then** $\ell := m$ **else** $r := m$
   return ℓ

Fig. 5.29. Multisequence selection. Split the sorted input sequences in S such that the sum of the resulting splitting positions is k and such that all elements up to the splitting positions are no larger than the elements to the right of the splitting positions.

i.e., the elements in the left parts are smaller than the elements in the undecided parts which in turn are smaller than elements in the right parts. Initially, all elements belong to the undecided parts. For simplicity, we assume that all elements are pairwise distinct.

The algorithm works iteratively and continues as long as one of the undecided parts is nonempty. In each iteration, it chooses a random element v from the union of the undecided parts[14] and locates it in all the undecided parts. For each i, we determine m_i such that $S_i[m_i] \le v < S_i[m_i + 1]$ by binary search. This takes time logarithmic in $r_i - \ell_i$. If $\sum_i m_i \le k$, we set ℓ to m; otherwise we set r to m. In either case, the invariant is maintained.

*Exercise 5.53. Give detailed pseudocode for a generalization of the function *multi-SequenceSelect* that allows keys to appear multiple times. Hint: There is a generic approach that makes keys unique by replacing a key x stored at PE i in position j of the local input array by the triple (x, i_{proc}, j). These triples are ordered lexicographically. You can emulate this ordering without explicitly considering triples in the binary searches. Suppose pivot v has been chosen on PE i at position j of the input. Then, if $i_{\text{proc}} < i$, the binary search should look for the rightmost element with key $\le v$. At PE i, no search is necessary, and we set $m_i := j$. If $i_{\text{proc}} > i$, the binary search should look for the largest key less than v.

Exercise 5.54. The function *smmSort* in Fig. 5.28 defines the input and output by local arrays. Reformulate your code as a program with explicit parallel loops where the input and output are a single global array.

We turn now to the analysis. Assume *smmSort* is run on p local input sequences of size n/p each. Local sorting takes time $O(\frac{n}{p} \log \frac{n}{p})$, for example using sequential mergesort. Multiway merging takes time $O(\frac{n}{p} \log p)$. Summing this gives time

---

[14] This can be done by choosing a random number $x \in 1..\sum_i(r_i - \ell_i)$ and by setting v to the element with *global number* x, where the global number of $S_i[y]$ is $y - \ell_i + \sum_{j < i} r_j - \ell_j$ for $y \in \ell_i + 1..r_i$. Pivot v can be found in time $O(|S|)$ by scanning the ranges until the first range i with $\sum_{j \le i} r_i - \ell_i \ge x$ is found.

$O(\frac{n}{p}(\log\frac{n}{p}+\log p))=O(\frac{n}{p}\log n)$, i.e., optimal speedup so far. The barrier synchronizations take time $O(\log p)$; see Sect. 13.4.2. One iteration of multisequence selection takes time $O(p\log n)$. From the analysis of quickselect, we know that the expected number of iterations is $O(\log n)$. However, we have to be careful here. We are running p multisequence selections in parallel and we are only finished when the last of them has finished, i.e., we are interested in the expected maximum execution time of p parallel multisequence selections.

Lemma 5.13. *After* $O(\log n+\log p)$ *expected iterations, all* p *multisequence selections are finished.*

Proof. We first argue as in the proof of Theorem 5.8. With probability at least $1/3$, an iteration is *good* in the sense that it reduces $\sum_i r_i - \ell_i$ by a factor of at least $2/3$. Hence, $k := \log_{3/2} n$ good iterations suffice to reduce the problem size to 1. We use the Chernoff bound (A.6) to show that it is unlikely that some particular PE will need a large number of iterations to see k good ones. The probability that *any* PE needs a larger number of iterations is at most p times that probability. In order to be able to use a tail bound, we rewrite the definition of the expected values of an integer random variable as $E[I] := \sum_{t\geq 0} t\,\mathrm{prob}(I=t) = \sum_{t\geq 0}\mathrm{prob}(I>t)$. So, let I denote the total number of iterations until all PEs have seen k good iterations. Let X^t denote the number of good iterations that a particular PE j has seen after t iterations. X^t can be written as $\sum_{i=1}^t X_i^t$, where X_i^t is an indicator random variable with $X_i^t = 1$ if and only if iteration i is good for PE j. We have $E[X^t]=t/3$. We use (A.6) for $\varepsilon = \frac{1}{2}$ and $t \geq 6k$, which yields

$$\mathrm{prob}\left(X^t < \left(1-\frac{1}{2}\right)E[X^t]\right) \leq e^{-\left(1-\frac{1}{2}\right)^2 E[X^t]/2} = e^{-E[X^t]/8} = e^{-t/24}.$$

It is now easy to complete the proof. We first observe

$$E[I] = \sum_{t\geq 0}\mathrm{prob}(I>t) \leq t_0 + \sum_{t>t_0}\mathrm{prob}(I>t) \leq t_0 + \sum_{t>t_0} p\cdot\mathrm{prob}(X^t < k),$$

where t_0 is any integer. For the first inequality, we used $\mathrm{prob}(I>t) \leq 1$ for all t. For $t_0 \geq 6k$, we conclude further

$$E[I] \leq t_0 + p\sum_{t\geq t_0} e^{-t/24} = t_0 + p\frac{e^{-t_0/24}}{1-e^{-1/24}},$$

using (A.14). For $t_0 \geq 24\ln p + 4$, the last expression is bounded by $t_0 + 1$. □

Overall we obtain the following result.

Theorem 5.14. *Parallel multiway mergesort takes time* $O\left(\dfrac{n}{p}\log n + p\log^2 p\right)$.

Once more, we can replace $\log n$ by $\log p$, since for $n = \Omega(p^2\log p)$, the term $(n/p)\log n$ dominates the term $p\log^2 p$ and for smaller n, $\log n = O(\log p)$. Multiway mergesort is efficient for $n = \Omega(p^2\log p)$.

*Exercise 5.55. Design a deterministic algorithm for multisequence selection that runs in time $O(p\log^2 n)$ and is a generalization of our algorithm for two-sequence selection in Sect. 5.4. Hint: The smallest or largest midpoint of a range can replace a range endpoint.

**Exercise 5.56. Varman et al. [322] gave an algorithm for multisequence selection that runs in time $O(p\log n)$. Develop detailed pseudocode that works for arbitrary n and p without requiring much additional memory (e.g., for padded arrays). Can you demonstrate in an implementation that it outperforms our algorithm in practice?

*Exercise 5.57. Show a lower bound of time $O(p\log\frac{n}{p})$ for multisequence selection in the comparison-based model.

5.14.1 Distributed-Memory Multiway Mergesort

Multiway mergesort is also attractive for a distributed-memory algorithm. Figure 5.30 shows pseudocode and Figure 5.31 gives an example. The algorithm is similar to sample sort; in particular, each element is communicated only once in a single all-to-all communication. The main difference is that the input is sorted locally up front, and this makes it possible to find perfect splitters efficiently. This results in perfect load balance. The main difference with respect to shared-memory multiway mergesort is that multisequence selection has to be implemented differently, since

Function $dmmSort(s : Sequence$ **of** $Element) : Sequence$ **of** $Element$
    $sort(s)$
    $x := dmmmSelect(s, \langle\left\lceil i\frac{\sum_i|s@i|}{p}\right\rceil : i \in 1..p\rangle)$ // find splitters
    $\langle s_1,\ldots,s_p\rangle := allToAll(\langle s[1..x_1], s[x_1+1..x_2],\ldots,s[x_{p-1}+1..|s|]\rangle)$
    return $multiwayMerge(\langle s_1,\ldots,s_p\rangle)$

Fig. 5.30. SPMD pseudocode for distributed-memory multiway mergesort

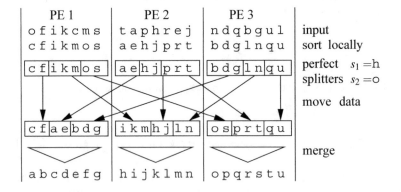

Fig. 5.31. Multiway mergesort of 21 characters on three PEs

Function *dmmmSelect*(*s* : *Sequence* **of** *Element; k* : *Array*[1..*p*] **of** ℕ) : *Array*[1..*p*] **of** ℕ
   ℓ, r, m, v, σ : *Array* [1..*p*] **of** ℕ
   for *i* := 1 **to** *p* **do** $(\ell_i, r_i) := (0, |s|)$ // initial search ranges
   while $\exists i, j : \ell_i @ j \neq r_i @ j$ **do**   // or-reduction
      $v := pickPivotVector(s, \ell, r)$ // reduction, prefix sum, broadcast
      for *i* := 1 **to** *p* **do** $m_i := binarySearch(v_i, s[\ell_i..r_i])$
      $\sigma := \sum_i m @ i$ // vector-valued reduction
      for *i* := 1 **to** *p* **do** **if** $\sigma_i \geq k_i$ **then** $r_i := m_i$ **else** $\ell_i := m_i$
   return ℓ

Fig. 5.32. SPMD pseudocode for distributed-memory multisequence multiselect with one sequence per PE and *p* ranks specified by vector *k*

multiSequenceSelect in Fig. 5.29 makes many remote memory accesses. Our strategy is to perform essentially the same computations but on different PEs. We apply the principle of "owner computes" (see also Sect. 3.1) – each PE is responsible for performing all the computations necessary for its local sequence. By doing this for all *p* desired ranks at once, we can make the communication very coarse-grained. Figure 5.32 shows pseudocode. The function runs *p* quickselect algorithms at once, and thus almost all of its local variables are vectors of dimension *p*. For example, the range $\ell_i..r_i$ now encloses the *i*th splitting position in *s*. The search can only terminate if all $p \times p$ ranges have unit size. To find out about that, all PEs have to communicate in an or-reduction collective communication operation; see Sect. 13.2. Picking pivots now also involves communication. This can be done analogously to pivot selection in parallel quicksort (see Sect. 5.7.1) except that the reduction, prefix sum, and broadcast operations involved work component-wise on *p*-dimensional vectors. Binary searches can be done locally. Adjusting the ranges once more requires a collective communication (reduction) to count the number of elements up to m_i. One iteration of the function *dmmmSelect* takes time $O(p \log n + p\beta + \alpha \log p)$ for *p* binary searches and a constant number of reduction/broadcast/prefix operations on vectors of length *p*. Overall, the selection takes time $O(p \log^2 n)$. It is also important to note that the number of startup overheads involved is much smaller – only $O(\log p \log n)$.

5.15 Parallel Sorting with Logarithmic Latency

With parallel mergesort (Sect. 5.4) and quicksort (Sect. 5.7), we have seen efficient parallel sorting algorithms that sort with span $O(\log^2 n)$. On the other hand, in Sect. 5.2 we have seen a fast, inefficient algorithm that sorts in logarithmic time. Here we want to outline a randomized asynchronous CRCW-PRAM algorithm that bridges this gap. For simplicity, we restrict ourselves to the case $n = p$, i.e., there is exactly one element per PE. The basic idea is to use a recursive variant of sample sort.

We use the fast, inefficient algorithm to sort a sample of size $\sqrt{p}$ in logarithmic time. We use an oversampling factor of $\Omega(\log p)$ to obtain an array of $k = \Omega\left(\sqrt{p}/\log p\right)$ splitters defining buckets of size $O(n/k)$ with high probability. Now, each PE searches the right bucket for its key using binary search in time $O(\log k)$.

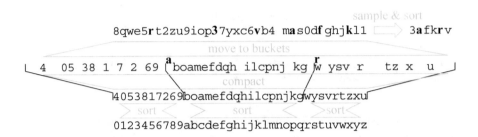

Fig. 5.33. Sorting 36 elements with logarithmic latency. The sample size is 6 and the number of buckets 3.

We then move elements to the right bucket. This is nontrivial. The trick is to allocate a target array for the buckets that is a constant factor larger than necessary with high probability. Each PE repeatedly attempts to copy its element into a random position in its bucket, using a CAS instruction. If this fails, the attempt is repeated until it suceeds. In each iteration, there is a constant success probability. Thus, with high probability, $O(\log p)$ iterations suffice until all PEs have placed their element.

Next, the bucket array is compressed. Denoting an empty entry by a 0 and a full entry by a 1, the position of a full entry in the compressed array is given by a prefix sum over these flags.

Finally, we recurse on the buckets such that again one PE is available for each element. An important technicality is how each PE learns about its bucket. Say that PE i is reponsible for element $A[i]$ where A is the compressed array. Then $A[i]$ is once more located among the splitters s using binary search, say $s[j] \leq A[i] < s[j+1]$. Now, by locating $s[j]$ and $s[j+1]$ in A, PE i can learn the left and right boundaries of its bucket. Figure 5.33 gives an example.

All activities in the first level of recursion run in time $O(\log p)$. Moreover, the number of PEs in a bucket shrinks by a factor $p^{\Omega(1)}$ with high probability. This means that the *logarithm* of the maximum number of PEs having to interact shrinks by a constant factor. Thus, summing over all levels of recursion, we obtain a geometric series summing to $O(\log p)$.

It is an interesting question to what extent this algorithm is practical. In its favor, when $n \gg p$, the algorithm can be generalized in such a way that the number of element comparisons approaches $n \log n$. Moreover, each element is moved only $O(\log \log p)$ times – the number of recursion levels. On the other hand, these element movements are implemented with very expensive random CAS operations, in contrast to the more cache-efficient operations used, for example, by parallel quicksort. Our conclusion is that, on currently predominant architectures, finding a sweet spot for the algorithm may be difficult. The algorithm might prove useful for large shared-memory machines that hide the cache miss latency by making extensive use of hardware threads. Such architectures have been built in the past [187, 250] and may reappear in the future.

5.16 Implementation Notes

Comparison-based sorting algorithms are usually available in standard libraries, and so you may not have to implement one yourself. Many libraries use tuned implementations of quicksort.

Canned noncomparison-based sorting routines are less readily available. Figure 5.34 shows an array-based implementation of *Ksort*. It works well for small to medium-sized problems. For large K and n, it suffers from the problem that the distribution of elements to the buckets may cause a cache fault for every element.

To fix this problem, one can use multiphase algorithms similar to MSD radix sort. The number K of output sequences should be chosen in such a way that one block from each bucket is kept in the cache; see also [221]. The distribution degree K can be larger when the subarray to be sorted fits into the cache. We can then switch to a variant of *uniformSort*; see Fig. 5.20.

Another important practical aspect concerns the type of elements to be sorted. Sometimes we have rather large elements that are to be sorted with respect to small keys. For example, you may want to sort an employee database by last name. In this situation, it makes sense to first extract the keys and store them in an array together with pointers to the original elements. Then, only the key–pointer pairs are sorted. If the original elements need to be brought into sorted order, they can be permuted accordingly in linear time using the sorted key–pointer pairs.

Procedure *KSortArray*(a,b : *Array* [1..n] **of** *Element*)
    $c = \langle 0,\dots,0 \rangle$: *Array* [0..$K-1$] **of** $\mathbb{N}$ // counters for each bucket
    for $i := 1$ **to** n **do** $c[key(a[i])]$++   // Count bucket sizes

    $C := 1$
    for $k := 0$ **to** $K-1$ **do** $(C,c[k]) := (C+c[k],C)$ // Store $\sum_{i<k} c[k]$ in $c[k]$.

    for $i := 1$ **to** n **do**   // Distribute $a[i]$
        $b[c[key(a[i])]] := a[i]$
        $c[key(a[i])]$++

Fig. 5.34. Array-based sorting with keys in the range $0..K-1$. The input is an unsorted array a. The output is b, containing the elements of a in sorted order. We first count the number of inputs for each key. Then we form the partial sums of the counts. Finally, we write each input element to the correct position in the output array

Multiway merging of a small number of sequences (perhaps up to eight) deserves special mention. In this case, the priority queue can be kept in the processor registers [263, 330].

5.16.1 C/C++

Sorting is one of the few algorithms that is part of the C standard library. However, the C sorting routine *qsort* is slower and harder to use than the C++ function *sort*. The main reason is that the comparison function is passed as a function pointer and is called for every element comparison. In contrast, *sort* uses the template mechanism of C++ to figure out at compile time how comparisons are performed so that the code generated for comparisons is often a single machine instruction. The parameters passed to *sort* are an iterator pointing to the start of the sequence to be sorted, and an iterator pointing after the end of the sequence. In our experiments using an Intel Pentium III and GCC 2.95, *sort* on arrays ran faster than our manual implementation of quicksort. One possible reason is that compiler designers may tune their code optimizers until they produce good code for the library version of quicksort. There is an efficient parallel-disk external-memory sorter in STXXL [88], an external-memory implementation of the STL. Efficient parallel sorters (parallel quicksort and parallel multiway mergesort) for multicore machines are available in the GNU standard library [212, 298]. On GPUs, radix sort [283], mergesort [83], and sample sort [199] have been used.

Exercise 5.58. Give a C or C++ implementation of the procedure *qSort* in Fig. 5.9. Use only two parameters: a pointer to the (sub)array to be sorted and its size.

5.16.2 Java

The Java 6 platform provides a method *sort* which implements a stable binary merge-sort for *Arrays* and *Collections*. One can use a customizable *Comparator*, but there is also a default implementation for all classes supporting the interface *Comparable*.
    The *Arrays* class provides a method *parallelSort*.

5.17 Historical Notes and Further Findings

In later chapters, we shall discuss several generalizations of sorting. Chapter 6 discusses priority queues, a data structure that supports insertions of elements and removal of the smallest element. In particular, inserting n elements followed by repeated deletion of the minimum amounts to sorting. Fast priority queues result in quite good sorting algorithms. A further generalization is the *search trees* introduced in Chap. 7, a data structure for maintaining a sorted list that allows searching, inserting, and removing elements in logarithmic time.
    We have seen several simple, elegant, and efficient randomized algorithms in this chapter. An interesting question is whether these algorithms can be replaced

by deterministic ones. Blum et al. [48] described a deterministic median selection algorithm that is similar to the randomized algorithm discussed in Sect. 5.8. This deterministic algorithm makes pivot selection more reliable using recursion: It splits the input set into subsets of five elements, determines the median of each subset, for example by sorting each five-element subset, then determines the median of the $n/5$ medians by calling the algorithm recursively, and finally uses the median of the medians as the splitter. The resulting algorithm has linear worst-case execution time, but the large constant factor makes the algorithm impractical. (We invite the reader to set up a recurrence for the running time and to show that it has a linear solution.)

There are quite practical ways to reduce the expected number of comparisons required by quicksort. Using the median of three random elements yields an algorithm with about $1.188 n \log n$ comparisons. The median of three medians of three-element subsets brings this down to $\approx 1.094 n \log n$ [41]. The number of comparisons can be reduced further by making the number of elements considered for pivot selection dependent on the size of the subproblem. Martínez and Roura [208] showed that for a subproblem of size m, the median of $\Theta(\sqrt{m})$ elements is a good choice for the pivot. With this approach, the total number of comparisons becomes $(1 + o(1)) n \log n$, i.e., it matches the lower bound of $n \log n - O(n)$ up to lower-order terms. Interestingly, the above optimizations can be counterproductive with respect to actual running time. Although fewer instructions are executed, it becomes impossible to predict when the inner while-loops of quicksort will be aborted. Since modern, deeply pipelined processors only work efficiently when they can predict the directions of branches taken, the net effect on performance can even be negative [173]. Therefore, in [280], a comparison-based sorting algorithm that avoids conditional branch instructions was developed. This algorithm is also cache-efficient, allows instruction parallelism, and can be made in-place [25]; see also Fig. 5.26. One can also implement quicksort [101] and mergesort [102] in such a way that conditional branches are avoided. An interesting deterministic variant of quicksort is proportion-extend sort [68].

A classical sorting algorithm of some historical interest is *Shell sort* [168, 288], a generalization of insertion sort, that gains efficiency by also comparing nonadjacent elements. It was open for a long time whether some variant of shell sort achieves $O(n \log n)$ average running time [168, 211]. Only recently was it shown that a randomized version of shell sort does so [132].

There are some interesting techniques for improving external multiway mergesort. The *snow plow* heuristic [185, Sect. 5.4.1] forms runs of expected size $2M$ using a fast memory of size M: Whenever an element is selected from the internal priority queue and written to the output buffer and the next element in the input buffer can extend the current run, we add it to the priority queue. Also, the use of *tournament trees* instead of general priority queues leads to a further improvement of multiway merging [185].

Multiway mergesort and distribution sort can be adapted to D parallel disks by *striping*, i.e., any D consecutive blocks in a run or bucket are evenly distributed over the disks. Using randomization, this idea can be developed into almost optimal algorithms that also overlap I/O and computation [89].

We have seen linear-time algorithms for highly structured inputs. A quite general model, for which the $n \log n$ lower bound does not hold, is the *word model*. In this model, keys are integers that fit into a single memory cell, say 32- or 64-bit keys, and the standard operations on words (bitwise AND, bitwise OR, addition, ...) are available in constant time. In this model, sorting is possible in deterministic time $O(n \log \log n)$ [16]. With randomization, even $O(n\sqrt{\log \log n})$ is possible [142]. *Flash sort* [242] is a distribution-based algorithm that works almost in-place.

There has been a huge amount of work on parallel sorting. On a CRCW-PRAM, sorting of n integers in the range $1..n$ is possible using logarithmic time and linear work [261]. However, since this algorithm is not stable, it cannot be extended to keys of polynomial size. Allowing a little more time changes the situation, however [32].

Sorting small inputs can be realized in hardware using *sorting networks*, which consist of wires and sorting gates which have two inputs a, b and two outputs $\max(a,b)$, $\min(a,b)$. Batcher's classical result [33] introduces a merging network which merges two n-element sorted sequences using $O(n \log n)$ gates and a critical path length of $O(\log n)$. A logarithmic number of merging stages then yields an n-element sorting network with $O(n \log^2 n)$ gates and a critical path length $O(\log^2 n)$. This algorithm has also been used as the base case of GPU sorting algorithms [83]. Ajtai et al. [11] gave a sorting network with $O(n \log n)$ gates and a critical path length $O(\log n)$. Unfortunately, the constant factor involved is prohibitively large. A recent improvement of these factors [133] still remains unpractical.

For practical sorting on large distributed-memory machines, there is a gap between algorithms such as sample sort (Sect. 5.13) or multiway mergesort (Sect. 5.14.1) on the one hand, which communicate their data only once but need $\Omega(p)$ message startups and, on the other hand, polylogarithmic time algorithms such as quicksort (Sect. 5.7) or binary mergesort (Sect. 5.4), which move all the data a logarithmic number of times. This gap can be filled using multilevel generalizations of sample sort and multiway mergesort that move the data k times and need $O(p^{1/k})$ message startups [23].

For very large data sets, one can combine techniques from external-memory and distributed-memory parallel processing [260].

Exercise 5.59 (Unix spellchecking). Assume you have a dictionary consisting of a sorted sequence of correctly spelled words. To check a text, you convert it to a sequence of words, sort it, scan the text and dictionary simultaneously, and output the words in the text that do not appear in the dictionary. Implement this spellchecker using Unix tools in a small number of lines of code. Can you do this in one line?

6

Priority Queues

The company TMG markets tailor-made first-rate garments. It organizes marketing, measurements, etc., but outsources the actual fabrication to independent tailors. The company keeps 20% of the revenue. When the company was founded in the 19th century, there were five subcontractors. Now it controls 15% of the world market and there are thousands of subcontractors worldwide.

Let us have a closer look at how orders are assigned to the subcontractors. The rule is that an order is assigned to the tailor who has so far (in the current year) been assigned the smallest total value of orders. The founders of TMG used a blackboard to keep track of the current total value of orders for each tailor; in computer science terms, they kept a list of values and spent linear time to find the correct tailor. The business has outgrown this solution. Can you come up with a more scalable solution where you have to look at only a small number of values to decide who will be assigned the next order?

Next year, the rules will be changed. In order to encourage timely delivery, orders will now be assigned to the tailor with the smallest value of unfinished *orders, i.e., whenever an order is assigned to a tailor, you have to increase the backlog of the tailor, and whenever a finished order arrives, you have to deduct the value of that order from the backlog of the tailor who executed it. In order to assign an order, you have to find the tailor with the smallest backlog. Is your strategy for assigning orders flexible enough to handle this efficiently?*

Priority queues[1] are the data structure required for the problem above and for many other applications. We start our discussion with the precise specification. (Nonaddressable) priority queues maintain a set M of *Element*s with *Key*s supporting the following operations:

---

[1] The photograph shows people queueing at the Eiffel Tower (Doods Dumaguing `www.flickr.com/photos/xianl2/8620507361`).

© Springer Nature Switzerland AG 2019

P. Sanders et al., *Sequential and Parallel Algorithms and Data Structures*,

https://doi.org/10.1007/978-3-030-25209-0_6

- $M.build(\{e_1,\ldots,e_n\})$: $M := \{e_1,\ldots,e_n\}$.
- $M.insert(e)$: $M := M \cup \{e\}$.
- $M.$ min: **return** $\min M$ (an element with minimum key).
- $M.deleteMin$: $e := \min M$; $M := M \setminus \{e\}$;   **return** e.

This is enough for the first part of our example. Each year, we build a new priority queue containing an *Element* with a *Key* of 0 for each contracted tailor. To assign an order, we delete the smallest *Element*, add the order value to its *Key*, and reinsert it. Section 6.1 presents a simple, efficient implementation of this basic functionality.

Addressable priority queues support additional operations. The elements in an addressable priority queue are accessible through a handle. The handle is established when the element is inserted into the queue. The additional operations are:

- $M.insert(e)$: $M := M \cup \{e\}$; Return a handle to e.
- $remove(h)$: Remove the element specified by the handle h.
- $decreaseKey(h,k)$: Decrease the key of the element at handle h to k.
- $M.merge(Q)$: $M := M \cup Q$; $Q := \emptyset$.

In our example, the operation *remove* might be helpful when a contractor is fired because he/she delivers poor quality. Using this operation together with *insert*, we can also implement the "new contract rules": When an order is assigned or delivered, we remove the *Element* for the contractor who executed the order, update its backlog value, and reinsert the *Element*. *DecreaseKey* streamlines the actions for a delivery to a single operation. In Sect. 6.2, we shall see that this is not just convenient but that decreasing keys can be implemented more efficiently than arbitrary element updates.

Priority queues have many applications. For example, in Sect. 12.2, we shall see that our introductory example can also be viewed as a greedy algorithm for a machine-scheduling problem. Also, the selection-sort algorithm of Sect. 5.1 can be implemented efficiently now: First, insert all elements into a priority queue, and then repeatedly delete the smallest element and output it. A tuned version of this idea is described in Sect. 6.1. The resulting *heapsort* algorithm is popular because it needs no additional space and is worst-case efficient.

In a *discrete-event simulation*, one has to maintain a set of pending events. Each event happens at some scheduled point in time and creates some number, maybe zero, of new events in the future. Pending events are kept in a priority queue. The main loop of the simulation deletes the next event from the queue, executes it, and inserts newly generated events into the priority queue. Note that the priorities (times) of the deleted elements (simulated events) increase monotonically during the simulation. It turns out that many applications of priority queues have this monotonicity property. Section 10.5 explains how to exploit monotonicity for integer keys.

Another application of monotone priority queues is the *best-first branch-and-bound* approach to optimization described in Sect. 12.4. Here, the elements are partial solutions of an optimization problem and the keys are optimistic estimates of the obtainable solution quality. The algorithm repeatedly removes the best-looking partial solution, refines it, and inserts zero or more new partial solutions.

Scheduling tasks of different importance in a parallel system is an application of parallel priority queues. For example, when parallelizing branch-and-bound, the partial solutions might be viewed as such tasks and the importance of a task might be a bound on its objective value.

We shall see two applications of addressable priority queues in the chapters on graph algorithms. In both applications, the priority queue stores nodes of a graph. Dijkstra's algorithm for computing shortest paths (Sect. 10.3) uses a monotone priority queue where the keys are path lengths. The Jarník–Prim algorithm for computing minimum spanning trees (Sect. 11.2) uses a (nonmonotone) priority queue where the keys are the weights of edges connecting a node to a partial spanning tree. In both algorithms, there can be a *decreaseKey* operation for each edge, whereas there is at most one *insert* and *deleteMin* for each node. Observe that the number of edges may be much larger than the number of nodes, and hence the implementation of *decreaseKey* deserves special attention.

Exercise 6.1. Show how to implement bounded nonaddressable priority queues using arrays. The size of the queue is bounded by w and when the queue has a size n, the first n entries of the array are used. Compare the complexity of the queue operations for two implementations: one by unsorted arrays and one by sorted arrays.

Exercise 6.2. Show how to implement addressable priority queues using doubly linked lists. Each list item represents an element in the queue, and a handle is a handle of a list item. Compare the complexity of the queue operations for two implementations: one by sorted lists and one by unsorted lists.

In Sect. 6.1 we begin with a very simple array-based data structure that is well suited to nonaddressable priority queues. We then discuss pointer-based addressable priority queues in Sect. 6.2. Section 6.3 shows how nonaddressable queues can be implemented efficiently in external memory. Section 6.4 discusses parallel priority queues. We shall describe a version with bulk operations that accesses many elements at once.

6.1 Binary Heaps

Heaps are a simple and efficient implementation of nonaddressable bounded priority queues [332]. They can be made unbounded in the same way as bounded arrays can be made unbounded; see Sect. 3.4. Heaps can also be made addressable, but we shall see better addressable queues in later sections.

We use an array $h[1..w]$ that stores the elements of the queue. The first n entries of the array are used. The array is *heap-ordered*, i.e.,

$$\text{for } j \text{ with } 2 \le j \le n: \quad h[\lfloor j/2 \rfloor] \le h[j].$$

What does "heap-ordered" mean? The key to understanding this definition is a bijection between positive integers and the nodes of a complete binary tree, as illustrated in Fig. 6.1. Node 1 is the root of the tree, and the children of node i are the nodes with numbers $2i$ and $2i + 1$. The parent of a node $i \ge 2$ is node $\lfloor i/2 \rfloor$. A heap of size

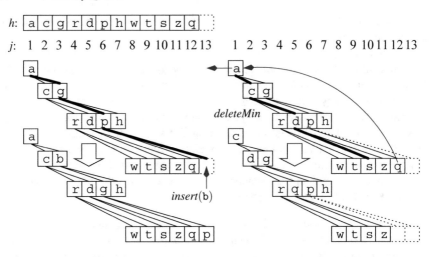

Fig. 6.1. The *top part* shows a heap with $n = 12$ elements stored in an array h with $w = 13$ entries. The root corresponds to index 1. The children of the root correspond to indices 2 and 3. The children of node i have indices $2i$ and $2i + 1$ (if they exist). The parent of a node $i, i \geq 2$, has index $\lfloor i/2 \rfloor$. If n elements are stored in the heap, they are stored in nodes 1 to n. The keys are characters, with their usual alphabetic order. The invariant states that the key of a parent is no larger than the keys of its children, i.e., the tree is heap-ordered. The *left part* shows the effect of inserting b. We add node 13 as a new leaf to the heap. The thick edges mark a path from node 13 to the root. The new element b is moved up this path until its parent is not larger. The remaining elements on the path are moved down to make room for b. The *right part* shows the effect of deleting the minimum. The thick edges mark the path p that starts at the root and always proceeds to the child with the smaller *Key*. The element q stored in the node n is provisionally moved to the root and then moves down p until its successor in p is not smaller anymore. The remaining elements move up to make room for q.

n uses nodes 1 to n. In a heap, the key of a parent is no larger than the keys of the children. In particular, a minimum element is stored in the root (= array position 1). Thus the operation min takes time $O(1)$. Creating an empty heap with space for w elements also takes constant time, as it only needs to allocate an array of size w:

Class *BinaryHeapPQ*$(w : \mathbb{N})$ **of** *Element*
    $h : Array\ [1..w]$ **of** *Element*   // The *heap h* is
    $n = 0 : \mathbb{N}$ // initially *empty* and has the
    invariant $\forall j \in 2..n : h[\lfloor j/2 \rfloor] \leq h[j]$ // *heap property*, which implies that
    Function min **assert** $n > 0$; **return** $h[1]$ // the *root* contains the *min*imum.

The minimum of a heap is stored in $h[1]$ and hence can be found in constant time; this is the same as for a sorted array. However, the heap property is much less restrictive than the property of being sorted. For example, there is only one sorted version of the set $\{1, 2, 3\}$, but both $\langle 1, 2, 3 \rangle$ and $\langle 1, 3, 2 \rangle$ are legal heap representations.

Exercise 6.3. Give all representations of $\{1, 2, 3, 4\}$ as a heap.

We shall next see that the increased flexibility permits efficient implementations of *insert* and *deleteMin*. We choose a description which is simple and easily proven correct. Section 6.5 gives some hints towards a more efficient implementation. An *insert* puts a new element *e* tentatively at the end of the heap *h*, i.e., it increments *n* and tentatively puts the new element into $h[n]$. This may violate the heap property at position *n*. To repair the heap property, we move *e* to an appropriate position on the path from leaf $h[n]$ to the root:

> **Procedure** *insert*(*e* : *Element*)
>   **assert** $n < w$
> n++; $h[n]:=e$
>   *siftUp*(*n*)

Here, *siftUp*(*n*) moves the contents of node *n* towards the root until either the root is reached or the key in the parent is no larger anymore; see Fig. 6.1. We have to prove that this restores the heap property. We write "heap except maybe at *i*" if *h* is a heap or $i > 1$, $h[i] < h[\lfloor i/2 \rfloor]$ and replacing $h[i]$ by $h[\lfloor i/2 \rfloor]$ turns *h* into a heap. When we put the new element into $h[n]$, *h* is a heap except maybe at *n*. Assume now that *h* is a heap except maybe at *i* when *siftUp*(*i*) is called. By the preceding sentence, this is true for the first call with $i = n$. If $i = 1$ or $h[\lfloor i/2 \rfloor] \leq h[i]$, *h* is a heap and we are done. If $i > 1$ and $h[\lfloor i/2 \rfloor] > h[i]$, we swap $h[i]$ and $h[\lfloor i/2 \rfloor]$. The heap property now holds for the children of *i*, since it sufficed to replace $h[i]$ by $h[\lfloor i/2 \rfloor]$ to restore the heap property. It clearly holds for *i*, and it holds for the sibling of *i* since we have replaced $h[\lfloor i/2 \rfloor]$ by something smaller. Hence *h* is a heap except maybe at $\lfloor i/2 \rfloor$, and we have established the invariant for the recursive call.

> **Procedure** *siftUp*(*i* : ℕ)
>   **assert** *h* is a heap except maybe at *i*.
>   **if** $i = 1 \vee h[\lfloor i/2 \rfloor] \leq h[i]$ **then return**
>   *swap*($h[i], h[\lfloor i/2 \rfloor]$)
>   *siftUp*($\lfloor i/2 \rfloor$)

Exercise 6.4. Show that the running time of *siftUp*(*n*) is O(log *n*) and hence an *insert* takes time O(log *n*). Reformulate *siftUp* as a while-loop.

A *deleteMin* returns the content of the root and replaces it by the content of node *n*. Since $h[n]$ might be larger than $h[2]$ or $h[3]$, this manipulation may violate the heap property at position 2 or 3. This possible violation is repaired using *siftDown*:

> **Function** *deleteMin* : *Element*
>   **assert** $n > 0$
>   *result* = $h[1]$: *Element*
> $h[1]:=h[n]$; n--
>   *siftDown*(1)
>   **return** *result*

The procedure *siftDown*(1) moves the new content of the root, which we call *e*, down the tree until the heap property holds. More precisely, consider the path *p* that starts at the root and always proceeds to the child with the smaller key; see Fig. 6.1. In the case of equal keys, the choice is arbitrary. We extend the path until all children of the last node of the path (there may be zero, one, or two) have a key no smaller than e. We put e into this position and move all elements on path p up by one position. In this way, the heap property is restored. Clearly, e is no larger than the elements stored in the children. Also moving the elements on p up by one position maintains the heap property because p is always extended to the child with the smaller key. The strategy is most easily formulated as a recursive procedure. A call of the following procedure, *siftDown(i)*, repairs the heap property in the subtree rooted at i, assuming that it holds already for the subtrees rooted at $2i$ and $2i + 1$; the heap property holds in the subtree rooted at i if we have $h[\lfloor j/2 \rfloor] \le h[j]$ for all proper descendants j of i:

Procedure *siftDown(i* : $\mathbb{N}$)
    assert the heap property holds for the trees rooted at $j = 2i$ and $j = 2i + 1$
    if $2i \le n$ **then**   // i is not a leaf
        if $2i + 1 > n \lor h[2i] \le h[2i+1]$ **then** $m := 2i$ **else** $m := 2i + 1$
        assert the sibling of m does not exist or it has a larger key than m
        if $h[i] > h[m]$ **then**   // the heap property is violated
            $swap(h[i], h[m])$
            $siftDown(m)$
    assert the heap property holds for the tree rooted at i

Exercise 6.5. Why is it important that the path is always extended to the child with the smaller key? Reformulate *siftDown* as a while-loop.

Exercise 6.6. Our current implementation of *siftDown* needs about $2 \log n$ element comparisons. Show how to reduce this to $\log n + O(\log \log n)$. Hint: Determine the path p first and then perform a binary search on this path to find the proper position for $h[1]$. Section 6.6 has more on variants of *siftDown*.

We can obviously build a heap from n elements by inserting them one after the other in $O(n \log n)$ total time. Interestingly, we can do better by establishing the heap property in a bottom-up fashion: *siftDown* allows us to establish the heap property for a subtree of height $k + 1$ provided the heap property holds for its subtrees of height k. The following exercise asks you to work out the details of this idea.

Exercise 6.7 (*buildHeap*). Assume that you are given an arbitrary array $h[1..n]$ and want to establish the heap property on it by permuting its entries. Consider two procedures for achieving this:

Procedure *buildHeapBackwards*
    for $i := \lfloor n/2 \rfloor$ **downto** 1 **do** *siftDown(i)*

Procedure *buildHeapRecursive(i* : $\mathbb{N}$)
    if $4i \le n$ **then**
        buildHeapRecursive(2i)
        buildHeapRecursive(2i + 1)
    siftDown(i)

(a) Show that both *buildHeapBackwards* and *buildHeapRecursive*(1) establish the heap property everywhere.
(b) Implement both algorithms efficiently and compare their running times for random integer keys and $n \in \{10^i : 2 \leq i \leq 8\}$. It will be important how efficiently you implement *buildHeapRecursive*. In particular, it might make sense to unravel the recursion for small subtrees.
*(c) For large n, the main difference between the two algorithms is in memory hierarchy effects. Analyze the number of I/O operations required by the two algorithms in the external-memory model described at the end of Sect. 2.2. In particular, show that if the block size is B and the fast memory has size $M = \Omega(B \log B)$, then *buildHeapRecursive* needs only $O(n/B)$ I/O operations.

The following theorem summarizes our results on binary heaps.

Theorem 6.1. *The heap implementation of nonaddressable priority queues realizes creating an empty heap and finding the minimum element in constant time, deleteMin and insert in logarithmic time* $O(\log n)$, *and build in linear time.*

Proof. The binary tree represented by a heap of n elements has height $k = \lfloor \log n \rfloor$. *insert* and *deleteMin* explore one root-to-leaf path and hence have logarithmic running time; min returns the content of the root and hence takes constant time. Creating an empty heap amounts to allocating an array and therefore takes constant time. *build* calls *siftDown* for at most 2^ℓ nodes of depth ℓ. Such a call takes time $O(k - \ell)$. Thus total the time is

$$O\left(\sum_{0 \leq \ell < k} 2^\ell (k - \ell)\right) = O\left(2^k \sum_{0 \leq \ell < k} \frac{k - \ell}{2^{k-\ell}}\right) = O\left(2^k \sum_{j \geq 1} \frac{j}{2^j}\right) = O(n).$$

The last equality uses (A.15). □

Heaps are the basis of *heapsort*. We first *build* a heap from the elements and then repeatedly perform *deleteMin*. Before the ith *deleteMin* operation, the ith smallest element is stored at the root $h[1]$. We swap $h[1]$ and $h[n - i + 1]$ and sift the new root down to its appropriate position. At the end, h stores the elements sorted in decreasing order. Of course, we can also sort in increasing order by using a *max-priority queue*, i.e., a data structure supporting the operations of *insert* and of deleting the maximum.

Heaps do not directly implement the addressable priority queues, since elements are moved around in the array h during insertion and deletion. Thus the array indices cannot be used as handles.

Exercise 6.8 (addressable binary heaps). Extend heaps to an implementation of addressable priority queues. How many additional pointers per element do you need? There is a solution with two additional pointers per element.

*Exercise 6.9 (bulk insertion). Design an algorithm for inserting k new elements into an n-element heap. Give an algorithm that runs in time $O(k \log k + \log n)$. Hint: Use a bottom-up approach similar to that for heap construction.

6.2 Addressable Priority Queues

Binary heaps have a rather rigid structure. All n elements are arranged into a single binary tree of height $\lfloor \log n \rfloor$. In order to obtain faster implementations of the operations *insert*, *decreaseKey*, *remove*, and *merge*, we now look at more flexible structures. The single, left-complete binary tree is replaced by a collection of trees (i.e., a forest) with arbitrary shape. Each tree is still *heap-ordered*, i.e., no child is smaller than its parent. In other words, the sequence of keys along any root-to-leaf path is nondecreasing. Here is an example of a heap-ordered forest for the set $\{0,1,3,4,5,7,8\}$:

The elements of the queue are now stored in *heap items* that have a persistent location in memory. Hence, pointers to heap items can serve as *handle* of priority queue elements. The tree structure is explicitly defined using pointers between items.

We shall discuss several variants of addressable priority queues. We start with the common principles underlying all of them. Figure 6.2 summarizes the commonalities. In order to keep track of the current minimum, we maintain a handle to the root containing it. We use *minPtr* to denote this handle. The forest is manipulated using three simple operations: adding a new tree (and keeping *minPtr* up to date), combining two trees into a single one, and cutting out a subtree, making it a tree on its own.

An *insert* adds a new single-node tree to the forest. So, a sequence of n *insert*s into an initially empty heap will simply create n single-node trees. The cost of an *insert* is clearly $O(1)$.

A *deleteMin* operation removes the node indicated by *minPtr*. This turns all children of the removed node into roots. We then scan the set of roots (old and new) to find the new minimum, a potentially very costly process. We also perform some rebalancing, i.e., we combine trees into larger ones. The details of this process distinguish different kinds of addressable priority queue and are the key to efficiency.

We turn now to *decreaseKey*(h,k), which decreases the key value at a handle h to k. Of course, k must not be larger than the old key stored with h. Decreasing the key associated with h may destroy the heap property because h may now be smaller than its parent. In order to maintain the heap property, we cut the subtree rooted at h and turn h into a root. This sounds simple enough, but may create highly skewed trees. Therefore, some variants of addressable priority queues perform additional operations to keep the trees in shape.

The remaining operations are easy. We can *remove* an item from the queue by first decreasing its key so that it becomes the minimum item in the queue, and then perform a *deleteMin*. To merge a queue o into another queue, we compute the union of *roots* and *o.roots*. To update *minPtr*, it suffices to compare the minima of the merged queues. If the root sets are represented by linked lists and no additional balancing is done, a merge needs only constant time.

In the remainder of this section, we shall discuss particular implementations of addressable priority queues.

Class *Handle* = **Pointer to** *PQItem*

Class *AddressablePQ*

    minPtr : *Handle*   // root that stores the minimum

    roots : *Set* **of** *Handle*   // pointers to tree roots

    Function min **return** element stored at *minPtr*

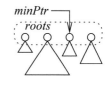

    Procedure *link*(*a*,*b* : *Handle*)

        assert $a \leq b$

        remove *b* from *roots*

        make *a* the parent of *b*   //

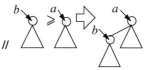

    Procedure *combine*(*a*,*b* : *Handle*)

        assert *a* and *b* are tree roots

        if $a \leq b$ **then** *link*(*a*,*b*) **else** *link*(*b*,*a*)

    Procedure *newTree*(*h* : *Handle*)

        roots := *roots* $\cup \{h\}$

        if $*h <$ min **then** *minPtr* := *h*

    Procedure *cut*(*h* : *Handle*)

        remove the subtree rooted at *h* from its tree   // *h*

        newTree(*h*)

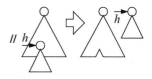

    Function *insert*(*e* : *Element*) : *Handle*

        i := a *Handle* for a new *PQItem* storing *e*

        newTree(*i*)

        return *i*

    Function *deleteMin* : *Element*

        e := the *Element* stored in *minPtr*

        foreach child *h* of the root at *minPtr* **do** *cut*(*h*)   //

        dispose *minPtr*

        perform some rebalancing and update *minPtr*   // uses *combine*

        return *e*

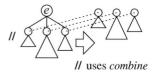

    Procedure *decreaseKey*(*h* : *Handle*, *k* : *Key*)

        change the key of *h* to *k*

        if *h* is not a root **then**

            cut(*h*);   possibly perform some rebalancing

    Procedure *remove*(*h* : *Handle*)   *decreaseKey*(*h*, $-\infty$);   *deleteMin*

    Procedure *merge*(*o* : *AddressablePQ*)

        if $*minPtr > *(o.minPtr)$ **then** *minPtr* := *o.minPtr*

        roots := *roots* $\cup$ *o.roots*

        o.roots := $\emptyset$;   possibly perform some rebalancing

Fig. 6.2. Addressable priority queues

6.2.1 Pairing Heaps

Pairing heaps [114] use a very simple technique for rebalancing. Pairing heaps are very efficient in practice. However, a full theoretical analysis is missing.

We present a simple variant of pairing heap that also has good provable bounds. There is always exactly one tree, and nodes may have an arbitrary number of children. Whenever an operation creates several roots, a rebalancing operation is necessary. The most complex rebalancing is done after a *deleteMin*. The root contains an element with a minimum key. After removal of the root, the children of the old root form a sequence $\langle r_1, \ldots, r_k \rangle$ of roots. They are combined into a single tree in the following *two-pass* process. In the first pass, the trees are combined in pairs, i.e., the trees with roots r_1 and r_2, r_3 and r_4, and so on, are joined by calls of *combine*. The resulting $\lceil k/2 \rceil$ trees are then combined into a single tree in the second pass. The last tree is joined with the next to last, the resulting tree is joined with the last tree but two, and so on. Figure 6.3 shows an example. The operations *insert*, *decreaseKey* and *merge* generate pairs of roots. They are simply combined into a single tree by a call of *combine*.

Exercise 6.10 (three-pointer items). Explain how to implement pairing heaps using three pointers per heap item i: one to the oldest child (i.e., the child linked first to i), one to the next younger sibling (if any), and one to the next older sibling. If there is no older sibling, the third pointer goes to the parent. Figure 6.6 gives an example.

*Exercise 6.11 (two-pointer items). Explain how to implement pairing heaps using two pointers per heap item: one to the oldest child and one to next younger sibling. If there is no younger sibling, the second pointer goes to the parent. Figure 6.6 gives an example.

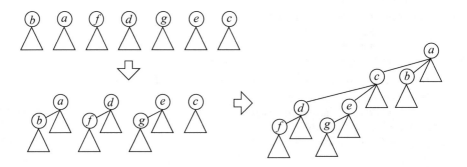

Fig. 6.3. The *deleteMin* operation for pairing heaps makes two passes over the nodes that became roots after the deletion of the old root. In the first pass, roots are combined pairwise. In the second pass, the roots are scanned sequentially from right to left and, in each step, the last two roots are joined. In this example, e becomes the child of c, then d becomes the child of c, and finally c becomes the child of a.

6.2.2 *Fibonacci Heaps

Fibonacci heaps [115] use more intensive balancing operations than do pairing heaps. This paves the way for a theoretical analysis. In particular, we obtain logarithmic amortized time for *remove* and *deleteMin* and worst-case constant time for all other operations.

Each item of a Fibonacci heap stores four pointers that link it to its parent, one child, and two siblings; see Fig. 6.6. The children of each node form a doubly linked circular list using the sibling pointers. The sibling pointers of the root nodes are used to represent the set *roots* of all roots in a similar way. Parent pointers of roots and child pointers of leaf nodes have a special value, for example a null pointer.

In addition, every heap item contains a field *rank*. The *rank* of an item is simply the number of its children. In Fibonacci heaps, *deleteMin* links only roots of equal rank r. The surviving root will then obtain a rank of $r + 1$. An efficient method to combine trees of equal rank is as follows. Let *maxRank* be an upper bound on the rank of any node. We shall prove below that *maxRank* is logarithmic in n. Maintain a set of buckets, initially empty and numbered from 0 to *maxRank*. Then scan the list of all roots. When scanning a root of rank i, inspect the ith bucket. If the ith bucket is empty, then put the root there. If the bucket is nonempty, then combine the two trees into one. This empties the ith bucket and creates a root of rank $i + 1$. Treat this root in the same way, i.e., try to throw it into the $(i + 1)$th bucket. If it is occupied, combine .... When all roots have been processed in this way, we have a collection of trees whose roots have pairwise distinct ranks; see Fig. 6.4.

A *deleteMin* can be very expensive if there are many roots. For example, a *deleteMin* following n insertions has a cost $\Omega(n)$. However, in an amortized sense, the cost of *deletemin* is O(*maxRank*). The reader must be familiar with the technique of amortized analysis (see Sect. 3.5) before proceeding further. For the amortized analysis, we postulate that each root holds one token. Tokens pay for a constant amount of computing time.

Lemma 6.2. *The amortized complexity of deleteMin is* O(*maxRank*).

Proof. A *deleteMin* first calls *newTree* at most *maxRank* times (since the degree of the root containing the old minimum is bounded by *maxRank*) and then initializes an array of size *maxRank*. So far, the running time is O(*maxRank*), and at most *maxRank* new tokens need to be created. The remaining time is proportional to the

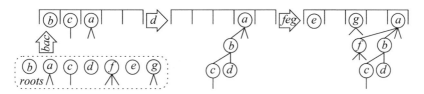

Fig. 6.4. An example of the development of the bucket array during execution of *deleteMin* for a Fibonacci heap. The arrows indicate the roots scanned. Note that scanning d leads to a cascade of three combine operations.

number of *combine* operations performed. Each *combine* turns a root into a nonroot and is paid for by the token associated with the node turning into a nonroot.   □

How can we guarantee that *maxRank* stays small? Let us consider a simple situation first. Suppose that we perform a sequence of insertions followed by a *deleteMin*. In this situation, we start with a certain number of single-node trees, and all trees formed by combining are *binomial trees*, as shown in Fig. 6.5. The binomial tree B_0 consists of a single node, and the binomial tree B_{i+1} is obtained by combining two copies of B_i. This implies that the root of B_i has rank i and that B_i contains exactly 2^i nodes. Thus the rank of a binomial tree is logarithmic in the size of the tree.

Unfortunately, *decreaseKey* may destroy the nice structure of binomial trees. Suppose an item v is cut out. We now have to decrease the rank of its parent w. The problem is that the size of the subtrees rooted at the ancestors of w has decreased but their rank has not changed, and hence we can no longer claim that the size of a tree stays exponential in the rank of its root. Therefore, we have to perform some rebalancing to keep the trees in shape. An old solution [325] is to keep all trees in the heap binomial. However, this causes logarithmic cost for a *decreaseKey*.

*Exercise 6.12 (binomial heaps). Work out the details of this idea. Assume that the key of v is decreased and becomes smaller than the key stored in its parent. Cut the following links. For each nonroot ancestor w of v (this includes v), cut the link to its parent. Moreover, for each such node w, cut the links from all siblings of w of rank higher than w to their parents. Show that all resulting trees are binomial. Then combine trees of equal rank until there is at most one tree of each rank. Argue that the cost of *decreaseKey* is logarithmic.

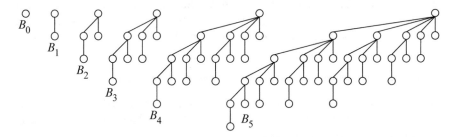

Fig. 6.5. The binomial trees of ranks 0 to 5

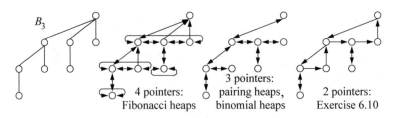

Fig. 6.6. Three ways to represent trees of nonuniform degree. The binomial tree of rank three, B_3, is used as an example.

Fibonacci heaps allow the trees to go out of shape but in a controlled way. The idea is surprisingly simple and is based on the amortized analysis of binary counters; see Sect. 3.4.3. We introduce an additional flag for each node. A node may or may not be marked. Roots are never marked. In particular, when *newTree*(h) is called in *deleteMin*, it removes the mark from h (if any). Thus when *combine* combines two trees into one, neither node is marked.

When a nonroot item x loses a child because *decreaseKey* has been applied to the child, x is marked; this assumes that x is not already marked. Otherwise, when x has already been marked, we cut x, remove the mark from x, and attempt to mark x's parent. If x's parent is already marked, then we continue in the same way. This technique is called *cascading cuts*. In other words, suppose that we apply *decreaseKey* to an item v and that the k nearest ancestors of v are marked. We turn v and the k nearest ancestors of v into roots, unmark them, and mark the $(k+1)$th nearest ancestor of v (if it is not a root). Figure 6.7 gives an example. Observe the similarity to carry propagation in binary addition.

For the amortized analysis, we postulate that each marked node holds two tokens and each root holds one token. Please check that this assumption does not invalidate the proof of Lemma 6.2.

Lemma 6.3. *The amortized complexity of decreaseKey is constant.*

Proof. Assume that we decrease the key of item v and that the k nearest ancestors of v are marked. Here, $k \geq 0$. The running time of the operation is $O(1+k)$. Each of the k marked ancestors carries two tokens, i.e., we have a total of $2k$ tokens available. We create $k+1$ new roots and need one token for each of them. Also, we mark one unmarked node and need two tokens for it. Thus we need a total of $k+3$ tokens. In other words, $k-3$ tokens are freed. They pay for all but $O(1)$ of the cost of *decreaseKey*. Thus the amortized cost of *decreaseKey* is constant.   □

How do cascading cuts affect the size of trees? We shall show that it stays exponential in the rank of the root. In order to do so, we need some notation. Recall the sequence $0, 1, 1, 2, 3, 5, 8, \ldots$ of Fibonacci numbers. These are defined by the recurrence $F_0 = 0$, $F_1 = 1$, and $F_i = F_{i-1} + F_{i-2}$ for $i \geq 2$. It is well known that $F_{i+2} \geq ((1+\sqrt{5})/2)^i \geq 1.618^i$ for all $i \geq 0$.

Exercise 6.13. Prove that $F_{i+2} \geq ((1+\sqrt{5})/2)^i \geq 1.618^i$ for all $i \geq 0$ by induction.

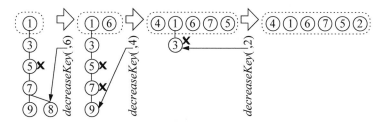

Fig. 6.7. Cascading cuts. Marks are drawn as crosses. Note that roots are never marked.

Lemma 6.4. *Let v be any item in a Fibonacci heap and let i be the rank of v. The subtree rooted at v then contains at least F_{i+2} nodes. In a Fibonacci heap with n items, all ranks are bounded by $1.4404 \log n$.*

Proof. Consider an arbitrary item v of rank i. Order the children of v by the time at which they were made children of v. Let w_j be the jth child, $1 \leq j \leq i$. When w_j was made a child of v, both nodes had the same rank. Also, since at least the nodes $w_1, \ldots, w_{j-1}$ were children of v at that time, the rank of v was at least $j - 1$ then. The rank of w_j has decreased by at most 1 since then, because otherwise w_j would no longer be a child of v. Thus the current rank of w_j is at least $j - 2$.

We can now set up a recurrence for the smallest number S_i of nodes in a tree whose root has rank i. Clearly, $S_0 = 1$ and $S_1 = 2$. Also, $S_i \geq 2 + S_0 + S_1 + \cdots + S_{i-2}$, since for $j \geq 2$ the number of nodes in the subtree with root w_j is at least S_{j-2}, and there are the nodes v and w_1. The recurrence above (with = instead of $\geq$) generates the sequence 1, 2, 3, 5, 8, $\ldots$, which is identical to the Fibonacci sequence (minus its first two elements).

Let us verify this by induction. Let $T_0 = 1$, $T_1 = 2$, and $T_i = 2 + T_0 + \cdots + T_{i-2}$ for $i \geq 2$. Then, for $i \geq 2$, $T_{i+1} - T_i = 2 + T_0 + \cdots + T_{i-1} - 2 - T_0 - \cdots - T_{i-2} = T_{i-1}$, i.e., $T_{i+1} = T_i + T_{i-1}$. This proves $T_i = F_{i+2}$.

For the second claim, we observe that $F_{i+2} \leq n$ implies $i \cdot \log((1 + \sqrt{5})/2) \leq \log n$, which in turn implies $i \leq 1.4404 \log n$.   □

This concludes our treatment of Fibonacci heaps. We have shown the following result.

Theorem 6.5. *The following time bounds hold for Fibonacci heaps: min, insert, and merge take worst-case constant time. decreaseKey takes amortized constant time. remove and deleteMin take an amortized time logarithmic in the size of the queue.*

Exercise 6.14. Describe a variant of Fibonacci heaps where all roots have distinct ranks. Hint: Whenever a new root comes into existence, immediately check whether there is already a root of the same rank. If so, combine.

6.3 *External Memory

We now go back to nonaddressable priority queues and consider their cache efficiency and I/O efficiency. A weakness of binary heaps is that the *siftDown* operation goes down the tree in an unpredictable fashion. This leads to many cache faults and makes binary heaps prohibitively slow when they do not fit into the main memory. We now outline a data structure for (nonaddressable) priority queues with more regular memory accesses. It is also a good example of a generally useful design principle: construction of a data structure out of simpler, known components and algorithms.

In this case, the components are internal-memory priority queues, sorting, and multiway merging; see also Sect. 5.12.1. Figure 6.8 depicts the basic design. The data structure consists of two priority queues Q and Q' (e.g., binary heaps) and k

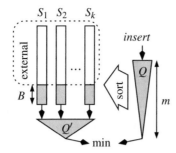

Fig. 6.8. Schematic view of an external-memory priority queue.

sorted sequences $S_1, \ldots, S_k$. Each element of the priority queue is stored either in the *insertion queue* Q, in the *deletion queue* Q', or in one of the sorted sequences. The size of Q is limited to a parameter m. The *deletion queue* Q' stores the smallest element of each sequence, together with the index of the sequence holding the element.

New elements are inserted into the insertion queue. If the insertion queue is full, it is first emptied. In this case, its elements form a new sorted sequence:

Procedure *insert*(e : *Element*)
    if $|Q| = m$ **then**
        k++; $S_k := sort(Q)$; $Q := \emptyset$; $Q'.insert((S_k.popFront, k))$
    $Q.insert(e)$

Q or Q' contains an element with the minimum key. We find it by comparing their minimum elements. If the minimum is in Q' and comes from sequence S_i, the next element in S_i is inserted into Q':

Function *deleteMin*
    if $\min Q \leq \min Q'$ **then** $e := Q.deleteMin$ // assume $\min \emptyset = \infty$
    else $(e, i) := Q'.deleteMin$
        if $S_i \neq \langle \rangle$ **then** $Q'.insert((S_i.popFront, i))$
    return e

It remains to explain how the ingredients of our data structure are mapped to the memory hierarchy. The queues Q and Q' are stored in internal memory. The size bound m for Q should be a constant fraction of the internal-memory size M and a multiple of the block size B. The sequences S_i are largely kept externally. Initially, only the B smallest elements of S_i are kept in an internal-memory buffer b_i. When the last element of b_i is removed, the next B elements of S_i are loaded. Note that we are effectively merging the sequences S_i. This is similar to our multiway merging algorithm described in Sect. 5.12.1. Each inserted element is written to external memory at most once and fetched back to internal memory at most once. Since all accesses to external memory transfer full blocks, the I/O requirement of our algorithm is at most n/B for n queue operations.

The total requirement for internal memory is at most the space for $m + kB + 2k$ elements. This is below the total fast-memory size M if $m = M/2$ and $k \leq \lfloor (M/2 - 2k)/B \rfloor \approx M/(2B)$. If there are many insertions and few deletions, the internal memory may eventually overflow. However, the earliest this can happen is after $m(1 + \lfloor (M/2 - 2k)/B \rfloor) \approx M^2/(4B)$ insertions. For example, if we have 8 GB of main memory, 8-byte elements, and 1 MB disk blocks, we have $M = 2^{30}$ and $B = 2^{17}$ (measured in elements). We can then perform about 2^{41} insertions – enough for 16 TB of data. Similarly to external mergesort, we can handle larger amounts of data by performing multiple phases of multiway merging; see [54, 272]. The data structure becomes considerably more complicated, but it turns out that the I/O requirement for n insertions and deletions is about the same as for sorting n elements. An implementation of this idea is two to three times faster than binary heaps for the hierarchy between cache and main memory [272]. There are also implementations for external memory [88].

6.4 Parallel Priority Queues

We first have to decide what a parallel priority queue should be. For example, are we allowing concurrent queue operations or are we only parallelizing single operations of an otherwise sequential queue? A simple answer is that we only want to parallelize single queue operations. There are indeed such data structures; see [55]. However, the maximum speedup we can hope for is $O(\log n)$, and the constant factor for PE interactions is likely to eat up much of that advantage. We are not aware of practical implementations achieving high speedups. In practice, a little bit of speedup can be obtained by parallelizing the sorting and multiway merging operations in priority queues based on the external queues described in Sect. 6.3; see [35, 45] for details.

Another view on parallel priority queues asks for concurrent access to a priority queue. However, as with the concurrent FIFOs discussed in Sect. 3.7, this raises two severe issues. First of all, it becomes nonobvious how to define the semantics of the queue operations. For example, if several PEs want to extract the minimum, should they all receive the same element? This definition restricts parallelism on the level of the program using the queue and – the second problem – leads to contention, as several PEs will have to access the same element.

We shall therefore concentrate on a third view of parallel priority queues – bulk parallel deletion on a distributed-memory machine. The operation $deleteMin^*(k)$ is executed collectively by all PEs and removes the k globally smallest elements from the queue.[2] Insertions still insert individual elements in an asynchronous fashion.

Our strategy for this kind of parallel priority queue is very simple. Each PE maintains a local priority queue Q. Inserted elements are sent asynchronously to a PE chosen uniformly at random, and inserted there. Thus, each PE has a representative

---

[2] We should point out that not all application programs can make use of bulk deletion. For example, Dijkstra's algorithm for computing shortest paths (Section 10.3) loses its label-setting property (see Theorem 10.6) when we scan several nodes at the same time.

Function *deleteMin**(k : ℕ) : *Sequence* **of** *Element*
   result:=⟨⟩
   $m := initialBufferSize(k,p)$ // choose initial # of removed elements such that with
   repeat   // high probability, one iteration of this loop is enough.
      for $i:=1$ **to** m **do** *result.pushBack(Q.deleteMin)*   // extract result candidates
      $(e_k, k_{here}) := parSelect2(result, k)$ // see Fig. 5.16
   until $\forall i : e_k \leq Q@i.\min$ // no result is missing. Needs all-reduce-and
   for $i:=1$ **to** $|result| - k_{here}$ **do** *Q.insert(result.popBack)*   // reinsert nonresult elements
   return *result*

Fig. 6.9. SPMD pseudocode for bulk *deleteMin* of the k globally smallest elements from a parallel priority queue.

sample of the overall data set. In particular, it can be shown that all globally small elements are among the locally small elements with high probability.

Lemma 6.6. *If $\ell \geq 1$ and $k = \Omega(p\log p)$, then with probability at least $1 - 1/k^\ell$, the k globally smallest elements are among the $O(\ell k/p)$ locally smallest elements of each queue.*

Exercise 6.15. Prove Lemma 6.6. Hint: Use the Chernoff bound (A.7).

Exercise 6.16. Refine Lemma 6.6 and derive the constant factors in the O-terms. What happens for $k = o(p\log p)$?

The operation *deleteMin** exploits Lemma 6.6; see also the pseudocode in Fig. 6.9 and the example in Fig. 6.10. It suffices to look at the locally smallest elements of each local queue to find the globally smallest ones. The function *initialBufferSize* makes an initial guess m of how many local elements will be needed. A simple choice that works well for $k = \Omega(p\log p)$ is $m = 2k/p$. For $k \gg p\log p$, we can use a value of m very close to k/p. Exercise 6.16 asks you to work out the constants in more detail. The m locally smallest elements are removed. They are tentativly moved to the result set. Then, parallel selection (see Fig. 5.16) is used to identify the k globally smallest elements among these result candidates. Let e_k denote the kth smallest result candidate. If there is no element smaller than e_k in any local queue, the result set contains the k globally smallest elements. Otherwise, we continue to remove elements from the local queues. We complete the operation by reinserting the result candidates that have not been selected for the final result.

Exercise 6.17. Our implementation of *deleteMin** delivers the results locally irrespective of load imbalance. Explain how to modify *deleteMin** such that every PE gets the same number of result elements up to rounding up or down. Hint: Use prefix sums.

Theorem 6.7. *For $k = \Omega(p\log p)$, the operation deleteMin** *works in expected time $O(\frac{k}{p}\log n)$, where n is the total queue size.*

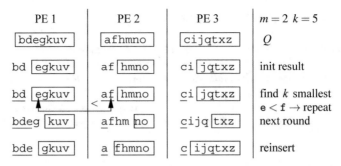

Fig. 6.10. Deleting the $k = 5$ smallest elements from a queue containing $\langle a,b,c,d,e,f,g,h,i,j,k,m,n,o,q,t,x,z\rangle$ distributed over three PEs

Proof. (Outline.) By Lemma 6.6, we can choose $m = \Theta(k/p)$ and achieve a constant expected number of iterations of the repeat loop. An iteration of the loop takes time $O(m\log n)$ for local *deleteMin* operations and expected time $O(k/p + \log p)$ for parallel selection; see Fig. 5.16. Testing for loop termination takes time $O(\log p)$ for an all-reduce-and operation; see Sect. 13.2. Reinserting unneeded elements takes no more time than deleting them in the first place. □

Exercise 6.18. Strengthen Theorem 6.7 and show a time bound of $O(\frac{k}{p}\log\frac{n}{p})$ for $n = \Omega(p\log p)$.

Exercise 6.19. Show that a *deleteMin** is communication-efficient in the sense that it has total communication cost $O(\alpha\log p)$ regardless of n and k. Are insertions equally efficient with respect to communication overhead?

6.4.1 Refinements

An inefficiency of our implementation of *deleteMin** is that it moves elements back and forth between the result buffer and Q. For $k = o(p\log p)$ this also becomes an issue asymptotically. For example, for $k = \Theta(p)$, it is known that we need $m = \Omega(\log p/\log\log p)$ locally removed elements, although on average only a constant number of elements per PE is actually returned. This problem can be avoided by keeping the result buffer around, emptying it only occasionally. It can be shown that this allows efficient operation all the way down to $k = O(p)$ [271].

We can reduce communication overhead by a significant constant factor if we allow some fluctuations in the size of the returned result set. Rather than running a full-fledged selection algorithm with several iterations of sample sorting and partitioning, we can just sort a single sample to determine a single pivot whose expected global rank is k. We then simply return all result candidates bounded by this rank; see also [158].

6.5 Implementation Notes

There are various places where *sentinels* (see Chap. 3) can be used to simplify or (slightly) accelerate the implementation of priority queues. Since sentinels may require additional knowledge about key values, this could make a reusable implementation more difficult, however:

- If $h[0]$ stores a *Key* no larger than any *Key* ever inserted into a binary heap, then *siftUp* need not treat the case $i = 1$ in a special way.
- If $h[n+1]$ stores a *Key* no smaller than any *Key* ever inserted into a binary heap, then *siftDown* need not treat the case $2i+1 > n$ in a special way. If such large keys are stored in $h[n+1..2n+1]$, then the case $2i > n$ can also be eliminated.
- Addressable priority queues can use a special dummy item rather than a null pointer.

For simplicity, we have formulated the operations *siftDown* and *siftUp* for binary heaps using recursion. It might be a little faster to implement them iteratively instead. Similarly, the *swap* operations could be replaced by unidirectional move operations, thus halving the number of memory accesses.

Exercise 6.20. Give iterative versions of *siftDown* and *siftUp*. Also, replace the *swap* operations.

Some compilers do the recursion elimination for you.

As with sequences, memory management for items of addressable priority queues can be critical for performance. Often, a particular application may be able to do this more efficiently than a general-purpose library. For example, many graph algorithms use a priority queue of nodes. In this case, items can be incorporated into nodes.

There are priority queues that work efficiently for integer keys. It should be noted that these queues can also be used for floating-point numbers. Indeed, the IEEE floating-point standard has the interesting property that for any valid floating-point numbers a and b, $a \leq b$ if and only if $bits(a) \leq bits(b)$, where $bits(x)$ denotes the reinterpretation of the bit string representing x as an integer.

6.5.1 C++

The STL class *priority_queue* offers nonaddressable priority queues implemented using binary heaps. The external-memory library STXXL [88] offers an external-memory priority queue. LEDA [195] and LEMON (Library for Efficient Modeling and Optimization in Networks) [201] implement a wide variety of addressable priority queues, including pairing heaps and Fibonacci heaps.

6.5.2 Java

The class *java.util.PriorityQueue* supports addressable priority queues to the extent that *remove* is implemented. However, *decreaseKey* and *merge* are not supported. Also, it seems that the current implementation of *remove* needs time $\Theta(n)$. JGraphT [167] offers an implementation of Fibonacci heaps.

6.6 Historical Notes and Further Findings

There is an interesting internet survey[3] of priority queues. It lists the following applications: (shortest-)path planning (see Chap. 10), discrete-event simulation, coding and compression, scheduling in operating systems, computing maximum flows, and branch-and-bound (see Sect. 12.4).

In Sect. 6.1 we saw an implementation of *deleteMin* by top-down search that needs about $2\log n$ element comparisons, and a variant using binary search that needs only $\log n + O(\log \log n)$ element comparisons. The latter is mostly of theoretical interest. Interestingly, a very simple "bottom-up" algorithm can be even better: The old minimum is removed and the resulting hole is sifted down all the way to the bottom of the heap. Only then, does the rightmost element fill the hole and it is subsequently sifted up. When used for sorting, the resulting *bottom-up heapsort* requires $\frac{3}{2}n\log n + O(n)$ comparisons in the worst case and $n\log n + O(1)$ in the average case [105, 284, 328]. While bottom-up heapsort is simple and practical, our own experiments indicate that it is not faster than the usual top-down variant (for integer keys). This surprised us at first. The explanation is that the bottom-up variant usually causes more cache faults that the standard variant and that the number of hard-to-predict branch operations is not reduced. Cache faults and incorrectly predicted branch operations have a larger influence on running time than does the number of comparisons; see [280] for more discussion. d-ary heaps, in which a node has d children instead of only two, outperform binary heaps in practice; see [193] for experiments.

The recursive *buildHeap* routine in Exercise 6.7 is an example of a *cache-oblivious algorithm* [116]. This algorithm is efficient in the external-memory model even though it does not explicitly use the block size or cache size.

Pairing heaps [114] have constant amortized complexity for *insert* and *merge* [160] and logarithmic amortized complexity for *deleteMin*. The best analysis is due to Pettie [253]. Fredman [112] has given operation sequences consisting of $O(n)$ insertions and *deleteMins* and $O(n\log n)$ *decreaseKeys* that require time $\Omega(n\log n\log\log n)$ for a family of addressable priority queues that includes all previously proposed variants of pairing heaps. Haeupler et al. [140] introduced a variant of pairing heaps that match the performance of Fibonacci heaps.

The family of addressable priority queues is large. Vuillemin [325] introduced binomial heaps, and Fredman and Tarjan [115] invented Fibonacci heaps. Høyer [157] described additional balancing operations that are akin to the operations used for search trees. One such operation yields *thin heaps* [174], which have performance guarantees similar to those of Fibonacci heaps and do without parent pointers and mark bits. It is likely that thin heaps are faster in practice than Fibonacci heaps. There are also priority queues with worst-case bounds asymptotically as good as the amortized bounds that we have seen for Fibonacci heaps [53]. The basic idea is to tolerate violations of the heap property and to continuously invest some work in

---

[3] www.leekillough.com/heaps/survey_results.html

reducing these violations. Other interesting variants are *fat heaps* [174] and *hollow heaps* [144].

Many applications need priority queues for integer keys only. For this special case, there are more efficient priority queues. The best theoretical bounds so far are constant time for *decreaseKey* and *insert* and $O(\log\log n)$ time for *deleteMin* [224, 314]. Using randomization, the time bound can even be reduced to $O(\sqrt{\log\log n})$ [142]. The algorithms are fairly complex. However, integer priority queues for operation sequences satisfying a *monotonicity property* are simple and practical. Section 10.3 gives examples. *Calendar queues* [58] are popular in the discrete-event simulation community. These are a variant of the *bucket queues* described in Sect. 10.5.1.

7

Sorted Sequences

*All of us spend a significant part of our time on searching, and so do computers:
They look up telephone numbers, balances of bank accounts, flight reservations, bills
and payments, ... In many applications, we want to search dynamic collections of
data. New bookings are entered into reservation systems, reservations are changed or
canceled, and bookings turn into actual flights. We have already seen one solution
to the problem, namely hashing. However, it is often desirable to keep a dynamic
collection sorted. The "manual data structure" used for this purpose is a filing-
card box. We can insert new cards at any position, we can remove cards, we can go
through the cards in sorted order, and we can use some kind of binary search to find
a particular card. Large libraries used to have filing-card boxes with hundreds of
thousands of cards.*[1]

Formally, we wish to maintain a set S of elements e, which are equipped with keys
$key(e)$ from a linearly ordered set Key. As before we write $e \leq e'$ if $key(e) \leq key(e')$.
This induces a linear preorder on S. Not only should searches be possible, but insertions and deletions as well. This leads to the following basic operations of a *sorted
sequence*:

- $S.locate(k : Key)$: **return** $\min \{e \in S : key(e) \geq k\}$
  (i.e., return an element e with $key(e) \geq k$ as small as possible).
- $S.remove(k : Key)$: $S := S \setminus \{e \in S : key(e) = k\}$
- $S.insert(e : Element)$: $S := (S \setminus \{e' \in S : key(e') = key(e)\}) \cup \{e\}$

The operation $locate(k)$ *locates* key k in S, i.e., it finds an element with key k
if it exists, and finds an element of S with a minimum key larger than S otherwise.
If k is larger than all keys that appear in S, no element is returned. The operation
$insert(e)$ inserts an element e with a new key or replaces an element e' in S that has
the same key as e. Note that this way of specifying $insert$ implies that all elements in
S have pairwise different keys. We shall reconsider this issue in Exercise 7.10. The
operation $delete(k)$ removes the element with key k, if S has such an element. We
shall show that these operations can be implemented to run in time $O(\log n)$, where

---

[1] The above photograph is from the library catalog of the University of Graz (Dr. M. Gossler).

© Springer Nature Switzerland AG 2019
P. Sanders et al., *Sequential and Parallel Algorithms and Data Structures*,
https://doi.org/10.1007/978-3-030-25209-0_7

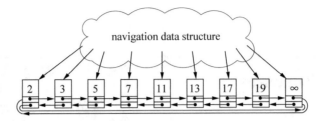

Fig. 7.1. A sorted sequence as a doubly linked list plus a navigation data structure.

n denotes the size of the sequence. How do sorted sequences compare with the data structures discussed in earlier chapters? They are more flexible than sorted arrays, because they efficiently support *insert* and *remove*. They are slower but also more powerful than hash tables, since *locate*(k) also works when there is no element with key k in S. Priority queues are a special case of sorted sequences; they can only locate and remove the smallest element.

Our basic realization of a sorted sequence consists of a sorted doubly linked list with an additional navigation data structure supporting *locate*. Figure 7.1 illustrates this approach. Recall that a doubly linked list for n elements consists of $n+1$ items, one for each element and one additional "dummy item". We use the dummy item to store a special key value ∞, which is larger than all keys that might possibly appear in an element of S. We can then define the result of *locate*(k) as the handle to the smallest list item $e \geq k$. If k is larger than all keys in S, *locate*(k) will return a handle to the dummy item. In Sect. 3.2.1, we saw that doubly linked lists support a large set of operations; most of them can also be implemented efficiently for sorted sequences. For example, we "inherit" constant-time implementations for *first*, *last*, *succ*, and *pred*. We shall see constant-amortized-time implementations for *remove*$(h : Handle)$ (note that here the object to be removed is given as a handle to a list item), *insertBefore*, and *insertAfter*, and logarithmic-time algorithms for concatenating and splitting sorted sequences. The indexing operator $[\cdot]$ and finding the position of an element in the sequence also take logarithmic time. Before we delve into a description of the navigation data structure, let us look at some concrete applications of sorted sequences.

Best-first heuristics. Assume that we want to pack some items into a set of bins. The items arrive one at a time and have to be put into a bin immediately. Each item i has a weight $w(i)$, and all bins have the same maximum capacity. The goal is to minimize the number of bins used. One successful heuristic solution to this problem is to put item i into the bin that fits best, i.e., the bin whose remaining capacity is the smallest among all bins that have a residual capacity at least as large as $w(i)$ [73]. To implement this algorithm, we can keep the bins in a sequence S sorted by their residual capacity. To place an item i, we call $S.locate(w(i))$, remove the bin that we have found, reduce its residual capacity by $w(i)$, and reinsert it into S. See also Exercise 12.9. Note that for this application we must allow distinct elements (bins) with identical keys, see Exercise 7.10.

Sweep-line algorithms. Assume that you have a set of horizontal and vertical line segments in the plane and want to find all points where two segments intersect. A sweep-line algorithm moves a vertical line over the plane from left to right and maintains, in a sorted sequence S, the set of horizontal line segments that intersect the sweep line. When the left endpoint of a horizontal segment is reached, it is inserted into S, with its y-coordinate as key, and when its right endpoint is reached, it is removed from S. (In the case where there are overlapping horizontal segments with identical y-coordinates, the sorted sequence must admit different elements with identical keys.) When a vertical line segment s is reached at a position x, we determine its vertical range $[y, y']$, call $S.locate(y)$, and, starting from the position thus found, scan S to the right until we reach an element with key y' or larger.[2] Exactly those horizontal line segments that are discovered during the scan have an intersection with s. The sweep-line algorithm can be generalized to arbitrary line segments [42], curved objects, and many other geometric problems [85].

Database indexes. A key problem in databases is to make large collections of data efficiently searchable. So-called *B-trees*, which are a variant of the (a, b)-tree data structure to be described in detail in Sect. 7.2, belong to the most important data structures used for databases.

The most popular navigation data structures belong to the class of *search trees*. Generally speaking, a search tree is a rooted ordered tree whose leaves correspond to the items in the doubly linked list. Internal nodes do not hold elements, but contain only keys to support navigation, and possibly other auxiliary information.[3] We shall frequently use the name of the navigation data structure to refer to the entire sorted-sequence data structure. Search tree algorithms will be introduced in three steps. As a warm-up, in Sect. 7.1 we consider (unbalanced) *binary search trees*, in which inner nodes have two children. They support *locate* in $O(\log n)$ time under certain favorable circumstances. Since binary search trees that guarantee this time bound are somewhat difficult to maintain under insertions and removals, we then switch to a generalization, namely (a, b)-trees, which allow search tree nodes of larger degree than 2. Section 7.2 explains how (a, b)-trees can be used to implement all three basic operations in logarithmic worst-case time. In Sects. 7.3 and 7.5, we augment search trees with additional mechanisms that support further operations. Section 7.4 takes a closer look at the (amortized) cost of update operations. Parallelizing search trees is briefly discussed in Sect. 7.6. As with priority queues, we shall concentrate on bulk operations that access many elements at once.

7.1 Binary Search Trees

Navigating a search tree is a bit like asking your way around in a foreign city. You ask a question, follow the advice given, ask again, follow the advice again, and so on, until you reach your destination.

---

[2] This *range query* operation is also discussed in Sect. 7.3.
[3] There are also variants of search trees where the elements are stored in all nodes of the tree.

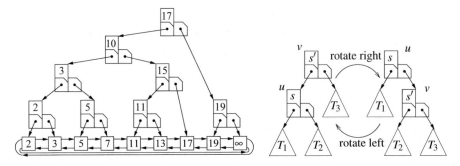

Fig. 7.2. *Left*: the sequence $\langle 2,3,5,7,11,13,17,19 \rangle$ represented by a binary search tree. In each node, we show the splitter key at the top and the pointers to the children at the bottom. *Right*: rotation in a binary search tree. The triangles indicate subtrees. Observe that the ancestor relationship between nodes u and v is interchanged.

A *binary search tree* is an ordered tree whose leaves store the elements of a sorted sequence in sorted order from left to right. In order to locate a key k, we would like to start at the root of the tree and follow the unique path to the appropriate leaf. How do we identify the correct path? To this end, the interior nodes of a search tree store keys (not elements!) that guide the search; we call these keys *splitter keys* or *splitters*. Every nonleaf node in a binary search tree with $n \geq 2$ leaves has exactly two children, a *left* child and a *right* child. The splitter key s associated with a nonleaf node v has the property that all elements stored in the left subtree of v have a key less than or equal to s, while all elements stored in the right subtree of v have a key larger than s.

Assume now that we are given a binary search tree T and a search key k. It is not hard to find the path to the correct leaf, i.e., the leaf with the entry e' that has the smallest key $k' \geq k$. Start at the root, and repeat the following step until a leaf is reached: If $k \leq s$ for the splitter key in the current node, go to the left child, otherwise go to the right child. If the data structure contains an element e with key k, it is quite easy to see that the process reaches the leaf that contains e. However, we also have to cover the case where no element with key k exists, and $k' = key(e') > k$. An easy case analysis shows that in this case the search procedure has the following invariant: Either the subtree rooted at the current node contains e' or this subtree contains the immediate predecessor of e', in its rightmost leaf. The search ends in a leaf with an element e'' and a key $k'' = key(e'')$, which is then compared with k. Because of the invariant, there are only two possibilities: If $k \leq k''$, then e'' is the desired element e'; if $k > k''$, then e' is the immediate successor of e'' in the doubly linked list. Figure 7.2 (left) shows an example of a binary search tree. You may want to test the behavior of the search procedure with keys such as 1, 9, 13, and 25. Recall that the height of a tree is the length of its longest root–leaf path. The height therefore tells us the maximum number of search steps needed to *locate* a search key k.

Exercise 7.1. Prove that, for every sorted sequence with $n \geq 1$ elements, there is a binary search tree with $n+1$ leaves and height $\lceil \log(n+1) \rceil$.

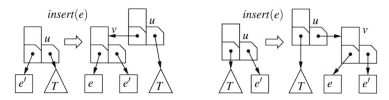

Fig. 7.3. Naive insertion into a binary search tree. A triangle indicates an entire subtree.

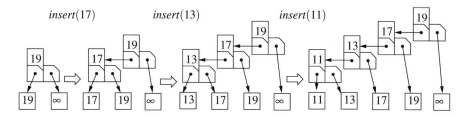

Fig. 7.4. Naively inserting elements in sorted order leads to a degenerate tree.

A binary search tree with $n+1$ leaves and height $\lceil\log(n+1)\rceil$ is called *perfectly balanced*. The resulting logarithmic search time is a dramatic improvement compared with the $\Omega(n)$ time needed for scanning a list. The bad news is that it is expensive to keep perfect balance when elements are inserted and removed. To understand this better, let us consider the "naive" insertion routine depicted in Fig. 7.3. We locate the key k of the new element e before its successor e', insert e into the list, and then introduce a new node v with left child e, right child e', and splitter key k. The old parent u of e' now points to v. In the worst case, every insertion operation will locate a leaf at the maximum depth so that the height of the tree increases with each insertion. Figure 7.4 gives an example: The tree may degenerate into a list, and we are back to scanning.

An easy solution to this problem is a healthy portion of optimism; perhaps it will not come to the worst. Indeed, if we insert n elements in *random* order, the expected height of the search tree is $\approx 2.99\log n$ [91]. We shall not prove this here, but will outline a connection to quicksort to make the result plausible. As an example, consider the tree in Fig. 7.2 (left), with the splitters 10 and 15 replaced by 7 and 13, respectively. This tree can be built by naive insertion. We first insert 17; this splits the set into subsets $\{2,3,5,7,11,13\}$ and $\{19\}$. From the elements in the left subset, we first insert 7; this splits the left subset into $\{2,3,5\}$ and $\{11,13\}$. In quicksort terminology, we would say that 17 is chosen as the splitter in the top-level call and that 7 is chosen as the splitter in the next recursive call for the left subarray. So building a binary search tree and quicksort are completely analogous processes; the same comparisons are made, but at different times. Every element of the set is compared with 17. In quicksort, these comparisons take place when the set is split in the top-level call. In building a binary search tree, these comparisons take place when the elements of the set are inserted. So the comparison between 17 and 11 takes place in the top-level call of quicksort and when 11 is inserted into the tree, respectively. We have seen (Theorem 5.6) that the expected number of comparisons in a randomized quicksort of n elements is $O(n\log n)$. By the above correspondence, the expected number of comparisons in building a binary tree by inserting n elements in random order is also $O(n\log n)$. Thus an insertion requires $O(\log n)$ comparisons on average. Even more is true. With high probability, each single insertion requires $O(\log n)$ comparisons, and the expected height is $\approx 2.99\log n$.

Can we guarantee that the height stays logarithmic in the worst case? The answer is yes, and there are many different ways to achieve logarithmic height. We shall survey these techniques in Sect. 7.8, and discuss two solutions in detail in Sect. 7.2. We shall first discuss a solution which allows nodes of varying degree, and then show how to balance binary trees using rotations.

Exercise 7.2. Figure 7.2 (right) indicates how the shape of a binary tree can be changed by transformations called (left and right) *rotations*. Apply rotations to subtrees of the tree in Fig. 7.2 so that the node labeled 11 becomes the root of the tree.

Exercise 7.3. Explain how to implement an *implicit* binary search tree, i.e., the tree is stored in an array using the same mapping of the tree structure to array positions as in the binary heaps discussed in Sect. 6.1. What are the advantages and disadvantages compared with a pointer-based implementation? Compare searching in an implicit binary tree with binary search in a sorted array.

7.2 (a,b)-Trees and Red–Black Trees

An (a,b)-tree is a search tree where all interior nodes, except for the root, have an outdegree between a and b. Here, a and b are constants. The root has degree 1 for a trivial tree with a single leaf. Otherwise, the root has a degree between 2 and b. For $a \geq 2$ and $b \geq 2a - 1$, the flexibility in node degrees allows us to efficiently maintain the invariant that *all leaves have the same depth*, as we shall see in a short while. Consider a node with outdegree d. With such a node we associate an array $c[1..d]$ of pointers to children and a sorted array $s[1..d-1]$ of $d-1$ splitter keys. The splitters guide the search. To simplify the notation, we additionally define $s[0] = -\infty$

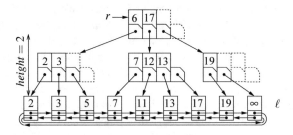

Fig. 7.5. Representation of $\langle 2,3,5,7,11,13,17,19 \rangle$ by a $(2,4)$-tree. The tree has height 2.

and $s[d] = \infty$. The keys of the elements e contained in the subtree rooted at the ith child $c[i]$, $1 \leq i \leq d$, lie between the $(i-1)$th splitter (exclusive) and the ith splitter (inclusive), i.e., we have $s[i-1] < key(e) \leq s[i]$. Figure 7.5 shows a $(2,4)$-tree storing the sequence $\langle 2,3,5,7,11,13,17,19 \rangle$.

Lemma 7.1. *An (a,b)-tree for n elements has height at most $1 + \lfloor \log_a(n+1)/2 \rfloor$, provided $n \geq 1$.*

Proof. The tree has $n+1$ leaves, where the "+1" accounts for the dummy leaf with key ∞. If $n = 0$, the root has degree 1 and there is a single leaf. So, assume $n \geq 1$. Let h be the height of the tree. Since the root has degree at least 2 and all other nodes have degree at least a, the number of leaves is at least $2a^{h-1}$. So $n+1 \geq 2a^{h-1}$, or $h \leq 1 + \log_a(n+1)/2$. Since the height is an integer, the bound follows.   □

Exercise 7.4. Prove that the height of an (a,b)-tree for n elements is at least $\lceil \log_b(n+1) \rceil$. Prove that this bound and the bound given in Lemma 7.1 are tight.

Searching in an (a,b)-tree is only slightly more complicated than searching in a binary tree. Let k be the search key and let k' be the smallest key in the doubly linked list that is at least as large as k. Instead of performing a single comparison at a nonleaf node, we have to find the correct child from among up to b choices, i.e., the child $c[i]$ with the property that $s[i-1] < k \leq s[i]$. Using binary search on the sorted sequence of the splitters, we need at most $\lceil \log b \rceil$ comparisons for each node on the search path. Correctness of the search procedure is proven using the same invariant as in the case of binary search: If key k is present in the tree, the search ends at the leaf with key k; otherwise, the subtree rooted at the current node either contains the element with key k' or contains the immediate predecessor of this element, in its rightmost leaf. Figure 7.6 gives pseudocode for initializing (a,b)-trees and for the *locate* operation. Recall that we use the search tree as a way to locate items of a doubly linked list and that the dummy list item is considered to have key value ∞. This dummy item is the rightmost leaf in the search tree. Hence, there is no need to treat the special case of root degree 0, and the handle of the dummy item can serve as a return value when one is locating a key larger than all values in the sequence.

Exercise 7.5. Prove that the total number of comparisons in a search is bounded by $\lceil \log b \rceil (1 + \log_a((n+1)/2))$. Assuming in addition that $b \leq 2a$, show that this number is $O(\log b) + O(\log n)$. What is the constant in front of the $\log n$ term?

To *insert* an element e with key $k = key(e)$, we first descend the tree recursively to locate k. This leads to a sequence element e' with key k'. Either k' is the smallest key $\geq k$ in the sequence or k' is the largest key strictly smaller than k. The latter situation can arise only if k is not present in the sequence. If $k = key(e')$, we simply replace e' by e in the list item. Note that this choice ensures that each key occurs at most once in the sorted sequence. If $k < k'$, we have to insert a new list item for e immediately before the item for e'. This case applies in particular if $k' = \infty$. We can avoid treating the case $k > k'$ separately, as follows: If this happens, we create

Class *ABHandle* : **Pointer** *to ABItem or Item*
// an ABItem (Item) is an item in the navigation data structure (doubly linked list)

Class *ABItem*(*splitters* : *Sequence* **of** *Key, children* : *Sequence* **of** *ABHandle*)
$d = |children| : 1..b$ // Degree
$s = splitters : Array\ [1..b-1]$ **of** *Key*
$c = children : Array\ [1..b]$ **of** *ABHandle*

Function *locateLocally*(k : *Key*) : $\mathbb{N}$
  return $\min\{i \in 1..d : k \leq s[i]\}$

Function *locateRec*(k : *Key, h* : $\mathbb{N}$) : *ABHandle*
  $i := locateLocally(k)$
  if $h = 1$ **then**
    if $c[i]{\rightarrow}e \geq k$ **then return** $c[i]$
    else return $c[i]{\rightarrow}next$
  else return $c[i]{\rightarrow}locateRec(k, h-1)$ //

Class *ABTree*($a \geq 2 : \mathbb{N}, b \geq 2a - 1 : \mathbb{N}$) **of** *Element*
$\ell = \langle\rangle$: *List* **of** *Element*
$r : ABItem(\langle\rangle, \langle\ell.head\rangle)$
$height = 1 : \mathbb{N}$ //

// Locate the item with the smallest key $k' \geq k$
Function *locate*(k : *Key*) : *ABHandle* **return** $r.locateRec(k, height)$

Fig. 7.6. (a,b)-trees. An *ABItem* is constructed from a sequence of keys and a sequence of handles to the children. The outdegree d is the number of children. We allocate space for the maximum possible outdegree b. There are two functions local to *ABItem*: *locateLocally*(k) locates key k among the splitters and *locateRec*(k,h) assumes that the *ABItem* has height h and descends h levels down the tree. The constructor for *ABTree* creates a tree for the empty sequence. The tree has a single leaf, the dummy element, and the root has degree 1. Locating a key k in an (a,b)-tree is solved by calling $r.locateRec(k,h)$, where r is the root and h is the height of the tree.

a new item for e, swap it with the e'-item in the linked list, and redirect the pointer to the e'-item in the navigation data structure to the e-item. Then the e'-item can be inserted immediately before the e-item, just as in the first case. Since we found e' when localizing k, and the splitter keys in the tree have not changed, we know that localizing k' in the changed structure would lead to the e-item. Now we focus on the case where the new e-item has to be inserted before the e'-item. If e' was the ith child $c[i]$ of its parent node v, then e becomes the new child $c[i]$ and $k = key(e)$ becomes the corresponding splitter $s[i]$. The old children $c[i..d]$ and their corresponding splitters $s[i..d-1]$ are shifted one position to the right. If the degree d of v was smaller than b before, d can be incremented by 1, and we are finished.

The difficult part is when a node v already has a degree $d = b$ and now would get a degree $b + 1$. Let $s'[1..b]$ denote the array of splitters of this illegal node, $c'[1..b+1]$ its child array, and u the parent of v (if it exists). The solution is to *split* v in the middle (see Fig. 7.7). More precisely, we create a new node t to the left of v and reduce the degree of v to $d = \lceil (b+1)/2 \rceil$ by moving the $b+1-d$ leftmost child pointers $c'[1..b+1-d]$ and the corresponding keys $s'[1..b-d]$ to this new node. The old node v keeps the d rightmost child pointers $c'[b+2-d..b+1]$ and the corresponding splitters $s'[b+2-d..b]$, but moved to the left by $b+1-d$ positions.

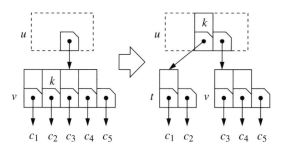

Fig. 7.7. Node splitting: The node v of degree $b+1$ (here 5) is split into a node of degree $\lfloor (b+1)/2 \rfloor$ and a node of degree $\lceil (b+1)/2 \rceil$. The degree of the parent increases by 1. The splitter key separating the two "parts" of v is moved to the parent.

The "leftover" middle key $k = s'[b+1-d]$ is an upper bound for the keys reachable from t. This key and the pointer to t are needed in the predecessor u of v. The situation for u is analogous to the situation for v before the insertion: If v was the ith child of u, t displaces it to the right; children $i+1,\ldots,d$ are pushed one position to the right as well. Now t becomes the ith child, and k is inserted as the ith splitter. The addition of t as an additional child of u increases the degree of u. If the degree of u becomes $b+1$, we split u. The process continues until either some ancestor of v has room to accommodate the new child or the root is split.

In the latter case, we allocate a new root node pointing to the two fragments of the old root. This is the only situation where the height of the tree can increase. In this case, the depth of all leaves increases by 1, i.e., the invariant that all leaves have the same depth is maintained. Since the height of the tree is $O(\log n)$ (see Lemma 7.1), we obtain a worst-case execution time of $O(\log n)$ for *insert*. Pseudocode is shown in Fig. 7.8.[4]

We still need to argue that *insert* leaves us with a correct (a,b)-tree. When we split a node of degree $b+1$, we create nodes of degree $d = \lceil (b+1)/2 \rceil$ and $b+1-d$. Both degrees are clearly at most b. Also, $b+1-\lceil(b+1)/2\rceil \geq a$ if $b \geq 2a-1$. Convince yourself that $b = 2a-2$ will not work.

Exercise 7.6. It is tempting to streamline *insert* by calling *locate* to replace the initial descent of the tree. Why does this not work? Would it work if every node had a pointer to its parent?

---

[4] We have borrowed the notation $C::m$ from C++ to define a method m for class C.

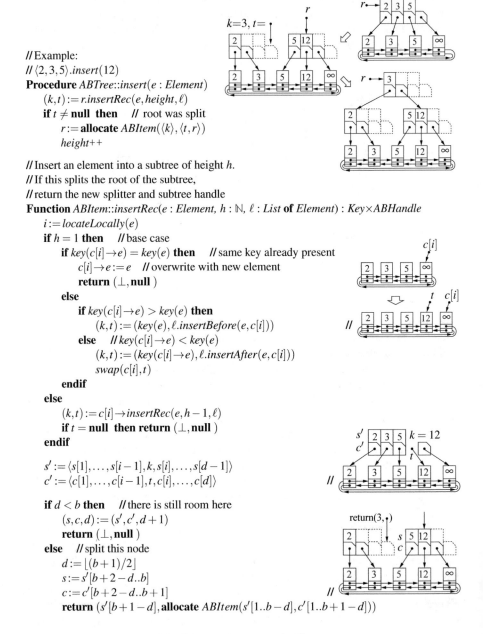

// Example:
// ⟨2, 3, 5⟩.insert(12)

Procedure *ABTree::insert(e : Element)*
   $(k,t) := r.insertRec(e, height, \ell)$
   if $t \neq$ **null then**   *// root was split*
      $r :=$ **allocate** *ABItem*$(\langle k \rangle, \langle t, r \rangle)$
      height++

// Insert an element into a subtree of height h.
// If this splits the root of the subtree,
// return the new splitter and subtree handle

Function *ABItem::insertRec(e : Element, h : ℕ, ℓ : List* **of** *Element) : Key×ABHandle*
   $i := locateLocally(e)$
   if $h = 1$ **then**   *// base case*
      if $key(c[i] \rightarrow e) = key(e)$ **then**   *// same key already present*
         $c[i] \rightarrow e := e$ *// overwrite with new element*
         return $(\bot, \textbf{null})$
      else
         if $key(c[i] \rightarrow e) > key(e)$ **then**
            $(k,t) := (key(e), \ell.insertBefore(e, c[i]))$
         else   *// key(c[i]→e) < key(e)*
            $(k,t) := (key(c[i] \rightarrow e), \ell.insertAfter(e, c[i]))$
            $swap(c[i], t)$
      endif
   else
      $(k,t) := c[i] \rightarrow insertRec(e, h-1, \ell)$
      if $t =$ **null then return** $(\bot, \textbf{null})$
   endif

   $s' := \langle s[1], \ldots, s[i-1], k, s[i], \ldots, s[d-1] \rangle$
   $c' := \langle c[1], \ldots, c[i-1], t, c[i], \ldots, c[d] \rangle$

   if $d < b$ **then**   *// there is still room here*
      $(s, c, d) := (s', c', d+1)$
      return $(\bot, \textbf{null})$
   else   *// split this node*
      $d := \lfloor (b+1)/2 \rfloor$
      $s := s'[b+2-d..b]$
      $c := c'[b+2-d..b+1]$
      return $(s'[b+1-d],$ **allocate** *ABItem*$(s'[1..b-d], c'[1..b+1-d]))$

Fig. 7.8. Insertion into an (a, b)-tree

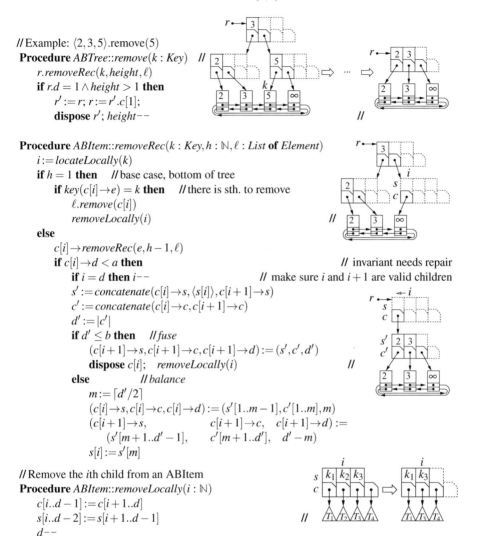

// Example: $\langle 2,3,5 \rangle$.remove(5)
Procedure *ABTree::remove(k : Key)*   //
  r.removeRec(k, height, ℓ)
  if $r.d = 1 \wedge height > 1$ **then**
    $r' := r; \; r := r'.c[1];$
    dispose r'; *height*--

Procedure *ABItem::removeRec(k : Key, h : ℕ, ℓ : List* **of** *Element)*
  $i := locateLocally(k)$
  if $h = 1$ **then**   // base case, bottom of tree
    if $key(c[i] \to e) = k$ **then**   // there is sth. to remove
      $\ell.remove(c[i])$
      $removeLocally(i)$
  else
    $c[i] \to removeRec(e, h-1, \ell)$
    if $c[i] \to d < a$ **then**   // invariant needs repair
      if $i = d$ **then** i--   // make sure i and $i+1$ are valid children
      $s' := concatenate(c[i] \to s, \langle s[i] \rangle, c[i+1] \to s)$
      $c' := concatenate(c[i] \to c, c[i+1] \to c)$
      $d' := |c'|$
      if $d' \le b$ **then**   // fuse
        $(c[i+1] \to s, c[i+1] \to c, c[i+1] \to d) := (s', c', d')$
        dispose $c[i]$; $removeLocally(i)$
      else   // balance
        $m := \lceil d'/2 \rceil$
        $(c[i] \to s, c[i] \to c, c[i] \to d) := (s'[1..m-1], c'[1..m], m)$
        $(c[i+1] \to s, \qquad\qquad c[i+1] \to c, \quad c[i+1] \to d) :=$
          $(s'[m+1..d'-1], \quad c'[m+1..d'], \quad d'-m)$
        $s[i] := s'[m]$

// Remove the ith child from an ABItem
Procedure *ABItem::removeLocally(i : ℕ)*
  $c[i..d-1] := c[i+1..d]$
  $s[i..d-2] := s[i+1..d-1]$
  d--

Fig. 7.9. Removal from an (a,b)-tree

We now turn to the operation *remove*. The approach is similar to what we already know from our study of *insert*. We locate the element to be removed, remove it from the sorted list, and repair possible violations of invariants on the way back up. Figure 7.9 shows pseudocode. When a parent u notices that the degree of its child $c[i]$ has dropped to $a-1$, it combines this child with one of its neighbors $c[i-1]$ or $c[i+1]$ to repair the invariant. There are two cases, illustrated in Fig. 7.10. If the neighbor has degree larger than a, we can *balance* the degrees by transferring one child or several children from that neighbor to $c[i]$. If the neighbor has degree a, balancing cannot help, since both nodes together have only $2a - 1$ children, so that we cannot give a children to both of them. However, in this case we can *fuse* them into a single node, since the requirement $b \geq 2a - 1$ ensures that the fused node has degree b at most.

To fuse a node $c[i]$ with its right neighbor $c[i + 1]$, we concatenate their child arrays. To obtain the new splitter array, we place the splitter $s[i]$ of the parent between the splitter arrays of $c[i]$ and $c[i + 1]$. The new arrays are stored in $c[i + 1]$, node $c[i]$ is deallocated, and the pointer to $c[i]$, together with the splitter $s[i]$, is removed from the parent node.

Exercise 7.7. Suppose a node v has been produced by fusing two nodes as described above. Prove that the ordering invariant is maintained: An element e reachable through child $v.c[i]$ has key $v.s[i - 1] < key(e) \leq v.s[i]$ for $1 \leq i \leq v.d$.

Balancing two neighbors is equivalent to first fusing them and then splitting the result, as in the operation *insert*. Since fusing two nodes decreases the degree of their parent, the need to fuse or balance might propagate up the tree. If the degree of the root drops to 1, we do one of two things. If the tree has height 1 and hence the linked list contains only a single item (the dummy node), there is nothing to do and we are finished. Otherwise, we deallocate the root and replace it by its sole child. The height of the tree decreases by 1.

The execution time of *remove* is proportional to the height of the tree and hence logarithmic in the size of the sorted sequence. We summarize the performance of (a,b)-trees in the following theorem.

Theorem 7.2. *For any integers a and b with $a \geq 2$ and $b \geq 2a - 1$, (a,b)-trees support the operations insert, remove, and locate on sorted sequences of size n in time $O(\log n)$.*

Exercise 7.8. Give a more detailed implementation of *locateLocally* based on binary search that needs at most $\lceil \log b \rceil$ comparisons. Your code should avoid both explicit use of infinite key values and special-case treatments for extreme cases.

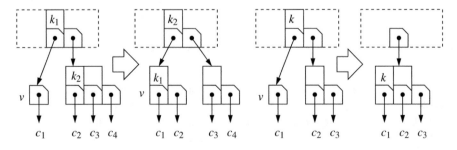

Fig. 7.10. Node balancing and fusing in (2,4)-trees: Node v has degree $a - 1$ (here 1). In the situation on the *left*, it has a sibling of degree $a + 1$ or more (here 3), and we *balance* the degrees. In the situation on the *right*, the sibling has degree a, and we *fuse* v and its sibling. Observe how keys are moved. When two nodes are fused, the degree of the parent decreases.

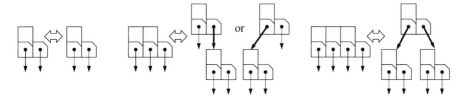

Fig. 7.11. The correspondence between (2,4)-trees and red–black trees. Nodes of degree 2, 3, and 4 as shown on the *left* correspond to the configurations on the *right*. Red edges are shown in **bold**.

Exercise 7.9. Suppose $a = 2^k$ and $b = 2a$. Show that $(1 + 1/k)\log n + 1$ element comparisons suffice to execute a *locate* operation in an (a,b)-tree. Hint: It is *not* quite sufficient to combine Exercise 7.4 with Exercise 7.8, since this would give you an additional term $+k$.

Exercise 7.10. Extend (a,b)-trees so that they can handle multiple occurrences of the same key. Elements with identical keys should be treated last-in first-out, i.e., *remove*(k) should remove the least recently inserted element with key k.

*Exercise 7.11 (red–black trees). A *red–black tree* is a binary search tree where the edges are colored either red or black. The *black-depth* of a node v is the number of black edges on the path from the root to v. The following invariants have to hold:

(a) All leaves have the same black-depth.
(b) Edges into leaves are black.
(c) No path from the root to a leaf contains two consecutive red edges.

Show that red–black trees and $(2,4)$-trees are isomorphic in the following sense: $(2,4)$-trees can be mapped to red–black trees by replacing nodes of degree three or four by two or three nodes, respectively, connected by red edges as shown in Fig. 7.11. Red–black trees can be mapped to $(2,4)$-trees using the inverse transformation, i.e., components induced by red edges are replaced by a single node. Now explain how to implement $(2,4)$-trees using a representation as a red–black tree.[5] Explain how the operations of adding a child to a node, removing a child from a node, splitting, fusing, and balancing nodes in the $(2,4)$-tree can be translated into recoloring and rotation operations (see Fig. 7.2) in the red–black tree. Colors are stored at the target nodes of the corresponding edges.

7.3 More Operations

Search trees support many operations in addition to *insert*, *remove*, and *locate*. We shall study them in two batches. In this section, we shall discuss operations directly supported by (a,b)-trees, and in Sect. 7.5 we shall discuss operations that require augmentation of the data structure:

---

[5] This may be more space-efficient than a direct representation if the keys are large.

- *min/max.* The constant-time operations *first* and *last* on a sorted list give us the smallest and the largest element in the sequence in constant time. For example, in Fig. 7.5, the dummy element of list ℓ gives us access to the smallest element, 2, and to the largest element, 19, via its *next* and *prev* pointers, respectively. In particular, search trees implement *double-ended priority queues*, i.e., sets that allow finding and removing both the smallest and the largest element; finding takes constant time and removing takes logarithmic time.
- *Range queries.* To retrieve all elements with keys in the range $[x,y]$, we first locate x and then traverse the sorted list until we see an element with a key larger than y. This takes time $O(\log n + \text{output size})$. For example, the range query $[4,14]$ applied to the search tree in Fig. 7.5 will find the 5, it subsequently outputs 7, 11, 13, and it stops when it sees the 17.
- *Build/rebuild.* Exercise 7.12 asks you to give an algorithm that converts a sorted list or array into an (a,b)-tree in linear time. Even if we first have to sort the elements, this operation is much faster than inserting the elements one by one. We also obtain a more compact data structure this way.

Exercise 7.12. Explain how to construct an (a,b)-tree from a sorted list in linear time. Which $(2,4)$-tree does your routine construct for the sequence $\langle 1,\ldots,17\rangle$? Next, remove the elements 4, 9, and 16.

7.3.1 *Concatenation

Two sorted sequences can be concatenated if the largest element of the first sequence is smaller than the smallest element of the second sequence. If sequences are represented as (a,b)-trees, two sequences S_1 and S_2 can be concatenated in time $O(\log(\max\{|S_1|,|S_2|\}))$. First, using the pointers to their dummy items, we find the beginning and end of both lists and make a note of the largest key in S_1. We remove the dummy item from S_1 and concatenate the underlying lists. Next, we fuse the root of one tree with an appropriate node of the other tree in such a way that the resulting tree remains sorted and balanced. More precisely, if $S_1.height \geq S_2.height$, we descend $S_1.height - S_2.height$ levels from the root of S_1 by following pointers to the rightmost children. The node v thus reached is then fused with the root of S_2. The new splitter key required is the largest key in S_1. If the degree of v now exceeds b, node v is split. From that point on, the concatenation proceeds like an *insert* operation, propagating splits up the tree until the invariant is fulfilled or a new root node is created. The case $S_1.height < S_2.height$ is a mirror image. We descend $S_2.height - S_1.height$ levels from the root of S_2 by following pointers to the leftmost children, fuse nodes, and propagate splits upwards, if necessary. Figure 7.12 gives an example of this case. The operation runs in time $O(\log(\max\{|S_1|,|S_2|\}))$, even if the heights of the trees have to be calculated. If the heights of the trees are known, time $O(1 + |S_1.height - S_2.height|)$ is sufficient.

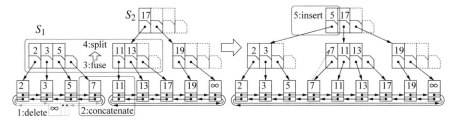

Fig. 7.12. Concatenating $(2,4)$-trees for $\langle 2,3,5,7\rangle$ and $\langle 11,13,17,19\rangle$.

7.3.2 *Splitting

We now show how to split a sorted sequence at a given element in logarithmic time. Consider a sequence $S = \langle w,\ldots,x,y,\ldots,z\rangle$. Splitting S at y results in the sequences $S_1 = \langle w,\ldots,x\rangle$ and $S_2 = \langle y,\ldots,z\rangle$. We implement splitting as follows. Consider the path p from the root to leaf y. We split each node v on p into two nodes, v_ℓ and v_r. Node v_ℓ gets the children of v that are to the left of p, and v_r gets the children that are to the right of p. Some of these nodes may get only one child or no children. Such a node with at least one child can be viewed as the root of an (a,b)-tree (slightly generalized by waiving the degree bound of 2 for the root). Note that we can record the heights of all these trees at practically no extra cost. The doubly linked list is split between the x-node and the y-node; the first part gets a new dummy item. Successively concatenating the left trees and the new dummy element yields an (a,b)-tree for the sorted list $S_1 = \langle w,\ldots,x\rangle$. Concatenating $\langle y\rangle$ and the right trees produces the (a,b)-tree for $S_2 = \langle y,\ldots,z\rangle$. The concatenation is simpler than in the general case, as we do not have to worry about the linked lists. These $O(\log n)$ concatenations can even be carried out in total time $O(\log n)$, by exploiting the fact that the heights of the left trees are strictly decreasing and the heights of the right trees are strictly increasing. Let us look at the trees to the left of p in more detail. Let $r_1, r_2,\ldots, r_k$ be the roots of these trees and let $h_1, h_2,\ldots, h_k$ be their heights. Then $h_1 > h_2 > \cdots > h_k$. Recall that these heights are known. We first concatenate the tree with root r_k with the new dummy element, which in time $O(1 + h_k)$ yields a new tree with height at most $h_k + O(1)$. This tree is concatenated with the tree with root r_{k-1}, which in time $O(1 + h_{k-1} - h_k)$ leads to a new tree with height at most $h_{k-1} + O(1)$. This is iterated: In the round for $j = k-2,\ldots,1$ the tree with root r_j is concatenated with the tree resulting from the previous rounds; the time needed is $O(1 + h_j - h_{j+1})$; the new tree has height at most $h_j + O(1)$ (see Exercise 7.13). After k rounds, only one tree is left. The total time needed is $O(1 + h_k + \sum_{1\le i<k}(1 + h_i - h_{i+1})) = O(k + h_1) = O(\log n)$. The trees on the right side are treated analogously, starting with the smallest tree and the single element y. Figure 7.13 gives an example.

Exercise 7.13. We glossed over one issue in the argument above. How can we bound the height of the tree resulting from concatenating the trees with roots r_k to r_i? Show (by induction) that the height is not larger than $h_i + 1$.

Exercise 7.14. Explain how to remove a subsequence $\langle e \in S : \alpha \le e \le \beta \rangle$ from an (a,b)-tree S in time $O(\log n)$.

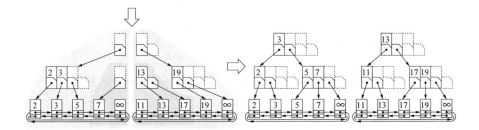

Fig. 7.13. Splitting the $(2,4)$-tree for $\langle 2,3,5,7,11,13,17,19 \rangle$ shown in Fig. 7.5 at element 11 produces the four (shaded) subtrees shown on the *left* and a single list node (with key 11). The list is split, which creates a new dummy item. Successively concatenating the trees indicated by the shaded areas leads to the $(2,4)$-trees shown on the *right*.

7.4 Amortized Analysis of Update Operations

Inserting or deleting an element in an (a,b)-tree costs $\Theta(\log n)$ time. Let us take a closer look at the best and the worst case. In the best case, we have to locate the affected element and to update the linked list and the bottommost internal node on the search path. In the worst case, *split* or *fuse* operations may propagate all the way up the tree. Since allocating (or deallocating) nodes is usually much slower than just reading them, these transformation steps make a significant difference in the running time.

Exercise 7.15. Exhibit a sequence of n operations on $(2,3)$-trees that requires $\Omega(n \log n)$ *split* and *fuse* operations.

In this section we study the cost of update operations in terms of the number of *split* and *fuse* operations. We show that, with repect to this cost measure, the *amortized* complexity is essentially equal to that of the best case if b is not at its minimum possible value but is at least $2a$. In Sect. 7.5.1, we shall see variants of *insert* and *remove* that, in the light of the analysis below, turn out to have constant amortized complexity even in the standard sense.

Theorem 7.3. *Consider a sorted sequence S organized as an (a,b)-tree with $b \ge 2a$. Assume that S is initially empty. Then, for any sequence of n insert or remove operations, the total number of split or fuse operations is $O(n)$.*

Proof. We give the proof for $(2,4)$-trees and leave the generalization to the reader (Exercise 7.16). We use the bank account method introduced in Sect. 3.5. The internal operations *split* and *fuse* are paid for by tokens; they cost one token each. We charge two tokens for each *insert* and one token for each *remove*, and claim that this suffices to pay for all *split* and *fuse* operations. Note that there is at most one *balance* operation for each *remove*, so we can account for the cost of *balance* directly without amortized analysis. In order to do the accounting, we associate the tokens with the nodes of the tree and show that the nodes can hold tokens according to the following table (*the token invariant*):

degree	1	2	3	4	5
tokens	∞	○		∞	○○○○

Note that we have included the cases of degree 1 and 5 which, (apart from the special case of a root of an empty sequence) occur temporarily during rebalancing. The purpose of splitting and fusing is to remove these exceptional degrees.

Creating an empty sequence makes a list with one dummy item and a root of degree 1. We charge two tokens for the *create* and put them on the root. Let us look next at insertions and removals. These operations add or remove a leaf and hence increase or decrease the degree of a node immediately above the leaf level. Increasing the degree of a node requires up to two additional tokens on the node (if the degree increases from 3 to 4 or from 4 to 5), and this is exactly what we charge for an insertion. If the degree grows from 2 to 3, we do not need additional tokens and we are overcharging for the insertion; there is no harm in this. Similarly, reducing the degree by 1 may require one additional token on the node (if the degree decreases from 3 to 2 or from 2 to 1). So, immediately after adding or removing a leaf, the token invariant is satisfied.

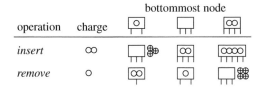

⊕ = leftover token

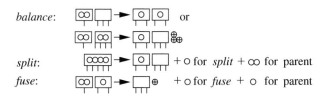

Fig. 7.14. The effect of (a,b)-tree operations on the token invariant. The *upper part* of the figure illustrates the addition or removal of a leaf. The two tokens charged for an insert are used as follows. When the leaf is added to a node of degree 3 or 4, the two tokens are put on the node. When the leaf is added to a node of degree 2, the two tokens are not needed, and the token on the node is also freed. The *lower part* illustrates the use of the tokens in *balance*, *split*, and *fuse* operations.

We need next to consider what happens during rebalancing. Figure 7.14 summarizes the following discussion graphically.

A *split* operation is performed on nodes of (temporary) degree 5 and results in a node of degree 3 and a node of degree 2. It also increases the degree of the parent by 1. The four tokens stored on the degree-5 node are spent as follows: One token pays for the *split*, one token is put on the new node of degree 2, and two tokens are used for the parent node. Again, we may not need the additional tokens for the parent node; in this case, we discard them.

A *balance* operation takes a node of degree 1 and a node of degree 3 or 4 and moves one child from the high-degree node to the node of degree 1. If the high-degree node has degree 3, two tokens are available and two tokens are needed; if the high-degree node has degree 4, four tokens are available and one token is needed. In either case, the available tokens are sufficient to maintain the token invariant.

A *fuse* operation fuses a degree-1 node with a degree-2 node into a degree-3 node and decreases the degree of the parent. Three tokens are available. We use one to pay for the operation and one to pay for the decrease in the degree of the parent. The third token is no longer needed, and we discard it.

Let us summarize. We charge two tokens for sequence creation, two tokens for each *insert*, and one token for each *remove*. These tokens suffice to pay one token each for every *split* or *fuse* operation. There is at most a constant amount of work for everything else done during an *insert* or *remove* operation. Hence, the total cost of n update operations is $O(n)$, and there are at most $2(n+1)$ *split* or *fuse* operations.   □

*Exercise 7.16. Generalize the above proof to arbitrary a and b with $b \geq 2a$. Show that n *insert* or *remove* operations cause only $O(n/(b-2a+1))$ *fuse* or *split* operations.

7.5 Augmented Search Trees

We show here that (a,b)-trees can support additional operations on sequences if we augment the data structure with additional information. However, augmentations come at a cost. They consume space and require time for keeping them up to date. Augmentations may also stand in each other's way.

Exercise 7.17 (reduction). Some operations on search trees can be carried out with the use of the navigation data structure alone and without the doubly linked list. Go through the operations discussed so far and discuss whether they require the *next* and *prev* pointers of linear lists. Range queries are a particular challenge.

7.5.1 Parent Pointers

Suppose we want to remove an element specified by the handle of a list item. In the basic implementation described in Sect. 7.2, the only thing we can do is to read the key k of the element and call *remove(k)*. This would take logarithmic time for the search, although we know from Sect. 7.4 that the amortized number of *fuse* operations required to rebalance the tree is constant. This detour is not necessary if each node v of the tree stores a handle indicating its *parent* in the tree (and perhaps an index i such that $v.parent.c[i] = v$).

Exercise 7.18. Assume $b \geq 2a$. Show that in (a,b)-trees with parent pointers, *remove*(h : Handle) and *insertAfter*(h : Handle) can be implemented to run in constant amortized time.

*Exercise 7.19 (avoiding augmentation). Assume $b \geq 2a$. Design an iterator class that allows one to represent a position in an (a,b)-tree that has no parent pointers. Creating an iterator is an extension of *locate* and takes logarithmic time. The class should support the operations *remove* and *insertAfter* in constant amortized time. Hint: Store the path to the current position.

*Exercise 7.20 (finger search). Augment search trees such that searching can profit from a "hint" given in the form of the handle of a *finger element* e'. If the sought element has rank r and the finger element e' has rank r', the search time should be $O(\log |r - r'|)$. Hint: One solution links all nodes at each level of the search tree into a doubly linked list.

*Exercise 7.21 (optimal merging). Section 5.3 discusses how to merge two sorted lists or arrays. Explain how to use finger search to implement merging of two sorted sequences in (optimal) time $O(n\log(m/n))$, where n is the size of the shorter sequence and m is the size of the longer sequence.

7.5.2 Subtree Sizes

Suppose that every nonleaf node t of a search tree stores its *size*, i.e., $t.size$ is the number of leaves in the subtree rooted at t. The kth smallest element of the sorted sequence can then be selected in a time proportional to the height of the tree. For simplicity, we shall describe this for binary search trees. Let t denote the current search tree node, which is initialized to the root. The idea is to descend the tree while maintaining the invariant that the kth element is contained in the subtree rooted at t. We also maintain the number i of elements that are to the *left* of t. Initially, $i = 0$. Let i' denote the size of the left subtree of t. If $i + i' \geq k$, then we set t to its left successor. Otherwise, t is set to its right successor and i is increased by i'. When a leaf is reached, the invariant ensures that the kth element is reached. Figure 7.15 gives an example.

Exercise 7.22. Generalize the above selection algorithm to (a,b)-trees. Develop two variants: one that needs time $O(b\log_a n)$ and stores only the subtree size, and another variant that needs only time $O(\log n)$ and stores $d - 1$ sums of subtree sizes in a node of degree d.

Exercise 7.23. Explain how to determine the rank of a sequence element with key k in logarithmic time.

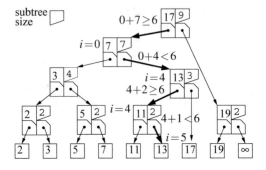

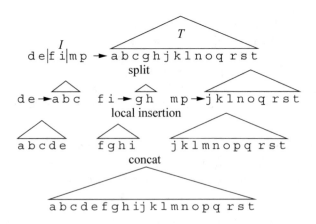

Fig. 7.15. Selecting the 6th smallest element from $\langle 2,3,5,7,11,13,17,19 \rangle$ represented by a binary search tree. The **thick** arrows indicate the search path.

Exercise 7.24. A colleague suggests supporting both logarithmic selection time and constant amortized update time by combining the augmentations described in Sects. 7.5.1 and 7.5.2. What will go wrong?

7.6 Parallel Sorted Sequences

Sorted sequences are difficult to parallelize for distributed-memory machines. Even on shared-memory machines, this is not easy. There has been a lot of work on sorted sequences that allow concurrent access [56, 134, 197]. However, it seems to be very difficult to achieve significant speedup compared with a tuned sequential implementation of a cache-friendly sorted sequence such as (a,b)-trees [292]. This is particularly difficult if there are a many *insert*, *remove*, and *update* operations and contention for the same keys.

If one needs only concurrent access, the (a,b)-trees introduced in Sect. 7.2 are a promising starting point [62, 134, 197], since their constant amortized update cost (see Sect. 7.4) requires concurrency control on only a constant number of tree nodes when averaged over all operations.

Fig. 7.16. Inserting $\langle d,e,f,i,m,p \rangle$ into $\langle a,b,c,g,h,j,k,l,n,o,q,r,s,t \rangle$ using three PEs.

We next discuss bulk operations on sorted sequences and show that simple algorithms can achieve sizable speedups in practice. The solution works for any sorted-sequence data structure with efficient support for concatenation (Sect. 7.3.1) and splitting (Sect. 7.3.2). We describe bulk insertion in detail. Bulk remove, bulk update, or arbitrary operation mixes are easy generalizations. One can also adapt the techniques described here to obtain algorithms for operations on two trees such as union, intersection, or difference [12, 46].

We first sort the insertion operations by their keys. Then, we split the sequence I of insertions and the sorted sequence T into corresponding pieces. We can then, independently and in parallel, perform the insertions in each piece of I into its corresponding piece of T. Concatenating the resulting sequences into a single sequence yields the desired result. Figure 7.16 gives an example. Figure 7.17 shows pseudocode for a parallel divide-and-conquer algorithm implementing the split–insert–concat approach. Not counting the cost of sorting I, we obtain a span of $\log k \log |T| + \frac{I}{k} \log |T|$; $\log k$ levels of recursion which take time $O(\log |T|)$ for splitting and concatenation in each level and time $O(\frac{|I|}{k} \log |T|)$ for the insertions. Again ignoring the time for sorting I, the overall work is $O(k \log |T|)$ for splitting and concatenation and $O(|I| \log |T|)$ for the insertions. Setting $k = p$, this yields a parallel execution time

$$O\left(\frac{|I|}{p} \log |T| + \log p \log |T| \right).$$

This result can be improved to $O((|I|/p) \log |T| + \log |T|)$ by splitting and concatenating with a more scalable parallel algorithm [12].

Procedure *bulkInsertPar*(T : *SortedSequence*, I : *Array* **of** *Element*, $k : \mathbb{N}$)
   assert I is sorted
   if $k = 1$ **then**   // base case
      foreach $e \in I$ **do** $T.insert(e)$
   else
      $m := \lfloor I/2 \rfloor$
      $(T_1, T_2) := T.split(I[m])$
      $bulkInsertRec(T_1, I[1..m], \lceil k/2 \rceil)$ || // parallel recursion
      $bulkInsertRec(T_2, I[m+1..|I|], \lfloor k/2 \rfloor)$
      $T := concat(T_1, T_2)$

Fig. 7.17. Task-parallel pseudocode for bulk insertion. The (presorted) insertion sequence I and the sorted sequence T are recursively split into k independent insertion tasks with an approximately equal number of insertions per task.

**Exercise 7.25. Show, using a more careful analysis, that with an appropriate choice of k, the algorithm in Fig. 7.17 actually only needs parallel execution time $O((|I|/p) \log(|T|/|I|) + \log^2 |T|)$. Hint: The *average* size of a subtree used for insertion is $|T|/k$. Argue that the overall work for insertions is maximized when all subtrees have average size.

7.7 Implementation Notes

Our pseudocode for (a,b)-trees is close to an actual implementation in a language such as C++ except for a few oversimplifications. The temporary arrays s' and c' in the procedures *insertRec* and *removeRec* can be avoided by appropriate case distinctions. In particular, a *balance* operation will not require calling the memory manager. A *split* operation of a node v might be slightly faster if v keeps the left half rather than the right half. We did not formulate the operation this way because then the cases of inserting a new sequence element and splitting a node would no longer be the same from the point of view of their parent.

For large b, *locateLocally* should use binary search. For small b, a linear search might be better. Furthermore, we may want to have a specialized implementation for small, fixed values of a and b that *unrolls*[6] all the inner loops.

Of course, the values of a and b are important. Let us start with the cost of *locate*. There are two kinds of operations that dominate the execution time of *locate*: Besides their inherent cost, element comparisons may cause branch mispredictions (see also Sect. 5.17). Furthermore, dereferencing pointers may cause cache faults. Exercise 7.9 indicates that element comparisons can be minimized by choosing a as a large power of two and $b = 2a$. Since the number of pointers that have to be dereferenced is proportional to the height of the tree (see Exercise 7.4), large values of a are also good for this measure. Taking this reasoning to the extreme, we would obtain the best performance for $a \geq n$, i.e., a single sorted array. This is not astonishing. We have concentrated on searches, and static data structures are best if updates are neglected.

Insertions and deletions have an amortized cost of one *locate* plus a constant number of node reorganizations (*split*, *balance*, or *fuse*) with cost $O(b)$ each. We obtain a logarithmic amortized cost for update operations as long as $b = O(\log n)$. A more detailed analysis (see Exercise 7.16) reveals that increasing b beyond $2a$ makes *split* and *fuse* operations less frequent and thus saves expensive calls to the memory manager associated with them. However, this measure has a slightly negative effect on the performance of *locate* and it clearly increases *space consumption*. Hence, b should remain close to $2a$.

Finally, let us take a closer look at the role of cache faults. A cache of size M can hold $\Theta(M/b)$ nodes. These are most likely to be the frequently accessed nodes close to the root. To a first approximation, the top $\log_a(M/b)$ levels of the tree are stored in the cache. Below this level, every time a pointer is dereferenced, a cache fault will occur, i.e., we will have about $\log_a(bn/\Theta(M))$ cache faults in each *locate* operation. Since the cache blocks of processor caches start at addresses that are a multiple of the block size, it makes sense to *align* the starting addresses of search tree nodes with a cache block, i.e., to make sure that they also start at an address that is a multiple of the block size. Note that (a,b)-trees might well be more efficient than binary search for large data sets because we may save a factor of $\log a$ in cache faults.

---

[6] *Unrolling* a loop "**for** $i := 1$ **to** K **do** $body_i$" means replacing it by the *straight-line program* "$body_1; \ldots; body_K$". This saves the overhead required for loop control and may give other opportunities for simplifications.

Very large search trees are stored on disks. Under the name *B-trees* [34], (a,b)-trees are the workhorse of the indexing data structures in databases. Internal nodes of B-trees have a size of several kilobytes. Furthermore, the items of the linked list are also replaced by entire data blocks that store between a' and b' elements, for appropriate values of a' and b' (see also Exercise 3.31). These leaf blocks will then also be subject to splitting, balancing, and fusing operations. For example, assume that we have $a = 2^{10}$, the internal memory is large enough (a few megabytes) to cache the root and its children, and the data blocks store between 16 and 32 KB of data. Then two disk accesses are sufficient to *locate* any element in a sorted sequence that takes 16 GB of storage. Since putting elements into leaf blocks dramatically decreases the total space needed for the internal nodes and makes it possible to perform very fast range queries, this measure can also be useful for a cache-efficient internal-memory implementation. However, note that update operations may now move an element in memory and thus will invalidate element handles stored outside the data structure. There are many more tricks for implementing (external-memory) (a,b)-trees. We refer the reader to [135] and [229, Chaps. 2 and 14] for overviews. A good free implementation of external B-trees is available in STXXL [88]. For an internal memory version, consider `github.com/bingmann/stx-btree`.

From the augmentations discussed in Sect. 7.5 and the implementation trade-offs discussed here, it becomes evident that *the* optimal implementation of sorted sequences does not exist, but depends on the hardware and the operation mix relevant to the actual application. We believe that (a,b)-trees with $b = 2^k = 2a = O(\log n)$, with a doubly linked list for the leaves, and augmented with parent pointers, are a sorted-sequence data structure that supports a wide range of operations efficiently.

Exercise 7.26. What choice of a and b for an (a,b)-tree guarantees that the number of I/O operations required for *insert*, *remove*, or *locate* is $O(\log_B(n/M))$? How many I/O operations are needed to *build* an n-element (a,b)-tree using the external sorting algorithm described in Sect. 5.12 as a subroutine? Compare this with the number of I/Os needed for building the tree naively using insertions. For example, try $M = 2^{29}$ bytes, $B = 2^{18}$ bytes[7], $n = 2^{32}$, and elements that have 8-byte keys and 8 bytes of associated information.

7.7.1 C++

The STL has four container classes *set*, *map*, *multiset*, and *multimap* for sorted sequences. The prefix *multi* means that there may be several elements with the same key. *Map*s offer the interface of an associative array (see also Chap. 4). For example, *someMap*$[k] := x$ inserts or updates the element with key k and sets the associated information to x.

The most widespread implementation of sorted sequences in STL uses a variant of red–black trees with parent pointers, where elements are stored in all nodes rather

---

[7] We are making a slight oversimplification here, since in practice one will use much smaller block sizes for organizing the tree than for sorting.

than only in the leaves. None of the STL data types supports efficient splitting or concatenation of sorted sequences.

LEDA [195] offers a powerful interface, *sortseq*, that supports all important operations on sorted sequences, including finger search, concatenation, and splitting. Using an implementation parameter, there is a choice between (a,b)-trees, red–black trees, randomized search trees, weight-balanced trees, and skip lists (for the last three, see Sect. 7.8).

7.7.2 Java

The Java library *java.util* offers the interface classes *SortedMap* and *SortedSet*, which correspond to the STL classes *set* and *map*, respectively. The corresponding implementation classes *TreeMap* and *TreeSet* are based on red–black trees.

7.8 Historical Notes and Further Findings

There is an entire zoo of sorted-sequence data structures. Just about any of them will do if you just want to support *insert*, *remove*, and *locate* in logarithmic time. Performance differences for the basic operations are often more dependent on implementation details than on the fundamental properties of the underlying data structures. The differences show up in the additional operations.

The first sorted-sequence data structure to support *insert*, *remove*, and *locate* in logarithmic time was AVL trees [5]. AVL trees are binary search trees which maintain the invariant that the heights of the subtrees of a node differ by at most one. Since this is a strong balancing condition, *locate* is probably a little faster than most competitors. On the other hand, AVL trees do *not* have constant amortized update costs. In each node, one has to store a "balance factor" from $\{-1, 0, 1\}$. In comparison, red–black trees have slightly higher costs for *locate*, but they have faster updates and the single color bit can often be squeezed in somewhere. For example, pointers to items will always store even addresses, so that their least significant bit could be diverted to storing color information.

$(2,3)$-trees were introduced in [7]. The generalization to (a,b)-trees and the amortized analysis in Sect. 3.5 come from [159]. There, it was also shown that the total number of splitting and fusing operations at the nodes of any given height decreases exponentially with height.

Splay trees [301] and some variants of randomized search trees [290] work even without any additional information besides one key and two successor pointers. A more interesting advantage of these data structures is their *adaptability* to nonuniform access frequencies. If an element e is accessed with probability p, these search trees will be reshaped over time to allow an access to e in time $O(\log(1/p))$. This can be shown to be asymptotically optimal for any comparison-based data structure. However, because of the large constants, this property leads to improved running times only for quite skewed access patterns.

Weight-balanced trees [244] balance the size of the subtrees instead of the height. They have the advantage that a node of weight w (= number of leaves of its subtree) is only rebalanced after $\Omega(w)$ insertions or deletions have passed through it [49].

There are so many *search tree* data structures for *sorted sequences* that these two terms are sometimes used as synonyms. However, there are also some equally interesting data structures for sorted sequences that are *not* based on search trees. Sorted arrays are a simple *static* data structure. Sparse tables [163] are an elegant way to make sorted arrays dynamic. The idea is to accept some empty cells to make insertion easier. Bender et al. [39] extended sparse tables to a data structure which is asymptotically optimal in an amortized sense. Moreover, this data structure is a crucial ingredient for a sorted-sequence data structure [39] that is *cache-oblivious* [116], i.e., it is cache-efficient on any two levels of a memory hierarchy without even knowing the size of the caches and cache blocks. The other ingredient is oblivious *static* search trees [116]; these are perfectly balanced binary search trees stored in an array such that any search path will exhibit good locality in any cache. We describe here the *van Emde Boas layout* used for this purpose, for the case where there are $n = 2^{2^k}$ leaves for some integer k. We store the top 2^{k-1} levels of the tree at the beginning of the array. After that, we store the $2^{2^{k-1}}$ subtrees of depth 2^{k-1}, allocating consecutive blocks of memory for them. The resulting $1 + 2^{2^{k-1}}$ subtrees of depth 2^{k-1} are stored recursively in the same manner. Static cache-oblivious search trees are practical in the sense that they can outperform binary search in a sorted array.

Skip lists [258] are based on another very simple idea. The starting point is a sorted linked list ℓ. The tedious task of scanning ℓ during *locate* can be accelerated by producing a shorter list ℓ' that contains only some of the elements in ℓ. If corresponding elements of ℓ and ℓ' are linked, it suffices to scan ℓ' and only descend to ℓ when approaching the searched element. This idea can be iterated by building shorter and shorter lists until only a single element remains in the highest-level list. This data structure supports all important operations efficiently in an expected sense. Randomness comes in because the decision about which elements to lift to a higher-level list is made randomly. Skip lists are particularly well suited for supporting finger search.

Yet another family of sorted-sequence data structures comes into play when we no longer consider keys as atomic objects. If keys are numbers given in binary representation, we can obtain faster data structures using ideas similar to the fast integer-sorting algorithms described in Sect. 5.10. For example, we can obtain sorted sequences with w-bit integer keys that support all operations in time $O(\log w)$ [216, 320]. At least for 32-bit keys, these ideas lead to a considerable speedup in practice [87]. Not astonishingly, string keys are also important. For example, suppose we want to adapt (a,b)-trees to use variable-length strings as keys. If we want to keep a fixed size for node objects, we have to relax the condition on the minimum degree of a node. Two ideas can be used to avoid storing long string keys in many nodes. *Common prefixes* of keys need to be stored only once, often in the parent nodes. Furthermore, it suffices to store the *distinguishing prefixes* of keys in inner nodes, i.e., just enough characters to be able to distinguish different keys in the current node [139]. Taking these ideas to the extreme results in *tries* [111], a search tree data structure specifically designed for string keys: Tries are trees whose edges are labeled by characters or strings. The characters along a root–leaf path represent a key. Using appropriate data structures for the inner nodes, a search in a trie for a string of size s can be carried out in time $O(s)$.

We close with three interesting generalizations of sorted sequences. The first generalization is *multidimensional objects*, such as intervals or points in d-dimensional space. We refer to textbooks on geometry for this wide subject [85]. The second generalization is *persistence*. A data structure is persistent if it supports nondestructive updates. For example, after the insertion of an element, there may be two versions of the data structure, the one before the insertion and the one after the insertion – both can be searched [99]. The third generalization is *searching many sequences* [66, 67, 217]. In this setting, there are many sequences, and searches need to locate a key in all of them or a subset of them.

8

Graph Representation

Scientific results are mostly available in the form of articles in journals and conference proceedings, and on various web[1] resources. These articles are not self-contained, but cite previous articles with related content. When you read an interesting article from 1975 today, you may ask yourself what the current state of the art is. In particular, you may want to know which newer articles cite the old article. Projects such as Google Scholar provide this functionality by analyzing the reference sections of articles and building a database of articles that efficiently supports looking up articles that cite a given article.

We can easily model this situation by a directed graph. The graph has a node for each article and an edge for each citation. An edge (u, v) from article u to article v means that u cites v. When an article is processed, the outgoing edges can be constructed directly from the list of references. In this way, every node (= article) stores all its outgoing edges (= the articles cited by it) but not the incoming edges (the articles citing it). If every node were also to store the incoming edges, it would be easy to find the citing articles. One of the main tasks of Google Scholar is to construct the reversed edges. This example shows that the cost of even a very basic elementary operation on a graph, namely finding all edges entering a particular node, depends heavily on the representation of the graph. If the incoming edges are stored explicitly, the operation is easy; if the incoming edges are not stored, the operation is nontrivial.

In this chapter, we shall give an introduction to the various possibilities for representing graphs in a computer. We focus mostly on directed graphs and assume that an undirected graph $G = (V, E)$ is represented as the corresponding (bi)directed graph $G' = (V, \bigcup_{\{u,v\} \in E} \{(u,v), (v,u)\})$. The top row of Fig. 8.1 shows an undirected graph and the corresponding bidirected graph. Most of the data structures presented also allow us to represent multiple parallel edges and self-loops. We start with a survey of the operations that we may want to support:

- *Accessing associated information.* Given a node or an edge, we frequently want to access information associated with it, for example the weight of an edge or the distance to a node. In many representations, nodes and edges are objects, and we can store this information directly as a member of these objects. If not

---

[1] The picture above shows a spider web (USFWS; see `commons.wikimedia.org/wiki/Image:Water_drops_on_spider_web.jpg`).

© Springer Nature Switzerland AG 2019
P. Sanders et al., *Sequential and Parallel Algorithms and Data Structures*,
https://doi.org/10.1007/978-3-030-25209-0_8

otherwise mentioned, we assume that $V = 1..n$ so that information associated with nodes can be stored in arrays. When all else fails, we can always store node or edge information in a hash table. Hence, accesses can be implemented to run in constant time. In the remainder of this book, we abstract from the various options for realizing access by using the data types *NodeArray* and *EdgeArray* to indicate array-like data structures that can be indexed by nodes and by edges, respectively.

- *Navigation.* Given a node, we may want to access its outgoing edges. This operation is at the heart of most graph algorithms. As we have seen in the example above, we sometimes also want to know the incoming edges.
- *Edge queries.* Given a pair of nodes (u,v), we may want to know whether this edge is in the graph. This can always be implemented using a hash table, but we may want to have something even faster. A more specialized but important query is to find the *reverse edge* (v,u) of a directed edge $(u,v) \in E$ if it exists. This operation can be implemented by storing additional pointers connecting edges with their reversals.
- *Construction, conversion and output.* The representation most suitable for the algorithmic problem at hand is not always the representation given initially. This is not a big problem, since most graph representations can be translated into each other in linear time.
- *Update.* Sometimes we want to add or remove nodes or edges. For example, the description of some algorithms is simplified if a node is added from which all other nodes can be reached (see, Fig. 10.10).

In Sect. 8.1 we begin with a very simple representation by a list of edges and continue in Sects. 8.2 and 8.3 with more structured representations that allow direct access from nodes to their incident edges. An interesting link to linear algebra comes from the matrix representation discussed in Sect. 8.4. For graphs with special structure, the *implicit* representations introduced in Sect. 8.5 can give extra performance. Parallel aspects are treated in Sect. 8.6.

8.1 Unordered Edge Sequences

Perhaps the simplest representation of a graph is as an unordered sequence of edges. Each edge contains a pair of node indices and, possibly, associated information such as an edge weight. Whether these node pairs represent directed or undirected edges is merely a matter of interpretation. Sequence representation is often used for input and output. It is easy to add edges or nodes in constant time. However, many other operations, in particular navigation, take time $\Theta(m)$, which is prohibitively slow. Only a few graph algorithms work well with the edge sequence representation; most algorithms require easy access to the edges incident to any given node. In this case the ordered representations discussed in the following sections are appropriate. In Chap. 11, we shall see two minimum-spanning-tree algorithms: One works well with an edge sequence representation and the other needs a more sophisticated data structure.

8.2 Adjacency Arrays – Static Graphs

To support easy access to the edges leaving any particular node, we can store the edges leaving any node in an array. If no additional information is stored with the edges, this array will just contain the indices of the target nodes. If the graph is *static*, i.e., does not change over time, we can concatenate all these little arrays into a single edge array E. An additional array V stores the starting positions of the subarrays, i.e., for any node v, $V[v]$ is the index in E of the first edge out of v. It is convenient to add a dummy entry $V[n+1]$ with $V[n+1] = m+1$. The edges out of any node v are then easily accessible as $E[V[v]], \ldots, E[V[v+1] - 1]$; the dummy entry ensures that this also holds true for node n. If a node v has no outgoing edge, $V[v] = V[v+1]$. Figure 8.1 (middle row, left side) shows an example.

The memory consumption for storing a directed graph using adjacency arrays is $n + m + \Theta(1)$ words. This is even more compact than the $2m$ words needed for an edge sequence representation.

Adjacency array representations can be generalized to store additional information: We may store information associated with edges in separate arrays or within the edge array. If we also need incoming edges, we may use additional arrays V' and E' to store the reversed graph.

Exercise 8.1. Design a linear-time algorithm for converting an edge sequence representation of a directed graph into an adjacency array representation. You should use only O(1) auxiliary space. Hint: View the problem as the task of sorting edges by their source node and adapt the integer-sorting algorithm shown in Fig. 5.34.

8.3 Adjacency Lists – Dynamic Graphs

Edge arrays are a compact and efficient graph representation. Their main disadvantage is that it is expensive to add or remove edges. For example, assume that we want to insert a new edge (u, v). Even if there is room in the edge array E to accommodate it, we still have to move the edges associated with nodes $u+1$ to n one position to the right, which takes time O(m).

In Chap. 3, we learned how to implement dynamic sequences. We can use any of the solutions presented there to produce a dynamic graph data structure. For each node v, we represent the sequence E_v of outgoing (or incoming, or both outgoing and incoming) edges by an unbounded array or by a (singly or doubly) linked list. We inherit the advantages and disadvantages of the respective sequence representations. Unbounded arrays are more cache-efficient. Linked lists allow constant-time insertion and deletion of edges at arbitrary positions. Most graphs arising in practice are sparse in the sense that every node has only a few incident edges. Adjacency lists for sparse graphs should be implemented without the dummy item introduced in Sect. 3.2, because an additional item per node would waste $\Theta(n)$ space. In the example in Fig. 8.1 (middle row, right side), we show circularly linked lists.

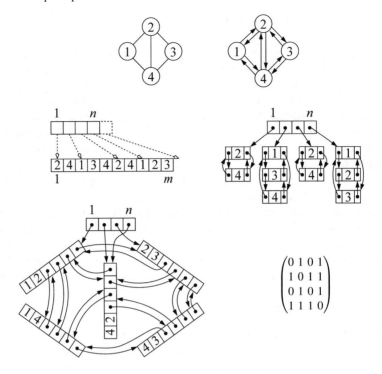

Fig. 8.1. The *first row* shows an undirected graph and the corresponding bidirected graph. The *second row* shows the adjacency array and adjacency list representations of this bidirected graph. The *third row* shows the linked-edge-objects representation and the adjacency matrix. In the former, there are five edge objects. Each object stores the names of its two endpoints and four pointers. The first two pointers point to the predecessor and successor edges of the first endpoint and the second two pointers do the same for the second endpoint. Thus, the third and fourth pointers of the edge $(1,2)$ point to the edges $(4,2)$ and $(2,3)$ respectively.

Exercise 8.2. Suppose the edges adjacent to a node u are stored in an unbounded array E_u, and an edge $e = (u,v)$ is specified by giving its position in E_u. Explain how to remove $e = (u,v)$ in constant amortized time. Hint: You do *not* have to maintain the relative order of the other edges.

Exercise 8.3. Explain how to implement the algorithm for testing whether a graph is acyclic discussed in Sect. 2.12 so that it runs in linear time, i.e., design an appropriate graph representation and an algorithm using it efficiently. Hint: Maintain a queue of nodes with outdegree 0.

Bidirected graphs arise frequently. Undirected graphs are naturally presented as bidirected graphs, and some algorithms that operate on directed graphs need access not only to outgoing edges but also to incoming edges. In these situations, we frequently want to store the information associated with an undirected edge or a directed edge and its reversal only once. Also, we may want to have easy access from an edge to its reversal.

We shall describe two solutions. The first solution simply associates two additional pointers with every directed edge. One points to the reversal, and the other points to the information associated with the edge.

The second solution has only one item for each undirected edge (or pair of directed edges) and makes this item a member of two adjacency lists. So, the item for an undirected edge $\{u, v\}$ would store the node names u and v (in no particular order) and be a member of lists E_u and E_v. If we want doubly linked adjacency information, the edge object for any edge $\{u, v\}$ stores four pointers: two are used for the doubly linked list representing E_u, and two are used for the doubly linked list representing E_v. We may use the convention that the first two pointers are for the node listed first in the edge item and the last two pointers for the node listed second. Any node stores a pointer to some edge incident to it. Starting from it, all edges incident to the node can be traversed. The bottom part of Fig. 8.1 gives an example. A small complication lies in the fact that finding the other end of an edge now requires some work. Note that the edge object for an edge $\{u, v\}$ stores the endpoints in no particular order. Hence, when we explore the edges out of a node u, we must inspect both endpoints and then choose the one which is different from u. An elegant alternative is to store $u \oplus v$ in the edge object [236]. An exclusive OR with either endpoint then yields the other endpoint. This representation saves space because only one node name has to be stored for each edge instead of two. However, we now need a different convention for how to interpret the four pointers in an edge object. We could, for example, say that the first two pointers are for the node with the smaller name and the last two pointers for the node with larger name.

If, in the case of a directed graph, one wants access to the incoming and the outgoing edges, again both solutions apply. In either solution, a node must store a pointer to one of its outgoing edges and to one of its incoming edges.

8.4 The Adjacency Matrix Representation

An n-node graph can be represented by an $n \times n$ *adjacency matrix A*. A_{ij} is 1 if $(i, j) \in E$ and 0 otherwise. Edge insertion or removal and edge queries work in constant time. It takes time $O(n)$ to obtain the edges entering or leaving a node. This is only efficient for very dense graphs with $m = \Omega(n^2)$. The storage requirement is n^2 bits. For very dense graphs, this may be better than the $n + m + O(1)$ words required for adjacency arrays. However, even for dense graphs, the advantage is small if additional edge information is needed.

Exercise 8.4. Explain how to represent an undirected graph with n nodes and without self-loops using $n(n-1)/2$ bits.

Perhaps more important than actually storing the adjacency matrix is the conceptual link between graphs and linear algebra introduced by the adjacency matrix. On the one hand, graph-theoretic problems can be solved using methods from linear algebra. For example, if $C = A^k$, then C_{ij} counts the number of paths from i to j with exactly k edges.

Exercise 8.5. Explain how to store an $n \times n$ matrix A with m nonzero entries using storage $O(m+n)$ such that a matrix–vector multiplication Ax can be performed in time $O(m+n)$. Describe the multiplication algorithm. Expand your representation so that products of the form $x^T A$ can also be computed in time $O(m+n)$.

On the other hand, graph-theoretic concepts can be useful for solving problems from linear algebra. For example, suppose we want to solve the matrix equation $Bx = c$, where B is a symmetric matrix. Now consider the corresponding adjacency matrix A, where $A_{ij} = 1$ if and only if $B_{ij} \neq 0$. If an algorithm for computing connected components finds that the undirected graph represented by A contains two distinct connected components, this information can be used to reorder the rows and columns of B such that we obtain an equivalent equation of the form

$$\begin{pmatrix} B_1 & 0 \\ 0 & B_2 \end{pmatrix} \begin{pmatrix} x_1 \\ x_2 \end{pmatrix} = \begin{pmatrix} c_1 \\ c_2 \end{pmatrix}.$$

This equation can now be solved by solving $B_1 x_1 = c_1$ and $B_2 x_2 = c_2$ separately. In practice, the situation is more complicated, since we rarely have matrices whose corresponding graphs are not connected. Still, more sophisticated graph-theoretic concepts such as cuts can help to discover structure in the matrix which can then be exploited to solve problems in linear algebra.

8.5 Implicit Representations

Many applications work with graphs of special structure. Frequently, this structure can be exploited to obtain simpler and more efficient representations. We shall give two examples.

The *grid graph* $G_{k\ell}$ with node set $V = [0..k-1] \times [0..\ell-1]$ and edge set

$$E = \{((i,j),(i,j')) \in V^2 : |j - j'| = 1\} \cup \{((i,j),(i',j)) \in V^2 : |i - i'| = 1\}$$

is completely defined by the two parameters k and ℓ. Figure 8.2 shows $G_{3,4}$. Edge weights could be stored in two two-dimensional arrays, one for the vertical edges and one for the horizontal edges.

An *interval graph* is defined by a set of intervals. For each interval, we have a node in the graph, and two nodes are adjacent if the corresponding intervals overlap. We may use open or closed intervals.

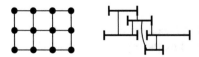

Fig. 8.2. The grid graph G_{34} (*left*) and an interval graph with five nodes and six edges (*right*).

Exercise 8.6 (representation of interval graphs).

(a) Show that, for any set of n intervals, there is a set of intervals whose endpoints are integers in $[1..2n]$, and that defines the same graph.
(b) Devise an algorithm that decides whether the graph defined by a set of n intervals is connected. Hint: Sort the endpoints of the intervals and then scan over the endpoints in sorted order. Keep track of the number of intervals that have started but not ended.
(c*) Devise a representation for interval graphs that needs $O(n)$ space and supports efficient navigation. Given an interval I, you need to find all intervals I' intersecting it; I' intersects I if I contains an endpoint of I' or $I \subseteq I'$. How can you find the former and the latter kinds of interval?

8.6 Parallel Graph Representation

Here, we discuss graph representations suitable for parallel processing. We do so first for shared memory and then for distributed memory.

8.6.1 Shared Memory

In a shared-memory machine, one may use the same graph representation as in the sequential case. As with other shared-memory data structures, read-only parallel access to the graph is efficient. However, we have to discuss how to do parallel construction and (bulk) updates. As already mentioned in Exercise 8.1, construction of an adjacency array can be viewed as the problem of sorting the edges by their endpoints. Hence, we can use the parallel algorithm presented in Sect. 5.11. At least in theory, it is relevant that we have very small keys – fewer than the elements. In this case, a CRCW-PRAM can perform the conversion using expected linear work and logarithmic time [261] (see also Sect. 5.17).

8.6.2 Distributed Memory

In a distributed-memory machine, one often *partitions* the graph by assigning each node to one PE. Another instance of the owner computes principle. For a partitioned graph, *cut* edges whose endpoints reside on different PEs require special attention. For cut edges, the graph representation needs to be able to identify the ID of the PEs responsible for the endpoints. During computations on a distributed graph, passing information along cut edges implies communication. Figure 8.3 gives an example.

PE 0 ○
PE 1 ◑
PE 2 ⊕
PE 3 ●

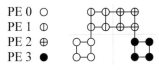

Fig. 8.3. A graph with 16 nodes partitioned between four PEs such that four edges are cut.

In order to minimize the communication required, it makes sense to choose the partition carefully. Hence, there has been intensive research on graph partitioning [59]. The best general-purpose graph-partitioning methods are quite sophisticated multilevel algorithms beyond the scope of this book. Since these methods are also relatively expensive, one also uses application-specific information to get a good partition fast.

For example, suppose we want to partition a web graph whose nodes represent web pages. We can obtain a reasonable partition by sorting the nodes by their URL. More precisely, the key used for a URL of the form d/f could be d^R/f, where d^R is the mirror image of d. For example, `algo2.iti.kit.edu/sanders.php` could be mapped to `ude.tik.iti.2ogla/sanders.php`. Then we cut the resulting ordering into balanced pieces. This partition has reasonable quality since many links are between nearby pages in this ordering.

Similarly, in many applications the nodes have a position in a geometrical space, for example geographical positions in road networks or three-dimensional coordinates in a graph describing a mathematical simulation. Interestingly, we can use the sorting idea from the URL example for multidimensional positions also. We map multidimensional coordinates to a single dimension. This is done using *space-filling curves* [27]. Figure 8.4 gives examples. A particularly simple such mapping is called *Z-order* or *Morton ordering*. It maps a d-tuple of k-bit integers $(x_1, \ldots, x_d)$ to dk bits by interleaving the bits. First come the first bits of all x_i's, then their second bits, and so on. Formally, bit $i \in 0..k-1$ of x_j is mapped to bit $id + j$ of the output. Figure 8.5 gives an example for $d = 3$ and $k = 4$.

Handling High Degree Vertices. The node based graph partitioning described above does not work for graphs which have nodes with very high degree. For nodes v with degree $\Omega(m/p)$, the work involved in handling just v can exceed the average work

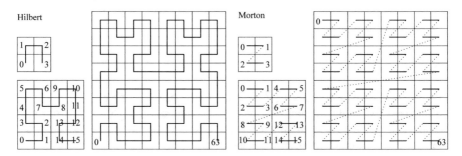

Fig. 8.4. Mapping two-dimensional points to a single dimension using space-filling curves.

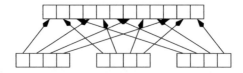

Fig. 8.5. Mapping of bits for 3D Morton ordering (3×4 bits $\rightarrow$ 12 bits).

per PE – leading to severe load imbalance. These nodes can be assigned to several PEs by partitioning their adjacency list between them. These PEs will then have to coordinate their work.

8.7 Implementation Notes

We have seen several representations of graphs in this chapter. They are suitable for different sets of operations on graphs, and can be tuned further for maximum performance in any particular application. The edge sequence representation is good only in specialized situations. Adjacency matrices are good for rather dense graphs. Adjacency lists are good if the graph changes frequently. Very often, some variant of adjacency arrays is fastest. This may be true even if the graph changes, because often there are only a few changes, all changes happen in an initialization phase of a graph algorithm, changes can be agglomerated into occasional rebuildings of the graph, or changes can be simulated by building several related graphs.

There are many variants of the adjacency array representation. Information associated with nodes and edges may be stored together with these objects or in separate arrays. A rule of thumb is that information that is frequently accessed should be stored with the nodes and edges. Rarely used data should be kept in separate arrays, because otherwise it would often be moved to the cache without being used. However, there can be other, more complicated reasons why separate arrays may be faster. For example, if both adjacency information and edge weights are read but only the weights are changed, then separate arrays may be faster because the amount of data written back to the main memory is reduced.

Unfortunately, no graph representation is best for all purposes. How can one cope with the zoo of graph representations? First, libraries such as LEDA and the Boost graph library offer several different graph data types, and one of them may suit your purposes. Second, if your application is not particularly time- or space-critical, several representations might do and there is no need to devise a custom-built representation for the particular application. Third, we recommend that graph algorithms should be written in the style of generic programming [119]. The algorithms should access the graph data structure only through a small interface – a set of operations such as iterating over the edges out of a node, accessing information associated with an edge, and proceeding to the target node of an edge. A graph algorithm that only accesses the graph using this interface can be run on any representation that realizes the interface. In this way, one can experiment with different representations. Fourth, if you have to build a custom representation for your application, make it available to others.

8.7.1 C++

LEDA [195, 218, 236] offers a powerful graph data type that supports a large variety of operations in constant time and is convenient to use, but is also space-consuming.

Therefore LEDA also implements several more space-efficient adjacency array representations.

The Boost graph library [50, 196] emphasizes a strict separation of representation and interface. In particular, Boost graph algorithms run on any representation that realizes the Boost interface. Boost also offers its own graph representation class *adjacency_list*. A large number of parameters allow one to choose between variants of graphs (directed and undirected graphs and multigraphs[2]), types of navigation available (in-edges, out-edges, ...), and representations of node and edge sequences (arrays, linked lists, sorted sequences, ...).

LEMON (Library for Efficient Modeling and Optimization in Networks) [201] also emphasizes the strict separation of representation and interface. LEMON offers a variety of general graph concepts, e.g., undirected graphs, directed graphs, and bipartite graphs, as well as special graph classes, e.g., grids and hypercubes. LEMON offers a richer class of graph algorithms than does the Boost library.

8.7.2 Java

JGraphT [167] offers rich support for graphs. It has a clear separation between interfaces, algorithms, and representations. It offers a rich class of algorithms.

8.8 Historical Notes and Further Findings

Special classes of graphs may result in additional requirements on their representation. An important example is *planar graphs* – graphs that can be drawn in the plane without edges crossing. Here, the ordering of the edges adjacent to a node should be in counterclockwise order with respect to a planar drawing of the graph. In addition, the graph data structure should efficiently support iterating over the edges along a *face* of the graph, a cycle that does not enclose any other node. LEDA offers representations for planar graphs.

Recall that *bipartite graphs* are special graphs where the node set $V = L \cup R$ can be decomposed into two disjoint subsets L and R such that the edges are only between nodes in L and nodes in R. All representations discussed here also apply to bipartite graphs. In addition, one may want to store the two sides L and R of the graph. A bipartite graph can be represented by an $|L| \times |R|$ matrix.

Hypergraphs $H = (V, E)$ are generalizations of graphs, where edges can connect more than two nodes. Hypergraphs are conveniently represented as the corresponding bipartite graph $B_H = (E \cup V, \{(e, v) : e \in E, v \in V, v \in e\})$.

Cayley graphs are an interesting example of implicitly defined graphs. Recall that a set V is a *group* if it has an associative multiplication operation $*$, a neutral element, and a multiplicative inverse operation. The *Cayley graph* (V, E) with respect to a set $S \subseteq V$ has the edge set $\{(u, u * s) : u \in V, s \in S\}$. Cayley graphs are useful because graph-theoretic concepts can be useful in group theory. On the other hand,

---

[2] Multigraphs allow multiple parallel edges.

group theory yields concise definitions of many graphs with interesting properties. For example, Cayley graphs have been proposed as interconnection networks for parallel computers [17].

In this book, we have concentrated on convenient data structures for *processing* graphs. There has also been a lot of work on *storing* graphs in a flexible, portable, space-efficient way. Significant compression is possible if we have a priori information about the graph. For example, the edges of a triangulation of n points in the plane can be represented with about $6n$ bits [74, 282].

9

Graph Traversal

Suppose you are working in the traffic planning department of a town with a nice medieval center.[1] An unholy coalition of shop owners, who want more street-side parking, and the Green Party, which wants to discourage car traffic altogether, has decided to turn most streets into one-way streets. You want to avoid the worst by checking whether the current plan maintains the minimum requirement that one can still drive from every point in town to every other point.

In the language of graphs (see Sect. 2.12), the question is whether the directed graph formed by the streets is strongly connected. The same problem comes up in other applications. For example, in the case of a communication network with unidirectional channels (e.g., radio transmitters), we want to know who can communicate with whom. Bidirectional communication is possible within the strongly connected components of the graph.

We shall present a simple, efficient algorithm for computing strongly connected components (SCCs) in Sect. 9.3.2. Computing SCCs and many other fundamental problems on graphs can be reduced to systematic graph exploration, inspecting each edge exactly once. We shall present the two most important exploration strategies: *breadth-first search* (BFS) in Sect. 9.1 and *depth-first search* (DFS) in Sect. 9.3. Both strategies construct forests and partition the edges into four classes: *Tree* edges comprising the forest, *forward* edges running parallel to paths of tree edges, *backward* edges running antiparallel to paths of tree edges, and *cross* edges. Cross edges are all remaining edges; they connect two different subtrees in the forest. Figure 9.1 illustrates the classification of edges.

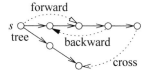

Fig. 9.1. Graph edges classified as tree edges, forward edges, backward edges, and cross edges.

---

[1] The copper engraving above shows part of Frankfurt around 1628 (M. Merian).

© Springer Nature Switzerland AG 2019
P. Sanders et al., *Sequential and Parallel Algorithms and Data Structures*,
https://doi.org/10.1007/978-3-030-25209-0_9

BFS can be parallelized to some extent and is discussed in Sect. 9.2. In contrast, DFS seems inherently difficult to parallelize. In Sect. 9.4, we therefore discuss only the traversal of directed acyclic graphs (DAGs) – a problem where DFS is used as the standard sequential solution. We also discuss several problems on undirected graphs, where a conversion to a DAG is an important step towards a parallel algorithm.

9.1 Breadth-First Search

A simple way to explore all nodes reachable from some node s is *breadth-first search* (BFS). BFS explores the graph (or, digraph) *layer by layer*. The starting node s forms layer 0. The direct neighbors (or, respectively, successors) of s form layer 1. In general, all nodes that are neighbors (or, successors) of a node in layer i but not neighbors (or, successors) of nodes in layers 0 to $i-1$ form layer $i+1$. Instead of saying that node v belongs to layer i, we also say that v has *depth i* or *distance i* from s.

The algorithm in Fig. 9.3 takes a node s and constructs the BFS tree rooted at s. This tree comprises exactly the nodes that are reachable from s. For each node v in the tree, the algorithm records its distance $d(v)$ from s, and the parent node $parent(v)$ from which v was first reached. The algorithm returns the pair $(d, parent)$. Initially, s has been reached and all other nodes store some special value $\bot$ to indicate that they have not been reached yet. Also, the depth of s is 0. The main loop of the algorithm builds the BFS tree layer by layer. We maintain two sets, Q and Q'; Q contains the nodes in the current layer, and we construct the next layer in Q'. The inner loops inspect all edges (u,v) leaving nodes u in the current layer, Q. Whenever v has no parent pointer yet, we put it into the next layer, Q', and set its parent pointer and distance appropriately. Figure 9.2 gives an example of a BFS tree and the resulting backward and cross edges.

BFS has the useful feature that its tree edges define paths from s that have a minimum number of edges. For example, you could use such paths to find railway connections that minimize the number of times you have to change trains or to find paths in communication networks with a smallest number of hops. An actual path from s to a node v can be found by following the parent references from v backwards.

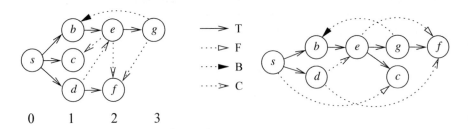

Fig. 9.2. BFS (*left*) and DFS (*right*) classify edges into tree (T), backward (B), cross(C), and forward edges (F). BFS visits the nodes in the order s, b, c, d, e, f, g and partitions them into layers $\{s\}$, $\{b,c,d\}$, $\{e,f\}$, and $\{g\}$. DFS visits the nodes in the order s, b, e, g, f, c, d.

Function *bfs*(*s* : *NodeId*) : (*NodeArray* **of** 0..*n*) × (*NodeArray* **of** *NodeId*)
    $d = \langle \infty, \ldots, \infty \rangle$: *NodeArray* **of** 0..*n*   // distance from root
    parent = $\langle \bot, \ldots, \bot \rangle$: *NodeArray* **of** *NodeId*
    $d[s] := 0$
    parent[*s*] := *s*   // self-loop signals root
    $Q = \langle s \rangle$: *Set* **of** *NodeId*   // current layer of BFS tree
    $Q' = \langle \rangle$: *Set* **of** *NodeId*   // next layer of BFS tree
    for $\ell := 0$ **to** ∞ **while** $Q \neq \langle \rangle$ **do**   // explore layer by layer
        invariant Q contains all nodes with distance ℓ from s
        foreach $u \in Q$ **do**
            foreach $(u, v) \in E$ **do**   // *scan* edges out of *u*
                if *parent*(*v*) = $\bot$ **then**   // found an unexplored node
                    parent(*v*) := *u*   // update BFS tree
                    $d[v] := \ell + 1$
                    $Q' := Q' \cup \{v\}$ // remember for next layer
      $(Q, Q') := (Q', \langle \rangle)$ // switch to next layer
    return (*d*, *parent*)   // the BFS tree is now $\{(v, w) : w \in V, v = parent(w)\}$

Fig. 9.3. Breadth-first search starting at a node *s*

Exercise 9.1. Show that BFS will never classify an edge as forward, i.e., there are no edges (u, v) with $d(v) > d(u) + 1$.

Exercise 9.2. What can go wrong with our implementation of BFS if *parent*[*s*] is initialized to $\bot$ rather than *s*? Give an example of an erroneous computation.

Exercise 9.3. BFS trees are not necessarily unique. In particular, we have not specified the order in which nodes are removed from the current layer. Give the BFS tree that is produced when *d* is removed before *b* when one performs a BFS from node *s* in the graph in Fig. 9.2.

Exercise 9.4 (FIFO BFS). Explain how to implement BFS using a single FIFO queue of nodes whose outgoing edges still have to be scanned. Prove that the resulting algorithm and our two-queue algorithm compute exactly the same tree if the two-queue algorithm traverses the queues in an appropriate order. Compare the FIFO version of BFS with Dijkstra's algorithm described in Sect. 10.3 and the Jarník–Prim algorithm described in Sect. 11.2. What do they have in common? What are the main differences?

Exercise 9.5 (graph representation for BFS). Give a more detailed description of BFS. In particular, make explicit how to implement it using the adjacency array representation described in Sect. 8.2. Your algorithm should run in time O(*n* + *m*).

Exercise 9.6 (BFS in undirected graphs). Assume the bidirected representation of undirected graphs. Show that edges are traversed in at most one direction, i.e., only the scanning of one of the directed versions (u, v) or (v, u) of an undirected edge $\{u, v\}$ can add a node to Q'. When does neither directed version add a node to Q'?

Exercise 9.7 (connected components). Explain how to modify BFS so that it computes a spanning forest of an undirected graph in time $O(m+n)$. In addition, your algorithm should select a *representative* node r for each connected component of the graph and assign it to *component*$[v]$ for each node v in the same component as r. Hint: Scan all nodes $s \in V$ in an outer loop and start BFS from any node s that it still unreached when it is scanned. Do not reset the parent array between different runs of BFS. Note that isolated nodes are simply connected components of size 1.

Exercise 9.8 (transitive closure). The *transitive closure* $G^+ = (V, E^+)$ of a graph $G = (V, E)$ has an edge $(u, v) \in E^+$ whenever there is a path of length 1 or more from u to v in E. Design an algorithm for computing transitive closures. Hint: Run *bfs(v)* for each node v to find all nodes reachable from v. Try to avoid a full reinitialization of the arrays d and *parent* at the beginning of each call. What is the running time of your algorithm?

9.2 Parallel Breadth-First Search

Here, we parallelize each iteration of the main loop of the BFS algorithm shown in Fig. 9.3. We first describe the parallelization for the shared-memory model and assume that it is sufficient to process the nodes in the current layer Q in parallel. Then Sect. 9.2.3 outlines what has to be changed for the distributed-memory model. Section 9.2.4 explains how to handle nodes with very high degree by also parallelizing the loop over the edges out of a single node u.

9.2.1 Shared-Memory BFS

Suppose the current layer Q is represented as a global array. We run the loop over Q in parallel. Several of the load-balancing algorithms presented in Chap. 14 can be used. We describe the variant using prefix sums given in Sect. 14.2. Since the work for a node is roughly proportional to its outdegree, we compute the prefix sum over the outdegrees of the nodes in Q, i.e., for node $Q[j]$, we compute $\sigma[j] := \sum_{k \leq j}$ outdegree$(Q[k])$. Let $m_\ell := \sigma[|Q|]$ be the total outdegree of the layer. Node $Q[j]$ is then assigned to PE $\lceil \sigma[j] p / m_\ell \rceil$. PE j finds the last node it has to process using binary search in σ, searching for jm_ℓ / p.

To parallelize the loop over Q, we avoid contention on Q' by splitting it into local pieces – each PE works on a local array Q' storing the nodes it enqueues. After the loop finishes, these local pieces are copied to the global array Q for the next iteration. Each PE copies its own piece. The starting addresses in Q can be computed as a prefix sum over the piece sizes.

A further dependence between computations on different nodes is caused by multiple edges between nodes in the current layer and a node v in the next layer. If these are processed at the same time by different PEs, there might be write contention for $d[v]$ and *parent*(v). In that case, v could also be inserted into Q' multiple times.

Let us first discuss a solution for CRCW-PRAMs with *arbitrary* semantics for concurrent writes; see Sect. 2.4.1. First, note that in iteration ℓ of the main loop, all competing threads will write the same value $\ell + 1$ to $d[v]$. Moreover, we do not care which node u will successfully set $parent(v) := u$. Finally, we can avoid duplicate entries in Q' by only inserting v into Q' if $parent(v) = u$. The lockstep synchronization of PRAMs will ensure that we read the right value here.

This PRAM algorithm yields the following bound.

Theorem 9.1. *On an arbitrary-CRCW-PRAM, BFS from s can be implemented to run in time*

$$O\left(\frac{m+n}{p} + D \cdot (\Delta + \log p)\right),$$

where D is the largest BFS distance from s and Δ is the maximum degree of a node.

Proof. Consider iteration ℓ of the main loop, working on n_ℓ nodes in Q with m_ℓ outgoing edges. The prefix sum for load balancing takes time $O(n_\ell/p + \log p)$. Each PE will be assigned at most $\lceil m_\ell/p \rceil + \Delta$ edges. Each edge can be processed in constant time. The prefix sum for finding positions in Q takes time $O(\log p)$. Copying Q' to Q takes time $O(m_\ell/p + \Delta)$. Summing over all iterations, we get the time bound

$$\sum_{\ell=0}^{D} O\left(\frac{m_\ell}{p} + \frac{n_\ell}{p} + \Delta + \log p\right) = O\left(\frac{m+n}{p} + D(\Delta + \log p)\right). \qquad \square$$

On an asynchronous shared-memory machine, a conservative solution is to use a CAS instruction to set parent pointers. $CAS(parent(v), \bot, u)$ could be used to make sure that exactly one predecessor u of v succeeds in establishing itself as the parent of v. Only when this CAS succeeds, are Q' and $d[v]$ subsequently updated. However, this is one of the rare cases where we can avoid a CAS instruction despite write conflicts. As long as the accesses to $d[v]$ and $parent(v)$ are atomic, the same discussion as for the PRAM algorithm will ensure that we get consistent values. The only complication that we cannot rule out is occasional duplicate entries in Q'. Figure 9.4 gives an example. This might incur additional work but does not render the computation incorrect. The only explicit synchronization we need is a barrier synchronization (see Sect. 13.4.2) after incrementing $\ell + 1$ in the main loop.

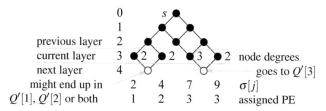

Fig. 9.4. Computations for layer 3 during a shared-memory BFS. Nodes 1 to 4 of the current layer are assigned to PEs $\lceil 2 \cdot 3/9 \rceil = 1$, $\lceil 4 \cdot 3/9 \rceil = 2$, $\lceil 7 \cdot 4/9 \rceil = 3$, and $\lceil 9 \cdot 3/9 \rceil = 3$ respectively. Since nodes 3 and 4 are handled by the same PE, the second node on the next layer goes to $Q'[3]$. Since nodes 1 and 2 are handled by distinct PEs, the first node on the next layer may end up in $Q'[1]$ or in $Q'[2]$ or in both.

9.2.2 Shared-Memory BFS Implementation

Here, we implement the algorithm presented in Sect. 9.2.1 in C++11. Listing 9.1 introduces a static graph data structure implemented as an adjacency array and a generic parallel prefix sum function. Listing 9.2 contains the shared-memory BFS implementation.

Listing 9.1. Utility functions for C++ BFS implementation

```
class Graph {                                                                    1
  int n;                                                                         2
  size_t m;                                                                      3
  std::vector<size_t> begin;                                                     4
  std::vector<int> adj; // adjacency array representation                        5
public:                                                                          6
  vector<int>::const_iterator beginNeighbor(const int v) const {                 7
    return adj.cbegin() + begin[v];                                              8
  }                                                                              9
  vector<int>::const_iterator endNeighbor(const int v) const {                  10
    return adj.cbegin() + begin[v + 1];                                         11
  }                                                                            12
  int getN() const { return n; } // number of nodes                           13
  size_t getM() const { return m; } // number of edges                        14
};                                                                             15
template <class Iterator, class F, class B>                                    16
void prefixSum(Iterator outBegin, Iterator outEnd, int iPE, int p,             17
    Iterator tmp, F f, B & barrier) {                                          18
  const size_t begin = (outEnd−outBegin)*iPE/p;                                19
  const size_t end = (outEnd−outBegin)*(iPE+1)/p;                              20
  size_t sum = 0, i = begin;                                                   21
  for (; i != end ; ++i) *(outBegin + i) = (sum += f(i));                      22
  *(tmp + iPE) = sum;                                                          23
  barrier.wait(iPE, p);                                                        24
  size_t a = 0;                                                                25
  for(i=0; i< iPE; ++i) a += *(tmp + i);                                       26
  for(i=begin; i!=end; ++i) *(outBegin + i) += a;                             27
}//SPDX−License−Identifier: BSD−3−Clause; Copyright(c) 2018 Intel Corporation 28
```

Since we are targeting a moderate number of threads, we do not use the prefix sum algorithm presented in Sect.13.3, whose asymptotic execution time is logarithmic in p, but a simpler code with linear execution time $O(p + n/p)$ but favorable constant factors. The *beginNeighbor* function returns a pointer to the first edge out of a node. The *endNeighbor* function returns a pointer to the first edge after the last edge out of a node. For a node with no outgoing edge, both functions return the same value. We do not provide graph initialization functions in this listing, as they are trivial. The *prefixSum* function takes the output range specified by pointer iterators *outBegin* and *outEnd*, the identifier of the PE (*iPE*), the total number of PEs p, the pointer to a temporary array *tmp*, and a function object f. The use of a function object allows user code to compute values on demand without requiring them to be stored in intermediate arrays. In lines 27–30 of Listing 9.2, we call *prefixSum* with a lambda function (line 29) that computes the outdegree of a node.

We come to the details of *prefixSum*. In lines 19 and 20, the array boundaries for *iPE*'s subrange are computed. Then the prefix sum within the local range is computed in line 22. The total sum of the range is saved in the *tmp* array (line 23). A barrier synchronization (Sect. 13.4.3) is executed in line 24 to ensure that the array *tmp* is completely filled before the PE accumulates the total sums of the PEs with smaller identifiers (lines 25 and 26) and adds the total aggregate to the local output items (line 27).

Listing 9.2 shows the bulk of the implementation. The parallel BFS routine *pBFS* accepts the total number of PEs *p*, the input graph *g*, the root of the BFS tree *s*, the array of output distances *d*, and the *parent* array. In lines 6–13, we initialize the arrays *d*, *parent*, *Q*, *Qp*, and *sigma* and/or preallocate space for them. These arrays correspond to *d*, *parent*, *Q*, *Q'*, and σ in the abstract algorithm in Sect. 9.2.1. *Qp* is an array of arrays storing the local *Qp* for each PE. The entries of *Qp* are *padded* to have the size of a full cache line (line 2) in order to avoid false sharing; see Sect. B.3. The atomic flag *done* and the barrier *barrier* are created and initialized in lines 14–15. Then the C++ threads are created and started. Recall that the constructor of a thread takes two arguments, a worker function passed as a function argument and a second argument that is passed to the worker function by the thread; see App C. The call of the constructor extends through lines 18–65. The worker function is defined in-place (lines 19–65). The second parameter is the loop index *i* (line 65) so that *iPE* is set to *i* in the *i*-th thread. The loop in line 66 waits until all *worker* threads are completed.

The body of the BFS worker function is given in lines 19–64. Before entering the main loop over BFS levels *l*, each PE initializes its part of the output arrays *d* and *parent* (lines 19–23). The PE which has the root *s* in its range initializes the distance *d* and *parent* of *s* in line 24. The main loop extends over lines 26–63. It first computes the prefix sum σ over outdegrees of the nodes in *Q* (lines 27–30). After the thread barrier on the next line, the local array *Qp* is cleared. Then some memory is preallocated for it to reduce future reallocations and copying on capacity overflows (lines 32–34).

Each PE determines in lines 36–38 the range of nodes (from *curQ* to *endQ* -1) assigned to it using the *upper_bound* binary search function from the C++ standard library. Edges (u, v) out of nodes belonging to the PE are traversed with two nested *for*-loops in lines 39–51. If the PE does not see a valid *parent* identifier for node *v*, then it updates its *parent* with *u*, the distance *d* with *l*, and adds *u* to the local *Qp* array (lines 45–48).

After a barrier, the PE computes the position (*outPos*) in *Q*, where it subsequently moves its local part of *Qp* (lines 52–54). To this end, it sums the sizes of the parts of *Qp* with smaller identifiers. The last PE resizes *Q* and *sigma* to accommodate $|Q'|$ elements. If there is nothing left to be done, this is signaled using the *done* flag (lines 55–59). After a further barrier, the local array *Qp* is copied to the global array *Q* (line 61). A final barrier ensures that *Q* is in a consistent state before the next iteration of the outermost loop is started.

Exercise 9.9. For each of the barrier synchronizations in Listing 9.2, give an example of what could go wrong if this barrier were omitted.

Listing 9.2. Shared-memory BFS implementation in C++

```cpp
// pad to avoid false sharing                                                    1
typedef pair<vector<int>, char [64 − sizeof(vector<int>)] > PaddedVector;        2

void pBFS(unsigned p, const Graph & g, const int s,                              3
                vector<int> & d, vector<int> & parent)                           4
{                                                                                5
  d.resize(g.getN());                                                            6
  parent.resize(g.getN());                                                       7
  vector<int> Q;                                                                 8
  Q.reserve(g.getN());                                                           9
  Q.push_back(s);                                                                10
  vector<PaddedVector> Qp(p);                                                    11
  vector<size_t> sigma(1), tmp(p);                                              12
  sigma.reserve(g.getN());                                                       13
  atomic<bool> done(false);                                                      14
  Barrier barrier(p);                                                            15
  vector<thread> threads(p);                                                     16
  for (unsigned i = 0; i < p; ++i)// go parallel                                 17
    threads[i] = thread([&](const unsigned iPE) { // worker function             18
      const size_t beginI = iPE*g.getN()/p;                                      19
      const size_t endI = (iPE + 1)*g.getN()/p;                                  20
      fill(d.begin() + beginI, d.begin() + endI,                                 21
        (numeric_limits<int>::max()));                                           22
      fill(parent.begin()+beginI, parent.begin()+endI,INVALID_NODE_ID);          23
      if(s >= beginI && s < endI) { d[s] = 0; parent[s] = s; }                   24
      int l = 1;                                                                 25
      for(; !done; ++l) {                                                        26
        prefixSum(sigma.begin(), sigma.end(), iPE, p, tmp.begin(),               27
            /* a lambda function */ [&] (int j)                                  28
            { return g.endNeighbor(Q[j]) − g.beginNeighbor(Q[j]);},              29
            barrier);                                                            30
        barrier.wait(iPE, p);                                                    31
        Qp[iPE].first.clear();                                                   32
        size_t ml = sigma.back();                                               33
        Qp[iPE].first.reserve(2*ml/p); // preallocate memory                     34
        size_t curQ = upper_bound(sigma.cbegin(), sigma.cend(),                 35
                        iPE*ml/p) − sigma.cbegin();                             36
        const size_t endQ = upper_bound(sigma.cbegin(), sigma.cend(),           37
            (iPE+1)*ml/p) − sigma.cbegin();                                     38
        for (; curQ != endQ; ++curQ) { // loop over nodes                       39
          const int u = Q[curQ];                                                40
          auto vIter = g.beginNeighbor(u);                                      41
          auto vEnd = g.endNeighbor(u);                                         42
          for (; vIter != vEnd; ++vIter) { // loop over edges                   43
```

```
        const int v = *vIter; // target of current edge                    44
        if(parent[v] == INVALID_NODE_ID) { // not visited yet              45
          parent[v] = u;                                                    46
          d[v] = l;                                                         47
          Qp[iPE].first.push_back(v); // queue for next layer               48
        }                                                                    49
      }                                                                      50
    }                                                                        51
    barrier.wait(iPE, p);                                                    52
    size_t outPos = 0;                                                       53
    for(int j=0; j < iPE; ++j) outPos += Qp[j].first.size();                 54
    if(iPE == p-1) {                                                         55
      Q.resize(outPos + Qp[iPE].first.size());                              56
      sigma.resize(Q.size());                                                57
      if (Q.empty()) done = true;                                            58
    }                                                                        59
    barrier.wait(iPE, p);                                                    60
    copy(Qp[iPE].first.cbegin(), Qp[iPE].first.cend(),                       61
      Q.begin() + outPos);                                                   62
    barrier.wait(iPE, p);                                                    63
    }                                                                        64
  }, i);                                                                     65
  for (auto & t : threads) t.join();                                         66
}//SPDX-License-Identifier: BSD-3-Clause; Copyright(c) 2018 Intel Corporation  67
```

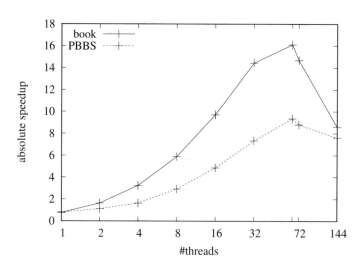

Fig. 9.5. Speedup of parallel BFS implementations over sequential BFS for the grid graph.

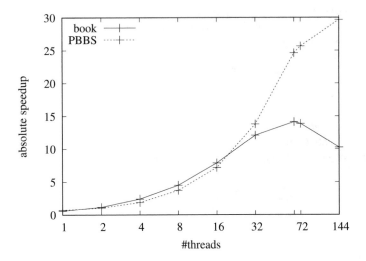

Fig. 9.6. Speedup of parallel BFS implementations over sequential BFS for the R-MAT graph

We measured the speedup of this implementation on the four-socket machine described in Appendix B and compared it with the implementation from the Problem Based Benchmark Suite (PBBS) [297]. Figures 9.5 and 9.6 show the speedup as a function of the number of threads for two graphs with $n = 10^7$ nodes – a three-dimensional grid and the R-MAT recursive matrix graph [63] with parameters $(.3,.1,.1,.5)$ (about $5n$ edges). Our implementation outperforms PBBS for the grid instance. For the R-MAT instance[2], our code is slightly better than PBBS for up to 16 threads, but does not scale to the full number of 72 available cores and gets even worse for 144 threads (using hyper-threading). The main difference between the two implementations is that we use static load balancing based on prefix sums, whereas PBBS uses the dynamic load balancing integrated into Cilk. This implementation difference may explain the performance difference. Static load balancing has less overhead and hence works well on instances with small average queue size such as the grid instance. Its layers are of size $O(\sqrt{n})$. The R-MAT graph has a much smaller diameter and hence a larger average queue size.

9.2.3 Distributed-Memory BFS

Our design for BFS on distributed-memory machines is based on the "owner computes" principle. We assign each node u of the graph to a PE $u.p$, which does the work related to node u. If u is in the set Q of nodes to be scanned, PE $u.p$ is responsible for doing this. However, when scanning edge (u,v), PE $i = u.p$ delegates this to PE $j = v.p$ since node v is the object actually affected. Hence, PE i sends the edge (u,v) to PE j. Figure 9.7 gives pseudocode. Of course, an efficient algorithm will

---

[2] The instance "local random" from [297] exhibited similar behavior.

handle local edges (u,v) with $u.p = v.p$ without explicit communication and it will deliver the messages in bulk fashion – gathering all local messages destined for the same PE in a message buffer and delivering this buffer as a whole or in large chunks. For example, a library based on the BSP model can be used here.

Function $dBFS(s : NodeId) : (NodeArray$ **of** $0..n) \times (NodeArray$ **of** $NodeId)$
  $d = \langle \infty, \ldots, \infty \rangle : NodeArray$ **of** $0..n$ // distance from root
  $parent = \langle \perp, \ldots, \perp \rangle : NodeArray$ **of** $NodeId$
  $Q = \langle \rangle : Set$ **of** $NodeId$ // current layer of BFS tree
  if $s.p = i_{\mathrm{proc}}$ **then** $d[s] := 0;\ parent[s] := s;\ Q := \langle s \rangle$
  for $\ell := 0$ **to** ∞ **while** $\exists i : Q@i \neq \langle \rangle$ **do**   // explore layer by layer
    invariant Q contains all local nodes with distance ℓ from s
    foreach $u \in Q$ **do**
      foreach $(u,v) \in E$ **do**   // *scan edges out of u*
        post message (u,v) to PE $v.p$
    deliver all messages
    $Q := \{\}$
    foreach *received message* (u,v) **do**
      if $parent(v) = \perp$ **then**   // found an unexplored node
        $parent(v) := u$ // update BFS tree
        $d[v] := \ell + 1$
        $Q := Q \cup \{v\}$ // remember for next layer
  return $(d, parent)$ // the BFS tree is now $\{(v,w) : w \in V, v = parent(w)\}$

Fig. 9.7. Distributed-memory BFS starting at a node s

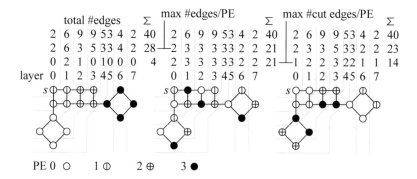

Fig. 9.8. Distributed-memory BFS with three different partitions. *Left*: four compact blocks – bad load balance but low communication. *Middle*: scattered nodes – good load balance but high communication. *Right*: eight scattered blocks of size two – a compromise. The light lines indicate the partition of the graph into layers. The top row of numbers shows the number of edges incident to the nodes of the current layer, the second row shows the maximum number of edges that have to be handled by a single PE, and the third row shows the maximum number of cut edges that have to be handled by a single PE. The column Σ contains the row sums.

Note that we are facing a trade-off between communication overhead and load balance here. Using a graph partition with a small overall cut as proposed in Sect. 8.6 will result in a small overall communication volume. However, in graphs with high locality, the BFS will initially run just on PE $s.p$, spreading only slowly over further PEs. On the other hand, assigning nodes to random PEs will destroy locality, but parallelism will spread very fast and the work will be well balanced. Perhaps compromises are in order here – working with a partition with $k \gg p$ pieces spread randomly over the PEs. Figure 9.8 illustrates this trade-off.

9.2.4 Handling High-Degree Nodes

We apply the general approach described on page 266. We use the same prefix sum σ over outdegrees for load balancing as in Sect. 9.2.1. However, we use it to actually distribute edges. The kth edge of node $Q[i]$ gets the number $\sigma[j-1]+k$. Edge i will be processed by PE $\lceil ip/m_\ell \rceil$. This means that PE j performs binary search for $k_j := jm_l/p$ in σ. So suppose $\sigma[i] \le k_j < \sigma[i+1]$. Then PE j starts scanning edges at edge $k_j - \sigma[i]$ of node i.

Exercise 9.10. Show that with the above modification of the algorithm, you can sharpen Theorem 9.1 to obtain running time

$$O\left(\frac{m+n}{p} + D\log p\right).$$

9.3 Depth-First Search

You may view breadth-first search as a careful, conservative strategy for systematic exploration that completely inspects known things before venturing into unexplored territory. In this respect, *depth-first search* (DFS) is the exact opposite: Whenever it finds a new node, it immediately continues to explore from it. It goes back to previously explored nodes only if it runs out of options to go forward. Although DFS leads to unbalanced exploration trees compared with the orderly layers generated by BFS, the combination of eager exploration with the perfect memory of a computer makes DFS very useful. Figure 9.9 gives an algorithm template for DFS. We can derive specific algorithms from it by specifying the subroutines *init*, *root*, *traverseTreeEdge*, *traverseNonTreeEdge*, and *backtrack*.

DFS uses node marks. Initially, all nodes are unmarked. Nodes are marked *active* when they are discovered and their exploration begins. Once the exploration of a node is completed, the mark is changed to *completed* and keeps this value until the end of the execution. The main loop of DFS looks for unmarked nodes s and calls $DFS(s,s)$ to grow a tree rooted at s. The recursive call $DFS(u,v)$ organizes the exploration out of v. The argument (u,v) indicates that v was reached via the edge (u,v) into v. For root nodes s, we use the "dummy" argument (s,s). We write $DFS(*,v)$ if the specific nature of the incoming edge is irrelevant to the discussion at hand.

Depth-first search of a directed graph $G = (V, E)$
unmark all nodes
init
foreach $s \in V$ **do**
  if s is not marked **then**
    $root(s)$ // Make s a root and grow
    $DFS(s, s)$ // a new DFS tree rooted at it.

Procedure $DFS(u, v : NodeId)$ // Explore v coming from u.
  mark v as active
  foreach $(v, w) \in E$ **do**
    if w is marked **then** *traverseNonTreeEdge*(v, w) // w was reached before
    else   *traverseTreeEdge*(v, w) // w was not reached before
          $DFS(v, w)$
  $backtrack(u, v)$ // Return from v along the incoming edge.
  mark v as completed

Fig. 9.9. A template for depth-first search of a graph $G = (V, E)$. We say that a call $DFS(*, v)$ explores v. The exploration is complete when we return from this call.

The call $DFS(*, v)$ first marks v as active and then inspects all edges (v, w) out of v. Assume now that we are exploring edge (v, w) with end node w. If w has been seen before (w is marked as either active or completed), w is already a node of the DFS forest. So (v, w) is not a tree edge, and hence we call *traverseNonTreeEdge*(v, w) and make no recursive call of DFS. If w has not been seen before (w is unmarked), (v, w) becomes a tree edge. We therefore call *traverseTreeEdge*(v, w) and make the recursive call of $DFS(v, w)$. When we return from this call, we explore the next edge out of v. Once all edges out of v have been explored, the procedure $backtrack(u, v)$ is called, where (u, v) is the edge in the call DFS for v; it performs summarizing and cleanup work. We then change the mark of v to "completed" and return.

At any point in time during the execution of DFS, there are a number of active calls. More precisely, there are nodes $v_1, v_2, \ldots, v_k$ such that we are currently exploring edges out of v_k, and the active calls are $DFS(v_1, v_1)$, $DFS(v_1, v_2)$, $\ldots$, $DFS(v_{k-1}, v_k)$. In this situation, precisely the nodes v_1 to v_k are marked *active* and the recursion stack contains the sequence $\langle (v_1, v_1), (v_1, v_2), \ldots, (v_{k-1}, v_k) \rangle$. More compactly, we say that the recursion stack contains $\langle v_1, \ldots, v_k \rangle$. A node is called active (or completed) if it is marked active (or completed, respectively). The node v_k is called the *current node*. We say that a node v has been *reached* when $DFS(*, v)$ has already been called. So the reached nodes are the active and the completed nodes.

Exercise 9.11. Give a nonrecursive formulation of DFS. There are two natural realizations. One maintains the stack of active nodes and, for each active node, the set of unexplored edges (it suffices to keep a pointer into the list or array of outgoing edges of the active node). The other maintains a stack of all unexplored edges emanating from active nodes. When a node is activated, all its outgoing edges are pushed onto this stack.

9.3.1 DFS Numbering, Completion Times, and Topological Sorting

DFS has numerous applications. In this section, we use it to number the nodes in two ways. As a by-product, we see how to detect cycles. We number the nodes in the order in which they are reached (array *dfsNum*) and in the order in which they are completed (array *compNum*). We have two counters, *dfsPos* and *compPos*, both initialized to 1. When we encounter a new root or traverse a tree edge, we set the *dfsNum* of the newly encountered node and increment *dfsPos*. When we backtrack from a node, we set its *compNum* and increment *compPos*. We use the following subroutines:

init:	$dfsPos = 1 : 1..n;$ $compPos = 1 : 1..n$
root(s):	$dfsNum[s] := dfsPos{+}{+}$
traverseTreeEdge(v,w):	$dfsNum[w] := dfsPos{+}{+}$
backtrack(u,v):	$compNum[v] := compPos{+}{+}$

The ordering by *dfsNum* is so useful that we introduce a special notation '$\prec$' (pronounced "precedes") for it. For any two nodes u and v, we define

$$u \prec v \Leftrightarrow dfsNum[u] < dfsNum[v].$$

The numberings *dfsNum* and *compNum* encode important information about the execution of *DFS*, as we shall show next. We shall first show that the DFS numbers increase along any path of the DFS tree, and then show that the numberings together classify the edges according to their type. They can also be used to encode the node marks during the execution of DFS as follows. We use *init* to initialize *dfsNum* and *compNum* with the all-zero vector. Then a node is unmarked if and only if its *dfsNum* is equal to 0. It is active if and only if its *dfsnum* is positive and *compNum* is 0. The node is completed if and only if its *compNum* is positive.

Lemma 9.2. *The nodes on the DFS recursion stack are ordered with respect to $\prec$.*

Proof. *dfsPos* is incremented after every assignment to *dfsNum*. Thus, when a node v is made active by a call $DFS(u,v)$ and is put on the top of the recursion stack, it has just been assigned the largest *dfsNum* so far.   □

*dfsNum*s and *compNum*s classify edges according to their type, as shown in Table 9.1. The argument is as follows. We first observe that two calls of DFS are either nested within each other, i.e., when the second call starts, the first is still active, or disjoint, i.e., when the second starts, the first is already completed. If $DFS(*,w)$ is nested in $DFS(*,v)$, the former call starts after the latter and finishes before it, i.e., $dfsNum[v] < dfsNum[w]$ and $compNum[w] < compNum[v]$. If $DFS(*,w)$ and $DFS(*,v)$ are disjoint and the former call starts before the latter, it also ends before the latter, i.e., $dfsNum[w] < dfsNum[v]$ and $compNum[w] < compNum[v]$.

Next we observe that the tree edges record the nesting structure of recursive calls. When a tree edge (v,w) is explored within $DFS(*,v)$, the call $DFS(v,w)$ is made and hence is nested within $DFS(*,v)$. Thus w has a larger DFS number and a smaller completion number than v. A forward edge (v,w) runs parallel to a path of tree edges

Table 9.1. The classification of an edge (v, w). The last column indicates the mark of w at the time when the edge (v, w) is explored.

type	$dfsNum[v] < dfsNum[w]$	$compNum[w] < compNum[v]$	Mark of w
tree	Yes	Yes	unmarked
forward	Yes	Yes	completed
backward	No	No	active
cross	No	Yes	completed

and hence w has a larger DFS number and a smaller completion number than v. We can distinguish tree and forward edges by the mark of w at the time when the edge (v, w) is inspected. If w is unmarked, the edge is a tree edge. If w is already marked, the edge is a forward edge. In the case of a forward edge, w is marked as completed; if w were active, it would be part of the recursion stack. Since v is the topmost node of the recursion stack, this would imply $dfsNum(w) < dfsNum(v)$, a contradiction.

A backward edge (v, w) runs antiparallel to a path of tree edges, and hence w has a smaller DFS number and a larger completion number than v. Furthermore, when the edge (v, w) is inspected, the call $DFS(*, v)$ is active and hence, by the nesting structure, so is the call $DFS(*, w)$. Thus w is active when the edge (v, w) is inspected.

Let us look, finally, at a cross edge (v, w). Since (v, w) is not a tree, forward, or backward edge, the calls $DFS(*, v)$ and $DFS(*, w)$ cannot be nested within each other. Thus they are disjoint. So w is completed either before $DFS(*, v)$ starts or after it ends. The latter case is impossible, since, in this case, w would be unmarked when the edge (v, w) was explored, and the edge would become a tree edge. So w is completed before $DFS(*, v)$ starts and hence $DFS(*, w)$ starts and ends before $DFS(*, v)$. Thus $dfsNum[w] < dfsNum[v]$ and $compNum[w] < compNum[v]$. The following lemma summarizes this discussion.

Lemma 9.3. *Table 9.1 shows the characterization of edge types in terms of dfsNum and compNum and the mark of the endpoint at the time of the inspection of the edge.*

Completion numbers have an interesting property for directed acyclic graphs.

Lemma 9.4. *The following properties are equivalent:*

(a) G is a DAG;
(b) DFS on G produces no backward edges;
(c) all edges of G go from larger to smaller completion numbers.

Proof. Backward edges run antiparallel to paths of tree edges and hence create cycles. Thus DFS on an acyclic graph cannot create any backward edges. This shows that (a) implies (b). All edges except backward edges run from larger to smaller completion numbers, according to Table 9.1. This shows that (b) implies (c). Finally, assume that all edges run from larger to smaller completion numbers. In this case the graph is clearly acyclic. This shows that (b) implies (a).   □

An order of the nodes of a DAG in which all edges go from earlier to later nodes is called a *topological sorting*. By Lemma 9.4, the ordering by decreasing completion number is a topological ordering. Many problems on DAGs can be solved efficiently by iterating over the nodes in a topological order. For example, in Sect. 10.2 we shall see a fast, simple algorithm for computing shortest paths in acyclic graphs.

Exercise 9.12. Modify DFS such that it labels the edges with their type.

Exercise 9.13 (topological sorting). Design a DFS-based algorithm that outputs the nodes in topological order if G is a DAG. Otherwise, it should output a cycle.

Exercise 9.14. Design a BFS-based algorithm for topological sorting.

Exercise 9.15. In a DFS on an undirected graph, it is convenient to explore edges in only one direction. When an undirected edge $\{v,w\}$ is inspected for the first time, say in the direction from v to w, it is "turned off" in the adjacency list of w and not explored in the opposite direction. Show that DFS (with this modification) on an undirected graph does not produce any cross edges or forward edges.

9.3.2 Strongly Connected Components

We now come back to the problem posed at the beginning of this chapter. Recall that two nodes belong to the same strongly connected component (SCC) of a graph if and only if they are reachable from each other. In undirected graphs, the relation "being reachable" is symmetric, and hence strongly connected components are the same as connected components. Exercise 9.7 outlines how to compute connected components using BFS, and adapting this idea to DFS is equally simple. For directed graphs, the situation is more interesting. Figure 9.10 shows a graph and its strongly connected components. It also illustrates the concept of the *shrunken graph* G^s corresponding to a directed graph G, which will turn out to be extremely useful for this section. The nodes of G^s are the SCCs of G. If C and D are distinct SCCs of G, we have an edge (C,D) in G^s if and only if there is an edge (u,v) in G with $u \in C$ and $v \in D$.

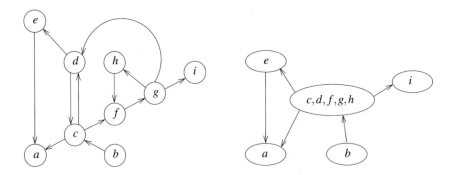

Fig. 9.10. A digraph G and the corresponding shrunken graph G^s. The SCCs of G have node sets $\{a\}$, $\{b\}$, $\{c,d,f,g,h\}$, $\{e\}$, and $\{i\}$.

Exercise 9.16. Show that the node sets of distinct SCCs are disjoint. Hint: Assume that SCCs C and D have a common node v. Show that any node in C can reach any node in D and vice versa.

Lemma 9.5. *The shrunken graph G^s with respect to a digraph G is acyclic.*

Proof. Assume otherwise, and let $C_1, C_2, \ldots, C_{k-1}, C_k$ with $C_k = C_1$ be a cycle in G^s. Recall that the C_i's are SCCs of G. By the definition of G^s, G contains an edge (v_i, w_{i+1}) with $v_i \in C_i$ and $w_{i+1} \in C_{i+1}$ for $0 \le i < k$. Define $v_k = v_1$. Since C_i is strongly connected, G contains a path from w_{i+1} to v_{i+1}, $0 \le i < k$. Thus all the v_i's belong to the same SCC, a contradiction. □

We shall show that the strongly connected components of a digraph G can be computed using DFS in linear time $O(n+m)$. More precisely, the algorithm outputs an array *component* indexed by nodes such that $component[v] = component[w]$ if and only if v and w belong to the same SCC. Alternatively, it could output the node set of each SCC.

The idea underlying the algorithm is simple. We imagine that the edges of G are added one by one to an initially edgeless graph. We use $G_c = (V, E_c)$ to denote the current graph, and keep track of how the SCCs of G_c evolve as edges are added. Initially, there are no edges and each node forms an SCC of its own. We use G_c^s to denote the shrunken graph of G_c.

How do the SCCs of G_c and G_c^s change when we add an edge e to G_c? There are three cases to consider. (1) Both endpoints of e belong to the same SCC of G_c. Then the shrunken graph and the SCCs do not change. (2) e connects nodes in different SCCs but does not close a cycle. The SCCs do not change, and an edge is added to the shrunken graph. (3) e connects nodes in different SCCs and closes one or more cycles. In this case, all SCCs lying on one of the newly formed cycles are merged into a single SCC, and the shrunken graph changes accordingly.

In order to arrive at an efficient algorithm, we need to describe how we maintain the SCCs as the graph evolves. If the edges are added in arbitrary order, no efficient simple method is known. However, if we use DFS to explore the graph, an efficient solution is fairly easy to obtain. Consider a depth-first search on G, let E_c be the set of edges already explored by DFS, and let $G_c = (V, E_c)$ be the current graph. Recall that a node is either unmarked, active, or completed. We distinguish between three kinds of SCCs of G_c: unreached, open, and closed. Unmarked nodes have indegree and outdegree 0 in G_c and hence form SCCs consisting of a single node. The corresponding node in the shrunken graph is isolated. We call these SCCs *unreached*. The other SCCs consist of marked nodes only. We call an SCC consisting of marked nodes *open* if it contains an active node, and *closed* if it contains only completed nodes. We call a marked node "open" if it belongs to an open component and "closed" if it belongs to a closed component. Observe that a closed node is always completed and that an open node may be either active or completed. For every SCC, we call the node with the smallest DFS number in the SCC the *representative* of the SCC. Figure 9.11 illustrates these concepts. We next state some important invariant properties of G_c; see also Fig. 9.12:

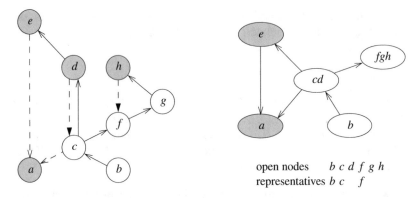

open nodes   *b c d f g h*
representatives *b c   f*

Fig. 9.11. A snapshot of DFS on the graph in Fig. 9.10 and the corresponding shrunken graph. The first DFS was started at node a and a second DFS was started at node b, the current node is g, and the recursion stack contains b, c, f, g. The edges (g, i) and (g, d) have not been explored yet. Edges (h, f) and (d, c) are back edges, (e, a) is a cross edge, and all other edges are tree edges. Finished nodes and closed components are shaded. There are closed components $\{a\}$ and $\{e\}$ and open components $\{b\}$, $\{c, d\}$, and $\{f, g, h\}$. The open components form a path in the shrunken graph with the current node g belonging to the last component. The representatives of the open components are the nodes b, c, and f, respectively. DFS has reached the open nodes in the order b, c, d, f, g, h. The representatives partition the sequence of open nodes into the SCCs of G_c.

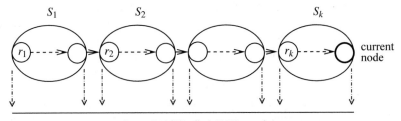

open nodes ordered by their DFS number

Fig. 9.12. The open SCCs are shown as ovals, and the current node is shown as a **bold** circle. The tree path to the current node is indicated. It enters each component at its representative. The horizontal line below represents the open nodes, ordered by *dfsNum*. Each open SCC forms a contiguous subsequence, with its representative as its leftmost element.

(a) All edges in G (not just G_c) out of closed nodes lead to closed nodes. In our example, the nodes a and e are closed.
(b) The tree path to the current node contains the representatives of all open components. Let S_1 to S_k be the open components as they are traversed by the tree path to the current node. There is then a tree edge from a node in S_{i-1} to the representative of S_i, and this is the only edge in G_c into S_i, $2 \leq i \leq k$. Also, there is no edge from an S_j to an S_i with $i < j$. Finally, all nodes in S_j are reachable

from the representative r_i of S_i for $1 \leq i \leq j \leq k$. In short, the open components form a path in the shrunken graph. In our example, the current node is g. The tree path $\langle b, c, f, g \rangle$ to the current node contains the open representatives b, c, and f.

(c) Consider the nodes in the open components ordered by their DFS numbers. The representatives partition the sequence into the open components. In our example, the sequence of open nodes is $\langle b, c, d, f, g, h \rangle$ and the representatives partition this sequence into the open components $\{b\}$, $\{c, d\}$, and $\{f, g, h\}$.

We shall show below that all three properties hold true generally, and not only for our example. The three properties will be invariants of the algorithm to be developed. The first invariant implies that the closed SCCs of G_c are actually SCCs of G, i.e., it is justified to call them closed. This observation is so important that it deserves to be stated as a lemma.

Lemma 9.6. *A closed SCC of G_c is an SCC of G.*

Proof. Let v be a closed vertex, let S be the SCC of G containing v, and let S_c be the SCC of G_c containing v. We need to show that $S = S_c$. Since G_c is a subgraph of G, we have $S_c \subseteq S$. So, it suffices to show that $S \subseteq S_c$. Let w be any vertex in S. There is then a cycle C in G passing through v and w. The invariant (a) implies that all vertices of C are closed. Since closed vertices are completed, all edges out of them have been explored. Thus C is contained in G_c, and hence $w \in S_c$. □

The invariants (b) and (c) suggest a simple method to represent the open SCCs of G_c. We simply keep a sequence *oNodes* of all open nodes in increasing order of DFS number, and the subsequence *oReps* of open representatives. In our example, we have *oNodes* $= \langle b, c, d, f, g, h \rangle$ and *oReps* $= \langle b, c, f \rangle$. We shall see later that both sequences are best kept as a stack (type *Stack* **of** *NodeId*).

Let us next see how the SCCs of G_c develop during DFS. We shall discuss the various actions of DFS one by one and show that the invariants are maintained. We shall also discuss how to update our representation of the open components.

When DFS starts, the invariants clearly hold: No node is marked, no edge has been traversed, G_c is empty, and hence there are neither open nor closed components yet. The sequences *oNodes* and *oReps* are empty.

Just before a new root is to be marked, i.e., the construction of a new DFS tree is started, all marked nodes are completed and hence there cannot be any open component. Therefore, both of the sequences *oNodes* and *oReps* are empty, and marking a new root s produces the open component $\{s\}$. The invariants are clearly maintained. We obtain the correct representation by adding s to both sequences.

If a tree edge $e = (v, w)$ is traversed and hence w is marked as active, $\{w\}$ becomes an open component on its own. All other open components are unchanged. The invariant (a) is clearly maintained, since v is active and hence open. The old current node is v and the new current node is w. The sequence of open components is extended by $\{w\}$. The open representatives are the old open representatives plus the node w. Thus the invariant (b) is maintained. Also, w becomes the open node with

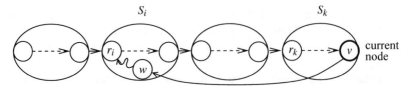

Fig. 9.13. The open SCCs are shown as ovals and their representatives as circles on the left side of the oval. All representatives lie on the tree path to the current node v. The nontree edge $e = (v, w)$ ends in an open SCC S_i with representative r_i. There is a path from w to r_i since w belongs to the SCC with representative r_i. Thus edge (v, w) merges S_i to S_k into a single SCC.

the largest DFS number and hence *oNodes* and *oReps* are both extended by w. Thus the invariant (c) is maintained.

Now suppose that a nontree edge $e = (v, w)$ out of the current node v is explored. If w is closed, the SCCs of G_c do not change when e is added to G_c, since, by Lemma 9.6, the SCC of G_c containing w is already an SCC of G *before* e is traversed. So, assume that w is open. Then w lies in some open SCC S_i of G_c. We claim that the SCCs S_i to S_k are merged into a single component and all other components are unchanged; see Fig. 9.13. Let r_i be the representative of S_i. We can then go from r_i to v along a tree path by invariant (b), then follow the edge (v, w), and finally return to r_i. The path from w to r_i exists, since w and r_i lie in the same SCC of G_c. We conclude that any node in an S_j with $i \leq j \leq k$ can be reached from r_i and can reach r_i. Thus the SCCs S_i to S_k become one SCC, and r_i is their representative. The S_j with $j < i$ are unaffected by the addition of the edge.

The invariant (c) tells us how to find r_i, the representative of the component containing w. The sequence *oNodes* is ordered by *dfsNum*, and the representative of an SCC has the smallest *dfsNum* of any node in that component. Thus $dfsNum[r_i] \leq dfsNum[w]$ and $dfsNum[w] < dfsNum[r_j]$ for all $j > i$. It is therefore easy to update our representation. We simply delete all representatives r with $dfsNum[r] > dfsNum[w]$ from *oReps*.

Finally, we need to consider completing a node v. When will this close an SCC? Completion of v will close the the SCC containing it if and only if v was the only remaining active node in the SCC. By invariant (b), all nodes in a component are tree descendants of the representative of the component, and hence the representative of a component is the last node to be completed in the component. In other words, we close a component if and only if we complete its representative. Since *oReps* is ordered by *dfsNum*, we close a component if and only if the last node of *oReps* completes. So, assume that we complete a representative v. Then, by invariant (c), the component S_k with representative $v = r_k$ consists of v and all nodes in *oNodes* following v. Completing v closes S_k. By invariant (a), there is no edge out of S_k into an open component. Thus invariant (a) holds after S_k is closed. If $k = 1$, the exploration of the entire DFS tree is completed and invariants (b) and (c) clearly hold. If $k \geq 2$, the new current node is the parent of v. By invariant (b), the parent of

v lies in S_{k-1}. Thus invariant (b) holds after S_k is closed. Invariant (c) holds after v is removed from *oReps*, and v and all nodes following it are removed from *oNodes*.

init:
    component : NodeArray of NodeId   // SCC representatives
    oReps = ⟨⟩ : *Stack* **of** *NodeId*   // representatives of open SCCs
    oNodes = ⟨⟩ : *Stack* **of** *NodeId*   // all nodes in open SCCs

root(w) or traverseTreeEdge(v,w):
    oReps.push(w)   // new open
    oNodes.push(w)   // component

traverseNonTreeEdge(v,w):
    if $w \in$ *oNodes* **then**
        while $w \prec$ *oReps.top* **do** *oReps.pop* // collapse components on cycle

backtrack(u,v):
    if $v =$ *oReps.top* **then**
        oReps.pop   // close
        repeat   // component
            $w :=$ *oNodes.pop*
            component$[w] := v$
        until $w = v$

Fig. 9.14. An instantiation of the DFS template that computes strongly connected components of a graph $G = (V,E)$

It is now easy to instantiate the DFS template. Figure 9.14 shows the pseudocode, and Fig. 9.15 illustrates a complete run. We use an array *component* indexed by nodes to record the result, and two stacks *oReps* and *oNodes*. When a new root is marked or a tree edge is explored, a new open component consisting of a single node is created by pushing this node onto both stacks. When a cycle of open components is created, these components are merged by popping all representatives from *oReps* having a larger DFS number than w. An SCC S is closed when its representative v finishes. At that point, all nodes of S are stored above v in *oNodes*. The operation *backtrack* therefore closes S by popping v from *oReps*, and by popping the nodes $w \in S$ from *oNodes* and setting their *component* to the representative v.

Note that the test $w \in$ *oNodes* in *traverseNonTreeEdge* can be done in constant time by storing information with each node that indicates whether the node is open. This indicator is set when a node v is first marked, and reset when the component of v is closed. We give implementation details in Sect. 9.5. Furthermore, the while-loop and the repeat loop can make at most n iterations during the entire execution of the algorithm, since each node is pushed onto the stacks exactly once. Hence, the execution time of the algorithm is $O(m+n)$. We have the following theorem.

Theorem 9.7. *The algorithm in Fig. 9.14 computes strongly connected components in time* $O(m+n)$.

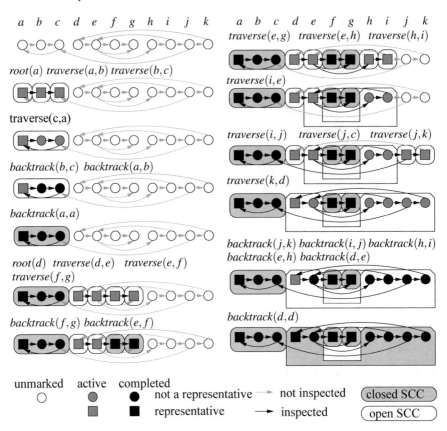

Fig. 9.15. An example of the development of open and closed SCCs during DFS. Unmarked nodes are shown as empty circles, active nodes are shown in gray, and completed nodes are shown in black. Nontraversed edges are shown in gray, and traversed edges are shown in black. Open SCCs are shown as unfilled closed curves, and closed SCCs are shaded gray. Representatives are shown as squares and nonrepresentatives are shown as circles. We start in the situation shown at the upper left. We make a a root and traverse the edges (a,b) and (b,c). This creates three open SSCs. The traversal of edge (c,a) merges these components into one. Next, we backtrack to b and then to a, and finally complete a. At this point, the component becomes closed. Exercise: Please complete the description.

Exercise 9.17 (certificates). Let G be a strongly connected graph and let s be a node of G. Show how to construct two trees rooted at s. The first tree proves that all nodes can be reached from s, and the second tree proves that s can be reached from all nodes. Can you modify the SCC algorithm so that it constructs both trees?

Exercise 9.18 (2-edge-connected components). An undirected graph is 2-edge-connected if its edges can be oriented so that the graph becomes strongly connected. The 2-edge-connected components are the maximal 2-edge-connected subgraphs; see Fig. 9.16. Modify the SCC algorithm shown in Fig. 9.14 so that it computes 2-

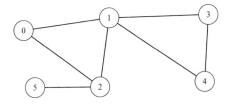

Fig. 9.16. The graph has two 2-edge-connected components, namely $\{0,1,2,3,4\}$ and $\{5\}$. The graph has three biconnected components, namely the subgraphs spanned by the sets $\{0,1,2\}$, $\{1,3,4\}$, and $\{2,5\}$. The vertices 1 and 2 are articulation points.

edge-connected components. Hint: Use the fact that DFS of an undirected graph does not produce either forward or cross edges (Exercise 9.15).

Exercise 9.19 (biconnected components). An *articulation point* in an undirected graph is a node whose removal disconnects the graph. An undirected graph without an articulation point is called *biconnected*. Two trivial cases are a single node and two nodes connected by an edge. Show that a graph with more than two nodes is biconnected if and only if every pair of distinct nodes is connected by two node-disjoint paths; see Fig. 9.16. A *biconnected component* (BCC) of an undirected graph is a maximal biconnected subgraph. Biconnected components are pairwise edge-disjoint. They may share nodes. The nodes that belong to more than one BCC are precisely the articulation points. Design an algorithm that computes the biconnected components of an undirected graph using a single pass of DFS. Hint: Adapt the strongly-connected-components algorithm. Define the representative of a BCC as the node with the second smallest *dfsNum* in the BCC. Prove that a BCC consists of the parent of the representative and all tree descendants of the representative that can be reached without passing through another representative. Modify *backtrack*. When you return from a representative v, output v, all nodes above v in *oNodes*, and the parent of v.

*Exercise 9.20 (open ear decomposition of biconnected graphs). An open ear decomposition of a graph is a sequence of paths $P_0, \ldots, P_k$ with the following properties: P_0 is a simple cycle and each P_i is a path whose endpoints lie on one of the preceding paths, whose endpoints are distinct, and which is internally disjoint from the preceding paths.

(a) Show: If a graph has an open ear decomposition, it is biconnected.
(b) Show: Every biconnected graph has an open ear decomposition. Hint: Consider a DFS on a biconnected graph. Decompose G into a set of paths $P_0, P_1, \ldots$ as follows. First consider any back edge (u, s) ending in the root s of the DFS tree. The first path P_0 consists of the back edge (u, s) plus the tree path from s to u. To construct P_i, choose a back edge (u, v) where v lies on $V_{i-1} := \cup_{j<i} P_j$ and then trace back tree edges from u until a node in V_{i-1} is encountered. In what order should one consider the back edges so that an open ear decomposition results?

9.4 Parallel Traversal of DAGs

In this section we describe a simple parallel algorithm for traversing the nodes of directed acyclic graphs (DAGs) in topological order. This immediately yields an algorithm for topological sorting but also serves as an algorithm template for other graph problems, including maximal independent sets and graph coloring. Indeed, the template can be viewed as a way to parallelize a class of greedy graph algorithms.

We have already seen a sequential algorithm for topological sorting in Sect. 2.12 and Exercise 8.3. We maintain the current indegree of each node and initialize a set with the nodes of indegree 0. Repeatedly, we remove all nodes in the set and their outgoing edges from the graph and add the nodes whose indegree becomes 0 to the set. This algorithm performs $D + 1$ iterations, where D is the length of the longest path in the network. Each iteration can be parallelized. We describe the distributed-memory version of the algorithm. Figure 9.17 gives pseudocode. As in parallel BFS (Sect. 9.2.3), we maintain a local array Q of local nodes that are ready to be processed. For DAG traversal, this means that they have indegree 0. Each PE iterates through the nodes u in its part of Q and sends messages to the PEs responsible for handling the nodes v reached by the edges out of u. Using prefix sums over the size of Q on each PE, we can assign unique numbers to the nodes which overall form a topological sorting[3] of the nodes of V. After all messages have been delivered, the incoming messages are processed. For topological sorting, the only thing that needs to be done for a message (u, v) is to decrement the indegree counter $\delta^-[v]$, and, if it has dropped to 0, to put v into Q.

Our DAG traversal algorithm can be generalized into an algorithm template whose main abstraction is sending messages along edges. More concretely, we get basic subroutines for initialization, sending messages, receiving a message, and processing the last message to a node. This is quite elegant because it completely abstracts from the parallel machine architecture used. Indeed, the basic approach was originally invented for external-memory graph algorithms, where it is known as time forward processing [72]. Using external-memory priority queues (see also Sect. 6.3) for message delivery, time forward processing yields algorithms with sorting complexity. The parallel complexity of the algorithm is a more complicated issue. To keep matters simple, we just consider a simple case in an exercise.

Exercise 9.21. Let Δ denote the maximum degree and D the length of the longest path in a DAG. Explain how to do topological sorting in time

$$O\left(\frac{m+n}{p} + D(\Delta + \log n)\right)$$

on a CRCW-PRAM model that allows fetch-and-decrement in constant time.

---

[3] Note that in each iteration, we process nodes that have the same length of the longest path from a source node of the graph. Thus there can be no edges between them. Hence, the relative numbering of these nodes is arbitrary.

Function *traverseDAG*   // let V denote the set of local nodes.
  $\delta^- = \langle \text{indegree}(v) : v \in V \rangle : NodeArray \text{ of } \mathbb{N}_0$
  $topOrder : NodeArray \text{ of } \mathbb{N}$
  $Q = \langle v \in V : \delta^-[v] = 0 \rangle : Set \text{ of } NodeId$ // and further *initializations*
  for $(pos:=0; \quad \sum_i |Q|@i > 0;)$ // explore layer by layer
    $offset := pos + \sum_{i < i_{\text{proc}}} |Q|@i$ // offset for local PE numbers using prefix sums
    foreach $u \in Q$ **do**
      $topOrder[u] := {+}{+}offset$
      foreach $(u,v) \in E$ **do** *post message* (u,v) *to PE v.p*
    $pos += \sum_i |Q|@i$ // advance to next layer
    deliver all messages
    $Q := \{\}$
    foreach *received message* (u,v) **do**   // process message
      if $-{-}\delta^-[v] = 0$ **then**   // process last message to v
        $Q := Q \cup \{v\}$ // remember for next layer
return *topOrder*

Fig. 9.17. Topological sorting using SPMD distributed-memory traversal of DAGs

We now give instantiations of the parallel DAG traversal template for two additional basic problems on *undirected graphs* – maximal independent sets and graph coloring. Another example (shortest paths) can be found in Sect. 10.2. The basic trick is to convert the undirected graph to a DAG by choosing an ordering of the nodes and then to convert an undirected edge $\{u,v\}$ to the directed edge $(\min(u,v),\max(u,v))$. In some cases it is a good idea to choose a random ordering of the nodes, since this keeps the longest path length D short, at least for graphs with small node degrees.

Lemma 9.8. *The DAG G' resulting from an undirected graph G with maximum degree Δ and a random ordering of the nodes has expected maximum path length $D = O(\Delta + \log n)$.*

Proof. There are fewer than $n\Delta^\ell$ simple paths of length ℓ in G; there are n choices for the first node of the path and at most Δ choices for each subsequent node. The probability that any particular one of these paths becomes a path in the DAG is $1/\ell!$. Hence, the probability that there is any path of length ℓ in G' is at most

$$p_\ell = n\frac{\Delta^\ell}{\ell!} \leq n\left(\frac{\Delta e}{\ell}\right)^\ell .$$

This estimate uses (A.9). We show that this probability is small for $\ell = \Omega(\Delta + \log n)$. First note that $(\Delta e/\ell)^\ell$ decreases with growing ℓ for $\ell \geq \Delta$. This holds since

$$\frac{d}{d\ell}\ln\left(\frac{\Delta e}{\ell}\right)^\ell = \frac{d}{d\ell}(\ell(\ln(\Delta e) - \ln\ell)) = \ln(\Delta e) - \ln\ell - 1 = \ln\Delta - \ln\ell.$$

For $\ell \geq \ell_0 = 5\max(\Delta, \ln n)$, we obtain

$$\left(\frac{\ell}{\Delta e}\right)^\ell \geq \left(\frac{\ell_0}{\Delta e}\right)^{\ell_0} \geq \left(\frac{5\Delta}{\Delta e}\right)^{5\ln n} \geq \left(\frac{5}{e}\right)^{5\ln n} = e^{\ln n \ln\left(\frac{5}{e}\right)^5} = n^{3.047\cdots}.$$

Hence, for $\ell \geq \ell_0$, $p_\ell \leq n \cdot 1/n^{3.047\cdots} = n^{-2.047\cdots}$. We can now bound the expectation of D, namely,

$$\mathrm{E}[D] \leq \ell_0 - 1 + \sum_{\ell_0 \leq \ell \leq n} p_\ell \ell \leq \ell_0 - 1 + n \cdot n^{-2.047\cdots} \leq \ell_0 - 1 + 1 = \ell_0,$$

where the last inequality holds for sufficiently large n.   □

Figure 9.18 gives an example. Note that the undirected graph on the left has a diameter of 8; the longest path is from node 4 to node b and has eight edges. In constrast, the longest path in the DAG obtained by randomly numbering the nodes is fairly short – only three edges.

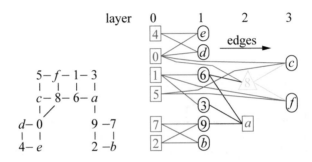

Fig. 9.18. An undirected graph with 16 nodes with randomly assigned node IDs. The DAG resulting from this node ordering ($0 < 1 < \ldots < 9 < a < \ldots < f$) has longest path length $D = 3$. The shapes of the boxes around the nodes on the right hand side indicate their color when the DAG is used to color the nodes; see Sect. 9.4.2.

9.4.1 Maximal Independent Sets

An independent set $I \subseteq V$ of an undirected graph $G = (V, E)$ is a set of nodes with no edges between them, i.e., $E \cap \binom{I}{2} = \emptyset$. Independent sets are an important concept in graph theory and are immediately relevant to parallel processing since several PEs can concurrently update the vertices of an independent set while being sure that the neighbors of the nodes in I will not change.

The following instantiation of the DAG traversal template yields an independent set I. We represent I by an array I of Boolean values and initialize it to 1 everywhere. At the end, $I[v]$ is true if and only if v is in the independent set. The only information we need to pass along an edge (u, v) is whether u is in I. A node v remains in I only if there is no edge (u, v) with $u \in I$. The following pseudocode summarizes this instantiation:

init: I = ⟨1,...,1⟩ : *NodeArray* **of** {0,1}
message sent for edge (u,v): I[u]
on receiving message x to v: I[v] := I[v] ∧ x

In the example graph in Fig. 9.18, nodes 0, 1, 2, 4, 5, and 7 are processed first and are all put into the independent set. Nodes 3, 6, 9, *b*, *d*, and *e* in layer 1, and nodes 8, *c*, and *f* in layers 2 and 3 receive messages from these nodes and hence stay out of the independent set. However, node *a* in layer 2 becomes part of the independent set since it receives no message – all its neighbors stay out of the independent set.

9.4.2 Graph Coloring

Recall, that graph coloring asks for colors to be assigned to nodes such that no two neighboring nodes have the same color; see also Sect. 12.5.2.1. A greedy heuristic for this task is only slightly more complicated than the one for computing a maximal independent set. We encode colors as positive integers. Nodes send their color along outgoing edges. Each node chooses the smallest color that is not already taken by one of its predecessors in the DAG. The following pseudocode shows how to implement the heuristic efficiently. Each node *v* pushes received messages onto a stack $S[v]$. Note that this is possible in constant time per incoming edge. When all messages have been received, we find the smallest color not in the stack $S[v]$ in time $O(|S[v]|)$. We do this using an auxiliary array *used*, recording the colors in $S[v]$. Since $S[v]$ can contain very large colors but we are guaranteed to find a free color in $1..|S[v]| + 1$, it suffices to record colors less than or equal to $|S[v]| + 1$:

init:c : NodeArray **of** ℕ;   *S : NodeArray* **of** *Stack* **of** ℕ
    foreach *v* ∈ *Q* **do** $c[v] := 1$
message sent for edge (u,v): c[v]
on receiving message x to v: S[v].push(x)
postprocess messages to v: used = ⟨0,...,0⟩ : *Array*$[1..|S[v]| + 1]$ **of** {0,1}
    while $S[v] \neq \emptyset$ **do** $x := S.pop$; **if** $x \leq |used|$ **then** $used[x] := 1$
    $c[v] := \min\{i \in 1..|used| : \neg used[i]\}$

Figure 9.18 gives an example. Nodes 0, 1, 2, 4, 5, and 7 have indegree 0 in the DAG and thus set their color to 1. The nodes in the subsequent layer, 3, 6, 9, *b*, *d*, and *e* receive only messages that color 1 is taken. Hence, color 2 is the first free color for all of them. Node *a* receives messages only from nodes 3, 6, and 9, which all have color 2. Hence, color 1 is the first free color for node *a*. In contrast, node 8 receives color 1 from node 0 and color 2 from node 6. Hence, its first free color is 3. Finally, nodes *c* and *f* receive colors 1 and 3, so that their first free color is 2.

Exercise 9.22. Show that this heuristic ensures that $\Delta + 1$ colors suffice where Δ is the maximum degree of the graph.

9.5 Implementation Notes

BFS is usually implemented by keeping unexplored nodes (with depths d and $d + 1$) in a FIFO queue. We chose a formulation using two separate sets for nodes at depth d and at depth $d + 1$ mainly because it allows a simple loop invariant that makes correctness immediately evident. However, our formulation might also turn out to be somewhat more efficient. If Q and Q' are organized as stacks, we shall have fewer cache faults than with a queue, in particular if the nodes of a layer do not quite fit into the cache. Memory management becomes very simple and efficient when just a single array a of n nodes is allocated for both of the stacks Q and Q'. One stack grows from $a[1]$ to the right, and the other grows from $a[n]$ to the left. When the algorithm switches to the next layer, the two memory areas switch their roles.

Our SCC algorithm needs to store four kinds of information for each node v: an indication of whether v is marked, an indication of whether v is open, something like a DFS number in order to implement "$\prec$", and, for closed nodes, the *NodeId* of the representative of its component. The array *component* suffices to keep this information. For example, if *NodeId*s are integers in $1..n$, $component[v] = 0$ could indicate an unmarked node. Negative numbers can indicate negated DFS numbers, so that $u \prec v$ if and only if $component[u] > component[v]$. This works because "$\prec$" is never applied to closed nodes. Finally, the test $w \in oNodes$ becomes $component[v] < 0$. With these simplifications in place, additional tuning is possible. We make *oReps* store *component* numbers of representatives rather than their IDs, and save an access to *component[oReps.top]*. Finally, the array *component* should be stored with the node data as a single array of records. The effect of these optimizations on the performance of our SCC algorithm is discussed in [219].

9.5.1 C++

LEDA [195] has implementations for topological sorting, reachability from a node (DFS), DFS numbering, BFS, strongly connected components, biconnected components, and transitive closure. BFS, DFS, topological sorting, and strongly connected components are also available in a very flexible implementation that separates representation and implementation, supports incremental execution, and allows various other adaptations.

The Boost graph library [50] and the LEMON graph library [201] use the *visitor concept* to support graph traversal. A visitor class has user-definable methods that are called at *event points* during the execution of a graph traversal algorithm. For example, the DFS visitor defines event points similar to the operations *init*, *root*, *traverse∗*, and *backtrack* used in our DFS template; there are more event points in Boost and LEMON.

9.5.2 Java

The JGraphT [167] library supports DFS in a very flexible way, not very much different from the visitor concept described for Boost and LEMON. There are also more specialized algorithms, for example for biconnected components.

9.6 Historical Notes and Further Findings

BFS and DFS were known before the age of computers. Tarjan [306] discovered the power of DFS and provided linear-time algorithms for many basic problems related to graphs, in particular biconnected and strongly connected components. Our SCC algorithm was invented by Cheriyan and Mehlhorn [70] and later rediscovered by Gabow [118]. Yet another linear-time SCC algorithm is that due to Kosaraju and Sharir [293]. It is very simple, but needs two passes of DFS. DFS can be used to solve many other graph problems in linear time, for example ear decomposition, planarity testing, planar embeddings, and triconnected components.

It may seem that problems solvable by graph traversal are so simple that little further research is needed on them. However, the bad news is that graph traversal itself is very difficult on advanced models of computation. In particular, DFS is a nightmare for both parallel processing [264] and memory hierarchies [215, 229]. Therefore alternative ways to solve seemingly simple problems are an interesting area of research. For example, in Sect. 11.9 we describe an approach to constructing minimum spanning trees using *edge contraction* that also works for finding connected components. Furthermore, the problem of finding biconnected components can be reduced to finding connected components [310]. The DFS-based algorithms for biconnected components and strongly connected components are almost identical. But this analogy completely disappears for advanced models of computation. Thus, parallel algorithms for strongly connected components remain an area of intensive (and sometimes frustrating) research (e.g., [104, 155]). More generally, it seems that problems for undirected graphs (such as finding biconnected components) are easier to solve than analogous problems for directed graphs (such as finding strongly connected components).

Parallel BFS has become a very popular benchmark for graph processing; see graph500.org/. Amazing performance values have been achieved by exploiting rather special properties of the benchmark graphs. In particular, most of the work is done in a very small number of layers.

10

Shortest Paths

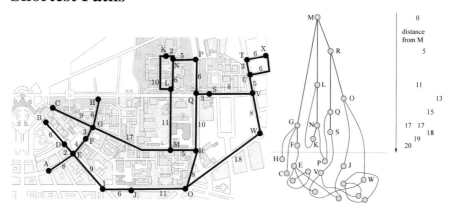

The problem of finding the shortest, quickest or cheapest path between two locations
is ubiquitous. You solve it daily. When you are in a location s and want to move to a
location t, you ask for the quickest path from s to t. The fire department may want to
compute the quickest routes from a fire station s to all locations in town – the single-
source all-destinations problem. Sometimes we may even want a complete distance
table from everywhere to everywhere – the all-pairs problem. In an old fashioned
road atlas, you will usually find an all-pairs distance table for the most important
cities.

 Here is a route-planning algorithm that requires a city map (all roads are assumed to be two-way roads) and a lot of dexterity but no computer. Lay thin threads
along the roads on the city map. Make a knot wherever roads meet, and at your
starting position. Now lift the starting knot until the entire net dangles below it. If
you have successfully avoided any tangles and the threads and your knots are thin
enough so that only tight threads hinder a knot from moving down, the tight threads
define the shortest paths. The illustration above shows a map of the campus of the
Karlsruhe Institute of Technology[1] and illustrates the route-planning algorithm for
the source node M.

Route planning in road networks is one of the many applications of shortest-path
computations. Many other problems profit from shortest-path computations, once
an appropriate graph model is defined. For example, Ahuja et al. [9] mention such
diverse applications as planning flows in networks, planning urban housing, inven-
tory planning, DNA sequencing, the knapsack problem (see also Chap. 12), production planning, telephone operator scheduling, vehicle fleet planning, approximating
piecewise linear functions, and allocating inspection effort on a production line.

---

[1] © KIT, Institut für Photogrammetrie und Fernerkundung.

© Springer Nature Switzerland AG 2019
P. Sanders et al., *Sequential and Parallel Algorithms and Data Structures*,
https://doi.org/10.1007/978-3-030-25209-0_10

The most general formulation of the shortest-path problem considers a directed graph $G = (V, E)$ and a cost function c that maps edges to arbitrary real numbers. It is fairly expensive to solve and many of the applications mentioned above do not need the full generality. For example, roads always have positive length. So we are also interested in various restrictions that allow simpler and more efficient algorithms: nonnegative edge costs, integer edge costs, and acyclic graphs. Note that we have already solved the very special case of unit edge costs in Sect. 9.1 – the breadth-first search (BFS) tree rooted at node s is a concise representation of all shortest paths from s. We begin in Sect. 10.1 with some basic concepts that lead to a generic approach to shortest-path algorithms. A systematic approach will help us to keep track of the zoo of shortest-path algorithms. As our first example of a restricted but fast and simple algorithm, we look at acyclic graphs in Sect. 10.2. In Sect. 10.3, we come to the most widely used algorithm for shortest paths: Dijkstra's algorithm for general graphs with nonnegative edge costs. The efficiency of Dijkstra's algorithm relies heavily on efficient priority queues. In an introductory course or on first reading, Dijkstra's algorithm might be a good place to stop. But there are many more interesting things about shortest paths in the remainder of the chapter. We begin with an average-case analysis of Dijkstra's algorithm in Sect. 10.4 which indicates that priority queue operations might dominate the execution time less than one might think based on the worst-case analysis. In Sect. 10.5, we discuss *monotone priority queues for integer keys* that take additional advantage of the properties of Dijkstra's algorithm. Combining this with average-case analysis leads even to a linear expected execution time. Section 10.6 deals with arbitrary edge costs, and Sect. 10.7 treats the all-pairs problem. We show that the all-pairs problem for general edge costs reduces to one general single-source problem plus n single-source problems with nonnegative edge costs. This reduction introduces the generally useful concept of node potentials. In Sect. 10.8, we go back to our original question about a shortest path between two specific nodes, in particular, in the context of road networks. Finally, we discuss parallel shortest path algorithms in Sect. 10.9.

10.1 From Basic Concepts to a Generic Algorithm

We extend the cost function to paths in the natural way. The cost of a path is the sum of the costs of its constituent edges, i.e., the cost of the path $p = \langle e_1, e_2, \dots, e_k \rangle$ is equal to $c(p) = \sum_{1 \le i \le k} c(e_i)$. The empty path has cost 0.

For a pair s and v of nodes, we are interested in a shortest path from s to v. We avoid the use of the definite article "the" here, since there may be more than one shortest path. Does a shortest path always exist? Observe that the number of paths from s to v may be infinite. For example, if $r = pCq$ is a path from s to v containing a cycle C, then we may go around the cycle an arbitrary number of times and still have a path from s to v; see Fig. 10.1. More precisely, assume p is a path leading from s to u, C is a path leading from u to u, and q is a path from u to v. Consider the path $r^{(i)} = pC^i q$ which first uses p to go from s to u, then goes around the cycle i times, and finally follows q from u to v. The cost of $r^{(i)}$ is $c(p) + i \cdot c(C) + c(q)$. If

C is a *negative cycle*, i.e., $c(C) < 0$, there is no shortest path from s to v, since the set $\{c(p) + c(q) - i \cdot |c(C)| : i \geq 0\}$ contains numbers smaller than any fixed number. We shall show next that shortest paths exist if there are no negative cycles.

Lemma 10.1. *If G contains no negative cycles and v is reachable from s, then a shortest path P from s to v exists. Moreover, P can be chosen to be simple.*

Proof. Let p_0 be a shortest *simple* path from s to v. Note that there are only finitely many simple paths from s to v. If p_0 is not a shortest path from s to v, there is a shorter path r from s to v. Then r is nonsimple. Since r is nonsimple we can, as in Fig. 10.1, write r as pCq, where C is a cycle and pq is a simple path. Then $c(p_0) \leq c(pq)$ and $c(pq) + c(C) = c(r) < c(p_0) \leq c(pq)$. So $c(C) < 0$ and we have shown the existence of a negative cycle, a contradiction to the assumption of the lemma. Thus there can be no path shorter than p_0. ☐

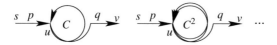

Fig. 10.1. A nonsimple path pCq from s to v.

Exercise 10.1. Strengthen the lemma above and show that if v is reachable from s, then a shortest path from s to v exists if and only if there is no negative cycle that is reachable from s and from which one can reach v.

For two nodes s and v, we define the *shortest-path distance*, or simply *distance*, $\mu(s, v)$ from s to v as

$$\mu(s,v) := \begin{cases} +\infty & \text{if there is no path from } s \text{ to } v, \\ -\infty & \text{if there is a path from } s \text{ to } v, \\ & \quad \text{but no shortest path from } s \text{ to } v, \\ c \text{ (a shortest path from } s \text{ to } v) & \text{otherwise.} \end{cases}$$

Since we use s to denote the source vertex most of the time, we also use the shorthand $\mu(v) := \mu(s, v)$. Observe that if v is reachable from s but there is no shortest path from s to v, then there are paths of arbitrarily large negative cost. Thus it makes sense to define $\mu(v) = -\infty$ in this case. Shortest paths have further nice properties, which we state as exercises.

Exercise 10.2 (subpaths of shortest paths). Show that subpaths of shortest paths are themselves shortest paths, i.e., if a path of the form pqr is a shortest path, then q is also a shortest path.

Exercise 10.3 (shortest-path trees). Assume that all nodes are reachable from s and that there are no negative cycles. Show that there is an n-node tree T rooted at s such that all tree paths are shortest paths. Hint: Assume first that shortest paths are unique and consider the subgraph T consisting of all shortest paths starting at s. Use the preceding exercise to prove that T is a tree. Extend this result to the case where shortest paths are not necessarily unique.

Exercise 10.4 (alternative definition). Show that

$$\mu(s,v) = \inf\{c(p) : p \text{ is a path from } s \text{ to } v\},$$

where the infimum of the empty set is $+\infty$.

Our strategy for finding shortest paths from a source node s is a generalization of the BFS algorithm shown in Fig. 9.2. We maintain two *NodeArrays* d and *parent*. Here, $d[v]$ contains our current knowledge about the distance from s to v, and *parent*$[v]$ stores the predecessor of v on the current shortest path to v. We usually refer to $d[v]$ as the *tentative distance* of v. Initially, $d[s] = 0$ and *parent*$[s] = s$. All other nodes have tentative distance "infinity" and no parent.

The natural way to improve distance values is to propagate distance information across edges. If there is a path from s to u of cost $d[u]$, and $e = (u,v)$ is an edge out of u, then there is a path from s to v of cost $d[u] + c(e)$. If this cost is smaller than the best previously known distance $d[v]$, we update d and *parent* accordingly. This process is called *edge relaxation*:

> **Procedure** *relax*$(e = (u,v) : Edge)$
> **if** $d[u] + c(e) < d[v]$ **then** $d[v] := d[u] + c(e);$ *parent*$[v] := u$

Arithmetic with ∞ is done in the natural way: $a < \infty$ and $\infty + a = \infty$ for all reals a, and $\infty \not< \infty$.

Lemma 10.2. *After any sequence of edge relaxations, if $d[v] < \infty$, then there is a path of length $d[v]$ from s to v.*

Proof. We use induction on the number of edge relaxations. The claim is certainly true before the first relaxation. The empty path is a path of length 0 from s to s, and all other nodes have infinite distance. Consider next a relaxation of an edge $e = (u,v)$. If the relaxation does not change $d[v]$, there is nothing to show. Otherwise, $d[u] + c(e) < d[v]$ and hence $d[u] < \infty$. By the induction hypothesis, there is a path p of length $d[u]$ from s to u. Then pe is a path of length $d[u] + c(e)$ from s to v.   □

The common strategy of the algorithms in this chapter is to relax edges until either all shortest paths have been found or a negative cycle has been discovered. For example, the (reversed) thick (solid and dashed) edges in Fig. 10.2 give us the *parent* information obtained after a sufficient number of edge relaxations: Nodes f, g, i, and h are reachable from s using these edges and have reached their respective $\mu(\cdot)$ values 2, -3, -1, and -3. Nodes b, j, and d form a negative-cost cycle reachable from s so that their shortest-path cost is $-\infty$. Node a is attached to this cycle, and

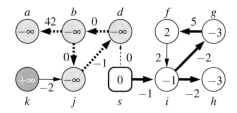

Fig. 10.2. A graph with source node s and shortest-path distances $\mu(v)$. Edge costs are shown as edge labels, and the distances are shown inside the nodes. The thick solid edges indicate shortest paths starting in s. The dashed edges and the light gray nodes belong to an infinite collection of paths with arbitrary small negative costs. The thick dashed edges (d,b), (b,j), and (j,d) form a negative cycle. The reversals of the thick solid and dashed edges indicate the parent function after a sufficient number of iterations. Node k is not reachable from s.

thus $\mu(a) = -\infty$. The edge (a,d) proves that d and hence d, b, j, and a are reachable from s, but this is not recorded in the parent information.

What is a good sequence of edge relaxations? Let $p = \langle e_1, \ldots, e_k \rangle$ be a path from s to v. If we relax the edges in the order e_1 to e_k, we have $d[v] \leq c(p)$ after the sequence of relaxations. If p is a shortest path from s to v, then $d[v]$ cannot drop below $c(p)$, by the preceding lemma, and hence $d[v] = c(p)$ after the sequence of relaxations.

Lemma 10.3 (correctness criterion). *After performing a sequence R of edge relaxations, we have $d[v] = \mu(v)$ if, for some shortest path $p = \langle e_1, e_2, \ldots, e_k \rangle$ from s to v, p is a subsequence of R, i.e., there are indices $t_1 < t_2 < \cdots < t_k$ such that $R[t_1] = e_1, R[t_2] = e_2, \ldots, R[t_k] = e_k$. Moreover, the parent information defines a path of length $\mu(v)$ from s to v.*

Proof. The following is a schematic view of R and p. The first row indicates the time. At time t_1, the edge e_1 is relaxed, at time t_2, the edge e_2 is relaxed, and so on:

$$
\begin{array}{llll}
& 1, 2, \ldots, & t_1, & \ldots, & t_2, & \ldots\ldots & ,t_k, & \ldots \\
R = \langle & \cdots & ,e_1, & \ldots, & e_2, & \ldots\ldots & ,e_k, & \ldots \rangle \\
p = & & \langle e_1, & & e_2, & \cdots & ,e_k \rangle
\end{array}
$$

We have $\mu(v) = \sum_{1 \leq j \leq k} c(e_j)$. For $i \in 1..k$, let v_i be the target node of e_i, and we define $t_0 = 0$ and $v_0 = s$. Then $d[v_i] \leq \sum_{1 \leq j \leq i} c(e_j)$ after time t_i, as a simple induction shows. This is clear for $i = 0$, since $d[s]$ is initialized to 0 and d-values are only decreased. After the relaxation of $e_i = R[t_i]$ for $i > 0$, we have $d[v_i] \leq d[v_{i-1}] + c(e_i) \leq \sum_{1 \leq j \leq i} c(e_j)$. Thus, after time t_k, we have $d[v] \leq \mu(v)$. Since $d[v]$ cannot go below $\mu(v)$, by Lemma 10.2, we have $d[v] = \mu(v)$ after time t_k and hence after performing all relaxations in R.

Let us prove next that the *parent* information traces out shortest paths. We shall do so under the additional assumption that shortest paths are unique, and leave the general case to the reader. After the relaxations in R, we have $d[v_i] = \mu(v_i)$ for $1 \leq$

$i \leq k$. So at some point in time, some operation $relax(u, v_i)$ sets $d[v_i]$ to $\mu(v_i)$ and $parent[v_i]$ to u. Note that this point of time may be before time t_i; it cannot be after t_i. By the proof of Lemma 10.2, there is a path from s to v_i of length $\mu(v_i)$ ending in the edge (u, v_i). Since, by assumption, the shortest path from s to v_i is unique, we must have $u = v_{i-1}$. So the *relax* operation sets $parent[v_i]$ to v_{i-1}. Later relax operations do not change this value since $d[v_i]$ is not decreased further.   □

Exercise 10.5. Redo the second paragraph in the proof above, but without the assumption that shortest paths are unique.

Exercise 10.6. Let S be the edges of G in some arbitrary order and let $S^{(n-1)}$ be $n-1$ copies of S. Show that $\mu(v) = d[v]$ for all nodes v with $\mu(v) \neq -\infty$ after the relaxations $S^{(n-1)}$ have been performed.

In the following sections, we shall exhibit more efficient sequences of relaxations for acyclic graphs and for graphs with nonnegative edge weights. We come back to general graphs in Sect. 10.6.

10.2 Directed Acyclic Graphs

In a directed acyclic graph (DAG), there are no directed cycles and hence no negative cycles. Moreover, we have learned in Sect. 9.3.1 that the nodes of a DAG can be topologically sorted into a sequence $\langle v_1, v_2, \ldots, v_n \rangle$ such that $(v_i, v_j) \in E$ implies $i < j$. A topological order can be computed in linear time $O(m+n)$ using depth-first search. The nodes on any path in a DAG increase in topological order. Thus, by Lemma 10.3, we can compute correct shortest-path distances if we first relax the edges out of v_1, then the edges out of v_2, etc.; see Fig. 10.3 for an example. In this way, each edge is relaxed only once. One may even ignore edges that emanate from nodes before s in the topological order. Since every edge relaxation takes constant time, we obtain a total execution time of $O(m+n)$.

Theorem 10.4. *Shortest paths in acyclic graphs can be computed in time* $O(m+n)$.

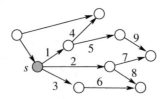

Fig. 10.3. Order of edge relaxations for the computation of the shortest paths from node s in a DAG. The topological order of the nodes is given by their x-coordinates. There is no need to relax the edges out of nodes "to the left of" s.

Exercise 10.7 (route planning for public transportation). Finding the quickest routes in public transportation systems can be modeled as a shortest-path problem for an acyclic graph. Consider a bus or train leaving a place p at time t and reaching its next stop p' at time t'. This connection is viewed as an edge connecting nodes (p,t) and (p',t'). Also, for each stop p and subsequent events (arrival and/or departure) at p, say at times t and t' with $t < t'$, we have the *waiting link* from (p,t) to (p,t'). (a) Show that the graph obtained in this way is a DAG. (b) You need an additional node that models your starting point in space and time. There should also be one edge connecting it to the transportation network. What should this edge be? (c) Suppose you have computed the shortest-path tree from your starting node to all nodes in the public transportation graph reachable from it. How do you actually find the route you are interested in? (d) Suppose there are minimum connection times at some of the stops. How can you incorporate them into the model? (e) How do you find the quickest connection with at most two intermediate stops?

Exercise 10.5. Instantiate the parallel DAG processing framework presented in Sect. 9.4 to compute shortest paths in DAGs.

10.3 Nonnegative Edge Costs (Dijkstra's Algorithm)

We now assume that all edge costs are nonnegative. Thus there are no negative cycles, and shortest paths exist for all nodes reachable from s. We shall show that if the edges are relaxed in a judicious order, every edge needs to be relaxed only once.

What is the right order? Along any shortest path, the shortest-path distances increase (more precisely, do not decrease). This suggests that we should scan nodes (to scan a node means to relax all edges out of the node) in order of increasing shortest-path distance. Lemma 10.3 tells us that this relaxation order ensures the computation of shortest paths, at least in the case where all edge costs are positive. Of course, in the algorithm, we do not know the shortest-path distances; we only know the *tentative distances* $d[v]$. Fortunately, for an unscanned node with smallest tentative distance, the true and tentative distances agree. We shall prove this in Theorem 10.6. We obtain the algorithm shown in Fig. 10.4. This algorithm is known as Dijkstra's shortest-path algorithm. Figure 10.5 shows an example run.

Dijkstra's Algorithm
declare all nodes unscanned and initialize d and *parent*
while there is an unscanned node with tentative distance $< +\infty$ **do**

    $u:=$ the unscanned node with smallest tentative distance
    relax all edges (u,v) out of u and declare u scanned

Fig. 10.4. Dijkstra's shortest-path algorithm for nonnegative edge weights

Operation	Queue
insert(s)	$\langle(s,0)\rangle$
deleteMin↝ $(s,0)$	$\langle\rangle$
relax $s \xrightarrow{2} a$	$\langle(a,2)\rangle$
relax $s \xrightarrow{10} d$	$\langle(a,2),(d,10)\rangle$
deleteMin↝ $(a,2)$	$\langle(d,10)\rangle$
relax $a \xrightarrow{3} b$	$\langle(b,5),(d,10)\rangle$
deleteMin↝ $(b,5)$	$\langle(d,10)\rangle$
relax $b \xrightarrow{2} c$	$\langle(c,7),(d,10)\rangle$
relax $b \xrightarrow{1} e$	$\langle(e,6),(c,7),(d,10)\rangle$
deleteMin↝ $(e,6)$	$\langle(c,7),(d,10)\rangle$
relax $e \xrightarrow{9} b$	$\langle(c,7),(d,10)\rangle$
relax $e \xrightarrow{8} c$	$\langle(c,7),(d,10)\rangle$
relax $e \xrightarrow{0} d$	$\langle(d,6),(c,7)\rangle$
deleteMin↝ $(d,6)$	$\langle(c,7)\rangle$
relax $d \xrightarrow{4} s$	$\langle(c,7)\rangle$
relax $d \xrightarrow{5} b$	$\langle(c,7)\rangle$
deleteMin↝ $(c,7)$	$\langle\rangle$

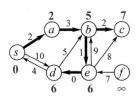

Fig. 10.5. Example run of Dijkstra's algorithm on the graph given on the *right*. The bold edges form the shortest-path tree, and the numbers in bold indicate shortest-path distances. The table on the *left* illustrates the execution. The *queue* contains all pairs $(v,d[v])$ with v reached and unscanned. A node is called *reached* if its tentative distance is less than $+\infty$. Initially, s is reached and unscanned. The actions of the algorithm are given in the first column. The second column shows the state of the queue after the action.

Note that Dijkstra's algorithm, when applied to undirected graphs is basically the thread-and-knot algorithm we saw in the introduction to this chapter. Suppose we put all threads and knots on a table and then lift the starting node. The other knots will leave the surface of the table in the order of their shortest-path distances.

Theorem 10.6. *Dijkstra's algorithm solves the single-source shortest-path problem for graphs with nonnegative edge costs.*

Proof. We proceed in two steps. In the first step, we show that all nodes reachable from s are scanned. In the second step, we show that the tentative and true distances agree when a node is scanned. In both steps, we argue by contradiction.

For the first step, assume the existence of a node v that is reachable from s, but never scanned. Consider a path $p = \langle s = v_1, v_2, \ldots, v_k = v \rangle$ from s to v, and let i be smallest such that v_i is not scanned. Then $i > 1$, since s is the first node scanned (in the first iteration, s is the only node whose tentative distance is less than $+\infty$). By the definition of i, v_{i-1} is scanned. When v_{i-1} is scanned, the edge (v_{i-1}, v_i) is relaxed. After this operation, we have $d[v_i] \le d[v_{i-1}] + c(v_{i-1}, v_i) < \infty$. So v_i must be scanned at some point during the execution, since the only nodes that stay unscanned are nodes u with $d[u] = +\infty$ at termination.

For the second step, consider the first point in time t at which a node v is scanned whose tentative distance $d[v]$ at time t is larger than its true distance $\mu[v]$. Consider a shortest path $p = \langle s = v_1, v_2, \ldots, v_k = v \rangle$ from s to v, and let i be smallest such that v_i is not scanned before time t. Then $i > 1$, since s is the first node scanned and $\mu(s) = 0 = d[s]$ when s is scanned. By the definition of i, v_{i-1} is scanned before time

t. Hence $d[v_{i-1}] \leq \mu(v_{i-1})$ when v_{i-1} is scanned. By Lemma 10.2, we then even have $d[v_{i-1}] = \mu(v_{i-1})$, and $d[v_{i-1}]$ does not change anymore afterwards. After v_{i-1} has been scanned, we always have $d[v_i] \leq d[v_{i-1}] + c(v_{i-1}, v_i) \leq \mu(v_{i-1}) + c(v_{i-1}, v_i)$, even if $d[v_i]$ should decrease. The last expression is at most $c(\langle v_1, v_2, \ldots, v_i \rangle)$, which in turn is at most $c(p)$, since all edge costs are nonnegative. Since $c(p) = \mu(v) < d[v]$, it follows that $d[v_i] < d[v]$. Hence at time t the algorithm scans some node different from v, a contradiction. □

Exercise 10.8. Let $v_1, v_2, \ldots$ be the order in which the nodes are scanned. Show that $\mu(v_1) \leq \mu(v_2) \leq \ldots$, i.e., the nodes are scanned in order of nondecreasing shortest-path distance.

Exercise 10.9 (checking of shortest-path distances (positive edge costs)). Assume that all edge costs are positive, that all nodes are reachable from s, and that d is a node array of nonnegative reals satisfying $d[s] = 0$ and $d[v] = \min_{(u,v) \in E} d[u] + c(u, v)$ for $v \neq s$. Show that $d[v] = \mu(v)$ for all v. Does the claim still hold in the presence of edges of cost 0?

Exercise 10.10 (checking of shortest-path distances (nonnegative edge costs)). Assume that all edge costs are non-negative, that all nodes are reachable from s, that *parent* encodes a tree rooted at s (for each node v, $parent(v)$ is the parent of v in this tree), and that d is a node array of nonnegative reals satisfying $d[s] = 0$ and $d[v] = d[parent(v)] + c(parent(v), v) = \min_{(u,v) \in E} d[u] + c(u, v)$ for $v \neq s$. Show that $d[v] = \mu(v)$ for all v.

We now turn to the implementation of Dijkstra's algorithm. We store all unscanned reached nodes in an addressable priority queue (see Sect. 6.2) using their tentative-distance values as keys. Thus, we can extract the next node to be scanned using the priority queue operation *deleteMin*. We need a variant of a priority queue where the operation *decreaseKey* addresses queue items using nodes rather than handles. Given an ordinary priority queue, such a *NodePQ* can be implemented using an additional *NodeArray* translating nodes into handles. If the items of the priority queue are objects, we may store them directly in a *NodeArray*. We obtain the algorithm given in Fig. 10.6.

Next, we analyze its running time in terms of the running times for the queue operations. Initializing the arrays d and *parent* and setting up a priority queue $Q = \{s\}$ takes time $O(n)$. Checking for $Q = \emptyset$ and loop control takes constant time per iteration of the while loop, i.e., $O(n)$ time in total. Every node reachable from s is removed from the queue exactly once. Every reachable node is also *insert*ed exactly once. Thus we have at most n *deleteMin* and *insert* operations. Since each node is scanned at most once, each edge is relaxed at most once, and hence there can be at most m *decreaseKey* operations. We obtain a total execution time of

$$T_{Dijkstra} = O\left(n + m + m \cdot T_{decreaseKey}(n) + n \cdot (T_{deleteMin}(n) + T_{insert}(n))\right),$$

where $T_{deleteMin}$, T_{insert}, and $T_{decreaseKey}$ denote the execution times for *deleteMin*, *insert*, and *decreaseKey*, respectively. Note that these execution times are a function of the queue size $|Q| = O(n)$.

Function *Dijkstra*(*s* : *NodeId*) : *NodeArray*×*NodeArray*   // returns (*d*, *parent*)
    $d = \langle \infty, \ldots, \infty \rangle$: *NodeArray* **of** $\mathbb{R} \cup \{\infty\}$ // tentative distance from root
    parent = $\langle \perp, \ldots, \perp \rangle$: *NodeArray* **of** *NodeId*
    parent[*s*] := *s*   // self-loop signals root
    Q : *NodePQ*   // unscanned reached nodes
    d[*s*] := 0;   *Q.insert*(*s*)
    while $Q \neq \emptyset$ **do**
        u := *Q.deleteMin*   // we have $d[u] = \mu(u)$

        foreach *edge* $e = (u, v) \in E$ **do**

            if $d[u] + c(e) < d[v]$ **then**   // relax
                $d[v] := d[u] + c(e)$
                parent[*v*] := *u*   // update tree
                if $v \in Q$ **then** *Q.decreaseKey*(*v*, *d*[*v*])

            else *Q.insert*(*v*)

    return (*d*, *parent*)

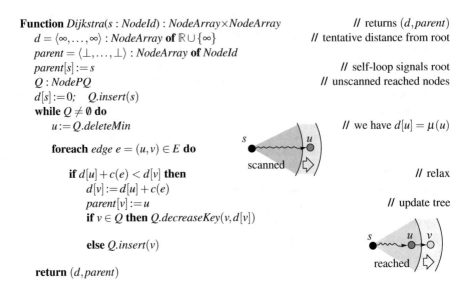

Fig. 10.6. Pseudocode for Dijkstra's algorithm

Exercise 10.11. Assume $n - 1 \leq m \leq n(n-1)/2$. Design a graph and a nonnegative cost function such that the relaxation of $m - (n-1)$ edges causes a *decreaseKey* operation.

In his original 1959 paper, Dijkstra proposed the following implementation of the priority queue: Maintain the number of reached unscanned nodes and two arrays indexed by nodes – an array *d* storing the tentative distances and an array storing, for each node, whether it is unreached or reached and unscanned or scanned. Then *insert* and *decreaseKey* take time O(1). A *deleteMin* takes time O(*n*), since it has to scan the arrays in order to find the minimum tentative distance of any reached unscanned node. Thus the total running time becomes

$$T_{Dijkstra59} = O(m + n^2).$$

Much better priority queue implementations have been invented since Dijkstra's original paper. Using the binary heap and Fibonacci heap priority queues described in Sect. 6.2, we obtain

$$T_{DijkstraBHeap} = O((m+n)\log n)$$

and

$$T_{DijkstraFibonacci} = O(m + n\log n),$$

respectively. Asymptotically, the Fibonacci heap implementation is superior except for sparse graphs with $m = O(n)$. In practice, Fibonacci heaps are usually not the fastest implementation, because they involve larger constant factors and the actual number of *decreaseKey* operations tends to be much smaller than what the worst case predicts. This experimental observation will be supported by theoretical analysis in the next section.

10.4 *Average-Case Analysis of Dijkstra's Algorithm

We shall show that the expected number of *decreaseKey* operations is $O(n\log(m/n))$.

Our model of randomness is as follows. The graph G and the source node s are arbitrary. Also, for each node v, we have an arbitrary multiset $C(v)$ of $indegree(v)$ nonnegative real numbers. So far, everything is arbitrary. The randomness comes now: We assume that, for each v, the costs in $C(v)$ are assigned randomly to the edges *into* v, i.e., our probability space consists of the $\prod_{v\in V} indegree(v)!$ possible assignments of edge costs to edges. Each such assignment has the same probability. We want to stress that this model is quite general. In particular, it covers the situation where edge costs are drawn independently from a common distribution.

Theorem 10.7. *Under the assumptions above, the expected number of decreaseKey operations is* $O(n\log(m/n))$.

Proof. We present a proof due to Noshita [245].

We need to start with a technical remark before we can enter the proof proper. For the analysis of the worst-case running time, it is irrelevant how nodes with the same tentative distance are treated by the priority queue. Any node with smallest tentative distance may be returned by the *deleteMin* operation. Indeed, the specification of the *deleteMin* operation leaves it open which element of smallest key is returned. For this proof, we need to be more specific and need to assume that nodes with equal tentative distance are returned in some *consistent* order. The detailed requirements will become clear in the proof. Consistency can, for example, be obtained by using the keys (tentative distance, node name) under lexicographic ordering instead of simply the tentative distances. Then nodes with the same tentative distance are removed in increasing order of node name.

Consider a particular node $v \neq s$ and assume that the costs of all edges not ending in v are already assigned. Only the assignment of the edge costs in $C(v)$ is open and will be determined randomly. We shall show that the expected number of *decreaseKey*$(v, *)$ operations is bounded by $\ln(indegree(v))$. We do so by relating the number of *decreaseKey* operations to the number of left-to-right maxima in a random sequence of length $indegree(v)$ (see Sect. 2.11).

The main difficulty in the proof is dealing with the fact that the order in which the edges into v are relaxed may depend on the assignment of the edge costs to these edges. The crucial observation is that up to the time when v is scanned, this order is fixed. Once v is scanned, the order may change. However, no further *decreaseKey* operations occur once v is scanned. In order to formalize this observation, we consider the execution of Dijkstra's algorithm on $G \setminus v$ (G without v and all edges incident to

v) and on G with the same assignment of costs to the edges not incident to v and an arbitrary assignment of costs to the edges incident to v. Before v is scanned in the run on G, exactly the same nodes are scanned in both executions and these scans are in the same order. This holds because the tentative distance of v has no influence on the other nodes before v is scanned. Also, the presence of v in the priority queue has no influence on the results of *deleteMin* operations before v is scanned in the run on G. This property is the consistency requirement. Of course, the time when v is scanned depends on the assignment of edge costs to the edges into v.

Let $u_1, \ldots, u_k$, where $k = indegree(v)$, be the source nodes of the edges into v in the order in which they are scanned in a run of Dijkstra's algorithm on $G \setminus v$ and let $\mu'(u_i)$ be the distance from s to u_i in $G \setminus v$. Nodes u_i that cannot be reached from s in $G \setminus v$ have infinite distance and come at the end of this ordering. According to Exercise 10.8, we have $\mu(u_1) \le \mu(u_2) \le \ldots \le \mu(u_k)$. In the run on G, the edges $e_1 = (u_1, v)$, $e_2 = (u_2, v)$, $\ldots$, $e_\ell = (u_\ell, v)$ are relaxed in that order until v is scanned; ℓ is a random variable. We do not know in what order the remaining edges into v are relaxed. However, none of them leads to a *decreaseKey*$(v, *)$ operation. We have $\mu'(u_i) = \mu(u_i)$ for $1 \le i \le \ell$. If e_i causes a *decreaseKey* operation, then $1 < i \le \ell$ (the relaxation of e_1 causes an *insert*(v) operation) and

$$\mu(u_i) + c(e_i) = \mu'(u_i) + c(e_i) < \min_{j<i} \mu'(u_j) + c(e_j) = \min_{j<i} \mu(u_j) + c(e_j).$$

Since $\mu(u_j) \le \mu(u_i)$, this implies

$$c(e_i) < \min_{j<i} c(e_j),$$

i.e., only left-to-right minima of the sequence $c(e_1), \ldots, c(e_k)$ can cause *decreaseKey* operations. Left-to-right minima are defined analogously to left-to-right maxima; see Sect. 2.11. We conclude that the number of *decreaseKey* operations on v is bounded by the number of left-to-right minima in the sequence $c(e_1), \ldots, c(e_k)$ minus 1; the "-1" accounts for the fact that the first element in the sequence counts as a left-to-right minimum but causes an *insert* and no *decreaseKey*. In Sect. 2.11, we have shown that the expected number of left-to-right maxima in a permutation of size k is bounded by H_k. The same bound holds for minima. Thus the expected number of *decreaseKey* operations is bounded by $H_k - 1$, which in turn is bounded by $\ln k = \ln indegree(v)$.

By the linearity of expectations, we may sum this bound over all nodes to obtain the following bound for the expected number of *decreaseKey* operations:

$$\sum_{v \in V} \ln indegree(v) \le n \ln \frac{m}{n},$$

where the last inequality follows from the concavity of the ln function (see (A.16)). □

We conclude that the expected running time is $O(m + n\log(m/n)\log n)$ with the binary heap implementation of priority queues. For sufficiently dense graphs ($m > n\log n \log\log n$), we obtain an execution time linear in the size of the input.

Exercise 10.12. Show that $E[T_{DijkstraBHeap}] = O(m)$ if $m = \Omega(n\log n \log\log n)$.

10.5 Monotone Integer Priority Queues

Dijkstra's algorithm scans nodes in order of nondecreasing distance values. Hence, a monotone priority queue (see Chap. 6) suffices for its implementation. It is not known whether monotonicity can be exploited in the case of general real edge costs. However, for integer edge costs, significant savings are possible. We therefore assume in this section that edge costs are integers in the range $0..C$ for some integer C, which we assume to be known when the queue is initialized.

Since a shortest path can consist of at most $n - 1$ edges, the shortest-path distances are at most $(n - 1)C$. The range of values in the queue at any one time is even smaller. Let min be the last value deleted from the queue (0 before the first deletion). Dijkstra's algorithm maintains the invariant that all values in the queue are contained in $min..min + C$. The invariant certainly holds after the first insertion. A $deleteMin$ may increase min. Since all values in the queue are bounded by C plus the old value of min, this is certainly true for the new value of min. Edge relaxations insert priorities of the form $d[u] + c(e) = min + c(e) \in min..min + C$.

10.5.1 Bucket Queues

A bucket queue is a circular array B of $C + 1$ doubly linked lists (see Figs. 10.7 and 3.12). We view the natural numbers as being wrapped around the circular array; all integers of the form $i + (C + 1)j$ map to the index i. A node $v \in Q$ with tentative distance $d[v]$ is stored in $B[d[v] \bmod (C + 1)]$. Since the priorities in the queue are always in $min..min + C$, all nodes in a bucket have the *same* tentative distance value.

Initialization creates $C + 1$ empty lists. An $insert(v)$ inserts v into $B[d[v] \bmod (C + 1)]$. A $decreaseKey(v)$ removes v from the list containing it and inserts v into $B[d[v] \bmod (C + 1)]$. Thus $insert$ and $decreaseKey$ take constant time if buckets are implemented as doubly linked lists.

A $deleteMin$ first looks at bucket $B[min \bmod (C + 1)]$. If this bucket is empty, it increments min and repeats. In this way, the total cost of all $deleteMin$ operations is $O(n + nC) = O(nC)$, since min is incremented at most nC times and at most n elements are deleted from the queue. Plugging the operation costs for the bucket queue implementation with integer edge costs in $0..C$ into our general bound for the cost of Dijkstra's algorithm, we obtain

$$T_{DijkstraBucket} = O(m + nC).$$

*Exercise 10.13 (Dinic's refinement of bucket queues [97]). Assume that the edge costs are positive real numbers in $[c_{min}, c_{max}]$. Explain how to find shortest paths in time $O(m + nc_{max}/c_{min})$. Hint: Use buckets of width c_{min}. Show that all nodes in the smallest nonempty bucket have $d[v] = \mu(v)$.

10.5.2 *Radix Heaps

Radix heaps [10] improve on the bucket queue implementation by using buckets of different widths. Narrow buckets are used for tentative distances close to min, and

wide buckets are used for tentative distances far away from *min*. In this subsection, we shall show how this approach leads to a version of Dijkstra's algorithm with running time
$$T_{DijkstraRadix} = O(m + n \log C).$$

Radix heaps exploit the binary representation of tentative distances. We need the concept of the *most significant distinguishing index* of two numbers. This is the largest index where the binary representations differ, i.e., for numbers a and b with binary representations $a = \sum_{i \geq 0} \alpha_i 2^i$ and $b = \sum_{i \geq 0} \beta_i 2^i$, we define the most significant distinguishing index $msd(a,b)$ as the largest i with $\alpha_i \neq \beta_i$, and let it be -1 if $a = b$. If $a < b$, then a has a 0-bit in position $i = msd(a,b)$ and b has a 1-bit.

A radix heap consists of an array of buckets $B[-1]$, $B[0]$, ..., $B[K]$, where $K = 1 + \lfloor \log C \rfloor$. The queue elements are distributed over the buckets according to the following rule:

queue element v is stored in bucket $B[i]$, where $i = \min(msd(min, d[v]), K)$.

We refer to this rule as the bucket queue invariant. Figure 10.7 gives an example. We remark that if *min* has a 1-bit in position $i \in 0..K - 1$, the corresponding bucket $B[i]$ is empty. This holds since any $d[v]$ with $i = msd(min, d[v])$ would have a 0-bit in position i and hence be smaller than *min*. But all keys in the queue are at least as large as *min*.

How can we compute $i := msd(a,b)$? We first observe that for $a \neq b$, the bitwise exclusive OR $a \oplus b$ of a and b has its most significant 1 in position i and hence represents an integer whose value is at least 2^i and less than 2^{i+1}. Thus $msd(a,b) =$

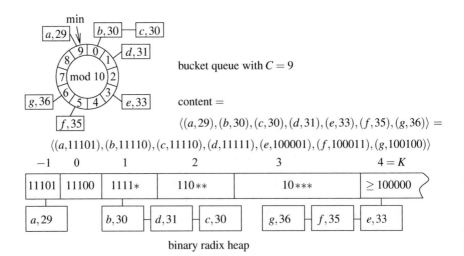

Fig. 10.7. Example of a bucket queue (*upper part*) and a radix heap (*lower part*). Since $C = 9$, we have $K = 1 + \lfloor \log C \rfloor = 4$. The bit patterns in the buckets of the radix heap indicate the set of keys they can accommodate.

$\lfloor \log(a \oplus b) \rfloor$, since $\log(a \oplus b)$ is a real number with its integer part equal to i and the floor function extracts the integer part. Many processors support the computation of *msd* by machine instructions.[2] Alternatively, we can use lookup tables or yet other solutions. From now on, we shall assume that *msd* can be evaluated in constant time.

Exercise 10.14. There is another way to describe the distribution of nodes over buckets. Let $min = \sum_j \mu_j 2^j$, let k be the smallest index greater than K with $\mu_k = 0$, and let $M_i = \sum_{j>i} \mu_j 2^j$. B_{-1} contains all nodes $v \in Q$ with $d[v] = min$, for $0 \le i < K$, $B_i = \emptyset$ if $\mu_i = 1$, and $B_i = \{v \in Q : M_i + 2^i \le d[x] < M_i + 2^{i+1} - 1\}$ if $\mu_i = 0$, and $B_K = \{v \in Q : M_k + 2^k \le d[x]\}$. Prove that this description is correct.

We turn now to the queue operations. Initialization, *insert*, and *decreaseKey* work completely analogously to bucket queues. The only difference is that bucket indices are computed using the bucket queue invariant.

A *deleteMin* first finds the minimum i such that $B[i]$ is nonempty. If $i = -1$, an arbitrary element in $B[-1]$ is removed and returned. If $i \ge 0$, the bucket $B[i]$ is scanned and *min* is set to the smallest tentative distance contained in the bucket. Since *min* has changed, the bucket queue invariant needs to be restored. A crucial observation for the efficiency of radix heaps is that only the nodes in bucket i are affected. We shall discuss below how they are affected. Let us consider first the buckets $B[j]$ with $j \ne i$. The buckets $B[j]$ with $j < i$ are empty. If $i = K$, there are no j's with $j > K$. If $i < j \le K$, any key a in bucket $B[j]$ will still have $msd(a, min) = j$, because the old and new values of *min* agree at bit positions greater than i.

What happens to the elements in $B[i]$? Its elements are moved to the appropriate new bucket. Thus a *deleteMin* takes constant time if $i = -1$ and takes time $O(i + |B[i]|) = O(K + |B[i]|)$ if $i \ge 0$. Lemma 10.8 below shows that every node in bucket $B[i]$ is moved to a bucket with a smaller index. This observation allows us to account for the cost of a *deleteMin* using amortized analysis. As our unit of cost (one token), we shall use the time required to move one node between buckets.

We charge $K + 1$ tokens for the *insert(v)* operation and associate these $K + 1$ tokens with v. These tokens pay for the moves of v to lower-numbered buckets in *deleteMin* operations. A node starts in some bucket j with $j \le K$, ends in bucket -1, and in between never moves back to a higher-numbered bucket. Observe that a *decreaseKey(v)* operation will also never move a node to a higher-numbered bucket. Hence, the $K + 1$ tokens can pay for all the node moves of *deleteMin* operations. The remaining cost of a *deleteMin* is $O(K)$ for finding a nonempty bucket. With amortized costs $K + 1 + O(1) = O(K)$ for an *insert* and $O(1)$ for a *decreaseKey*, we obtain a total execution time of $O(m + n \cdot (K + K)) = O(m + n \log C)$ for Dijkstra's algorithm, as claimed.

It remains to prove that *deleteMin* operations move nodes to lower-numbered buckets.

Lemma 10.8. *Let i be the smallest index such that $B[i]$ is nonempty, and assume $i \ge 0$. Let min be the smallest element in $B[i]$. Then $msd(min, x) < i$ for all $x \in B[i]$.*

---

[2] $\oplus$ is a direct machine instruction, and $\lfloor \log x \rfloor$ is the exponent in the floating-point representation of x.

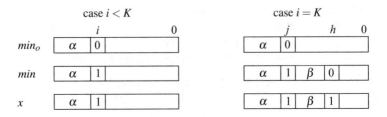

Fig. 10.8. The structure of the keys relevant to the proof of Lemma 10.8. In the proof, it is shown that β starts with $j - K$ 0's.

Proof. Observe first that the case $x = min$ is easy, since $msd(x,x) = -1 < i$. For the nontrivial case $x \neq min$, we distinguish the subcases $i < K$ and $i = K$. Let min_o be the old value of min. Figure 10.8 shows the structure of the relevant keys.

Case $i < K$. The most significant distinguishing index of min_o and any $x \in B[i]$ is i, i.e., min_o has a 0 in bit position i, and all $x \in B[i]$ have a 1 in bit position i. They agree in all positions with an index larger than i. Thus the most significant distinguishing index for min and x is smaller than i.

Case $i = K$. Consider any $x \in B[K]$. Let $j = msd(min_o, min)$. Since $min \in B[K]$, we have $j \geq K$. Let $h = msd(min, x)$. We want to show that $h < K$. Let α comprise the bits in positions larger than j in min_o, and let A be the number obtained from min_o by setting the bits in positions 0 to j to 0. The binary representation of A is α followed by $j + 1$. Since the jth bit of min_o is 0 and that of min is 1 (j is the most significant distinguishing index and $min_0 < min$), we have $min_o < A + 2^j \leq min$. Also, $x \leq min_o + C < A + 2^j + C \leq A + 2^j + 2^K$. So

$$A + 2^j \leq min \leq x < A + 2^j + 2^K,$$

and hence the binary representations of min and x consist of α followed by a 1, followed by $j - K$ many 0's, followed by some bit string of length K. Thus min and x agree in all bits with index K or larger, and hence $h < K$.

In order to aid intuition, we give a second proof for the case $i = K$. We first observe that the binary representation of min starts with α followed by a 1. We next observe that x can be obtained from min_o by adding some K-bit number. Since $min \leq x$, the final carry in this addition must run into position j. Otherwise, the jth bit of x would be 0 and hence $x < min$. Since min_o has a 0 in position j, the carry stops at position j. We conclude that the binary representation of x is equal to α followed by a 1, followed by $j - K$ many 0's, followed by some K-bit string. Since $min \leq x$, the $j - K$ many 0's must also be present in the binary representation of min.   □

*Exercise 10.15. Radix heaps can also be based on number representations with base b for any $b \geq 2$. In this situation we have buckets $B[i, j]$ for $i = -1, 0, 1, \ldots, K$ and $0 \leq j \leq b$, where $K = 1 + \lfloor \log C / \log b \rfloor$. An unscanned reached node x is stored in bucket $B[i, j]$ if $msd(min, d[x]) = i$ and the ith digit of $d[x]$ is equal to j. We also store, for each i, the number of nodes contained in the buckets $\cup_j B[i, j]$. Discuss the implementation of the priority queue operations and show that a shortest-path algorithm with running time $O(m + n(b + \log C/\log b))$ results. What is the optimal choice of b?

If the edge costs are random integers in the range $0..C$, a small change to Dijkstra's algorithm with radix heaps guarantees linear running time [129, 227]. For every node v, let $c_{min}^{in}(v)$ denote the minimum cost of an incoming edge. We divide Q into two parts, a set F which contains unscanned nodes whose tentative distance is known to be equal to their exact distance from s (we shall see below how one can learn this), and a part B which contains all other reached unscanned nodes. B is organized as a radix heap. We also maintain a value min. We scan nodes as follows.

When F is nonempty, an arbitrary node in F is removed and the outgoing edges are relaxed. When F is empty, the minimum node is selected from B and min is set to its distance label. When a node is selected from B, the nodes in the first nonempty bucket $B[i]$ are redistributed if $i \geq 0$. There is a small change in the redistribution process. When a node v is to be moved, and $d[v] \leq min + c_{min}^{in}(v)$, we move v to F. Observe that any future relaxation of an edge into v cannot decrease $d[v]$, and hence $d[v]$ is known to be exact at this point.

We call this algorithm ALD (average-case linear Dijkstra). The algorithm ALD is correct, since it is still true that $d[v] = \mu(v)$ when v is scanned. For nodes removed from F, this was argued in the previous paragraph, and for nodes removed from B, this follows from the fact that they have the smallest tentative distance among all unscanned reached nodes.

Theorem 10.9. *Let G be an arbitrary graph and let c be a random function from E to $0..C$. The algorithm ALD then solves the single-source problem in expected time $O(m + n + \log C)$.*

Proof. We still need to argue the bound on the running time. To do this, we modify the amortized analysis of plain radix heaps. Initialization of the heap takes time $O(K) = O(\log C)$. Consider now an arbitrary node v and how it moves through the buckets. It starts out in some bucket $B[j]$ with $j \leq K$. When it has just been moved to a new bucket but not yet to F, $d[v] \geq min + c_{min}^{in}(v) + 1$, and hence the index i of the new bucket satisfies $i \geq \lfloor \log(c_{min}^{in}(v) + 1) \rfloor + 1$. Therefore, in order to pay for all moves of node v between buckets, it suffices to charge $K - (\lfloor \log(c_{min}^{in}(v) + 1) \rfloor + 1) + 1 = K - \lfloor \log(c_{min}^{in}(v) + 1) \rfloor$ tokens to v. Summing over all nodes, we obtain a total payment of

$$\sum_v \left(K - \lfloor \log(c_{min}^{in}(v) + 1) \rfloor \right) = n + \sum_v \left(K - \lceil \log(c_{min}^{in}(v) + 1) \rceil \right).$$

We need to estimate this sum. For each vertex, we have one incoming edge contributing to this sum. We therefore bound the sum from above if we sum over all edges, i.e.,

$$\sum_v \left(K - \lceil \log(c_{min}^{in}(v) + 1) \rceil \right) \leq \sum_e (K - \lceil \log(c(e) + 1) \rceil).$$

Now, $K - \lceil \log(c(e)+1) \rceil$ is the number of leading 0's in the binary representation of $c(e)$ when written as a K-bit number. Our edge costs are uniform random numbers in $0..C$, and $K = 1 + \lfloor \log C \rfloor$. Thus

$$\text{prob}(K - \lceil \log(c(e)+1) \rceil \geq k) \leq \frac{|0..2^{K-k}-1|}{C+1} = \frac{2^{K-k}}{C+1} \leq 2^{-(k-1)}.$$

The last inequality follows from $C \geq 2^{K-1}$. Using (A.2) and (A.15), we obtain for each edge e

$$E[K - \lceil \log(c(e)+1) \rceil]] = \sum_{k \geq 1} \text{prob}(K - \lceil \log(c(e)+1) \rceil \geq k) \leq \sum_{k \geq 1} 2^{-(k-1)} = 2.$$

By the linearity of expectations, we obtain further

$$E\left[\sum_e (K - \lceil \log(c(e)+1) \rceil)\right] = \sum_e E[K - \lceil \log(c(e)+1) \rceil]] \leq \sum_e 2 = 2m = O(m).$$

Thus the total expected cost of the *deleteMin* operations is $O(m+n)$. The time for all *decreaseKey* operations is $O(m)$, and the time spent on all other operations is also $O(m+n)$.   □

Observe that the preceding proof does not require edge costs to be independent. It suffices that the cost of each edge is chosen uniformly at random in $0..C$. Theorem 10.9 can be extended to real-valued edge costs.

**Exercise 10.16. Explain how to adapt the algorithm ALD to the case where c is a random function from E to the real interval $(0, 1]$. The expected time should be $O(m+n)$. What assumptions do you need about the representation of edge costs and about the machine instructions available? Hint: You may first want to solve Exercise 10.13. The narrowest bucket should have a width of $\min_{e \in E} c(e)$. Subsequent buckets have geometrically growing widths.

10.6 Arbitrary Edge Costs (Bellman–Ford Algorithm)

For acyclic graphs and for nonnegative edge costs, we got away with m edge relaxations. For arbitrary edge costs, no such result is known. However, it is easy to guarantee the correctness criterion of Lemma 10.3 using $O(n \cdot m)$ edge relaxations: the Bellman–Ford algorithm [38, 108] given in Fig. 10.9 performs $n - 1$ rounds. In each round, it relaxes all edges. Since simple paths consist of at most $n - 1$ edges, every shortest path is a subsequence of this sequence of relaxations. Thus, after the relaxations are completed, we have $d[v] = \mu(v)$ for all v with $-\infty < d[v] < \infty$, by Lemma 10.3. Moreover, *parent* encodes the shortest paths to these nodes. Nodes v unreachable from s will still have $d[v] = \infty$, as desired.

It is not so obvious how to find the nodes w with $\mu(w) = -\infty$. The following lemma characterizes these nodes.

Function *BellmanFord(s : NodeId) : NodeArray×NodeArray*
$d = \langle \infty, \ldots, \infty \rangle : NodeArray$ **of** $\mathbb{R} \cup \{-\infty, \infty\}$ // distance from root
parent $= \langle \bot, \ldots, \bot \rangle : NodeArray$ **of** *NodeId*
$d[s] := 0;\quad parent[s] := s$ // self-loop signals root
for $i := 1$ **to** $n - 1$ **do**
$\quad$ **forall** $e \in E$ **do** *relax(e)*   // round i
forall $e = (u, v) \in E$ **do**   // postprocessing
$\quad$ **if** $d[u] + c(e) < d[v]$ **then** *infect(v)*
return $(d, parent)$

Procedure *infect(v)*
if $d[v] > -\infty$ **then**
$\quad d[v] := -\infty$
$\quad$ **foreach** $(v, w) \in E$ **do** *infect(w)*

Fig. 10.9. The Bellman–Ford algorithm for shortest paths in arbitrary graphs

Lemma 10.10. *After $n - 1$ rounds of edge relaxations, we have $\mu(w) = -\infty$ if and only if there is an edge $e = (u, v)$ with $d[u] + c(e) < d[v]$ such that w is reachable from v.*

Proof. If $d[u] + c(e) < d[v]$, then $d[u] < \infty$ and hence u and v are reachable from s. A further relaxation of e would further decrease $d[v]$ and hence $\mu(v) < d[v]$. Since $\mu(v) = d[v]$ also for all nodes v with $\mu(v) > -\infty$, we must have $\mu(v) = -\infty$. Then $\mu(w) = -\infty$ for any node reachable from v.

Assume conversely that $\mu(w) = -\infty$. Then there is a cycle $C = \langle v_0, v_1, \ldots, v_k \rangle$ with $v_k = v_0$ of negative cost that is reachable from s and from which w can be reached. Since C can be reached from s, we have $d[v_i] < \infty$ for all i. We claim that there must be at least one i, with $d[v_i] + c(v_i, v_{i+1}) < d[v_{i+1}]$. Assume otherwise, i.e., $d[v_i] + c(v_i, v_{i+1}) \geq d[v_{i+1}]$ for $0 \leq i < k$. Summing over all i yields $\sum_{0 \leq i < k} d[v_i] + \sum_{0 \leq i < k} c(v_i, v_{i+1}) \geq \sum_{0 \leq i < k} d[v_{i+1}]$. Since $v_0 = v_k$, the two summations over tentative distances are equal and we conclude that $c(C) = \sum_{0 \leq i < k} c(v_i, v_{i+1}) \geq 0$, a contradiction to the fact that C is a cycle of negative cost.   □

The pseudocode implements the lemma using a recursive function *infect(v)*. For any edge $e = (u, v)$ with $d[u] + c(e) < d[v]$, it sets the d-value of v and all nodes reachable from it to $-\infty$. If *infect* reaches a node w that already has $d[w] = -\infty$, it breaks the recursion because previous executions of *infect* have already explored all nodes reachable from w.

Exercise 10.17. Show that the postprocessing runs in time $O(m)$. Hint: Relate *infect* to *DFS*.

Exercise 10.18. Someone proposes an alternative postprocessing algorithm: Set $d[v]$ to $-\infty$ for all nodes v for which following parents does not lead to s. Give an example where this method overlooks a node with $\mu(v) = -\infty$.

Exercise 10.19 (arbitrage). Consider a set of currencies C with an exchange rate of r_{ij} between currencies i and j (you obtain r_{ij} units of currency j for one unit of currency i). A *currency arbitrage* is possible if there is a sequence of elementary currency exchange actions (*transactions*) that starts with one unit of a currency and ends with more than one unit of the same currency. (a) Show how to find out whether a matrix of exchange rates admits currency arbitrage. Hint: $\log(xy) = \log x + \log y$. (b) Refine your algorithm so that it outputs a sequence of exchange steps that maximizes the average profit *per transaction*.

Section 10.11 outlines further refinements of the Bellman–Ford algorithm that are necessary for good performance in practice.

10.7 All-Pairs Shortest Paths and Node Potentials

The all-pairs problem is tantamount to n single-source problems and hence can be solved in time $O(n^2m)$. A considerable improvement is possible. We shall show that it suffices to solve one general single-source problem plus n single-source problems with nonnegative edge costs. In this way, we obtain a running time of $O(nm + n(m + n\log n)) = O(nm + n^2\log n)$. We need the concept of node potentials.

A *(node) potential function* assigns a number $pot(v)$ to each node v. For an edge $e = (v,w)$, we define its *reduced cost* $\bar{c}(e)$ as

$$\bar{c}(e) = pot(v) + c(e) - pot(w).$$

Lemma 10.11. *Let p and q be paths from v to w. Then $\bar{c}(p) = pot(v) + c(p) - pot(w)$ and $\bar{c}(p) \leq \bar{c}(q)$ if and only if $c(p) \leq c(q)$. In particular, the shortest paths with respect to $\bar{c}$ are the same as those with respect to c.*

Proof. The second and the third claim follow from the first. For the first claim, let $p = \langle e_0, \ldots, e_{k-1} \rangle$, where $e_i = (v_i, v_{i+1})$, $v = v_0$, and $w = v_k$. Then

$$\bar{c}(p) = \sum_{i=0}^{k-1} \bar{c}(e_i) = \sum_{0 \leq i < k} (pot(v_i) + c(e_i) - pot(v_{i+1}))$$

$$= pot(v_0) + \sum_{0 \leq i < k} c(e_i) - pot(v_k) = pot(v_0) + c(p) - pot(v_k). \qquad \square$$

Exercise 10.20. Node potentials can be used to generate graphs with negative edge costs but no negative cycles: Generate a (random) graph, assign to every edge e a (random) nonnegative cost $c(e)$, assign to every node v a (random) potential $pot(v)$, and set the cost of $e = (u,v)$ to $\bar{c}(e) = pot(u) + c(e) - pot(v)$. Show that this rule does not generate negative cycles.

Lemma 10.12. *Assume that G has no negative cycles and that all nodes can be reached from s. Let $pot(v) = \mu(v)$ for $v \in V$. With these node potentials, the reduced edge costs are nonnegative.*

All-Pairs Shortest Paths in the Absence of Negative Cycles

add a new node s and zero-length edges (s,v) for all v	// no new cycles, time $O(m)$
compute $\mu(v)$ for all v with Bellman–Ford	// time $O(nm)$
set $pot(v) = \mu(v)$ and compute reduced costs $\bar{c}(e)$ for $e \in E$	// time $O(m)$
forall nodes x **do**	// time $O(n(m+n\log n))$
use Dijkstra's algorithm to compute the reduced shortest-path distances $\bar{\mu}(x,v)$	
using source x and the reduced edge costs $\bar{c}$	
// translate distances back to original cost function	// time $O(m)$
forall $e = (v,w) \in V \times V$ **do** $\mu(v,w) := \bar{\mu}(v,w) + pot(w) - pot(v)$	// use Lemma 10.11

Fig. 10.10. Algorithm for all-pairs shortest paths in the absence of negative cycles

Proof. Since all nodes are reachable from s and since there are no negative cycles, $\mu(v) \in \mathbb{R}$ for all v. Thus the reduced costs are well defined. Consider an arbitrary edge $e = (v,w)$. We have $\mu(v) + c(e) \geq \mu(w)$, and thus $\bar{c}(e) = \mu(v) + c(e) - \mu(w) \geq 0$.   □

Theorem 10.13. *The all-pairs shortest-path problem for a graph without negative cycles can be solved in time* $O(nm + n^2 \log n)$.

Proof. The algorithm is shown in Fig. 10.10. We add an auxiliary node s and zero-cost edges (s,v) for all nodes of the graph. This does not create negative cycles and does not change $\mu(v,w)$ for any of the existing nodes. Then we solve the single-source problem for the source s, and set $pot(v) = \mu(v)$ for all v. Next we compute the reduced costs and then solve the single-source problem for each node x by means of Dijkstra's algorithm. Finally, we translate the computed distances back to the original cost function. The computation of the potentials takes time $O(nm)$, and the n shortest-path calculations take time $O(n(m+n\log n))$. The preprocessing and postprocessing take linear time $O(n+m)$.   □

The assumption that G has no negative cycles can be removed [220].

Exercise 10.21. The *diameter D* of a graph G is defined as the largest distance between any two of its nodes. We can easily compute it using an all-pairs computation. Now we want to consider ways to *approximate* the diameter of a strongly connected graph using a constant number of single-source computations. (a) For any starting node s, let $D'(s) := \max_{u \in V} \mu(s,u)$ be the maximum distance from s to any node u. Show that $D'(s) \leq D \leq 2D'(s)$ for undirected graphs. Also, show that no such relation holds for directed graphs. Let $D''(s) := \max_{u \in V} \mu(u,s)$ be the maximum distance from any node u to s. Show that $\max(D'(s), D''(s)) \leq D \leq D'(s) + D''(s)$ for both undirected and directed graphs. (b) How should a graph be represented to support shortest-path computations for source nodes s as well as target node s? (c) Can you improve the approximation by considering more than one node s?

10.8 Shortest-Path Queries

We are often interested in the shortest path from a specific source node s to a specific target node t; route planning in a traffic network is one such scenario. We shall explain some techniques for solving such *shortest-path queries* efficiently and argue for their usefulness for the route-planning application. Edge costs are assumed to be nonnegative in this section.

We start with a technique called *early stopping*. We run Dijkstra's algorithm to find shortest paths starting at s. We stop the search as soon as t is removed from the priority queue, because at this point in time the shortest path to t is known. This helps except in the unfortunate case where t is the node farthest from s. On average, assuming that every target node is equally likely, early stopping saves a factor of two in scanned nodes. In practical route planning, early stopping saves much more because modern car navigation systems have a map of an entire continent but are mostly used for distances of up to a few hundred kilometers.

Another simple and general heuristic is *bidirectional search*, from s forward and from t backward until the search frontiers meet. More precisely, we run two copies of Dijkstra's algorithm side by side, one starting from s and one starting from t (and running on the reversed graph). The two copies have their own queues, say Q_s and Q_t, respectively. We grow the search regions at about the same speed, for example by removing a node from Q_s if $\min Q_s \le \min Q_t$ and a node from Q_t otherwise.

It is tempting to stop the search once the first node u has been removed from both queues, and to claim that $\mu(t) = \mu(s,u) + \mu(u,t)$. Observe that execution of Dijkstra's algorithm on the reversed graph with a starting node t determines $\mu(u,t)$. This is not quite correct, but almost so.

Exercise 10.22. Give an example where u is *not* on the shortest path from s to t.

However, we have collected enough information once some node u has been removed from both queues. Let d_s and d_t denote the tentative distances at the time of termination in the runs with source s and source t, respectively. We show that $\mu(t) < \mu(s,u) + \mu(u,t)$ implies the existence of a node $v \in Q_s$ with $\mu(t) = d_s[v] + d_t[v]$.

Let $p = \langle s = v_0, \ldots, v_i, v_{i+1}, \ldots, v_k = t \rangle$ be a shortest path from s to t. Let i be the largest index such that v_i has been removed from Q_s. Then $d_s[v_{i+1}] = \mu(s, v_{i+1})$ and $v_{i+1} \in Q_s$ when the search stops. Also, $\mu(s,u) \le \mu(s, v_{i+1})$, since u has already been removed from Q_s, but v_{i+1} has not. Next, observe that

$$\mu(s, v_{i+1}) + \mu(v_{i+1}, t) = c(p) < \mu(s, u) + \mu(u, t),$$

since p is a shortest path from s to t. By subtracting $\mu(s, v_{i+1})$, we obtain

$$\mu(v_{i+1}, t) < \mu(s, u) - \mu(s, v_{i+1}) + \mu(u, t) \le \mu(u, t)$$

and hence, since u has been scanned from t, v_{i+1} must also have been scanned from t, i.e., $d_t[v_{i+1}] = \mu(v_{i+1}, t)$ when the search stops. So we can determine the shortest distance from s to t by inspecting not only the first node removed from both queues, but also all nodes in, say, Q_s. We iterate over all such nodes v and determine the minimum value of $d_s[v] + d_t[v]$.

Dijkstra's algorithm scans nodes in order of increasing distance from the source. In other words, it grows a disk centered at the source node. The disk is defined by the shortest-path metric in the graph. In the route-planning application for a road network, we may also consider geometric disks centered on the source and argue that shortest-path disks and geometric disks are about the same. We can then estimate the speedup obtained by bidirectional search using the following heuristic argument: A disk of a certain diameter has twice the area of two disks of half the diameter. We could thus hope that bidirectional search will save a factor of two compared with unidirectional search.

Exercise 10.23 (bidirectional search). (a) Consider bidirectional search in a grid graph with unit edge weights. How much does it save compared with unidirectional search? (b) Give an example where bidirectional search in a real road network takes *longer* than unidirectional search. Hint: Consider a densely inhabited city with sparsely populated surroundings. (c) Design a strategy for switching between forward (starting in s) and backward (starting in t) search such that bidirectional search will *never* inspect more than twice as many nodes as unidirectional search. (*d) Try to find a family of graphs where bidirectional search visits exponentially fewer nodes on average than does unidirectional search. Hint: Consider random graphs or hypercubes.

We shall next describe several techniques that are more complex and less generally applicable. However, if they are applicable, they usually result in larger savings. These techniques mimic human behavior in route planning. The most effective variants of these speedup techniques are based on storing *preprocessed* information that depends on the graph but not on the source and target nodes. Note that with extreme preprocessing (compute the complete distance table using an all-pairs shortest-path computation), queries can answered very fast (time O(path length)). The drawback is that the distance table needs space $\Theta(n^2)$. Running Dijkstra's algorithm for every query needs no extra space, but is slow. We shall discuss compromises between these extremes.

10.8.1 Goal-Directed Search

The idea is to bias the search space such that Dijkstra's algorithm does not grow a disk but a region protruding towards the target; see Fig. 10.11. Assume we know a function $f : V \to \mathbb{R}$ that estimates the distance to the target, i.e., $f(v)$ estimates $\mu(v,t)$ for all nodes v. We use the estimates to modify the distance function. For each $e = (u,v)$, let[3] $\bar{c}(e) = c(e) + f(v) - f(u)$. We run Dijkstra's algorithm with the modified distance function (assuming for the moment that it is nonnegative). We know

---

[3] In Sect. 10.7, we added the potential of the source and subtracted the potential of the target. We do exactly the opposite now. The reason for changing the sign convention is that in Lemma 10.12, we used $\mu(s,v)$ as the node potential. Now, f estimates $\mu(v,t)$.

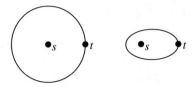

Fig. 10.11. The standard Dijkstra search grows a circular region centered on the source; goal-directed search grows a region protruding towards the target.

already (Lemma 10.11) that node potentials do not change shortest paths, and hence correctness is preserved. Tentative distances are related via $\bar{d}[v] = d[v] + f(v) - f(s)$. An alternative view of this modification is that we run Dijkstra's algorithm with the original distance function but remove the node with smallest value $d[v] + f(v)$ from the queue. The algorithm just described is known as A^*-*search*. In this context, f is frequently called the *heuristic* or the *heuristic distance estimate*.

Before we state requirements on the estimate f, let us see one specific example that illustrates the potential usefulness of heuristic information. Assume, in a thought experiment, that $f(v) = \mu(v,t)$. Then $\bar{c}(e) = c(e) + \mu(v,t) - \mu(u,t)$ and hence edges on a shortest path from s to t have a modified cost equal to 0. Thus any node ever removed from the queue has a modified tentative distance equal to 0. Consider any node that does not lie on a shortest path from s to t. Any shortest path from s to such a node must contain an edge $e = (u,v)$ such that u lies on a shortest path from s to t but e does not. Then $\mu(v,t) + c(e) > \mu(u,t)$ and hence $\bar{c}(e) > 0$. Thus nodes not on a shortest path from s to t have positive tentative distance, and hence Dijkstra's algorithm followsonly shortest paths, without looking left or right.

The function f must have certain properties to be useful. First, we want the modified distances to be nonnegative. So, we need $c(e) + f(v) \geq f(u)$ for all edges $e = (u,v)$. In other words, our estimate of the distance from u should be at most our estimate of the distance from v plus the cost of going from u to v. This property is called *consistency of estimates*. It has an interesting consequence. Consider any path $p = \langle v = v_0, v_1, \ldots, v_k = t \rangle$ from a node v to t. Adding the consistency relation $f(v_i) + c((v_{i-1}, v_i)) \geq f(v_{i-1})$ for all edges of the paths yields $f(t) + c(p) \geq f(v)$. Thus $c(p) \geq f(v) - f(t)$ and, further, $\mu(v,t) \geq f(v) - f(t)$. We may assume without loss of generality that $f(t) = 0$; otherwise, we simply subtract $f(t)$ from all f-values. Then $f(v)$ is a lower bound for $\mu(v,t)$.

It is still true that we can stop the search when t is removed from the queue? Consider the point in time when t is removed from the queue, and let p be any path from s to t. If all edges of p have been relaxed at termination, $d[t] \leq c(p)$. If not all edges of p have been relaxed at termination, there is a node v on p that is contained in the queue at termination. Then $d[t] + f(t) \leq d[v] + f(v)$, since t was removed from the queue but v was not, and hence

$$d[t] = d[t] + f(t) \leq d[v] + f(v) \leq d[v] + \mu(v,t) \leq c(p).$$

In either case, we have $d[t] \leq c(p)$, and hence the shortest distance from s to t is known as soon as t is removed from the queue.

What is a good heuristic function for route planning in a road network? Route planners often give a choice between *shortest* and *fastest* connections. In the case of shortest paths, a feasible lower bound $f(v)$ is the straight-line distance between v and t. Speedups by a factor of roughly four are reported in the literature. For fastest paths, we may use the geometric distance divided by the speed assumed for the best kind of road. This estimate is extremely optimistic, since targets are frequently in the center of a town, and hence no good speedups have been reported. More sophisticated methods for computing lower bounds are known which are based on preprocessing; we refer the reader to [130] for a thorough discussion.

10.8.2 Hierarchy

Road networks are structured into a hierarchy of roads, for example residential roads, urban roads, highways of different categories, and motorways. Roads of higher status typically support higher average speeds and are therefore to be preferred for a long-distance journey. A typical fastest path for a long distance journey first changes to roads of higher and higher status and then travels on a road of highest status for most of the journey until it descends again towards the target. Early industrial route planners used this observation as a heuristic acceleration technique. However, this approach sacrifices optimality. The second author frequently drives from Saarbrücken to Bonn. The fastest route uses the motorway for most of the journey but descends to a road of lower status for about 10 kilometers near the middle of the route. The reason is that one has to switch motorways and the two legs of motorway form an acute angle. It is then better to take a shortcut on a route of lesser status in the official hierarchy of German roads. Thus, for fastest-path planning, the shortcut road should be in the top-level of the hierarchy. Such misclassification can be fixed manually. However, manually classifying the importance of roads is expensive and error-prone.

An algorithmic classification is called for. Several algorithmic approaches have been developed in the last two decades which have not only achieved better and better performance but also, eventually, have become simpler and simpler. We refer our readers to [30] for an overview. We discuss two approaches, an early one and a recent one.

The early approach [276] is quite intuitive. We call a road *level one* if there is a shortest path between some source and some target that includes that road outside (!!!) the initial and final segments of, say, 10 kilometers. The level-one roads form the level-one network. The level-one network will contain vertices of degree 2. For example, if a slow road forms a three-way junction with a fast road, it is likely that the two legs of the fast road belong to the level-one network, but the slow road does not. Nodes of degree 2 may be removed by replacing the two incident edges by a single edge passing over that node of degree 2. Once the simplified level-one network is constructed, the same strategy is used to form the level-two network. And so on.

We next describe *contraction hierarchies* (CHs) [123], which are less intuitive, but very effective (up to four orders of magnitude faster than Dijkstra's algorithm) and yet fairly easy to implement (a basic implementation is possible with a few hundred lines of code).

Suppose for now that the nodes are already ordered by increasing importance. We describe below how this is done. The idea is that nodes that lie on many shortest paths are late in the ordering and nodes that are intermediate nodes of only a few shortest paths are early in the ordering. CH preprocessing goes through $n-1$ steps of *node contraction* – removing one node after another until only the most important node is left. CHs maintain the invariant that the shortest-path distances between the remaining nodes are preserved in each step. This is achieved by inserting *shortcut edges*. Suppose node v is to be removed but the path $\langle u,v,w \rangle$ is a shortest path. Then inserting a shortcut edge (u,w) with weight $c(u,v)+c(v,w)$ preserves the shortest-path distance between u and w and hence all shortest-path distances. Deciding whether $\langle u,v,w \rangle$ is a shortest path amounts to a local shortest-path search from u, looking for *witness* paths showing that the shortcut (u,w) is not needed. We can also speed up preprocessing by stopping the witness search early – if in doubt, simply insert a shortcut. For the removal of v, it suffices to inspect all pairs of edges entering and leaving v and to insert a shortcut if they form a shortest path. CHs are effective on road networks since, with a good node ordering, few shortcuts are needed.

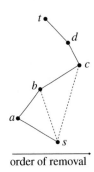

order of removal

Fig. 10.12. The path $\langle s,a,b,c,d,t \rangle$ is a shortest path from s to t. The nodes are removed in the order a, b, t, s, d, c. When a is removed, the shortcut (s,b) is added. When b is removed, the shortcut (s,c) is added. When s and t are removed, no shortcut edges relevant to the shortest path from s to t are added. When d is removed, there is no shortcut edge (c,t) added since t is no longer part of the graph at this point in time. All solid and dashed edges are part of the graph H. The path $\langle s,c,d,t \rangle$ is a shortest path from s to t in H. It consists of the up-path $\langle s,c \rangle$ followed by the down-path $\langle c,d,t \rangle$. Expansion of $\langle s,c \rangle$ first yields $\langle s,b,c \rangle$ and then $\langle s,a,b,c \rangle$.

The result of preprocessing is the original graph plus the shortcuts inserted during contraction. We call this graph H; see Fig. 10.12 for an example. The crucial property of H is that, for arbitrary s and t, it contains a shortest path P from s to t that is an *up–down path*. We call an edge (u,v) *upward* if u is removed before v and *downward* otherwise. An up–down path consists of a first segment containing only upward edges and a second segment containing only downward edges. The existence of such a path is easily seen. Let $\langle \dots, u,v,w, \dots \rangle$ be a shortest path in H such that v was removed before u and w. Then the short-cut (u,w) was introduced when v was removed and $\langle \dots, u,w, \dots \rangle$ is also a path in H. Continuing in this way, we obtain an up–down path. The *up–down property* leads to a very simple and fast query algorithm. We do bidirectional search in H using a modification of Dijkstra's algorithm where forward search considers only upward edges and backward search considers only downward edges. The computed path contains shortcuts that have to be expanded. To support unpacking in time linear in the output size, it suffices to store the *midpoints* of shortcuts with every shortcut, i.e., the node that was contracted when the shortcut was inserted. Expanding then becomes a recursive procedure, with edges in the input graph as the base case.

We still need to discuss the order of removal. Of course, if the ordering is bad, we will need a huge number of shortcuts, resulting in large space requirements and slow queries. Many heuristics for node ordering have been considered. Even simple ones yield reasonable performance. These heuristics compute the ordering online by identifying unimportant nodes in the remaining graph to be contracted next. To compute the importance of v, a contraction of v is simulated. In particular, it is determined how many shortcuts would be inserted. This number and other statistics gathered during the simulated contraction are condensed into a score. Nodes with low score are contracted first.

We close with a comparison of CHs with the heuristic based on the official hierarchy of roads. The official hierarchy consists of not more than 10 layers. Rather than using a small, manually defined set of layers, CHs use n different layers. This means that CHs can exploit the hierarchy inherent in the input in a much more fine-grained way. The CH query algorithm is also much more aggressive than the heuristic algorithm – it switches to a higher layer in every step. CHs can afford this aggressiveness since shortcuts repair any "error" made in defining the hierarchy.

10.8.3 Transit Node Routing

Another observation from daily life can be used to obtain very fast query times. When you drive to somewhere "far away", you will leave your start location via one of only a few "important" road junctions. For example, the second author lives in Scheidt,

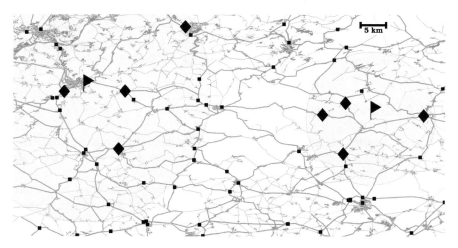

Fig. 10.13. Finding the optimal travel time between two points (the flags) somewhere between Saarbrücken and Karlsruhe amounts to retrieving the 2×4 *access nodes* (diamonds), performing 16 table lookups between all pairs of access nodes, and checking that the two disks defining the *locality filter* do not overlap. The small squares indicate further transit nodes.

a small village near Saarbrücken. For trips to the south-east to north-east, he enters the motorway system in Sankt Ingbert. For trips to the south, he enters the system at Grossblittersdorf, and so on. So all shortest paths from his home to distances far away go through a small number of transit nodes. Moreover, these transit nodes are the same for the entire population of Scheidt. As a consequence, the total number of transit nodes for all locations in Germany (or Europe, for that matter) is small. Figure 10.13 gives another example.

The notion of a transit node can be made precise and algorithmic; see [31]. In their scheme, about $\sqrt{n}$ of the most important road junctions are selected as transit nodes. Here, n is the number of nodes in the road network (a few million for the German network). The algorithmic strategy is now as follows:

(a) Select the set A of transit nodes.
(b) Compute the complete distance table between the nodes in A.
(c) For each node v, determine its transit nodes A_v conntecting v to the long-distance network. The sets A_v are typically small, usually no more than a handful of elements.
(d) For each node v, compute the distances to the nodes in A_v.

Shortest-path distances between faraway nodes can then be computed as

$$\mu(s,t) = \min_{u \in A_s} \min_{v \in A_t} \mu(s,u) + \mu(u,v) + \mu(v,t). \qquad (10.1)$$

In this way, shortest-path queries are reduced to a small number of table lookups. This can be more than $1\,000\,000$ times faster than Dijkstra's algorithm. Unfortunately, (10.1) does not work for "local" shortest paths that do not touch any transit node. Therefore, a full implementation of transit node routing needs additionally to define a *locality filter* that detects local paths and a routine for finding the local paths. This local algorithm could, for example, use CHs, the labeling method, or further layers of local transit node routing. The precomputation of the distance tables can also use all methods discussed in the preceding sections; see [184, 3, 21] for details.

10.9 Parallel Shortest Paths

Different shortest-path computations (Dijkstra, Bellman–Ford, contraction hierarchies, all-pairs) pose different challenges for parallelization. We progress from the easy to the difficult.

Computing all-pairs shortest paths is embarrassingly parallel for $p \leq n$, just assign n/p starting nodes to each PE. We only need to be careful about space consumption. We only want to replicate the distance array of the shortest-path tree currently computed and we should use a priority queue implementation that only consumes space proportional to the actual queue size rather than $\Theta(n)$.

The basic Bellman–Ford algorithm allows for a lot of fine-grained parallelism. All the edge relaxations in an iteration can be performed in parallel. We do not even have to use CAS instructions, as long as write operations are atomic. Suppose two

PEs try to update $d[v]$ to x and y, respectively, where $x < y$. The bad case is when x is written first and then overwritten by the worse value y. The overall algorithm will remain correct, since the value x will be tried again in the next iteration. Hence, as long as the bad case happens rarely, we pay only a small performance penalty. Moreover, in many situations distances decrease only rarely (see also Sect. 10.4). Therefore contention is much lower than in other situations. This *priority update principle* is also useful in many other shared memory algorithms [296]. Unfortunately, the optimizations discussed in Sect. 10.11 are more difficult to parallelize.

Many of the preprocessing techniques discussed in Sect. 10.8 can be parallelized. We outline an approach for contraction hierarchies [182, 324]. The algorithm computes a score for each node of how attractive it is to contract the node. Then it contracts in parallel all nodes which are less important than all nodes reachable within BFS distance 2. Distance 2 is necessary in order to avoid concurrent updates of adjacency lists when inserting shortcuts. Parallelizing over many CH queries is also easy. This even works on distributed memory machines where each PE knows only about a small part of the road network [182]: One stores important nodes redundantly such that the search spaces of the forward and the backward search can be computed locally. Then the forward search space is sent to the PE responsible for the backward search, which intersects the two search spaces yielding the result. In [182], preprocessing was also parallelized for distributed memory. The graph is partitioned as explained in Sect. 8.6.2, but the approach is generalized to store a *halo* of order ℓ locally: Suppose v is in some local block B according to the partition. Then it is ensured that every node within BFS distance ℓ is stored locally. This way, the local searches done during contraction can be done locally. The halo is adapted after every iteration in order to take newly inserted shortcut edges into account.

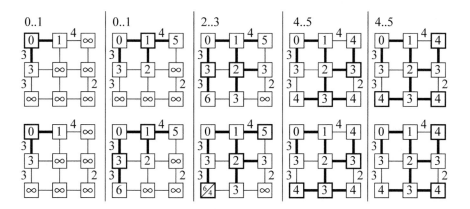

Fig. 10.14. Example execution of Δ-stepping with $\Delta = 2$ (*top*) and $\Delta = \infty$ (Bellman–Ford, *bottom*). Edges have unit weight unless a different value is explicitly given. Thick edges denote the current approximation to the shortest-path tree. Bold squares indicate nodes currently being scanned. Note that Bellman–Ford scans 11 nodes overall, whereas 2-stepping scans only 9 nodes. The range $a..b$ given in the top row indicates the bucket currently being scanned.

Unfortunately, the "bread-and-butter" shortest-path problem with a single source and nonnegative edge weights is not easy to parallelize. We may get a little bit of parallelism by scanning all nodes with smallest distance value (min Q) in parallel. This can be generalized slightly by observing that a node v also has its final distance value if $d[v] \leq \min Q + \min_{(u,v) \in E} c(u,v)$ [80]. If we exploit more parallelism, for example using a parallel priority queue, it might happen that we now scan nodes v with $d[v] > \mu(v)$. We compensate for this "mistake" by rescanning nodes that are reinserted into the priority queue. This preserves correctness albeit with an increase in the work done.

We can develop this idea further. Once we abandon the nice properties of Dijkstra's algorithm, it makes sense to switch to an approximate priority queue that has less overhead. A simple way to do this is to use a bucket priority queue where key x is mapped to bucket $\lfloor x/\Delta \rfloor$, where Δ is a tuning parameter. This queue is a slight generalization of the simple integer bucket queue from Sect. 10.5.1. The idea behind the Δ-*stepping* algorithm is to achieve parallelism by scanning in parallel all nodes in the first nonempty bucket of the bucket queue. This algorithm can be shown to be work efficient at least for random edge weights. Adding further measures, that involve different treatments of edges with weight $< \Delta$, one can obtain the following bound.

Theorem 10.14 ([228]). $1/d$-*stepping with random edge weights from* $[0,1]$ *runs in time*

$$O\left(\frac{n+m}{p} + dL\log n + \log^2 n\right)$$

on a CRCW-PRAM, where d is the maximum degree and L is the largest shortest-path distance from the starting node.

Δ-stepping can also be viewed as a generalization of both the Bellman–Ford algorithm and Dijkstra's algorithm. For $\Delta \leq \min\{e \in E : c(e)\}$, we get Dijkstra's algorithm, and for $\Delta \geq \max\{e \in E : c(e)\}$, we get the Bellman–Ford algorithm.

Implementing a basic variant of Δ-stepping in parallel is a slight generalization of the parallel BFS described in Sect. 9.2. We now describe the distributed memory version in more detail. Each PE i maintains a local bucket priority queue Q storing reached but unscanned nodes assigned to PE i. One iteration of the main loop first finds the first bucket j that is nonempty on some PE. All PEs then work in parallel on $Q[j]$. For a node $u \in Q[j]$ and an outgoing edge $e = (u,v)$, a relaxation request $(v, d[u] + c(e))$ is sent to the PE responsible for node v. The iteration is finished by processing incoming requests. For a request (v,c), if $d[v] > c$, $d[v]$ is set to c and the queue is updated accordingly, i.e., if v is already in Q, a *decreaseKey* is performed and otherwise an insertion. Note that in Δ-stepping, a node may be reinserted after it has been scanned. Figure 10.14 gives an example.

10.10 Implementation Notes

Shortest-path algorithms work over the set of extended reals $\mathbb{R} \cup \{+\infty, -\infty\}$. We may ignore $-\infty$, since it is needed only in the presence of negative cycles and, even there, it is needed only for the output; see Sect. 10.6. We can also get rid of $+\infty$ by noting that $parent(v) = \perp$ if and only if $d[v] = +\infty$, i.e., when $parent(v) = \perp$, we assume that $d[v] = +\infty$ and ignore the number stored in $d[v]$.

A refined implementation of the Bellman–Ford algorithm [218, 308] explicitly maintains a current approximation T to the shortest-path tree. Nodes still to be scanned in the current iteration of the main loop are stored in a set Q. Consider the relaxation of an edge $e = (u, v)$ that reduces $d[v]$. All descendants of v in T will subsequently receive a new d-value. Hence, there is no reason to scan these nodes with their current d-values and one may remove them from Q and T. Furthermore, negative cycles can be detected by checking whether v is an ancestor of u in T.

Implementations of contraction hierarchies are available at github.com under RoutingKit/RoutingKit and Project-OSRM/osrm-backend.

10.10.1 C++

LEDA [195] has a special priority queue class *node_pq* that implements priority queues of graph nodes. Both LEDA and the Boost graph library [50] have implementations of the Dijkstra and Bellman–Ford algorithms and of the algorithms for acyclic graphs and the all-pairs problem. There is a graph iterator based on Dijkstra's algorithm that allows more flexible control of the search process. For example, one can use it to search until a given set of target nodes has been found. LEDA also provides a function that verifies the correctness of distance functions (see Exercise 10.9).

The Boost graph library [50] and the LEMON graph library [201] use the *visitor concept* to support graph traversal. A visitor class has user-definable methods that are called at *event points* during the execution of a graph traversal algorithm. For example, the DFS visitor defines event points similar to the operations *init*, *root*, *traverse∗*, and *backtrack* used in our DFS template; there are more event points in Boost and LEMON.

10.10.2 Java

The JGraphT [167] library supports DFS in a very flexible way, not very much different from the visitor concept described for Boost and LEMON. There are also more specialized algorithms, for example for biconnected components.

10.11 Historical Notes and Further Findings

Dijkstra [96], Bellman [38], and Ford [108] found their algorithms in the 1950s. The original version of Dijkstra's algorithm had a running time $O(m + n^2)$ and there is a long history of improvements. Most of these improvements result from better data structures for priority queues. We have discussed binary heaps [332], Fibonacci heaps [115], bucket heaps [92], and radix heaps [10]. Experimental comparisons can be found in [71, 218]. For integer keys, radix heaps are not the end of the story. The best theoretical result is $O(m + n \log \log n)$ time [313]. For *undirected* graphs, linear time can be achieved [311]. The latter algorithm still scans nodes one after the other, but not in the same order as in Dijkstra's algorithm.

Meyer [227] gave the first shortest-path algorithm with linear average-case running time. The algorithm ALD was found by Goldberg [129].

Integrality of edge costs is also of use when negative edge costs are allowed. If all edge costs are integers greater than $-N$, a *scaling algorithm* achieves a time $O(m\sqrt{n}\log N)$ [128].

In Sect. 10.8, we outlined a small number of speedup techniques for route planning. Many other techniques exist. In particular, we have not done justice to advanced goal-directed techniques, combinations of different techniques, etc. A recent overview can be found in [30]. Theoretical performance guarantees beyond Dijkstra's algorithm are more difficult to achieve. Positive results exist for special families of graphs such as planar graphs and when approximate shortest paths suffice [103, 313, 315].

There is a generalization of the shortest-path problem that considers several cost functions at once. For example, your grandfather may want to know the fastest route for visiting you but he only wants routes where he does not need to refuel his car, or you may want to know the fastest route subject to the condition that the road toll does not exceed a certain limit. Constrained shortest-path problems are discussed in [143, 223].

Shortest paths can also be computed in geometric settings. In particular, there is an interesting connection to optics. Different materials have different refractive indices, which are related to the speed of light in the material. Astonishingly, the laws of optics dictate that a ray of light always travels along a quickest path.

Exercise 10.24. An ordered semigroup is a set S together with an associative and commutative operation $+$, a neutral element 0, and a linear ordering $\leq$ such that for all x, y, and z, $x \leq y$ implies $x + z \leq y + z$. Which of the algorithms in this chapter work when the edge weights are from an ordered semigroup? Which of them work under the additional assumption that $0 \leq x$ for all x?

11

Minimum Spanning Trees

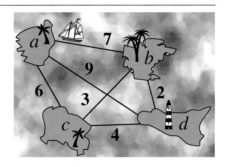

The atoll of Taka-Tuka-Land in the South Seas asks you for help.[1] The people want to connect their islands by ferry lines. Since money is scarce, the total cost of the connections is to be minimized. It needs to be possible to travel between any two islands; direct connections are not necessary. You are given a list of possible connections together with their estimated costs. Which connections should be opened?

More generally, we want to solve the following problem. Consider a connected undirected graph $G = (V, E)$ with real edge costs $c : E \to \mathbb{R}$. A *minimum spanning tree* (MST) of G is a set $T \subseteq E$ of edges such that the graph (V, T) is a tree minimizing $c(T) := \sum_{e \in T} c(e)$. In our example, the nodes are islands, the edges are possible ferry connections, and the costs are the costs of opening a connection. Throughout this chapter, G denotes an undirected connected graph.

Minimum spanning trees are perhaps the simplest variant of an important family of problems known as *network design problems*. Because MSTs are such a simple concept, they also show up in many seemingly unrelated problems such as clustering, finding paths that minimize the maximum edge cost used, and finding approximations for harder problems, for example the Steiner tree problem and the traveling salesman tour problem. Sections 11.7 and 11.9 discuss this further. An equally good reason to discuss MSTs in a textbook on algorithms is that there are simple, elegant, and fast algorithms to find them. We shall derive two simple properties of MSTs in Sect. 11.1. These properties form the basis of most MST algorithms. The Jarník–Prim algorithm grows an MST starting from a single node and will be discussed in Sect. 11.2. Kruskal's algorithm grows many trees in unrelated parts of the graph at once and merges them into larger and larger trees. This will be discussed in Sect. 11.3. An efficient implementation of the algorithm requires a data structure for maintaining a partition of a set into subsets. Two operations have to be supported by this data structure: "determine whether two elements are in the same subset" and "join two subsets". We shall discuss this union–find data structure in Sect. 11.4. It has many applications besides the construction of minimum spanning trees. Sec-

---

[1] The figure above was drawn by A. Blancani.

© Springer Nature Switzerland AG 2019
P. Sanders et al., *Sequential and Parallel Algorithms and Data Structures*,
https://doi.org/10.1007/978-3-030-25209-0_11

tions 11.5 and 11.6 discuss external-memory and parallel algorithms, respectively, for minimum spanning trees.

Exercise 11.1. If the input graph is not connected, we may ask for a *minimum spanning forest* – a set of edges that defines an MST for each connected component of G. Develop a way to find minimum spanning forests using a single call of an MST routine. Do not find connected components first. Hint: Insert $n-1$ additional edges.

Exercise 11.2 (spanning sets). A set T of edges *spans* a connected graph G if (V,T) is connected. Is a minimum-cost spanning set of edges necessarily a tree? Is it a tree if all edge costs are positive? For a cost function $c : E \to \mathbb{R}$, let $m = 1 + \max_{e \in E} |c(e)|$ and define a cost function c' by $c'(e) = m + c(e)$. Show that a minimum-cost spanning tree with respect to the cost function c is a minimum-cost spanning set of edges with respect to the cost function c' and vice versa.

Exercise 11.3. Reduce the problem of finding *maximum*-cost spanning trees to the minimum-spanning-tree problem.

Exercise 11.4. Suppose you have a highly tuned library implementation of an MST algorithm that you would like to reuse. However, this routine accepts only 32-bit integer edge weights while your graph uses 64-bit floating-point values. Show how to use the library routine provided that $m < 2^{32}$. Your preprocessing and postprocessing may take time $O(m \log m)$.

11.1 Cut and Cycle Properties

We shall prove two simple lemmas which allow one to add edges to an MST and to exclude edges from consideration for an MST. We need the concept of a cut in a graph. A *cut* is a partition $(S, V \setminus S)$ of the node set V into two nonempty parts. A cut

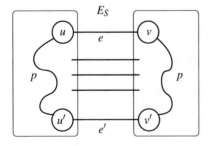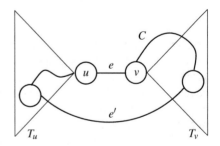

Fig. 11.1. Cut and cycle properties. The *left* part illustrates the proof of the cut property. Edge e has minimum cost in the cut E_S, and p is a path in the MST connecting the endpoints of e; p must contain an edge in E_S. The figure on the *right* illustrates the proof of the cycle property. C is a cycle in G, e is an edge of C of maximum weight, and T is an MST containing e. T_u and T_v are the components of $T \setminus e$, and e' is an edge in C connecting T_u and T_v.

determines the set $E_S = \{\{u,v\} \in E : u \in S, v \in V \setminus S\}$ of edges that connect S with $V \setminus S$. Figure 11.1 illustrates the proofs of the following lemmas.

Lemma 11.1 (cut property). *Let $(S, V \setminus S)$ be a cut and let e be a minimum cost edge in E_S. Consider a set T' of edges that is contained in some MST and contains no edge from E_S. Then $T' \cup \{e\}$ is also contained in some MST. In particular, there is an MST containing e.*

Proof. Consider any MST T of G with $T' \subseteq T$. Let $u \in S$ and $v \in V \setminus S$ be the endpoints of e. Since T is a spanning tree, it contains a unique path p from u to v. Since p connects $u \in S$ with $v \in V \setminus S$, it must contain an edge $e' = \{u', v'\}$ with $u' \in S$ and $v' \in V \setminus S$, i.e., $e' \in E_S$. Note that the case $e' = e$ is possible. Recall that we assume $e' \notin T'$. Now, $T'' := (T \setminus \{e'\}) \cup \{e\}$ is also a spanning tree, because removal of e' splits T into two subtrees, which are then joined together by e. Since $c(e) \le c(e')$, we have $c(T'') \le c(T)$, and hence T'' is also an MST. Obviously, $T' \cup \{e\} \subseteq T''$.
The second claim is a special case of the first. Set $T' = \emptyset$.   □

Lemma 11.2 (cycle property). *Consider any cycle $C \subseteq E$ and an edge $e \in C$ with maximum cost among all edges of C. Then any MST of $G' = (V, E \setminus \{e\})$ is also an MST of G.*

Proof. Note first that since the edge e that is removed is on a cycle in G, the graph G' is connected. Consider any MST T' of G' and assume that it is not an MST of G. Then there must be an MST T of G with $c(T) < c(T')$. If $e \notin T$, T is a spanning tree of G' cheaper than T', a contradiction. So T contains e. Let $e = \{u, v\}$. Removing e from T splits (V, T) into two subtrees (V_u, T_u) and (V_v, T_v) with $u \in V_u$ and $v \in V_v$. Since C is a cycle, there must be another edge $e' = \{u', v'\}$ in C such that $u' \in V_u$ and $v' \in V_v$. Replacing e by e' in T yields a spanning tree $T'' := (T \setminus \{e\}) \cup \{e'\}$ which does not contain e and for which $c(T'') = c(T) - c(e) + c(e') \le c(T) < c(T')$. So T'' is a spanning tree of G' cheaper than T', a contradiction.   □

The cut property yields a simple greedy algorithm for finding an MST; see Fig. 11.2. We initialize T to the empty set of edges. As long as T is not a spanning tree, let $(S, V \setminus S)$ be a cut such that E_S and T are disjoint (S is the union of some but not all connected components of (V, T)), and add a minimum-cost edge from E_S to T.

Function *genericMST(V, E, c)* : *Set of Edge*
    $T := \emptyset$
    while $|T| < n - 1$ **do**   *// T is extendible to an MST, but no spanning tree yet*
         let $(S, V \setminus S)$ be a cut such that T and E_S are disjoint;
         let e be a minimum-cost edge in E_S;
         $T := T \cup \{e\};$ *// enlarge T*
    return T

Fig. 11.2. A generic MST algorithm

Lemma 11.3. *The generic MST algorithm is correct.*

Proof. The algorithm maintains the invariant that T is a subset of some minimum spanning tree of G. The invariant is clearly true when T is initialized to the empty set. When an edge is added to T, the invariant is maintained by the cut property (Lemma 11.1). When T has reached size $n - 1$, it is a spanning tree contained in an MST, so it is itself an MST.   □

Different choices of the cut $(S, V \setminus S)$ lead to different algorithms. We discuss three approaches in detail in the following sections. For each approach, we need to explain how to find a minimum-cost edge in the cut.

The cycle property also leads to a simple algorithm for finding an MST: Set E' to the set of all edges. As long as E' is not a spanning tree, find a cycle in E and delete an edge of maximum cost from T. No efficient implementation of this approach is known however.

Exercise 11.5. Show that the MST is uniquely defined if all edge costs are different. Show that in this case the MST does not change if each edge cost is replaced by its rank among all edge costs.

*Exercise 11.6. We discuss how to check the cycle property for all non-tree edges. Let T be any spanning tree. Construct a tree whose leaves correspond to the vertices of the graph and whose inner nodes correspond to the edges of T. Start with a forest consisting of n trees, one for each node of the graph. Then process the edges of T in order of increasing cost. In order to process $e = \{u, v\}$ create a new node labelled e and make the roots of the trees containing u and v in the current forest the children of the new node. See Fig. 11.3 for an example. Let C be the resulting tree.

(a) The *lowest common ancestor* $lca_C(x, y)$ of two nodes x and y of C is the lowest node (= maximum depth) having x and y as descendants. Show that for any two nodes x and y of G, the edge associated with the node $lca_C(x, y)$ of C is the heaviest edge on the tree path in T connecting x and y.
(b) Show that T is an MST if for every non-tree edge $e' = \{x, y\}$, we have $c(e') \geq c(lca_C(x, y))$. Remark: C can be preprocessed in time $O(n)$ such that lca-queries can be answered in constant time [40].

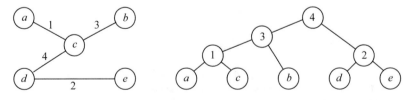

Fig. 11.3. A spanning tree T and the corresponding certification tree C. Observe that $c(lca_C(a, b)) = 3$ is the cost of the most costly edge on the path connecting a and b in T. Check that the analogous statement holds for any pair of nodes of T.

11.2 The Jarník–Prim Algorithm

The Jarník–Prim (JP) algorithm [96, 165, 257] for MSTs is very similar to Dijkstra's algorithm for shortest paths.[2] It grows a tree starting with an arbitrary source node. In each iteration of the algorithm, the cheapest edge connecting a node in the tree with a node outside the tree is added to the tree. The resulting spanning tree is returned. In the notation of the generic algorithm, T is a tree and S is the set of nodes of the tree. Initially, $S = \{s\}$, where s is an arbitrary node, and $T = \emptyset$. Generally, T and S are the edge and node sets of a tree, and E_S is the set of edges having exactly one endpoint in S. Let e be a cheapest such edge and assume $u \notin S$. Then u is added to S and e is added to T.

The main challenge is to find this edge e efficiently. To this end, the algorithm maintains a shortest edge between any node $v \in V \setminus S$ and a node in S in a priority queue Q. The key of v is the minimum cost of any edge connecting v to a node in S. If there is no edge connecting v to S, the key is ∞. The smallest element in Q then gives the desired edge and the node to be added to S.

Assume now that a node u is added to S. We inspect its incident edges $\{u,v\}$. If v lies outside S and the edge is a better connection for v to S, the key value of v is decreased accordingly. Figure 11.4 illustrates the operation of the JP algorithm, and Fig. 11.5 shows the pseudocode. The code uses two auxiliary arrays d and $parent$, where $d[v]$ stores the cost of the shortest edge from $v \in V \setminus S$ to a node in S and $parent[v]$ stores the endpoint in S of the corresponding edge. A value $d[v] = \infty$ means that no connection to S is available. Exploiting our assumption that all edge weights are positive, $d[v] = 0$ indicates that $v \in S$. Note that this convention for encoding membership in S allows us to combine two necessary tests in the inner loop of the pseudocode. Namely, an edge (u,v) only affects the priority queue if it connects S and $V \setminus S$ *and* improves the best connection found. Both conditions are true if and only if $c(e) < d[v]$. The test "**if** $w \in Q$" can be implemented by comparing the old value of $d[w]$ with ∞.

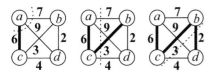

Fig. 11.4. A sequence of cuts (dotted lines) corresponding to the steps carried out by the Jarník–Prim algorithm with starting node a. The edges (a,c), (c,b), and (b,d) are added to the MST.

The only important difference from Dijkstra's algorithm is that the priority queue stores edge costs rather than path lengths. The analysis of Dijkstra's algorithm carries over to the JP algorithm, i.e., the use of a Fibonacci heap priority queue yields a running time $O(n \log n + m)$.

---

[2] Dijkstra also described this algorithm in his seminal 1959 paper on shortest paths [96]. Since Prim described the same algorithm two years earlier, it is usually named after him. However, the algorithm actually goes back to a paper from 1930 by Jarník [165].

Function *jpMST* : Set **of** *NodeId*
  $d = \langle \infty, \ldots, \infty \rangle : NodeArray[1..n]$ **of** $\mathbb{R} \cup \{\infty\}$ // $d[v]$ is the distance of v from the tree
  parent : *NodeArray* **of** *Edge*   // $(v, parent[v])$ is shortest edge between S and v
  $Q : NodePQ$ // uses $d[\cdot]$ as priority
  $Q.insert(s)$ for some arbitrary $s \in V$
  while $Q \neq \emptyset$ **do**
    $u := Q.deleteMin$
    $d[u] := 0$ // $d[u] = 0$ encodes $u \in S$
    foreach *edge* $e = \{u, v\} \in E$ **do**
      if $c(e) < d[v]$ **then**   // $c(e) < d[v]$ implies $d[v] > 0$ and hence $v \notin S$
        $d[v] := c(e)$
        $parent[v] := u$
        if $v \in Q$ **then** $Q.decreaseKey(v)$ **else** $Q.insert(v)$
    invariant $\forall v \in Q : d[v] = \min\{c((u,v)) : (u,v) \in E \wedge u \in S\}$
  return $\{(v, parent[v]) : v \in V \setminus \{s\}\}$

Fig. 11.5. The Jarník–Prim MST algorithm. Positive edge costs are assumed.

Exercise 11.7. Dijkstra's algorithm for shortest paths can use monotone priority queues. Show that monotone priority queues do *not* suffice for the JP algorithm.

*Exercise 11.8 (average-case analysis of the JP algorithm). Assume that the edge costs $1, \ldots, m$ are assigned randomly to the edges of G. Show that the expected number of *decreaseKey* operations performed by the JP algorithm is then bounded by $O(n \log(m/n))$. Hint: The analysis is very similar to the average-case analysis of Dijkstra's algorithm in Theorem 10.7.

11.3 Kruskal's Algorithm

The JP algorithm is a good general-purpose MST algorithm. Nevertheless, we shall now present an alternative algorithm, Kruskal's algorithm [191]. It also has its merits. In particular, it does not need a sophisticated graph representation, but works even when the graph is represented by its sequence of edges. For sparse graphs with $m = O(n)$, its running time is competitive with the JP algorithm.[3]

Kruskal's algorithm is also an instantiation of the generic algorithm. It grows a forest, i.e., in contrast to the JP algorithm, it grows several trees. In any iteration it adds the cheapest edge connecting two distinct components of the forest. In the notation of the generic algorithm, T is the set of edges already selected. Initially, T is the empty set. Let e be a cheapest edge connecting nodes in distinct subtrees of T. We let $(S, V - S)$ be any cut such that exactly one endpoint of e belongs to S. Then e is the cheapest edge in E_S. We add e to T.

---

[3] Kruskal's algorithm can be improved so that we get a very good algorithm for denser graphs also [246]. This *filterKruskal* algorithm needs average time $O(m + n \log n \log(m/n))$.

Function *kruskalMST(V, E, c)* : *Set* **of** *Edge*
    $T := \emptyset$
    invariant T *is a subforest of an MST*
    foreach $(u, v) \in E$ in ascending order of cost **do**
       if u and v are in different subtrees of T **then**
          $T := T \cup \{(u, v)\}$ // join two subtrees
    return T

Fig. 11.6. Kruskal's MST algorithm

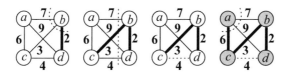

Fig. 11.7. In this example, Kruskal's algorithm first proves that (b, d) and (b, c) are MST edges using the cut property. Then (c, d) is excluded because it is the heaviest edge in the cycle $\langle b, c, d \rangle$, and, finally, (a, c) completes the MST.

How can we find the cheapest edge connecting two components of (V, T)? A first approach would be to first filter out the edges connecting two components and then to find the cheapest such edge. It is much simpler to combine both tasks. We iterate over the edges of G in order of increasing cost. When an edge is considered and its endpoints are in the same component of the current forest, the edge is discarded (filtering step), if its endpoints belong to distinct components, it is added to the forest (selection step). Figure 11.6 gives the pseudocode and Fig. 11.7 gives an example.

In an implementation of Kruskal's algorithm, we have to find out whether an edge connects two components of (V, T). In the next section, we shall see that this can be done so efficiently that the main cost factor is sorting the edges. This takes time $O(m \log m)$ if we use an efficient comparison-based sorting algorithm. The constant factor involved is rather small, so that for $m = O(n)$ we can hope to do better than the $O(m + n \log n)$ JP algorithm.

Exercise 11.9 (streaming MST). Suppose the edges of a graph are presented to you only once (for example over a network connection) and you do not have enough memory to store all of them. The edges do *not* necessarily arrive in sorted order.

(a) Outline an algorithm that nevertheless computes an MST using space $O(n)$.
(*b) Refine your algorithm to run in time $O(m \log n)$. Hint: Process batches of $O(n)$ edges (or use the *dynamic tree* data structure described by Sleator and Tarjan [300]).

11.4 The Union–Find Data Structure

A *partition* of a set M is a collection $M_1, \ldots, M_k$ of subsets of M with the property that the subsets are disjoint and cover M, i.e., $M_i \cap M_j = \emptyset$ for $i \neq j$ and $M = M_1 \cup \cdots \cup M_k$. The subsets M_i are called the *blocks* of the partition. For example, in Kruskal's algorithm, the forest T partitions V. The blocks of the partition are the connected components of (V, T). Some components may be trivial and consist of a single isolated node. Initially, all blocks are trivial. Kruskal's algorithm performs two operations on the partition: testing whether two elements are in the same subset (subtree) and joining two subsets into one (inserting an edge into T).

The *union–find data structure* maintains a partition of the set $1..n$ and supports these two operations. Initially, each element is a block on its own. Each block has a representative. This is an element of the block; it is determined by the data structure and not by the user. The function *find(i)* returns the representative of the block containing i. Thus, testing whether two elements are in the same block amounts to comparing their respective representatives. An operation *union(r, s)* applied to representatives of different blocks joins these blocks into a single block. The new block has r or s as its representative.

Exercise 11.10 (*union* versus *link*). In some other books *union* allows arbitrary parameters from $1..n$ and our restricted operation is called *link*. Explain how this generalized *union* operation can be implemented using *link* and *find*.

To implement Kruskal's algorithm using the union–find data structure, we refine the procedure shown in Fig. 11.6. Initially, the constructor of the class *UnionFind* initializes each node to represent its own block. The **if** statement is replaced by

$r := find(u)$; $s := find(v)$;
if $r \neq s$ **then** $T := T \cup \{\{u, v\}\}$; *union(r, s)*;

The union–find data structure is simple to implement as follows. Each block is represented as a rooted tree[4], with the root being the representative of the block. Each element stores its parent in this tree (stored in an array *parent*). A root of such a tree has itself as a *parent* (a self-loop).

The implementation of both operations is simple. We shall first describe unoptimized versions and later discuss optimizations that lead to the pseudocode shown in Fig. 11.8. For *find(i)*, we follow parent pointers starting at i until we encounter a self-loop. The self-loop is located at the representative of i, which we return. The implementation of *union(r, s)* is equally simple. We simply make one representative the child of the other. The root of the resulting tree is the representative of the combined block. What we have described so far yields a correct but inefficient union–find data structure. The *parent* references could form long chains that are traversed again and again during *find* operations. In the worst case, each operation may take linear time $\Omega(n)$.

---

[4] Note that this tree may have a structure very different from the corresponding subtree in Kruskal's algorithm.

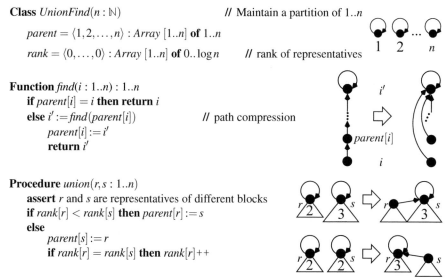

Class *UnionFind*(*n* : ℕ)   *// Maintain a partition of 1..n*

    parent = ⟨1,2,...,*n*⟩ : *Array* [1..*n*] **of** 1..*n*

    rank = ⟨0,...,0⟩ : *Array* [1..*n*] **of** 0..log*n*   *// rank of representatives*

Function *find*(*i* : 1..*n*) : 1..*n*
    if *parent*[*i*] = *i* **then return** *i*
    else *i'* := *find*(*parent*[*i*])   *// path compression*
       parent[*i*] := *i'*
       return *i'*

Procedure *union*(*r*,*s* : 1..*n*)
    assert *r* and *s* are representatives of different blocks
    if *rank*[*r*] < *rank*[*s*] **then** *parent*[*r*] := *s*
    else
       parent[*s*] := *r*
       if *rank*[*r*] = *rank*[*s*] **then** *rank*[*r*]++

Fig. 11.8. An efficient union–find data structure that maintains a partition of the set {1,...,*n*}

Exercise 11.11. Give an example of an *n*-node graph with O(*n*) edges where a naive implementation of the union–find data structure as described so far would lead to quadratic execution time for Kruskal's algorithm.

Therefore, Fig. 11.8 introduces two optimizations. The first optimization leads to a limit on the maximum depth of the trees representing blocks. Every representative stores a nonnegative integer, which we call its *rank*. Initially, every element is a representative and has rank 0. When the *union* operation is applied to two representatives with different rank, we make the representative of smaller rank a child of the representative of larger rank. When the two representatives have the same rank, the choice of the parent is arbitrary; however, we increase the rank of the new root. We refer to the first optimization as *union by rank*.

Exercise 11.12. Assume that no *find* operations are called. Show that in this case, the rank of a representative is the height of the tree rooted at it.

The second optimization is called *path compression*. This ensures that a chain of parent references is never traversed twice. Rather, all nodes visited during an operation *find*(*i*) redirect their parent pointers directly to the representative of *i*. In Fig. 11.8, we have formulated this rule as a recursive procedure. This procedure first traverses the path from *i* to its representative and then uses the recursion stack to traverse the path back to *i*. While the recursion stack is being unwound, the parent pointers are redirected. Alternatively, one can traverse the path twice in the forward direction. In the first traversal, one finds the representative, and in the second traversal, one redirects the parent pointers.

Exercise 11.13. Describe a nonrecursive implementation of *find*.

Theorem 11.4. *Union by rank ensures that the depth of no tree exceeds* $\log n$.

Proof. Without path compression, the rank of a representative is equal to the height of the tree rooted at it. Path compression does not increase heights and does not change the ranks of roots. It therefore suffices to prove that ranks are bounded by $\log n$. We shall show inductively that a tree whose root has rank k contains at least 2^k elements. This is certainly true for $k = 0$. The rank of a root grows from $k - 1$ to k when it receives a child of rank $k - 1$. Thus, by the induction hypothesis, the root had at least 2^{k-1} descendants before the union operation, and it receives a child which also had at least 2^{k-1} descendants. So the root has at least 2^k descendants after the union operation. $\square$

Union by rank and path compression make the union–find data structure "breathtakingly" efficient – the amortized cost of any operation is almost constant.

Theorem 11.5. *The union–find data structure of Fig. 11.8 performs m find and $n - 1$ union operations in time* $O(m\alpha(m,n))$. *Here,*

$$\alpha(m,n) = \min\left\{i \geq 1 : A(i, \lceil m/n \rceil) \geq \log n\right\},$$

where

$$
\begin{aligned}
A(1,j) &= 2^j && \text{for } j \geq 1, \\
A(i,1) &= A(i-1,2) && \text{for } i \geq 2, \\
A(i,j) &= A(i-1, A(i,j-1)) && \text{for } i \geq 2 \text{ and } j \geq 2.
\end{aligned}
$$

You will probably find the formulae overwhelming. The function[5] A grows extremely rapidly. We have $A(1,j) = 2^j$, $A(2,1) = A(1,2) = 2^2 = 4$, $A(2,2) = A(1, A(2,1)) = 2^4 = 16$, $A(2,3) = A(1, A(2,2)) = 2^{16}$, $A(2,4) = 2^{2^{16}}$, $A(2,5) = 2^{2^{2^{16}}}$, $A(3,1) = A(2,2) = 16$, $A(3,2) = A(2, A(3,1)) = A(2,16)$, and so on.

Exercise 11.14. Estimate $A(5,1)$.

For all practical n, we have $\alpha(m,n) \leq 5$, and union–find with union by rank and path compression essentially guarantees constant amortized cost per operation.

The proof of Theorem 11.5 is beyond the scope of this introductory text. We refer the reader to [291, 307]. Here, we prove a slightly weaker result that is equally useful for all practical purposes. In order to be able to state the result, we first define the numbers T_k, $k \geq 0$: $T_0 = 1$, and $T_k = 2^{T_{k-1}}$ for $k \geq 1$. The first terms of this rapidly growing sequence of numbers are:

k	0	1	2	3	4	5	$\ldots$	k
T_k	1	2	$4 = 2^2$	$16 = 2^{2^2}$	$65536 = 2^{2^{2^2}}$	$2^{65536} = 2^{2^{2^{2^2}}}$	$\ldots$	$2^{2^{\cdot^{\cdot^{\cdot^2}}}}$ of height k.

---

[5] The usage of the letter A is a reference to the logician Ackermann [4], who first studied a variant of this function in the late 1920s.

Note that $T_k = A(2, k-1)$ for $k \geq 2$ and that T_k is a "tower of twos" of height k.

For $x > 0$, we define $\log^* x$ as $\min\{k : T_k \geq x\}$. This is also the smallest non-negative integer k such that $\log^{(k)} x := \log(\log(\ldots \log(x) \ldots))$ (taking the logarithm k times) is less than or equal to 1. The function $\log^* x$ grows extremely slowly. For example, for all $x < 2^{65536}$, we have $\log^* x \leq 5$.

Theorem 11.6. *The union–find data structure with path compression and union by rank completes m find operations and $n-1$ union operations in $\mathrm{O}((m+n)\log^* n)$ time.*

Proof. (This proof is based on [156].) Consider an arbitrary sequence of $n-1$ *union* and m *find* operations starting with the initialization of the set $1..n$. Since *union* operations take constant time, we can concentrate the analysis on the *find* operations.

The rank of a root can grow while the sequence of operations is being executed. Once a node ceases to be a root, its rank no longer changes. In fact, its rank is no longer important for the execution of the algorithm. However, we shall use it in the analysis. We refer to the rank of a node v at the end of the execution as its *final rank* $fr(v)$. If v ever becomes a child, its final rank is the rank at the time when it first became a child. We make the following observations:

(a) Along paths defined by *parent* pointers, the values of *fr* strictly increase.
(b) When a node v obtains its final rank h, its subtree has as least 2^h nodes.
(c) There are at most $n/2^h$ nodes with *fr* value h.

Proof of the observations: (a) This is an invariant of the data structure. Initially, there are no (non-self-loop) edges at all. When v becomes a child of u during a *union* operation, we have $fr(v) = rank(v) < rank(u)$ right after the operation. Also $rank(u) \leq fr(u)$. The path compression in *find* operations shortcuts paths which can only increase the difference between *fr*-values. (b) is already implied by our proof of Theorem 11.4. Note that a node may lose descendants by path compresssion. However, this happens only when the node is no longer a root. (c) For a fixed *fr* value h and any node v with $fr(v) = h$, let M_v denote the set of children of v just before the moment when v becomes the child of some other node (or at the end of the execution, when v never becomes a child). We prove that the sets M_v are disjoint (this implies observation (c) since, according to (b), each M_v has at least 2^h elements and since there are only n nodes overall). Assume otherwise; say, node w belongs to M_{v_1} and M_{v_2} for distinct nodes v_1 and v_2 with final rank equal to h. Then they cannot both be roots at the end of the execution. Say, v_1 becomes a child of some node u at some point. Then $fr(u) > h$ by (a). By subsequent *union* operations, w can obtain further ancestors. However, by (a), these ancestors all have *fr* values larger than h. Hence, w can never become a descendant of another node with *fr* value h.

We partition the nodes with positive final rank into *rank groups* $G_0, G_1, \ldots$. Rank group G_k contains all nodes v with $T_{k-1} < fr(v) \leq T_k$ (defining $T_{-1} := 0$). For example, G_4 contains all nodes with final ranks between 17 and 65536. Since, by Theorem 11.4, ranks can never exceed $\log n$, it becomes apparent that only rank groups up to G_4 will be nonempty for practical values of n. Formally, for a node $v \in G_k$

with $k > 0$, $T_{k-1} < fr(v) \leq \log n$. Hence, $T_k = 2^{T_{k-1}} < 2^{\log n} = n$, or, equivalently, $k < \log^* n$. Thus, there are at most $\log^* n$ nonempty rank groups.

We are now ready for an amortized analysis of the cost of *find* operations. For an operation *find(v)*, we charge r units of cost, where r is the number of nodes on the path $\langle v = v_1, \ldots, v_{r-1}, v_r = s \rangle$ from v to its representative s. Note that the cost of the operation, including setting new *parent* pointers, is $\Theta(r)$ and hence we are covering the asymptotic cost. We distribute these r cost units as follows. We charge one unit to each node on the path with the following exceptions: Nodes v_1, v_{r-1}, v_r, and the nodes v_i whose parent is in a higher rank group are not charged. Note that the final rank of all nodes except maybe v_1 is positive and that all but nodes v_{r-1} and v_r get a new parent by path compression. Since, by observation (a), the *fr* values strictly increase along the path and since there are at most $\log^* n$ nonempty rank groups, the number of exceptions is $3 + \log^* n = O(\log^* n)$. We charge the exceptions directly to the *find* operation. In this way, each *find* is charged $O(\log^* n)$ for a total of $O(m \log^* n)$ for all all *find* operations.

Now we have to take care of the costs charged to nodes. Consider a node v belonging to rank group G_k. When v is charged during a *find* operation, v has a parent u (also belonging to G_k) that is not the root s. Hence, v gets s as a new parent by path compression. By observation (a), $fr(u) < fr(s)$, i.e., whenever node v is charged, it gets a new parent with a larger final rank. Since the ranks in group G_k are bounded by T_k, v is charged at most T_k times. Once its parent is in a higher rank group, v is never charged again. Therefore, the overall cost charged to v is at most T_k, and the total cost charged to nodes in rank group G_k is at most $|G_k| \cdot T_k$.

The final ranks of the nodes in G_k are $T_{k-1} + 1, \ldots, T_k$. By observation (3), there are at most $n/2^h$ nodes of final rank h. Therefore,

$$|G_k| \leq \sum_{T_{k-1} < h \leq T_k} \frac{n}{2^h} < \frac{n}{2^{T_{k-1}}} = \frac{n}{T_k},$$

by the definition of T_k. Equivalently, $|G_k| \cdot T_k < n$, i.e., the total cost charged to rank group k is at most n. Since there are at most $\log^* n$ nonempty rank groups, the total charge to all nodes is at most $n \log^* n$.

Adding the charges of the *find* operations and the nodes, we get $O(m \log^* n) + n \log^* n = O((n + m) \log^* n)$.   □

11.5 *External Memory

The MST problem is one of the very few graph problems that are known to have an efficient external-memory algorithm. We shall give a simple, elegant algorithm that exemplifies many interesting techniques that are also useful for other external-memory algorithms and for computing MSTs in other models of computation. Our algorithm is a composition of techniques that we have already seen: external sorting, priority queues, and internal union–find. More details can be found in [90].

11.5.1 A Semiexternal Kruskal Algorithm

We begin with an easy case. Suppose we have enough internal memory to store the union–find data structure of Sect. 11.4 for n nodes. This is enough to implement Kruskal's algorithm in the external-memory model. We first sort the edges using the external-memory sorting algorithm described in Sect. 5.12. Then we scan the edges in order of increasing weight, and process them as described by Kruskal's algorithm. If an edge connects two subtrees, it is an MST edge and can be output; otherwise, it is discarded. External-memory graph algorithms that require $\Theta(n)$ internal memory are called *semiexternal* algorithms.

11.5.2 Edge Contraction

If the graph has too many nodes for the semiexternal algorithm of the preceding subsection, we can try to reduce the number of nodes. This can be done using *edge contraction*. Suppose we know that $e = (u,v)$ is an MST edge, for example because e is the least-weight edge incident to v. We add e to the output, and need to remember that u and v are already connected in the MST under construction. Above, we used the union–find data structure to record this fact; now we use edge contraction to encode the information into the graph itself. We identify u and v and replace them by a single node. For simplicity, we again call this node u. In other words, we delete v and *relink* all edges incident to v to u, i.e., any edge (v,w) now becomes an edge (u,w). Figure 11.9 gives an example. In order to keep track of the origin of relinked edges, we associate an additional attribute with each edge that indicates its *original* endpoints. With this additional information, the MST of the contracted graph is easily translated back to the original graph. We simply replace each edge by its original.

We now have a blueprint for an external MST algorithm: Repeatedly find MST edges and contract them. Once the number of nodes is small enough, switch to a semiexternal algorithm. The following subsection gives a particularly simple implementation of this idea.

11.5.3 Sibeyn's Algorithm

Suppose $V = 1..n$. Consider the following simple strategy for reducing the number of nodes from n to n' [90]:

for *current* := 1 **to** $n - n'$ **do**
    find the lightest edge incident to current and contract it

Figure 11.9 gives an example, with $n = 4$ and $n' = 2$. The strategy looks deceptively simple. We need to discuss how we find the cheapest edge incident to *current* and how we relink the other edges incident to it, i.e., how we inform its neighbors that they are receiving additional incident edges. We can use a priority queue for both purposes. For each edge $e = \{u,v\}$, we store the item

$$(\min(u,v), \max(u,v), \text{weight of } e, \text{origin of } e)$$

in the priority queue. The ordering is lexicographic by the first and third components, i.e., edges are ordered first by the lower-numbered endpoint and then according to weight. The algorithm operates in phases. In each phase, we process all edges incident to the *current* node, i.e., a phase begins when the first edge incident to *current* is selected from the queue and ends when the last such edge is selected. The edges incident to *current* are selected from the queue in increasing order of weight. Let $(current, relinkTo, *, \{u_0, v_0\})$ be the lightest edge (= first edge delivered by the queue in the phase) incident to *current*. We add its original $\{u_0, v_0\}$ to the MST. Consider any other edge $(current, z, c, \{u_0', v_0'\})$ incident to *current*. If $z = RelinkTo$, we discard the edge because relinking would turn it into a self-loop. If $z \neq RelinkTo$, we add $(\min(z, RelinkTo), \max(z, RelinkTo), c, \{u_0', v_0'\})$ to the queue.

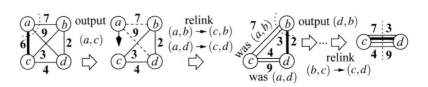

Fig. 11.9. An execution of Sibeyn's algorithm with $n' = 2$. The edge $(c, a, 6)$ is the cheapest edge incident to a. We add it to the MST and merge a into c. The edge $(a, b, 7)$ becomes an edge $(c, b, 7)$ and $(a, d, 9)$ becomes $(c, d, 9)$. In the new graph, $(d, b, 2)$ is the cheapest edge incident to b. We add it to the spanning tree and merge b into d. The edges $(b, c, 3)$ and $(b, c, 7)$ become $(d, c, 3)$ and $(d, c, 7)$, respectively. The resulting graph has two nodes that are connected by four parallel edges of weights 3, 4, 7, and 9.

Function *sibeynMST(V, E, c)* : *Set* **of** *Edge*
    let π be a random permutation of $1..n$
    Q: *priority queue*   // Order: *min node*, then *min edge weight*
    foreach $e = (u, v) \in E$ **do**
        $Q.insert(\min\{\pi(u), \pi(v)\}, \max\{\pi(u), \pi(v)\}, c(e), (u, v))$
    $current := 0$ // we are just before processing node 1
    loop
        $(u, v, c, \{u_0, v_0\}) := \min Q$ // next edge
        if $current \neq u$ **then**   // new node
            if $u = n - n' + 1$ **then** *break loop*   // node reduction completed
            $Q.deleteMin$
            output (u_0, v_0) // the original endpoints define an MST edge
            $(current, relinkTo) := (u, v)$ // prepare for relinking remaining u-edges
        else if $v \neq relinkTo$ **then**
            $Q.insert((\min\{v, relinkTo\}, \max\{v, relinkTo\}, c, \{u_0, v_0\})$ // relink
    $S := sort(Q)$ // sort by increasing edge weight
    apply semiexternal Kruskal to S

Fig. 11.10. Sibeyn's MST algorithm

Exercise 11.15. Let T be the partial MST just before the edges incident to *current* are inspected. Characterize the content of the queue, i.e., which edges $\{u_0, v_0\}$ have a representative in the queue and what is this representative? Hint: Show that every component of (V, T) contains exactly one node v with $v \geq current$. Call this node the representative of the component and use $rep(u_0)$ to denote the representative of the component containing u_0; then $\{u_0, v_0\}$ is represented by $\{rep(u_0), rep(v_0)\}$ if the two representatives are distinct and is not represented otherwise.

Figure 11.10 gives the details. For reasons that will become clear in the analysis, we renumber the nodes randomly before starting the algorithm, i.e., we chose a random permutation of the integers 1 to n and rename node v as $\pi(v)$. For any edge $e = \{u, v\}$ we store $(\min\{\pi(u), \pi(v)\}, \max\{\pi(u), \pi(v)\}, c(e), e)$ in the queue. The main loop stops when the number of nodes is reduced to n'. We complete the construction of the MST by sorting the remaining edges and then running the semiexternal Kruskal algorithm on them.

Theorem 11.7. *Let* $\mathrm{sort}(x)$ *denote the I/O complexity of sorting x items. The expected number of I/O steps needed by the algorithm sibeynMST is* $O(\mathrm{sort}(m \ln(n/n')))$.

Proof. From Sect. 6.3, we know that an external-memory priority queue can execute K queue operations using $O(\mathrm{sort}(K))$ I/Os. Also, the semiexternal Kruskal step requires $O(\mathrm{sort}(m))$ I/Os. Hence, it suffices to count the number of operations in the reduction phases. Besides the m insertions during initialization, the number of queue operations is proportional to the sum of the degrees of the nodes encountered. Let the random variable X_i denote the degree of node i when it is processed. When i is processed, the contracted graph has $n - i + 1$ remaining nodes and at most m edges. Hence the average degree of each remaining node is at most $2m/(n - i + 1)$. Owing to the random permutation of the nodes, each remaining node has the same probability of being removed next. Hence, $E[X_i]$ coincides with the average degree. By the linearity of expectations, we have $E[\sum_{1 \leq i \leq n-n'} X_i] = \sum_{1 \leq i \leq n-n'} E[X_i]$. We obtain

$$E\left[\sum_{1 \leq i \leq n-n'} X_i\right] = \sum_{1 \leq i \leq n-n'} E[X_i] \leq \sum_{1 \leq i \leq n-n'} \frac{2m}{n-i+1}$$

$$= 2m\left(\sum_{1 \leq i \leq n} \frac{1}{i} - \sum_{1 \leq i \leq n'} \frac{1}{i}\right) = 2m(H_n - H_{n'})$$

$$= 2m(\ln n - \ln n') + O(1) = 2m \ln \frac{n}{n'} + O(1),$$

where $H_n := \sum_{1 \leq i \leq n} 1/i = \ln n + \Theta(1)$ is the nth harmonic number (see (A.13)). $\square$

Note that we could do without switching to the semiexternal Kruskal algorithm. However, then the logarithmic factor in the I/O complexity would become $\ln n$ rather than $\ln(n/n')$ and the practical performance would be much worse. Observe that $n' = \Theta(M)$ is a large number, say 10^8. For $n = 10^{12}$, $\ln n$ is three times $\ln(n/n')$.

Exercise 11.16. For any n, give a graph with n nodes and $O(n)$ edges where Sibeyn's algorithm *without random renumbering* would need $\Omega(n^2)$ relink operations.

11.6 *Parallel Algorithms

The MST algorithms presented in the preceding sections add one edge after another to the MST and thus do not directly yield parallel algorithms. We need an algorithm that indentifies many MST edges at once. *Borůvka's algorithm* [51, 241] does just that. Interestingly, going back to 1926, it is also the oldest MST algorithm. For simplicity, let us assume that all edges have different weights. The algorithm is a recursive multilevel algorithm similar to the list-ranking algorithm presented in Sect. 3.3.2. Borůvka's algorithm applies the cut property to the simple cuts $\{v\}, V \setminus \{v\}$ for all $v \in V$, i.e., it finds the lightest edge incident to each node. Note that these edges cannot contain a cycle as the heaviest edge on the cycle is not the lightest edge incident to either endpoint. These edges are added to the MST and then contracted (see also Sect. 11.5.2). Recursively finding the MST of the contracted graph completes the MST.

It is relatively easy to see how to implement this algorithm sequentially such that each level of recursion takes linear time. We can also easily show that there are at most $\log n$ such levels: Each node finds one MST edge (recall that we assume the graph to be connected). Each new MST edge is found at most twice (once from each of its endpoints). Hence, at least $n/2$ distinct MST edges are identified. Contracting them at least halves the number of nodes.

An attractive feature of Borůvka's algorithm is that finding the new MST edges in each level of recursion is easy to parallelize. The contraction step is more complicated to parallelize, though. The difficulty is that we are not contracting single, unrelated edges but, instead, that the edges to be contracted form a graph. More concretely, consider the directed graph $H = (V, C)$, where $(u, v) \in C$ if $\{u, v\} \in E$ is the lightest edge incident to u. All nodes in H have outdegree 1. If you follow a path in H, the visited edges have nonincreasing weights. Since we assume the edge weights of G to be unique, this can only mean that any path in H ends in a cycle $u \rightleftharpoons v$ of length two, where the two edges of H, (u, v) and (v, u) represent the same edge $\{u, v\}$ of G. In other words, the components of H are "almost" rooted trees, except that there are two nodes pointing to each other instead of a single root. The most difficult step of our parallel algorithm is to transform these *pseudotrees* to *rooted stars*, i.e., rooted trees where all nodes point directly to the root.

The first, easy substep is to convert a pseudotree to a rooted tree. Assuming any ordering on the nodes, consider a cycle $u \rightleftharpoons v$ with $u < v$. We designate u as the root of the tree by replacing (u, v) by the self-loop (u, u).

Next, we have to "flatten" these trees. The pseudocode in Fig. 11.11 uses the doubling algorithm tha we have already seen for list ranking in Sect. 3.3.[6] The while-loop executing the doubling algorithm will perform $\log L$ iterations where $L < n$ is the longest path in H.

The resulting rooted trees are easy to contract. An edge $e = \{u, v\}$ in the input graph is transformed into an edge $\{u.R, v.R\}$ in the contracted graph. The weight remains the same. The endpoints of an edge in the input graph are seperately stored

---

[6] A work-efficient variant, using ideas similar to the independent-set algorithm for list ranking, is also possible [230].

so that they can be used for outputting the result. Some of the edges defined above are not required for the contracted graph: Those connecting nodes in the same component of H are not needed at all. Among parallel edges, only the lightest has to be kept. This can, for example, be achieved by inserting all edges into a hash table using their endpoints as keys. Then, within each bucket of the hash table, a minimum reduction is performed. Figure 11.12 gives an example.

// Input: edges are triples $(\{u,v\},c,\{u,v\})$ where c is the weight
// Output: triples $(\{u,v\},c,\{u',v'\})$ where u', v' are the original endpoints
Function *boruvkaMST(V, E)* : *Set* **of** *Edge*
    if $|V| = 1$ **then return** $\emptyset$ // base case
    foreach $u \in V$ **do**|| // find lightest incident edges
        $u.\underline{e} := \min\{(\{u,v\},c,o) : \{u,v\} \in E\}$ // minimum c; also in parallel
        $u.R :=$ other end of $u.\underline{e}$ // define pseudotrees – $u.R$ is the sole successor of u
    $T := \{u.\underline{e} : u \in V\}$ // remember MST edges
    foreach $u \in V$ **do**|| // pseudotrees → rooted trees
        if $u.R = v \wedge v.R = u \wedge u < v$ **then** $u.R := u$
    while $\exists u : u.R \neq u.R.R$ **do**   // trees → stars
        foreach $u \in V$ **do**||
            if $u.R \neq u.R.R$ **then** $u.R := u.R.R$ // doubling
    // contract
    $V' := \{u \in V : u.R = u\}$ // roots
    $E' := \{(\{u.R,v.R\},c,o) : (\{u,v\},c,o) \in E \wedge u.R \neq v.R\}$ // intertree edges
    optional: among parallel edges in (V',E'), *remove all but the lightest ones*

    return $T \cup boruvkaMST(V',E')$

Fig. 11.11. Loop-parallel CREW-PRAM pseudocode for Borůvka's MST algorithm

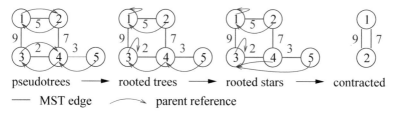

pseudotrees $\longrightarrow$ rooted trees $\longrightarrow$ rooted stars $\longrightarrow$ contracted
—— MST edge   parent reference

Fig. 11.12. One level of recursion of Borůvka's algorithm.

*Exercise 11.17. Refine the pseudocode in Fig. 11.11. Make explicit how the recursive instance can work with consecutive integers for node identifiers. Hint: see Sect. 3.3.2. Also, work out how to identify parallel edges in parallel (use hashing or sorting) and how to balance load even when nodes can have very high degree. Each level of recursion should work in expected time $O(m/p + \log p)$.

On a PRAM, Borůvka's algorithm can be implemented to run in polylogarithmic time.

Theorem 11.8. *On a CRCW-PRAM, the algorithm in Fig. 11.11 can be implemented to run in expected time*

$$O\left(\frac{m}{p}\log n + \log^2 n\right).$$

Proof. We analyze the algorithm without removal of parallel edges. There are at most $\log n$ levels of recursion since the number of nodes is at least halved in each level.

In each level, at most m edges are processed. Analogously to the BFS algorithm in Sect. 9.2.4, we can assign PEs to nodes using prefix sums such that each PE works on $O(m/p)$ edges. Using local minimum reductions, we can then compute the locally lightest edges. All this takes time $O(m/p + \log p)$ per level of recursion.

Transforming pseudotrees to rooted trees is possible in constant time. The while-loop for the doubling algorithm executes at most $\log n$ iterations. In recursion level i, there are at most $k = n/2^i$ nodes left. Performing one doubling step takes time $O(\lceil k/p \rceil)$. Hence, the total time for the while-loops is bounded by

$$\sum_{i=1}^{\log n} O\left(\left\lceil \frac{n}{p2^i} \right\rceil \log n\right) = O\left(\frac{n}{p}\log n + \log^2 n\right).$$

Building the recursive instances in adjacency array representation can be done using expected linear work and logarithmic time; see Sect. 8.6.1.

Overall we get expected time

$$O\left(\frac{m+n}{p}\log n + \log^2 n\right) = O\left(\frac{m}{p}\log n + \log^2 n\right).$$

For the last simplification, we exploit the fact that the graph is connected and hence $m \geq n-1$. □

The optional removal of superfluous parallel edges can be implemented using bulk operations on a hash table where the endpoints of the edge form the key. The asymptotic work needed is comparable to that for Kruskal's algorithm.

Exercise 11.18 (graphs staying sparse under edge contractions). We say that a class of graphs stays sparse under edge contractions if there is a constant C such that for every graph $G = (V,E)$ in the class and every graph $G' = (V',E')$ that can be obtained from G by edge contractions (and keeping only one copy of a set of parallel edges), we have $|E'| \leq C \cdot |V'|$. The class of planar graphs is sparse under edge contractions. Show that the running time of the parallel MST algorithm improves to $O(m/p + \log^2 n)$ for such graphs. Hint: Replace the sentence "In each level, at most m edges are processed" by "In level i of the recursion, there are at most $n/2^i$ nodes and hence $Cn/2^i$ edges left". Removal of parallel edges now becomes essential. Use the result of Exercise 11.17.

11.7 Applications

The MST problem is useful in attacking many other graph problems. We shall discuss the Steiner tree problem and the traveling salesman problem.

11.7.1 The Steiner Tree Problem

We are given a nonnegatively weighted undirected graph $G = (V, E)$ and a set S of nodes. The goal is to find a minimum-cost subset T of the edges that connects the nodes in S. Such a T is called a minimum Steiner tree. It is a tree connecting a set U with $S \subseteq U \subseteq V$. The challenge is to choose U so as to minimize the cost of the tree. The minimum-spanning-tree problem is the special case where S consists of all nodes. The Steiner tree problem arises naturally in our introductory example. Assume that some of the islands in Taka-Tuka-Land are uninhabited. The goal is to connect all the inhabited islands. The optimal solution may have some of the uninhabited islands in the solution.

The Steiner tree problem is **NP**-complete (see Sect. 2.13). We shall show how to construct a solution which is within a factor of two of the optimum. We construct an auxiliary complete graph H with node set S: For any pair u and v of nodes in S, the cost of the edge $\{u, v\}$ in H is their shortest-path distance in G. Let T_A be an MST of H and let c_H be its cost. Replacing every edge of T_A by the path it represents in G yields a subgraph of G connecting all the nodes in S. The resulting subgraph of G has cost c_H and may contain parallel edges and cycles. We remove parallel edges and delete edges from cycles until the remaining subgraph is cycle-free. The cost of the resulting Steiner tree is at most c_H.

Theorem 11.9. *The algorithm above constructs a Steiner tree which has at most twice the cost of an optimal Steiner tree.*

Proof. Let c_H be the cost of the MST in H. Recall that H is the complete graph on the node set S and that the cost of any edge $\{u, v\}$ of H is the cost of a shortest path connecting u and v in G. The algorithm constructs a Steiner tree for S of cost at most c_H. We show that $c_H \leq 2c(T_{\mathrm{opt}})$, where T_{opt} is a minimum Steiner tree for S in G. To this end, it suffices to show that the auxiliary graph H has a spanning tree of cost at most $2c(T_{\mathrm{opt}})$. Fig. 11.13 indicates how to construct such a spanning tree. "Walking once around the Steiner tree" defines a cycle C in G of cost $c(C) = 2c(T_{\mathrm{opt}})$; observe that every edge in T_{opt} occurs exactly twice in this cycle. Deleting the nodes outside S in this cycle gives us a cycle in H. The cost of this cycle in H is at most $2c(T_{\mathrm{opt}})$, because edge costs in H are shortest-path distances in G. More precisely, consider subpaths $P = \langle s, \dots, t \rangle$ of C whose endpoints s and t are in S and whose interior nodes are outside S. The cost of the edge $\{s, t\}$ of H is the shortest-path distance between s and t in G and hence at most the cost of the path P in G. We have now shown the existence of a cycle through all nodes of H of cost at most $2c(T_{\mathrm{opt}})$. Removing any edge of this cycle gives us a spanning tree of H. □

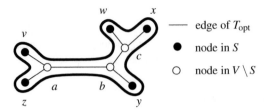

Fig. 11.13. "Once around the tree". The tree T_{opt} shown is a minimum Steiner tree for $S = \{v, w, x, y, z\}$. It also contains the nodes a, b, and c in $V \setminus S$. Walking once around the tree yields the cycle $\langle v, a, b, c, w, c, x, c, b, y, b, a, z, a, v \rangle$ in G, of cost $2c(T_{\text{opt}})$. Removal of the nodes outside S yields the cycle $\langle v, w, x, y, z, v \rangle$ in the auxiliary graph H. The cost of this cycle is at most $2c(T_{\text{opt}})$. Removal of any edge of the cycle yields a spanning tree of the auxiliary graph.

Exercise 11.19. Improve the above bound to $2(1 - 1/|S|)$ times the optimum.

The algorithm can be implemented to run in time $O(m + n \log n)$ [213]. Algorithms with better approximation ratios exist [266].

Exercise 11.20. Outline an implementation of the algorithm above and analyze its running time. Aim for running time $O(|S|(m + n \log n))$.

11.7.2 Traveling Salesman Tours

The traveling salesman problem is one of the most intensively studied optimization problems [18, 194, 318]: Given an undirected complete graph on a node set V with edge weights $c(e)$, the goal is to find the minimum-weight simple cycle passing through all nodes (also called a *tour*). This is the path a traveling salesman would want to take if his goal was to visit all nodes of the graph exactly once. We assume in this section that the edge weights satisfy the triangle inequality, i.e., $c(u, v) + c(v, w) \geq c(u, w)$ for all nodes u, v, and w. An important consequence of this assumption is that nonsimple cycles cannot lead to better tours than simple cycles, as dropping repeated nodes from a cycle does not increase its weight. There is a simple relation between the cost of MSTs and that of travaling salesman tours.

Theorem 11.10. *Let C_{opt} and C_{MST} be the costs of an optimal tour and of an MST, respectively. Then*
$$C_{\text{MST}} \leq C_{\text{opt}} \leq 2C_{\text{MST}}.$$

Proof. Let C be an optimal tour. Deleting any edge from C yields a spanning tree. Thus $C_{\text{MST}} \leq C_{\text{opt}}$. Conversely, let T be an MST. Walking once around the tree as shown in Fig. 11.13 gives us a cycle of cost at most $2C_{\text{MST}}$, passing through all nodes. It may visit nodes several times. Deleting an extra visit to a node does not increase the cost, owing to the triangle inequality. More precisely, replacing the cycle edges (u, v) and (v, w), where v is a node visited more than once, by (u, w) does not increase the cost. □

In the remainder of this section, we shall briefly outline a technique for improving the lower bound of Theorem 11.10. We need two additional concepts: 1-trees and node potentials. Let G' be obtained from G by deleting node 1 and the edges incident to it. A minimum 1-tree consists of the two cheapest edges incident to node 1 and an MST of G'. Since deleting the two edges incident to node 1 from a tour C yields a spanning tree of G', we have $C_1 \le C_{opt}$, where C_1 is the minimum cost of a 1-tree. A node potential is any real-valued function π defined on the nodes of G. We have also used node potentials in Sect. 10.7. A node potential π yields a modified cost function c_π defined as

$$c_\pi(u,v) = c(u,v) + \pi(v) + \pi(u)$$

for any pair u and v of nodes. For any tour C, the costs under c and c_π differ by $2S_\pi := 2\sum_v \pi(v)$, since a tour uses exactly two edges incident to any node. Let T_π be a minimum 1-tree with respect to c_π. Then

$$c_\pi(T_\pi) \le c_\pi(C_{opt}) = c(C_{opt}) + 2S_\pi,$$

and hence

$$c(C_{opt}) \ge \max_\pi (c_\pi(T_\pi) - 2S_\pi).$$

This lower bound is known as the Held–Karp lower bound [147, 148]. The maximum is over all node potential functions π. It is hard to compute the lower bound exactly. However, there are fast iterative algorithms for approximating it. The idea is as follows, and we refer the reader to the original papers for details. Assume we have a potential function π and the optimal 1-tree T_π with respect to it. If all nodes of T_π have degree two, we have a traveling salesman tour and stop. Otherwise, we make the edges incident to nodes of degree larger than two a little more expensive and the edges incident to nodes of degree 1 a little cheaper. This can be done by modifying the node potential of v as follows. We define a new node potential π' by

$$\pi'(v) = \pi(v) + \varepsilon \cdot (\deg(v, T_\pi) - 2),$$

where ε is a parameter which goes to 0 with increasing iteration number, and $\deg(v, T_\pi)$ is the degree of v in T_π. We next compute an optimal 1-tree with respect to π' and hope that it will yield a better lower bound.

11.8 Implementation Notes

The minimum-spanning-tree algorithms discussed in this chapter are so fast that the running time is usually dominated by the time required to generate the graphs and appropriate representations. The JP algorithm works well for all m and n if an adjacency array representation (see Sect. 8.2) of the graph is available. Pairing heaps [231] are a robust choice for the priority queue. Kruskal's algorithm may be faster for sparse graphs, in particular if only a list or array of edges is available or if we know how to sort the edges very efficiently.

The union–find data structure can be implemented more space-efficiently by exploiting the observation that only representatives need a rank, whereas only nonrepresentatives need a parent. We can therefore omit the array *rank* in Fig. 11.6. Instead, a root of rank g stores the value $n + 1 + g$ in *parent*. Thus, instead of two arrays, only one array with values in the range $1..n + 1 + \lceil \log n \rceil$ is needed. This is particularly useful for the semiexternal algorithm [90].

11.8.1 C++

LEDA [195] uses Kruskal's algorithm for computing MSTs. The union–find data structure is called *partition* in LEDA. The Boost graph library [50] and LEMON graph library [201] give choices of algorithms for computing MSTs. They also provide the union–find data structure.

11.8.2 Java

The JGraphT [167] library gives a choice between Kruskal's and the JP algorithm. It also provides the union–find data structure.

11.9 Historical Notes and Further Findings

There is a randomized linear-time MST algorithm that uses phases of Borůvka's algorithm to reduce the number of nodes [176, 183]. The second building block of this algorithm reduces the number of edges to about $2n$: We sample $O(m/2)$ edges randomly, find an MST T' of the sample, and remove edges $e \in E$ that are the heaviest edge in a cycle in $e \cup T'$. The last step is difficult to implement efficiently. But, at least for rather dense graphs, this approach can yield a practical improvement [179]. An adaptation for the external-memory model [2] saves a factor $\ln(n/n')$ in the asymptotic I/O complexity compared with Sibeyn's algorithm but is impractical for currently interesting values of n owing to its much larger constant factor in the O-notation.

The theoretically best *deterministic* MST algorithm [65, 254] has the interesting property that it has optimal worst-case complexity, although it is not known exactly what this complexity is. Hence, if you were to come up with a completely different deterministic MST algorithm and prove that your algorithm runs in linear time, then we would know that the old algorithm also runs in linear time.

There has been a lot of work on parallel MST algorithms. In particular, the linear-work algorithm [176] can be parallelized [76, 141]. Bader and Cong [26] achieved speedup in practice using a hybrid of Borůvka's algorithm and Prim's algorithm which starts growing trees from multiple source nodes in parallel. Zhou [334] gave an efficient shared-memory implementation of Borůvka's algorithm by exploiting the *priority update principle* [296]. The *filterKruskal* algorithm mentioned above can also be partially parallelized [246]. When a graph is partitioned into blocks, it suffices to consider the cut edges and the MST edges of the blocks. Wassenberg et al. developed a parallel version of Kruskal's algorithm for image segmentation based on this observation [327].

Minimum spanning trees define a single path between any pair of nodes. Interestingly, this path is a *bottleneck shortest path* [9, Application 13.3], i.e., it minimizes the maximum edge cost for all paths connecting the nodes in the original graph. Hence, finding an MST amounts to solving the all-pairs bottleneck-shortest-path problem in much less time than that for solving the all-pairs shortest-path problem.

A related and even more frequently used application is clustering based on the MST [9, Application 13.5]: By dropping $k - 1$ edges from the MST, it can be split into k subtrees. The nodes in a subtree T' are far away from the other nodes in the sense that all paths to nodes in other subtrees use edges that are at least as heavy as the edges used to cut T' out of the MST.

Many applications lead to MST problems on complete graphs. Frequently, these graphs have a compact description, for example if the nodes represent points in the plane and the edge costs are Euclidean distances (these MSTs are called Euclidean minimum spanning trees). In these situations, it is an important concern whether one can rule out most of the edges as too heavy without actually looking at them. This is the case for Euclidean MSTs. It can be shown that Euclidean MSTs are contained in the Delaunay triangulation [85] of the point set. This triangulation has linear size and can be computed in time $O(n \log n)$. This leads to an algorithm of the same time complexity for Euclidean MSTs.

We have discussed the application of MSTs to the Steiner tree and the traveling salesman problem. We refer the reader to the books [9, 18, 190, 194, 323] for more information about these and related problems.

12

Generic Approaches to Optimization

Fig. 3.

A smuggler in the mountainous region of Profitania has n items in his cellar. If he sells an item i across the border, he makes a profit p_i. However, the smuggler's trade union only allows him to carry knapsacks with a maximum weight of M. If item i has weight w_i, which of the items should he pack into the knapsack to maximize the profit from his next trip[1]?

This problem, usually called the *knapsack problem*, has numerous other applications, many of which are described in the books [180, 207]. For example, an investment bank might have an amount M of capital to invest and a set of possible investments. Each investment i has an expected profit p_i for an investment of cost w_i. In this chapter, we use the knapsack problem as an example to illustrate several generic approaches to optimization. These approaches are quite flexible and can be adapted to complicated situations that are ubiquitous in practical applications.

In the previous chapters we considered very efficient specific solutions for frequently occurring simple problems such as finding shortest paths or minimum spanning trees. Now we look at generic solution methods that work for a much larger range of applications. Of course, the generic methods do not usually achieve the same efficiency as specific solutions. However, they save development time.

Formally, an instance I of an optimization problem can be described by a set $\mathcal{U}_I$ of *potential* solutions, a set $\mathcal{L}_I$ of *feasible* solutions, and an *objective function* f_I with $f_I \colon \mathcal{L}_I \to \mathbb{R}$. For simplicity, we will mainly drop the subscript I and write $\mathcal{U}$, $\mathcal{L}$, and f. In a *maximization* problem, we are looking for a feasible solution $x^* \in \mathcal{L}$ that maximizes the value of the objective function over all feasible solutions. In a *minimization* problem, we are looking for a solution that minimizes the value of the objective function. In a *search* problem, the objective function is irrelevant; the task is to find any feasible solution $x \in \mathcal{L}$, if this set is not empty. Similarly, in an *existence problem*, the objective function is irrelevant and the question is whether $\mathcal{L}$ is nonempty.

For example, in the knapsack problem an instance I specifies the maximum weight M, the number n of objects as well as profits and weights of these objects, as vectors $p = (p_1, \dots, p_n)$ and $w = (w_1, \dots, w_n)$. Figure 12.1 gives an example in-

---

[1] The illustration above shows a 19th century American knapsack commons.wikimedia.org/wiki/File:19th_century_knowledge_hiking_and_camping_sheepskin_knapsack_sleeping_bag_rolled_up.jpg

© Springer Nature Switzerland AG 2019
P. Sanders et al., *Sequential and Parallel Algorithms and Data Structures*,
https://doi.org/10.1007/978-3-030-25209-0_12

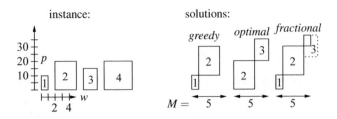

Fig. 12.1. The *left part* shows a knapsack instance with $p = (10, 20, 15, 20)$, $w = (1, 3, 2, 4)$, and $M = 5$. The items are indicated by rectangles whose width and height correspond to the weight and profit, respectively. The *right part* shows three solutions: the one computed by the *greedy* algorithm in Sect. 12.2, an *optimal* solution computed by the dynamic programming algorithm in Sect. 12.3, and the *fractional* solution of the linear relaxation (Sect. 12.1.2). The optimal solution has weight 5 and profit 35.

stance. A potential solution for I is simply a vector $x = (x_1, \ldots, x_n)$ with $x_i \in \{0, 1\}$. Here $x_i = 1$ indicates that item i is put into the knapsack and $x_i = 0$ indicates that item i is left out. Thus $\mathscr{U} = \{0, 1\}^n$. A potential solution x is feasible if its total weight does not exceed the capacity of the knapsack, i.e., if $\sum_{1 \le i \le n} w_i x_i \le M$. The dot product $w \cdot x$ is a convenient shorthand for $\sum_{1 \le i \le n} w_i x_i$. We can then say that $\mathscr{L} = \{x \in \mathscr{U} : w \cdot x \le M\}$ is the set of feasible solutions, and $f(x) = p \cdot x$ is the objective function.

The distinction between minimization and maximization problems is not essential because setting $f = -f$ converts a maximization problem into a minimization problem and vice versa. We shall use maximization as our default simply because our example problem is more naturally viewed as a maximization problem.[2]

In this chapter we shall present seven generic approaches to solving optimization problems. We start out with black-box solvers that can be applied to any optimization problem whose instances can be expressed in the problem specification language of the solver. In such a case, the only task of the user is to formulate the given problem in the language of the black-box solver. Section 12.1 introduces this approach using *linear programming* and *integer linear programming* as examples. The *greedy approach*, which we have already met in Chap. 11, is reviewed in Sect. 12.2. The approach of *dynamic programming* discussed in Sect. 12.3 is a more flexible way to construct solutions. We can also systematically explore the entire set of potential solutions, as described in Sect. 12.4. *Constraint programming*, *SAT solvers*, and *ILP solvers* are special cases of *systematic search*. Finally, we consider two very flexible approaches to exploring only a subset of the solution space. *Local search*, discussed in Sect. 12.5, modifies a single solution until it has the desired quality. *Evolutionary algorithms*, described in Sect. 12.6, simulate a population of candidate solutions. Most of the methods described above can be parallelized. We outline basic approaches within each section.

---

[2] Be aware that most of the literature uses minimization as the default.

12.1 Linear Programming – Use a Black-Box Solver

The easiest way to solve an instance of an optimization problem is to write down a specification of the space of feasible solutions and of the objective function and then use an existing software package to find an optimal solution. Such software packages are often called "black-box solvers" since the user only needs to know their interface and not the methods used for finding an optimal solution. Of course, the question is for which kinds of specification general solvers are available. In this section, we introduce a particularly large class of problem instances for which such black-box solvers are available. In *Linear Programming* one specifies the set of feasible solutions as a set of vectors in $\mathbb{R}^m$ by linear inequalities and the objective function as a linear function with values in $\mathbb{R}$. There are software packages that solve such instances, which can be termed "efficient" in a theoretical or in a practical sense. In *(Mixed) Integer Linear Programming* (some or) all the components of the feasible solutions are restricted to be integers. Black-box software packages are available for this type of specification as well, which can solve such instances quickly in many cases, although the algorithms are not guaranteed to run in polynomial time.

12.1.1 Linear Programming

Definition 12.1. *A* linear program[3] *(LP) with n* variables *and m* constraints *is an instance of a maximization problem that is specified in the following way. The possible solutions are vectors* $x = (x_1,\ldots,x_n)$ *with real components* x_j, $j \in 1..n$, *which are called* variables. *The objective function is a linear function f of x, i.e.,* $f\colon \mathbb{R}^n \to \mathbb{R}$ *with* $f(x) = c \cdot x$, *where* $c = (c_1,\ldots,c_n)$ *is called the* cost *or* profit vector[4]. *The variables are constrained by m linear constraints of the form* $a_i \cdot x \bowtie_i b_i$, *where* $\bowtie_i \in \{\le,\ge,=\}$, $a_i = (a_{i1},\ldots,a_{in}) \in \mathbb{R}^n$, *and* $b_i \in \mathbb{R}$ *for* $i \in 1..m$. *The set of feasible solutions is given by*

$$\mathscr{L} = \left\{ x \in \mathbb{R}^n : \forall i \in 1..m : a_i \cdot x \bowtie_i b_i \text{ and } \forall j \in 1..n : x_j \ge 0 \right\}.$$

Figure 12.2 shows a simple example. A classical application of linear programming is the *diet problem*. A farmer wants to mix food for his cows. There are n different kinds of food on the market, say, corn, soya, fish meal, ... One kilogram of food j costs c_j euros. There are m requirements for healthy nutrition. For example, the cows should get enough calories, protein, vitamin C, and so on. One kilogram of food j contains a_{ij} percent of a cow's daily requirement with respect to requirement i. A solution to the following linear program gives a cost-optimal diet that satisfies the health constraints: Let x_j denote the amount (in kilogram) of food j used by the farmer. The ith nutritional requirement is modeled by the inequality $\sum_j a_{ij} x_j \ge 100$. The cost of the diet is given by $\sum_j c_j x_j$. The goal is to minimize the cost of the diet.

---

[3] The term "linear programming" stems from the 1940s [81] and has nothing to do with the modern meaning of "program" as in "computer program".

[4] If all c_j are positive, it is common to use the term "profit" in maximization problems and "cost" in minimization problems.

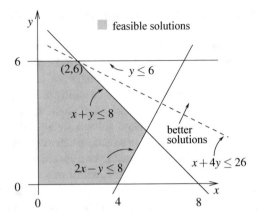

Fig. 12.2. A simple two-dimensional linear program in variables x and y, with three constraints and the objective function "maximize $x+4y$". The feasible region is shaded, and $(x,y) = (2,6)$ is the optimal solution. Its objective value is 26. The vertex $(2,6)$ is optimal because the half-plane described by $x+4y \le 26$ contains the entire feasible region and has $(2,6)$ in its boundary.

Exercise 12.1. How do you model supplies that are available only in limited amounts, for example food produced by the farmer himself? Also, explain how to specify additional constraints such as "no more than 0.01 mg cadmium contamination per cow per day".

Can the knapsack problem be formulated as a linear program? Probably not. Each item either goes into the knapsack as a whole or not at all. There is no possibility of adding only a part of an item. In contrast, it is assumed in the diet problem that any arbitrary amount of any food can be purchased, for example 3.7245 kg and not just 3 kg or 4 kg. Integer linear programs (see Sect. 12.1.2) are the suitable method for formulating the knapsack problem.

We next connect linear programming to a problem that was studied earlier in the book. We show how to formulate the single-source shortest-path problem with nonnegative edge weights in the language of linear programming. Let $G = (V,E)$ be a directed graph, let $s \in V$ be the source node, and let $c\colon E \to \mathbb{R}_{\ge 0}$ be the cost function on the edges of G. In the linear program that corresponds to this instance there is a variable d_v for each vertex v in G. The intention is that in an optimal solution d_v denotes the cost of the shortest path from s to v. Somewhat surprisingly, we formulate the *shortest* path problem as a *maximization* problem. Consider

maximize $\qquad \sum_{v \in V} d_v$

subject to $\qquad d_s = 0,$

$\qquad d_w \le d_v + c(e) \quad$ for all $e = (v,w) \in E.$

In order to gain intuition why this might be a suitable formulation, the reader should recall the string model for the single-source shortest path problem that we discussed at the beginning of Chap. 10. In this model every node sits *as far below* the start node s *as possible*, without any edge being overstretched. That is, for vertices v and w that form an edge $e = (v,w)$ and their distances d_v and d_w from s we must have the relation $d_w \leq d_v + c(e)$. Maximizing the sum is a weak version of expressing the goal that each d_v should be as large as possible. We prove that solving the LP above is equivalent to solving the given shortest-path instance.

Theorem 12.2. *Let $G = (V,E)$ be a directed graph, $s \in V$ a designated vertex, and $c: E \to \mathbb{R}_{\geq 0}$ a nonnegative cost function. If all vertices of G are reachable from s, the shortest-path distances in G are the unique optimal solution to the linear program above.*

Proof. Let $\mu(v)$ be the distance from s to v. Then $\mu(v) \in \mathbb{R}_{\geq 0}$, since edge costs are nonnegative and all nodes are reachable from s, and hence no vertex can have a distance $-\infty$ or $+\infty$ from s. We observe first that the choice $d_v := \mu(v)$ for all v satisfies the constraints of the LP. Indeed, $\mu(s) = 0$ and $\mu(w) \leq \mu(v) + c(e)$ for any edge $e = (v,w)$.

We next show that if $(d_v)_{v \in V}$ satisfies all constraints of the LP above, then $d_v \leq \mu(v)$ for all v. Consider any v, and let $\langle s = v_0, v_1, \ldots, v_k = v \rangle$ be a shortest path from s to v. Then $\mu(v) = \sum_{0 \leq i < k} c(v_i, v_{i+1})$. We shall show by induction on j that $d_{v_j} \leq \sum_{0 \leq i < j} c(v_i, v_{i+1})$, for $j = 0, \ldots, k$. For $j = 0$, this follows from $d_s = 0$ by the first constraint. For $j > 0$, we have

$$d_{v_j} \leq d_{v_{j-1}} + c(v_{j-1}, v_j) \leq \sum_{0 \leq i < j-1} c(v_i, v_{i+1}) + c(v_{j-1}, v_j) = \sum_{0 \leq i < j} c(v_i, v_{i+1}),$$

where the first inequality follows from the second set of constraints of the LP and the second inequality comes from the induction hypothesis.

We have now shown that $(\mu(v))_{v \in V}$ is a feasible solution, and that $d_v \leq \mu(v)$ for all v for all feasible solutions $(d_v)_{v \in V}$. Since the objective of the LP is to maximize the sum of the d_v's, we must have $d_v = \mu(v)$ for all v in the optimal solution to the LP. □

Exercise 12.2. Where does the proof above fail when not all nodes are reachable from s or when there are negative edge costs? Does it still work in the absence of negative cycles?

The proof that the LP above actually captures the given instance of the shortest-path problem is nontrivial. When you formulate a problem instance as an LP, you should always prove that the LP is indeed a correct description of the instance that you are trying to solve.

Exercise 12.3. Let $G = (V,E)$ be a directed graph and let s ("source") and t ("sink") be two nodes. Let $cap: E \to \mathbb{R}_{\geq 0}$ and $c: E \to \mathbb{R}_{\geq 0}$ be nonnegative functions on the edges of G. For an edge e, we call $cap(e)$ and $c(e)$ the capacity and cost, respectively, of e. A *flow* is a function $f: E \to \mathbb{R}_{\geq 0}$ with $0 \leq f(e) \leq cap(e)$ for all e and flow conservation at all nodes except s and t, i.e., for all $v \neq s,t$, we have

$$\text{flow into } v = \sum_{e=(u,v)} f(e) = \sum_{e=(v,w)} f(e) = \text{flow out of } v.$$

The *value* of the flow is the net flow out of s, i.e., $\sum_{e=(s,v)} f(e) - \sum_{e=(u,s)} f(e)$. The *maximum-flow problem* asks for a flow of maximum value, given G, s, t, and *cap*. Show that such an instance of the flow problem can be formulated as an LP.

If also c is given, the *cost* of a flow is $\sum_e f(e)c(e)$. The *minimum-cost maximum-flow problem* asks for a maximum flow of minimum cost. Show how to formulate instances of this problem as an LP.

Linear programs are of central importance because they combine expressive power with efficient solution algorithms. For discussing efficiency, we have to explain how to measure the size of the input. The coefficients a_{ij}, b_i, and c_j of the linear inequalities and the objective function are assumed to be rational numbers. The size of an LP is then specified by m, n, and L, where L is an upper bound on the number of bits needed to write (in binary) the numerators and denominators of these coefficients. We say an algorithm *solves* Linear Programming if, when presented with an arbitrary LP, it either (1) returns an optimal feasible solution or (2) correctly asserts that there is no feasible solution or (3) correctly asserts that there are feasible solutions with arbitrarily large values of the objective function.

Theorem 12.3. *Linear programs can be solved in polynomial time [177, 181].*

The two polynomial-time algorithms are the Ellipsoid method [181] and the interior point method [177]. A compact account of the interior point method can be found in [222]. The worst-case running time of the best interior point algorithm is $O(\max(m,n)^{7/2}L)$. Fortunately, the worst case rarely arises. Most linear programs can be solved relatively quickly by any one of several procedures. One, the simplex algorithm, is briefly outlined in Sect. 12.5.1. For now, the reader should remember two facts: First, many optimization problems can be formulated as linear programs, and second, there are efficient LP solvers that can be used as black boxes. In fact, although LP solvers are used on a routine basis, very few people in the world know exactly how to implement a highly efficient LP solver.

The simplex algorithm is notoriously difficult to parallelize, since an efficient implementation performs very little work in each step. The polynomial-time algorithms perform fewer steps with more work in each step and have been parallelized successfully in state-of-the-art LP solvers.

12.1.2 Integer Linear Programming

The expressive power of linear programming grows when some or all of the variables can be designated to be integral. Such variables can then take on only integer values, and not arbitrary real values. If all variables are constrained to be integral, the formulation of the problem is called an *integer linear program* (ILP). If some but not all variables are constrained to be integral, the formulation is called a *mixed integer linear program* (MILP). For example, our knapsack problem is tantamount to the following 0-1 integer linear program:

$$\text{maximize } p \cdot x$$

subject to

$$w \cdot x \leq M \quad \text{and} \quad x_j \in \{0,1\} \text{ for } j \in 1..n.$$

In a 0-1 integer linear program, the variables are constrained to the values 0 and 1.

Exercise 12.4. Explain how to replace any ILP by a 0-1 ILP, assuming that an upper bound U on the value of any variable in the optimal solution is known. Hint: Replace each variable of the original ILP by a set of $O(\log U)$ many 0-1 variables.

Unfortunately, solving ILPs and MILPs is **NP**-hard. Indeed, even the knapsack problem is **NP**-hard. Nevertheless, ILPs can often be solved in practice using linear-programming packages. In Sect. 12.4, we shall outline how this is done. When an exact solution would be too time-consuming, linear programming can help to find approximate solutions. The *linear-program relaxation* of an ILP is the LP obtained by omitting the integrality constraints on the variables. For example, in the knapsack problem we would replace the constraint $x_j \in \{0,1\}$ by the constraint $x_j \in [0,1]$.

An LP relaxation can be solved by an LP solver. In many cases, the solution to the relaxation teaches us something about the underlying ILP. One observation always holds true (for maximization problems): The objective value of the relaxation is at least as large as the objective value of the underlying ILP. This claim is trivial, because any feasible solution to the ILP is also a feasible solution to the relaxation. The optimal solution to the LP relaxation will in general be *fractional*, i.e., variables will take on rational values that are not integral. However, it might be the case that only a few variables have nonintegral values. By appropriate rounding of fractional variables to integer values, we can often obtain good integer feasible solutions.

We shall give an example. The linear relaxation of the knapsack problem is given by

$$\text{maximize } p \cdot x$$

subject to

$$w \cdot x \leq M \quad \text{and} \quad x_j \in [0,1] \text{ for } j \in 1..n.$$

This has a natural interpretation. It is no longer required to add items to the knapsack as a whole; one can now take any fraction of an item. In our smuggling scenario, the *fractional knapsack problem* corresponds to a situation involving divisible goods such as liquids or powders.

The fractional knapsack problem is easy to solve in time $O(n \log n)$; there is no need to use a general-purpose LP solver. We renumber (sort) the items by *profit density* p_j/w_j such that

$$\frac{p_1}{w_1} \geq \frac{p_2}{w_2} \geq \cdots \geq \frac{p_n}{w_n}.$$

We find the smallest index ℓ such that $\sum_{j=1}^{\ell} w_j > M$ (if there is no such index, we can take all knapsack items). The item ℓ with fractional value x_ℓ is also called *critical item*. Now we set

$$x_1 = \cdots = x_{\ell-1} = 1, \quad x_\ell = \left(M - \sum_{j=1}^{\ell-1} w_j\right)/w_\ell, \quad \text{and } x_{\ell+1} = \cdots = x_n = 0.$$

Figure 12.1 gives an example. The fractional solution above is the starting point for many good algorithms for the knapsack problem. We shall see more of this later.

Exercise 12.5 (linear relaxation of the knapsack problem).

(a) Prove that the above routine computes an optimal solution. Hint: You may want to use an *exchange argument* similar to the one used to prove the cut property of minimum spanning trees in Sect. 11.1.
(b) Outline an algorithm that computes an optimal solution in linear expected time. Hint: Use a variant of *quickSelect*, described in Sect. 5.8.
(c) Parallelize your algorithm from (b).

A solution to the fractional knapsack problem is easily converted to a feasible solution to the knapsack problem. We simply take the fractional solution and round the sole fractional variable x_ℓ to 0. We call this algorithm *roundDown*.

Exercise 12.6. Formulate the following *set-covering* problem as an ILP. Given a set M, subsets $M_i \subseteq M$ for $i \in 1..n$ with $\bigcup_{i=1}^n M_i = M$, and a cost $c_i \in \mathbb{N}$ for each M_i, select $F \subseteq 1..n$ such that $\bigcup_{i \in F} M_i = M$ and $\sum_{i \in F} c_i$ is minimized.

Boolean formulae. These provide another powerful description language for search and decision problems. Here, variables range over the Boolean values 1 and 0, and the connectors $\wedge$, $\vee$, and $\neg$ are used to build formulae. A Boolean formula is *satisfiable* if there is an assignment of Boolean values to the variables such that the formula evaluates to 1. As an example, we consider the *pigeonhole principle*: It is impossible to pack $n+1$ items into n bins such that every bin contains at most one item. This principle can be formulated as the statement that certain formulae are unsatisfiable. Fix n. We have variables x_{ij} for $1 \leq i \leq n+1$ and $1 \leq j \leq n$. So i ranges over items and j ranges over bins. Variable x_{ij} represents the statement "item i is in bin j". The constraint that every item must be put into (at least) one bin is formulated as $x_{i1} \vee \cdots \vee x_{in}$, for $1 \leq i \leq n+1$. The constraint that no bin should hold more than one item is expressed by the subformulas $\neg(\bigvee_{1 \leq i < h \leq n+1} x_{ij} \wedge x_{hj})$, for $1 \leq j \leq n$. The conjunction of these $n+m+1$ formulae is unsatisfiable, since from a satisfying assignment for the variables one could calculate a way of distributing the $n+1$ items into the n bins, which does not exist. SAT solvers decide if a given Boolean formula is *satisfiable* or not, and in the positive case calculate a satisfying assignment. Although the satisfiability problem is **NP**-complete, there are now solvers that can solve real-world instances that involve hundreds of thousands of variables.[5]

Exercise 12.7. Formulate the pigeonhole principle for $n+1$ items and n bins as an integer linear program.

---

[5] See www.satcompetition.org/.

SAT instances can be solved in parallel using the *portfolio approach* – run p SAT solvers in parallel and (almost) independently. Two measures are needed to make this successful. *Diversification* tries to make these SAT solvers to behave differently. For example, we can randomize decisions and vary tuning parameters. Furthermore, information "learned" during the search has to be *exchanged* in a judicious way. The resulting speedups fluctuate wildly – between occasional slowdown and huge superlinear speedup. This naturally leads to some debate about how effective this is. However, several natural ways to average speedups seem to indicate that the portfolio approach scales surprisingly well up to at least hundreds of PEs [28].

The portfolio approach can also be used for other systematic-search problems (see also Sect. 12.4); for example for solving ILPs or for other logical inference problems.

12.2 Greedy Algorithms – Never Look Back

The term *greedy algorithm* is used for a problem-solving strategy where the items under consideration are inspected in some order, usually some carefully chosen order. When an item is considered, a decision about this item is made; for example, whether it is included into the solution. Decisions are never reversed. The algorithm for the fractional knapsack problem given in the preceding section follows the greedy strategy; we consider the items in decreasing order of profit density. The algorithms for shortest paths in acyclic graphs and for the case of nonnegative edge weights in Sects. 10.2 and 10.3 and those for minimum spanning trees in Chap. 11 also follow the greedy strategy. For the single-source shortest-path problem with nonnegative edge weights, we considered the edges in the order of the tentative distances of their source nodes. For the latter two problems, the greedy approach led to an optimal solution.

Usually, greedy algorithms yield only suboptimal solutions. Let us consider the knapsack problem again. The greedy approach from above scans the items in order of decreasing profit density and includes items that will still fit into the knapsack. We shall give this algorithm the name *greedy*. Figures 12.1 and 12.3 give examples. Observe that *greedy* always gives solutions at least as good as *roundDown*

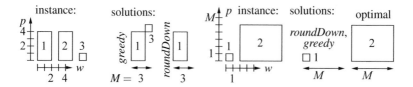

Fig. 12.3. Two instances of the knapsack problem. *Left*: For $p = (4, 4, 1)$, $w = (2, 2, 1)$, and $M = 3$, *greedy* performs better than *roundDown*. *Right*: For $p = (1, M - 1)$ and $w = (1, M)$, both *greedy* and *roundDown* are far from optimal.

gives. Once *roundDown* encounters an item that it cannot include, it stops. However, *greedy* keeps on looking and often succeeds in including additional items of less weight. Although the example in Fig. 12.1 gives the same result for both *greedy* and *roundDown*, the results generally *are* different. For example, consider the two instances in Fig. 12.3. With profits $p = (4,4,1)$, weights $w = (2,2,1)$, and $M = 3$, *greedy* includes the first and third items, yielding a profit of 5, whereas *roundDown* includes just the first item and obtains only a profit of 4. Both algorithms may produce solutions that are far from optimal. For example, for any capacity M, consider the two-item instance with profits $p = (1, M - 1)$ and weights $w = (1, M)$. Both *greedy* and *roundDown* include only the first item, which has a profit density slightly higher than that of the second item, but a very small absolute profit. In this case it would be much better to include just the second item.

We can turn this observation into an algorithm, which we call *round*. This computes two solutions: the solution x^d proposed by *roundDown* and the solution x^c obtained by choosing exactly the critical item x_ℓ.[6] It then returns the better of the two.

We can give an interesting performance guarantee for algorithm *round*. It always achieves at least 50% of the profit of the optimal solution. More generally, we say that an algorithm achieves an *approximation ratio* of α if, for all inputs, its solution is at most a factor α worse than the optimal solution.

Theorem 12.4. *The algorithm round achieves an approximation ratio of 2.*

Proof. Consider an instance p, w, x. Let x^* denote any optimal solution, and let x^f be the optimal solution for the same input when fractional solutions are admitted. Then $p \cdot x^* \leq p \cdot x^f$. The value of the objective function is increased by including the critical item, i.e., by setting $x_\ell = 1$ in the fractional solution. We obtain

$$p \cdot x^* \leq p \cdot x^f \leq p \cdot x^d + p \cdot x^c \leq 2 \max \left\{ p \cdot x^d, p \cdot x^c \right\},$$

and hence the profit achieved by algorithm *round* is at least half the optimum.   □

There are many ways to refine algorithm *round* without sacrificing this approximation guarantee. We can replace x^d by the greedy solution. We can similarly augment x^c with any greedy solution for a smaller instance where item j is removed and the capacity is reduced by w_j.

We now turn to another important class of optimization problems, called *scheduling problems*. Consider the following scenario, known as the *scheduling problem for independent weighted jobs on identical machines*. We are given m identical machines, on which we want to process n jobs; the execution of job j takes t_j time units. An assignment $x : 1..n \rightarrow 1..m$ of jobs to machines is called a *schedule*. Thus the load ℓ_j assigned to machine j is $\sum_{i:\, x(i)=j} t_i$. The goal is to minimize the *makespan* $L_{\max} = \max_{1 \leq j \leq m} \ell_j$ of the schedule.

---

[6] We assume here that "unreasonably large" items with $w_i > M$ have been removed from the problem in a preprocessing step.

This is a fundamental load-balancing problem for parallel processing and we shall consider it further in Chap. 14. One application scenario is as follows. We have a video game processor with several identical processor cores. The jobs are the tasks executed in a video game such as audio processing, preparing graphics objects for the image-processing unit, simulating physical effects, and simulating the intelligence of the game. The makespan will then determine the required time between two time steps of the game and hence the frame rate at which changes in the game can be displayed. Users of a game expect a frame rate which guarantees pleasant viewing.

We give next a simple greedy algorithm for the problem above [136] that has the additional property that it does not need to know the sizes of the jobs in advance. We can even assume that the jobs are presented one after the other, and we assign them on the basis of the knowledge we have so far. Algorithms with this property ("unknown future") are called *online* algorithms. When job i arrives, we assign it to the machine with the smallest load. Formally, we compute the loads $\ell_j = \sum_{h<i \wedge x(h)=j} t_h$ of all machines j and assign the new job to the least loaded machine, i.e., $x(i) := j_i$, where j_i is such that $\ell_{j_i} = \min_{1 \le j \le m} \ell_j$. This *shortest-queue algorithm* does not guarantee optimal solutions, but always computes nearly optimal solutions.

Theorem 12.5. *The list-scheduling algorithm ensures*

$$L_{\max} \le \frac{1}{m} \sum_{i=1}^{n} t_i + \frac{m-1}{m} \max_{1 \le i \le n} t_i.$$

Proof. In the schedule generated by the shortest-queue algorithm, some machine j^* has a load $L_{\max}$. We focus on the last job i^* that is assigned to machine j^*. When job i^* is assigned to j^*, all m machines have a load of at least $L_{\max} - t_{i^*}$, i.e.,

$$\sum_{i \ne i^*} t_i \ge (L_{\max} - t_{i^*}) \cdot m.$$

Solving this for $L_{\max}$ yields

$$L_{\max} \le \frac{1}{m} \sum_{i \ne i^*} t_i + t_{i^*} = \frac{1}{m} \sum_i t_i + \frac{m-1}{m} t_{i^*} \le \frac{1}{m} \sum_{i=1}^{n} t_i + \frac{m-1}{m} \max_{1 \le i \le n} t_i. \qquad \square$$

We next observe that $\frac{1}{m} \sum_i t_i / m$ and $\max_i t_i$ are lower bounds on the makespan of any schedule and hence also of the optimal schedule. We obtain the following corollary.

Corollary 12.6. *The approximation ratio of the shortest-queue algorithm is $2 - 1/m$.*

Proof. Let $L_1 = \frac{1}{m} \sum_i t_i$ and $L_2 = \max_i t_i$. The makespan L^* of the optimal solution is at least $\max(L_1, L_2)$. The makespan of the shortest-queue solution is bounded by

$$L_1 + \frac{m-1}{m} L_2 \le L^* + \frac{m-1}{m} L^* = \left(2 - \frac{1}{m}\right) L^*. \qquad \square$$

The shortest-queue algorithm is not better than claimed above. Consider an instance with $n = m(m-1)+1, t_i = 1$ for $i = 1,\ldots,n-1$, and $t_n = m$. The optimal solution has a makespan $L^* = m$, whereas the shortest-queue algorithm produces a solution with a makespan $L_{max} = 2m - 1$. The shortest-queue algorithm is an online algorithm. It produces a solution which is at most a factor $2 - 1/m$ worse than the solution produced by an algorithm that knows the entire input. In such a situation, we say that the online algorithm has a *competitive ratio* of $\alpha = 2 - 1/m$.

*Exercise 12.8. Show that the shortest-queue algorithm achieves an approximation ratio of $4/3$ if the jobs are sorted by decreasing size.

*Exercise 12.9 (bin packing). Suppose a smuggler has perishable goods in her cellar. She has to hire enough porters to ship all items tonight. Develop a greedy algorithm that tries to minimize the number of people she needs to hire, assuming that each one can carry a weight M. Try to obtain an approximation ratio for your *bin packing* algorithm.

Parallel Greedy Algorithms

In general, greedy algorithms are difficult to parallelize. However, we may be able to exploit special properties of the application. When the priorities of the items can be computed up front, we can order them using parallel sorting. Perhaps we can even compute the decisions in parallel, for example using a prefix sum. This was the case for the greedy knapsack algorithm, where the decision just depends on the total weight of the preceding objects.

Sometimes at least a certain subset of items can be processed independently. We saw an example in the parallel DAG traversal algorithm in Sect. 9.4 and for parallel shortest paths in Sect. 10.9. If all else fails, we can also relax the ordering of the objects, approximating a greedy algorithm. We have seen an example in Sect. 10.9 for the Δ-stepping algorithm for shortest paths.

12.3 Dynamic Programming – Build It Piece by Piece

The first idea in dynamic programming is to expand a given problem instance into a system of auxiliary instances, which are called *subproblems*. These subproblems are then solved systematically. When solving a subproblem, one assumes that the solutions for all its subproblems have been computed before. The second idea is that for many optimization problems the following *principle of optimality* holds: An *optimal* solution for a subproblem is composed of *optimal* solutions for some of its subproblems. If a subproblem has several optimal solutions, it does not matter which one is used. A recurring phenomenon is that there is a choice among the possible sets of subproblems that are to be combined to obtain an optimal solution. An algorithm that follows the principle of dynamic programming uses the optimality principle to systematically build up a table of optimal solutions for all subproblems. Finally, an optimal solution for the original problem instance is constructed from the solutions for the subproblems.

Again, we shall use the knapsack problem as an example. Consider an instance I, consisting of weight vector w, profit vector c, and capacity M. For each $i \in 0..n$ and each C between 0 and M we define a *subproblem*, as follows: $P(i,C)$ is the maximum profit possible when only items 1 to i can be put in the knapsack and the total weight is at most C. Our goal is to compute $P(n,M)$. (We shall see below that solutions for all these subproblems can be used for calculating an optimal selection.) We start with trivial cases and work our way up. The trivial cases are "no items" and "total weight 0". In both of these cases, the maximum profit is 0. So

$$P(0,C) = 0 \text{ for all } C \quad \text{and} \quad P(i,0) = 0 \text{ for all } i.$$

Consider next the case $i > 0$ and $C > 0$. In the solution that maximizes the profit, we either use item i or do not use it. In the latter case, the maximum achievable profit is $P(i-1,C)$. In the former case, the maximum achievable profit is $P(i-1,C-w_i)+p_i$, since we obtain a profit of p_i for item i and must use an optimal solution for the first $i-1$ items under the constraint that the total weight is at most $C - w_i$. Of course, the former alternative is only feasible if $C \geq w_i$. We summarize this discussion in the following recurrence for $P(i,C)$:

$$P(i,C) = \begin{cases} 0, & \text{if } i = 0 \text{ or } C = 0, \\ \max(P(i-1,C),P(i-1,C-w_i)+p_i) & \text{if } i \geq 1 \text{ and } w_i \leq C, \quad (12.1) \\ P(i-1,C) & \text{if } i \geq 1 \text{ and } w_i > C. \end{cases}$$

Exercise 12.10. Show that the case distinction for w_i in the definition of $P(i,C)$ can be avoided by defining $P(i,C) = -\infty$ for $C < 0$.

Using the above recurrence, we can compute $P(n,M)$ by filling a table P with one column for each possible capacity C and one row for each item set $1..i$. Table 12.1 gives an example. There are many possible orders in which to fill out the table, for example row by row. In order to construct an optimal solution for the original instance from the table, we work our way backwards, starting with entry $P(n,M)$ at the bottom right-hand corner of the table. We set $i = n$ and $C = M$. If $P(i,C) = P(i-1,C)$, we set $x_i = 0$ and continue to row $i-1$ and column C. Otherwise, we set $x_i = 1$. We have $P(i,C) = P(i-1,C-w_i)+p_i$, and therefore we continue to row $i-1$ and column $C - w_i$. We continue in this fashion until we arrive at row 0. At this point $(x_1,\ldots,x_n)$ is an optimal solution for the original knapsack instance.

Exercise 12.11. The dynamic programming algorithm for the knapsack problem, as just described, needs to store a table containing $\Theta(nM)$ integers. Give a more space-efficient solution that at any given time stores only a single bit in each table entry as well as two full rows of table P. What information is stored in the bit? How is the information in the bits used to construct a solution? Can you get down to storing the bits and only *one* full row of the table? Hint: Exploit the freedom you have in the order of filling in table values.

Table 12.1. A dynamic-programming table for the knapsack instance with $p = (10, 20, 15, 20)$, $w = (1, 3, 2, 4)$, and $M = 5$. Entries that are inspected when the optimal solution is constructed are in **boldface**.

	C = 0	1	2	3	4	5
i = 0	0	0	0	0	0	0
1	**0**	10	10	10	10	10
2	0	10	10	**20**	30	30
3	0	10	15	25	30	**35**
4	0	10	15	25	30	**35**

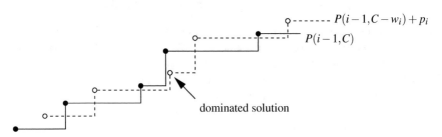

Fig. 12.4. The step function $C \mapsto P(i-1,C)$ is drawn with a solid line, and the step function $C \mapsto P(i-1, C-w_i) + p_i$ with a dashed line. The function $P(i,C)$ is the pointwise maximum of these two functions. The solid step function is stored as the sequence of solid points. The representation of the dashed step function is obtained by adding (w_i, p_i) to every solid point. The representation of $C \mapsto P(i,C)$ is obtained by merging the two representations and deleting all dominated elements.

We shall next describe an important improvement with respect to space consumption and speed. Instead of computing $P(i,C)$ for all i and all C, the *Nemhauser–Ullmann algorithm* [36, 239] computes only *Pareto-optimal* solutions. A solution x is Pareto-optimal if there is no solution that *dominates* it, i.e., has a greater profit and no greater cost or the same profit and less cost. In other words, since $P(i,C)$ is an increasing function of C, only the pairs $(C, P(i,C))$ with $P(i,C) > P(i, C-1)$ are needed for an optimal solution. We store these pairs in a list L_i sorted by the value of C. So $L_0 = \langle (0,0) \rangle$, indicating that $P(0,C) = 0$ for all $C \geq 0$, and $L_1 = \langle (0,0), (w_1, p_1) \rangle$, indicating that $P(1,C) = 0$ for $0 \leq C < w_1$ and $P(1,C) = p_1$ for $C \geq w_1$.

How can we go from L_{i-1} to L_i? The recurrence for $P(i,C)$ paves the way; see Fig. 12.4. We have the list representation L_{i-1} for the function $C \mapsto P(i-1,C)$. We obtain the representation L'_{i-1} for $C \mapsto P(i-1, C-w_i) + p_i$ by shifting every point in L_{i-1} by (w_i, p_i). We merge L_{i-1} and L'_{i-1} into a single list by order of first component and delete all elements that are dominated by another value, i.e., we delete all elements that are preceded by an element with a higher second component, and, for each fixed value of C, we keep only the element with the largest second component.

Exercise 12.12. Give pseudocode for the above merge. Show that the merge can be carried out in time $O(|L_{i-1}|)$. Conclude that the running time of the Nemhauser–Ullmann algorithm is proportional to the number of all Pareto-optimal solutions produced in the course of the algorithm for all i taken together.

Both the basic dynamic-programming algorithm for the knapsack problem and its improved (Nemhauser–Ullmann) version require $\Theta(nM)$ worst-case time. This is quite good if M is not too large. Since the running time is polynomial in n and M, the algorithm is called *pseudo-polynomial*. The "pseudo" means that it is not necessarily polynomial in the *input size* measured in bits; however, it is polynomial in the natural parameters n and M. There is, however, an important difference between the basic and the refined approach. The basic approach has best-case running time $\Theta(nM)$. The best case for the refined approach is $O(n)$. The *average-case* complexity of the refined algorithm is polynomial in n, independent of M. This holds even if the averaging is done only over perturbations of an arbitrary instance by a small amount of random noise. We refer the reader to [36] for details.

Exercise 12.13 (dynamic programming by profit). Assume an instance of the knapsack problem is given. Explore the following alternative way of defining subproblems: Let $W(i,P)$ be the smallest weight bound that makes it possible to achieve a profit of at least P, using knapsack items $1..i$. Obviously we have $W(i,P) = 0$ for $1 \le i \le n$ and all $P \le 0$. Let $W(0,P) = \infty$ for all $P > 0$.

(a) Show that $W(i,P) = \min\{W(i-1,P), W(i-1,P-p_i)+w_i\}$, for $1 \le i \le n$ and $P \ge 0$.
(b) Develop a table-based dynamic-programming algorithm using the above recurrence that computes optimal solutions for the given instance in time $O(np^*)$, where p^* is the profit of an optimal solution. Hint: Assume first that p^*, or at least a good upper bound for it, is known. Then explain how to achieve the goal without this assumption.

Exercise 12.14 (making change). Suppose you have to program a vending machine that should give exact change using a minimum number of coins.

(a) Develop an optimal greedy algorithm that works in the Euro zone with coins worth 1, 2, 5, 10, 20, 50, 100, and 200 cents and in Canada with coins worth (1,) 5, 10, 25, (50,) 100, and 200 cents[7].
(b) Show that this algorithm would not be optimal if there were also a 4 cent coin.
(c) Develop a dynamic-programming algorithm that gives optimal change for any currency system.

Exercise 12.15 (chained matrix products). We want to compute the matrix product $M_1 M_2 \cdots M_n$, where M_i is a $k_{i-1} \times k_i$ matrix. Assume that a pairwise matrix product is computed in the straightforward way using mks element multiplications to obtain the product of an $m \times k$ matrix with a $k \times s$ matrix. Exploit the associativity

---

[7] In Canada, the 50 cent coin is legal tender, but rarely used. Production of 1 cent coins was discontinued in 2012.

of matrix products to minimize the number of element multiplications needed. Use dynamic programming to find an optimal evaluation order in time $O(n^3)$. For example, the product of a 4×5 matrix M_1, a 5×2 matrix M_2, and a 2×8 matrix M_3 can be computed in two ways. Computing $M_1(M_2M_3)$ takes $5 \cdot 2 \cdot 8 + 4 \cdot 5 \cdot 8 = 240$ multiplications, whereas computing $(M_1M_2)M_3$ takes only $4 \cdot 5 \cdot 2 + 4 \cdot 2 \cdot 8 = 104$ multiplications.

Exercise 12.16 (edit distance). The *edit distance* (or *Levenshtein distance*) $L(s,t)$ between two strings s and t is the minimum number of character deletions, insertions, and replacements ("editing steps") one has to apply to s to produce the string t. For example, $L(\text{graph}, \text{group}) = 3$ (delete h, replace a by o, insert u before p). Let n be the length of s and m be the length of t. We define subproblems as follows: $d(i,j) = L(\langle s_1, \ldots, s_i \rangle, \langle t_1, \ldots, t_j \rangle)$, for $0 \leq i \leq n$, $0 \leq j \leq m$. Show that

$$d(i,j) = \min \left\{ d(i-1,j) + 1, d(i,j-1) + 1, d(i-1,j-1) + [s_i \neq t_j] \right\},$$

where $[s_i \neq t_j]$ is 1 if s_i and t_j are different and is 0 otherwise. Use these optimality equations to formulate a dynamic-programming algorithm to calculate the edit distance of $L(s,t)$. What is the running time? Explain how one subsequently finds an optimal sequence of editing steps to transform s into t.

Exercise 12.17. Does the principle of optimality hold for minimum spanning trees? Check the following three possibilities for definitions of subproblems: subsets of nodes, arbitrary subsets of edges, and prefixes of the sorted sequence of edges.

Exercise 12.18 (constrained shortest path). Consider a directed graph $G = (V, E)$ where edges $e \in E$ have a *length* $\ell(e)$ and a *cost* $c(e)$. We want to find a path from node s to node t that minimizes the total length subject to the constraint that the total cost of the path is at most C. Show that subpaths from s' to t' of optimal solutions are *not* necessarily shortest paths from s' to t'.

Parallel Dynamic Programming

Roughly speaking, dynamic-programming algorithms spend most of their time filling table entries. If several of these entries can be computed independently, this can be done in parallel. For example, for the knapsack problem, (12.1) tells us that all entries in row i depend only on row $i-1$ and hence can be computed in parallel.

Exercise 12.19. Explain how to parallelize one step of the Nemhauser–Ullmann algorithm for the knapsack problem using linear work and polylogarithmic time. Hint: You can use parallel merging and prefix sums.

Sometimes, in order to expose parallelism, we need to fill the tables in an order that we may not think of in the sequential algorithm. For example, when computing the edit distance in Exercise 12.16, we get some parallelism if we fill the table $d(\cdot, \cdot)$ along the diagonals, i.e., for fixed k, the entries $d(i, k-i)$ are independent and depend only on the preceding two diagonals.

12.4 Systematic Search – When in Doubt, Use Brute Force

In many optimization problems, for each given instance I the universe $\mathcal{U}$ of possible solutions is finite, so that we can in principle solve the optimization problem by trying all possibilities.[8] Applying this idea in the naive way does not lead very far. For many problems, the "search space" $\mathcal{U}$ grows rapidly with the input size $|I|$. But we can frequently restrict the search to *promising* candidates, and then the concept carries a lot further.

We shall explain the concept of systematic search using the knapsack problem and a specific approach to systematic search known as *branch-and-bound*. In Exercises 12.22 and 12.21, we outline systematic-search routines following a somewhat different pattern.

Branch-and-bound can be used when (feasible) solutions can be represented as vectors (or, more generally, sequences) whose components attain only finitely many values. The set of all these vectors (the *search space*) is searched systematically. Systematic search views the search space as a tree. Figure 12.5 gives pseudocode for a systematic-search routine *bbKnapsack* for the knapsack problem. The tree representing the search space is generated and traversed using a recursive procedure. Fig. 12.6 shows a sample run. A tree node corresponds to a partial solution in which some components have been fixed and the others are still free. Such a node can be regarded as a subinstance. The root is the vector in which all components are free. *Branching* is the elementary step in a systematic-search routine. The children of some node, or subinstance, are generated by inserting all sensible values for some free component of the partial solution. For each of the resulting new nodes the procedure is called recursively. Within the recursive call, the chosen value is fixed. In case of *bbKnapsack* the potential solutions are vectors in $\{0,1\}^n$, and in partial solutions some of the components are fixed, which means that for some objects it has been decided whether to include them or not. In principle an arbitrary component can be chosen to be fixed next. The routine *bbKnapsack* fixes the components x_i one after another in order of decreasing profit density. When treating x_i, it first tries including item i by setting $x_i := 1$, and then excluding it by setting $x_i := 0$. In both cases, components $x_{i+1}, \ldots, x_n$ are treated by recursion. The assignment $x_i := 1$ is not feasible and is not considered if the total weight of the included objects exceeds M. To organize this, a "remaining capacity" M' is carried along, which is reduced by the weight of any object when it is included. So the choice $x_i := 1$ is left out if $w_i > M'$. With these definitions, after all variables have been set, in the nth level of the recursion, *bbKnapsack* will have found a feasible solution. Indeed, without the *bounding rule*, which is discussed below, the algorithm would systematically explore a tree of partial solutions whose leaves correspond to all feasible solutions. *Branching* occurs at nodes at which both subtrees corresponding to the choices $x_i := 0$ and $x_i := 1$ are explored. Because of the order in which the components are fixed, the first feasible solution encountered would be the solution found by the algorithm *greedy*.

---

[8] The sentence "When in doubt, use brute force." is attributed to Ken Thompson.

Function $bbKnapsack((p_1,\ldots,p_n),(w_1,\ldots,w_n),M) : \{0,1\}^n$
    assert $p_1/w_1 \geq p_2/w_2 \geq \cdots \geq p_n/w_n$ // assume input sorted by profit density
    $\hat{x} = heuristicKnapsack((p_1,\ldots,p_n),(w_1,\ldots,w_n),M) : \{0,1\}^n$ // best solution so far
    $x : \{0,1\}^n$ // current partial solution $(x_1,\ldots,x_n)$
    $recurse(1,M,0)$
    return $\hat{x}$

    // *recurse* finds solutions assuming $x_1,\ldots,x_{i-1}$ are fixed, $M' = M - \sum_{j<i} x_j w_j, P = \sum_{j<i} x_j p_j$.
    Procedure $recurse(i,M',P : \mathbb{N})$
        $u := P + upperBound((p_i,\ldots,p_n),(w_i,\ldots,w_n),M')$
        if $u > p \cdot \hat{x}$ **then**   // possibly better solution than $\hat{x}$
           if $i > n$ **then** $\hat{x} := x$
           else   // branch on variable x_i
                if $w_i \leq M'$ **then** $x_i := 1; recurse(i+1,M'-w_i,P+p_i)$
                if $u > p \cdot \hat{x}$ **then** $x_i := 0; recurse(i+1,M',P)$

Fig. 12.5. A branch-and-bound algorithm for the knapsack problem. An initial feasible solution is constructed by the function *heuristicKnapsack* using some heuristic algorithm. The function *upperBound* computes an upper bound for the possible profit.

    Bounding is a method for pruning subtrees that cannot contain better solutions than those already known. A branch-and-bound algorithm keeps the best feasible solution found so far in a global variable, in our case the variable $\hat{x}$; this solution is often called the *incumbent* solution. It is initialized to a trivial solution or one determined by a heuristic routine. At all times, the value $p \cdot \hat{x}$ provides a lower bound on the value of the objective function that can be obtained. Whenever a partial solution x is processed, this lower bound is complemented by an upper bound u for the value of the objective function obtainable by extending x to a full feasible solution. In our example, the upper bound could be the the sum of the profit of the partial solution and the profit for the fractional knapsack problem with items $i..n$ and capacity $M' = M - \sum_{j<i} x_j w_j$. Branch-and-bound stops expanding the current branch of the search tree when $u \leq p \cdot \hat{x}$, i.e., when there is no hope of finding a better solution in the subtree rooted at the current node.

Exercise 12.20. Explain how to implement the function *upperBound* in Fig. 12.5 so that it runs in time $O(\log n)$. Hint: Precompute the prefix sums $\sum_{j \leq i} w_i$ and $\sum_{j \leq i} p_i$, for $1 \leq i \leq n$, and use binary search.

Exercise 12.21 (constraint programming and the eight-queens problem). Consider a chessboard. The task is to place eight queens on the board so that they do not attack each other, i.e., no two queens should be placed in the same row, column, diagonal, or antidiagonal. So each row contains exactly one queen. Let x_i be the position of the queen in row i. Then $x_i \in 1..8$. The solution must satisfy the following constraints: $x_i \neq x_j$, $i + x_i \neq j + x_j$, and $x_i - i \neq x_j - j$ for $1 \leq i < j \leq 8$. What do these conditions express? Show that they are sufficient. A systematic search can use

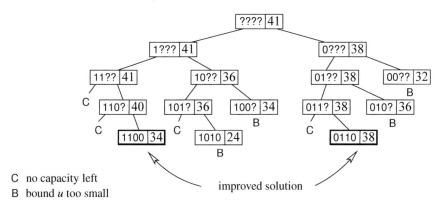

Fig. 12.6. The part of the search space explored by *bbKnapsack* for a knapsack instance with $p = (10, 24, 14, 24)$, $w = (1, 3, 2, 4)$, and $M = 5$ is processed. As initial solution we use the empty $\hat{x} = (0, 0, 0, 0)$ with profit 0. The function *upperBound* is computed by rounding down the optimal value of the objective function for the fractional knapsack problem. The nodes of the search tree contain the components $x_1 \cdots x_{i-1}$ fixed so far and the upper bound u. Left subtrees are explored first; they correspond to setting $x_i := 1$. There are two for not exploring the subtrees of a nonleaf node: Either there is not enough capacity left for the choice $x_i := 1$ (indicated by "C") or it turns out that a feasible solution with a profit at least as large as the upper bound is already known (indicated by "B", the recursive call does not have an effect).

the following optimization: When a variable x_i is fixed at some value, this excludes some values for variables that are still free. Modify the systematic search so that it keeps track of the values that are still available for free variables. Stop exploration as soon as there is a free variable that has no value available to it anymore. This technique of eliminating values is basic to *constraint programming*.

Exercise 12.22 (the 15-puzzle).
   The 15-puzzle is a popular sliding-block puzzle. Fifteen square tiles marked with numbers 1, 2, ..., 15, sit in a 4×4 frame. This leaves one "hole". For an example see the top of the figure on the left. You are supposed to move the tiles into the right order, by performing *moves*. A move is defined as the action of interchanging a square and the hole in the array of tiles. Design an algorithm that finds a shortest sequence of moves from a given starting configuration to the ordered configuration shown at the bottom of the figure on the left. Assume that there is a solution. Use *iterative deepening depth-first search* [189]: Try all one-move sequences first, then all two-move sequences, and so on. This should work for the simpler 8-puzzle, with eight tiles in a 3×3 frame. For the 15-puzzle, use the following optimizations: Never undo the immediately preceding move. Use the number of moves that would be needed if all pieces could move freely to their target position as a lower bound, and stop exploring a subtree if this bound proves that the current search depth is too small. Decide beforehand whether the number of moves is odd or even. Implement your algorithm to run in constant time per move tried.

12.4.1 Solving Integer Linear Programs

In Sect. 12.1.2, we have seen how to formulate the knapsack problem as a 0-1 ILP. We shall now indicate how the branch-and-bound procedure developed for the knapsack problem can be applied to any 0-1 ILP. Recall that in a 0-1 ILP the values of the variables are constrained to 0 and 1. Our discussion will be brief, and we refer the reader to a textbook on integer linear programming [240, 287] for more information.

The main change is that the function *upperBound* now solves a general linear program that has variables $x_i, \ldots, x_n$ with range $[0,1]$. The constraints for this LP come from the input ILP, with the variables x_1 to x_{i-1} replaced by their values. In the remainder of this section, we shall simply refer to this linear program as "the LP".

If the LP has a feasible solution, *upperBound* returns the optimal value for the LP. If the LP has no feasible solution, *upperBound* returns $-\infty$ so that the ILP solver will stop exploring this branch of the search space. We shall next describe several generalizations of the basic branch-and-bound procedure that sometimes lead to considerable improvements:

Branch selection. We may pick any unfixed variable x_j for branching. In particular, we can make the choice depend on the solution of the LP. A commonly used rule is to branch on a variable whose fractional value in the LP is closest to $1/2$.

Order of search tree traversal. In the knapsack example, the search tree was traversed depth-first, and the 1-branch was tried first. In general, we are free to choose any order of tree traversal. There are at least two considerations influencing the choice of strategy. If no good feasible solution is known, it is good to use a depth-first strategy so that complete solutions are explored quickly. Otherwise, it is better to use a *best-first* strategy that explores those search tree nodes that are most likely to contain good solutions. Search tree nodes are kept in a priority queue, and the next node to be explored is the most promising node in the queue. The priority could be the upper bound returned by the LP. However, since the LP is expensive to evaluate, one sometimes settles for an approximation.

Finding solutions. We may be lucky in that the solution of the LP turns out to assign integer values to all variables. In this case there is no need for further branching. Application-specific heuristics can additionally help to find good solutions quickly.

Branch-and-cut. When an ILP solver branches too often, the size of the search tree explodes and it becomes too expensive to find an optimal solution. One way to avoid branching is to add constraints to the linear program that *cut* away solutions with fractional values for the variables without changing the solutions with integer values.

12.4.2 *Parallel Systematic Search

Systematic search is time-consuming and hence parallelization is desirable. Since the search procedure tries many solution options, there is considerable potential for parallelization. However, there are also challenges:

- The computations can be very fine-grained. This means that we may not be able to afford communication or other PE interactions for every individual computation.
- It is difficult to predict the total execution time involved in a subtree of the search. Hence, load balancing has to be dynamic.
- The computations are not completely independent. For example, in Fig. 12.5, when a new best solution is found, this may prune subtrees currently being explored by other PEs. Hence, this information has to be disseminated quickly. This also implies that parallel processing may explore more candidates than the sequential code does.

We first outline how to parallelize the branch-and-bound framework for ILP presented in Sect. 12.4.1 on a shared-memory system. Then we describe in more detail a distributed-memory parallelization of the branch-and-bound solver for the knapsack problem shown in Fig. 12.5. The basic idea is simple. When branching on a variable x, we can spawn several tasks – one for each value tried. This gives only very limited parallelism for a single variable, but by recursively applying the same idea to the spawned tasks, we quickly get a large set of tasks that can all be executed in parallel. An attractive way to manage the tasks is to use a bulk parallel priority queue (see Sect. 6.4). We may want to mix this with a depth-first strategy – occasionally, a task switches to depth-first search in order to obtain concrete solutions. The best solution value found so far should be a global variable that can be used to prune tasks that cannot possibly yield an improvement.

For the knapsack solver in Fig. 12.5, this best-first approach will not work well – even creating a task already costs time $\Omega(n)$, whereas we have seen in Exercise 12.20 that exploring a single node in the depth-first search approach can be done in logarithmic time. On the other hand, strict depth-first search is inherently sequential. The work-stealing approach described in Sect. 14.5 offers a way out. Working on a problem sequentially means depth-first exploration using backtracking. Splitting a problem means splitting off a part of the explored tree. This is made possible by managing the stack for backtracking explicitly so that splitting amounts to manipulating entries of the stack. A simple and effective splitting strategy is to search the stack top-down for the first item i where the branch $x_i = 0$ has not been tried yet. One of the resulting parts is the old stack, except that the branch $x_i = 0$ is now marked as already tried. The split-off part is a stack which copies the old stack for positions $1..i$ and initializes the search to explore the corresponding subtree with $x_i = 0$. Since each split sets one x_i to 0, the splitting depth from Sect. 14.5 can become at most n. If the search really requires work exponential in n, this means that the splitting depth is only logarithmic in the work to be done.

In order to do bounding effectively, the value $p \cdot \hat{x}$ has to be globally known. On a shared-memory machine we simply would use a global variable that is atomically updated whenever a larger value is found. On a distributed-memory machine, we need to emulate that approach in a scalable way. This is possible by embedding a spanning tree of bounded degree (e.g., maximum degree 3) and diameter $O(\log p)$ into the processor network. A PE that finds or receives an improved value forwards it to those neighbors in the tree which may not have seen that value.[9] This approach ensures that a new bound spreads over all PEs in time $O(\log p)$ without causing significant contention anywhere.

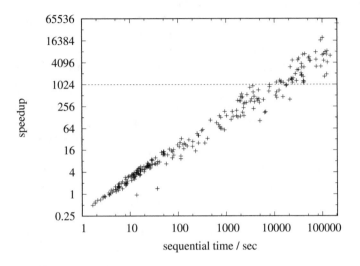

Fig. 12.7. Speedup as function of sequential execution time for 256 random instances with $p = 1024$, $n = 2000$, $w_i \in [0.01, 1.01]$, $p_i \in [w_i + 0.125,]$, and $M = \sum_i w_i/2$. See [268] for more details.

What speedup can we expect from this parallel knapsack searcher? When we use it just to validate an already known optimal solution $\hat{x}$, the analysis in Sect. 14.5 with a splitting depth and communication cost of $O(n)$ yields $O(T_{\text{seq}}/p + n^2)$, i.e., linear speedup for sufficiently difficult instances. For finding a solution, no such analysis is known. One might expect that the efficiency would be somewhat lower than for validating a solution since the delay in propagating new bounds will cause superfluous computations. However, superlinear speedup over the Algorithm in Fig. 12.5 for particular instances is also possible: Some PE might find the optimal solution very early, causing parts of the search space of the sequential algorithm to be pruned. Figure 12.7 shows speedups for 256 random difficult instances on 1024 PEs (on a rather slow machine even for the mid-1990s). As is to be expected, for instances solved quickly by the sequential solver, little speedup is obtained. However, for the most difficult instances, considerable superlinear speedup is observed. The sum of the sequential execution times is 1410 times larger than the sum of the parallel execution times, i.e., in some sense, the overall speedup is superlinear. This is surprising but not uncommon in parallel exploration of search spaces. Parallel search is more robust

---

[9] A similar, somewhat more complicated mechanism involving rooted trees is described in [172, 268].

than sequential search, since the sequential exploration of the search space might get bogged down in a large fruitless part of the search space.

For an even simpler approach to parallel systematic search that does not explicitly split the search space, refer to the discussion of portfolio-based SAT solvers (Page 365).

12.5 Local Search – Think Globally, Act Locally

The optimization algorithms we have seen so far are applicable only in special circumstances. Dynamic programming needs a special structure of the problem and may require a lot of space and time. Systematic search is usually too slow for large inputs. Greedy algorithms are fast but often yield only low-quality solutions. *Local search* is a widely applicable iterative procedure. It starts with some feasible solution and then moves from feasible solution to feasible solution by local modifications. Figure 12.8 gives the basic framework. We shall refine it later.

Local search maintains a current feasible solution x and the best solution $\hat{x}$ seen so far. In each step, local search moves from the current solution to a neighboring solution. What are neighboring solutions? Any solution that can be obtained from the current solution by making small changes to it. For example, in the case of the knapsack problem, we might remove up to two items from the knapsack and replace them by up to two other items. The precise definition of the neighborhood depends on the application and the algorithm designer. We use $\mathcal{N}(x)$ to denote the *neighborhood* of x. The second important design decision is which solution is chosen from the neighborhood. Finally, some heuristic decides when the search should stop.

In the rest of this section, we shall tell you more about local search.

12.5.1 Hill Climbing

Hill climbing is the greedy version of local search. It moves only to neighbors that are better than the currently best solution. This restriction further simplifies the local search. The variables $\hat{x}$ and x are the same, and we stop when there are no improved solutions in the neighborhood $\mathcal{N}$. The only nontrivial aspect of hill climbing is the choice of the neighborhood. We shall give two examples where hill climbing works quite well, followed by an example where it fails badly.

find some feasible solution $x \in \mathcal{L}$
$\hat{x} := x$ *// $\hat{x}$ is best solution found so far*
while not satisfied with $\hat{x}$ **do**
    $x :=$ some heuristically chosen element from $\mathcal{N}(x) \cap \mathcal{L}$
    if $f(x) > f(\hat{x})$ **then** $\hat{x} := x$

Fig. 12.8. Local search

Our first example is the traveling salesman problem described in Sect. 11.7.2. Given an undirected graph and a distance function on the edges satisfying the triangle inequality,
the goal is to find a shortest tour that visits all nodes of the
graph. We define the neighbors of a tour as follows. Let (u,v)
and (w,y) be two edges of the tour, i.e., the tour has the form
$(u,v), p, (w,y), q$, where p is a path from v to w and q is a
path from y to u. We remove these two edges from the tour,
and replace them by the edges (u,w) and (v,y). The new tour
first traverses (u,w), then uses the reversal of p back to v,
then uses (v,y), and finally traverses q back to u; see the Figure on the right for an illustration. This move is known as
a 2-exchange, and a tour that cannot be improved by a 2-
exchange is said to be 2-optimal. In many instances of the
traveling salesman problem, 2-optimal tours come quite close to optimal tours.

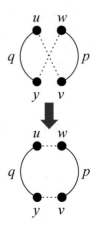

Exercise 12.23. Describe a scheme where three edges are removed and replaced by
new edges.

An interesting example of hill climbing with a clever choice of the neighborhood
function is the *simplex algorithm* for linear programming (see Sect. 12.1). This is
the most widely used algorithm for linear programming. The set of feasible solutions $\mathcal{L}$ of a linear program is defined by a set of linear equalities and inequalities
$a_i \cdot x \bowtie b_i$, $1 \le i \le m$. The points satisfying a linear equality $a_i \cdot x = b_i$ form a *hyperplane* in R^n, and the points satisfying a linear inequality $a_i \cdot x \le b_i$ or $a_i \cdot x \ge b_i$ form a
half-space. Hyperplanes are the n-dimensional analogues of planes and half-spaces
are the analogues of half-planes. The set of feasible solutions is an intersection of
m half-spaces and hyperplanes and forms a *convex polytope*. We have already seen
an example in two-dimensional space in Fig. 12.2. Figure 12.9 shows an example

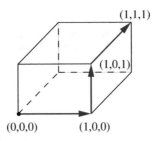

Fig. 12.9. The three-dimensional unit cube is defined by the inequalities $x \ge 0$, $x \le 1$, $y \ge 0$,
$y \le 1$, $z \ge 0$, and $z \le 1$. At the vertices $(1,1,1)$ and $(1,0,1)$, three inequalities are tight, and on
the edge connecting these vertices, the inequalities $x \le 1$ and $z \le 1$ are tight. For the objective
"maximize $x+y+z$", the simplex algorithm starting at $(0,0,0)$ may move along the path
indicated by the arrows. The vertex $(1,1,1)$ is optimal, since the half-space $x+y+z \le 3$
contains the entire feasible region and has $(1,1,1)$ in its boundary.

in three-dimensional space. Convex polytopes are the n-dimensional analogues of convex polygons. In the interior of the polytope, all inequalities are strict (= satisfied with inequality); on the boundary, some inequalities are tight (= satisfied with equality). The vertices and edges of the polytope are particularly important parts of the boundary. We shall now sketch how the simplex algorithm works. We assume that there are no equality constraints. Observe that an equality constraint c can be solved for any one of its variables; this variable can then be removed by substituting into the other equalities and inequalities. Afterwards, the constraint c is redundant and can be dropped.

The simplex algorithm starts at an arbitrary vertex of the feasible region. In each step, it moves to a neighboring vertex, i.e., a vertex reachable via an edge, with a larger objective value. If there is more than one such neighbor, a common strategy is to move to the neighbor with the largest objective value. If there is no neighbor with a larger objective value, the algorithm stops. *At this point, the algorithm has found the vertex with the maximum objective value.* In the examples in Figs. 12.2 and 12.9, the captions argue why this is true. The general argument is as follows. Let x^* be the vertex at which the simplex algorithm stops. The feasible region is contained in a cone with apex x^* and spanned by the edges incident to x^*. All these edges go to vertices with smaller objective values and hence the entire cone is contained in the half-space $\{x : c \cdot x \le c \cdot x^*\}$. Thus no feasible point can have an objective value larger than x^*. We have described the simplex algorithm as a walk on the boundary of a convex polytope, i.e., in geometric language. It can be described equivalently using the language of linear algebra. Actual implementations use the linear-algebra description.

In the case of linear programming, hill climbing leads to an optimal solution. In general, however, hill climbing will not find an optimal solution. In fact, it will not even find a near-optimal solution. Consider the following example. Our task is to find the highest point on earth, i.e., Mount Everest. A feasible solution is any point on earth. The local neighborhood of a point is any point within a distance of 10 km. So the algorithm would start at some point on earth, then go to the highest point within a distance of 10 km, then go again to the highest point within a distance of 10 km, and so on. If one were to start from the second author's home (altitude 206 meters), the first step would lead to an altitude of 350 m, and there the algorithm would stop, because there is no higher hill within 10 km of that point. There are very few places in the world where the algorithm would continue for long, and even fewer places where it would find Mount Everest.

Why does hill climbing work so nicely for linear programming, but fail to find Mount Everest? The reason is that the earth has many local optima, hills that are the highest point within a range of 10 km. In contrast, a linear program has only one local optimum (which then, of course, is also a global optimum). For a problem with many local optima, we should expect *any* generic method to have difficulties. Observe that increasing the size of the neighborhoods in the search for Mount Everest does not really solve the problem, except if the neighborhoods are made to cover the entire earth. But finding the optimum in a neighborhood is then as hard as the full problem.

12.5.2 Simulated Annealing – Learning from Nature

If we want to ban the bane of local optima in local search, we must find a way to escape from them. This means that we sometimes have to accept moves that decrease the objective value. What could "sometimes" mean in this context? We have contradictory goals. On the one hand, we must be willing to take many downhill steps so that we can escape from wide local optima. On the other hand, we must be sufficiently target-oriented in order to find a global optimum at the end of a long, narrow ridge. A very popular and successful approach to reconciling these contradictory goals is *simulated annealing*; see Fig. 12.10. This works in phases that are controlled by a parameter T, called the *temperature* of the process. We shall explain below why the language of physics is used in the description of simulated annealing. In each phase, a number of moves are made. In each move, a neighbor $x' \in \mathcal{N}(x) \cap \mathcal{L}$ is chosen uniformly at random, and the move from x to x' is made with a certain probability. This probability is 1 if x' improves upon x. It is less than 1 if the move is to an inferior solution. The trick is to make the probability depend on T. If T is large, we make the move to an inferior solution relatively likely; if T is close to 0, we make such a move relatively unlikely. The hope is that, in this way, the process zeroes in on a region containing a good local optimum in phases of high temperature and then actually finds a near-optimal solution in the phases of low temperature. The exact choice of the transition probability in the case where x' is an inferior solution is given by $\exp((f(x') - f(x))/T)$. Observe that T is in the denominator and that $f(x') - f(x)$ is negative. So the probability decreases with T and also with the absolute loss in objective value.

find some feasible solution $x \in \mathcal{L}$
$T :=$ some positive value   // initial temperature of the system
while T is still sufficiently large **do**
      perform a number of steps of the following form
          pick x' from $\mathcal{N}(x) \cap \mathcal{L}$ uniformly at random
          with probability $\min(1, \exp(\frac{f(x') - f(x)}{T}))$ **do** $x := x'$
      decrease T // make moves to inferior solutions less likely

Fig. 12.10. Simulated annealing

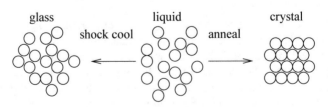

Fig. 12.11. Annealing versus shock cooling

Why is the language of physics used, and why this apparently strange choice of transition probabilities? Simulated annealing is inspired by the physical process of *annealing*, which can be used to minimize[10] the global energy of a physical system. For example, consider a pot of molten silica (SiO_2); see Fig. 12.11. If we cool it very quickly, we obtain a glass – an amorphous substance in which every molecule is in a local minimum of energy. This process of shock cooling has a certain similarity to hill climbing. Every molecule simply drops into a state of locally minimum energy; in hill climbing, we accept a local modification of the state if it leads to a smaller value of the objective function. However, a glass is not a state of global minimum energy. A state of much lower energy is reached by a quartz crystal, in which all molecules are arranged in a regular way. This state can be reached (or approximated) by cooling the melt very slowly. This process is called *annealing*. How can it be that molecules arrange themselves into a perfect shape over a distance of billions of molecular diameters although they feel only local forces extending over a few molecular diameters?

Qualitatively, the explanation is that local energy minima have enough time to dissolve in favor of globally more efficient structures. For example, assume that a cluster of a dozen molecules approaches a small perfect crystal that already consists of thousands of molecules. Then, with enough time, the cluster will dissolve and its molecules can attach to the crystal. Here is a more formal description of this process, which can be shown to hold for a reasonable model of the system: If cooling is sufficiently slow, the system reaches *thermal equilibrium* at every temperature. Equilibrium at temperature T means that a state x of the system with energy E_x is assumed with probability

$$\frac{\exp(-E_x/T)}{\sum_{y\in\mathscr{L}}\exp(-E_y/T)},$$

where T is the temperature of the system and $\mathscr{L}$ is the set of states of the system. This energy distribution is called the *Boltzmann distribution*. When T decreases, the probability of states with minimum energy grows. In fact, in the limit $T \to 0$, the probability of states with minimum energy approaches 1.

The same mathematics works for abstract systems corresponding to a maximization problem. We identify the cost function f with the energy of the system, and a feasible solution with the state of the system. It can be shown that the system approaches a Boltzmann distribution for a quite general class of neighborhoods and the following rules for choosing the next state:

pick x' from $\mathscr{N}(x)\cap\mathscr{L}$ uniformly at random;
with probability $\min\left(1,\exp((f(x')-f(x))/T)\right)$ **do** $x := x'$.

The physical analogy gives some idea of why simulated annealing might work,[11] but it does not provide an implementable algorithm. We have to get rid of two infinities: For every temperature, we wait infinitely long to reach equilibrium, and we do that for infinitely many temperatures. Simulated-annealing algorithms therefore

---

[10] Note that we are talking about *minimization* now.
[11] Note that we have written "might work" and not "works".

have to decide on a *cooling schedule*, i.e., how the temperature T should be varied over time. A simple schedule chooses a starting temperature T_0 that is supposed to be just large enough that all neighbors are accepted. Furthermore, for a given problem instance, there is a fixed number N of iterations to be used at each temperature. The idea is that N should be as small as possible but still allow the system to get close to equilibrium. After every N iterations, T is decreased by multiplying it by a constant α less than 1. Typically, α is between 0.8 and 0.99. When T has become so small that moves to inferior solutions have become highly unlikely (this is the case when T is comparable to the smallest difference in objective value between any two feasible solutions), T is finally set to 0, i.e., the annealing process concludes with a hill-climbing search.

Better performance can be obtained with *dynamic schedules*. For example, the initial temperature can be determined by starting with a low temperature and increasing it quickly until the fraction of transitions accepted approaches 1. Dynamic schedules base their decision about how much T should be lowered on the actually observed variation in $f(x)$ during the local search. If the temperature change is tiny compared with the variation, it has too little effect. If the change is too close to or even larger than the variation observed, there is a danger that the system will be forced prematurely into a local optimum. The number of steps to be made until the temperature is lowered can be made dependent on the actual number of moves accepted. Furthermore, one can use a simplified statistical model of the process to estimate when the system is approaching equilibrium. The details of dynamic schedules are beyond the scope of this exposition. Readers are referred to [1] for more details on simulated annealing.

Exercise 12.24. Design a simulated-annealing algorithm for the knapsack problem. The local neighborhood of a feasible solution is all solutions that can be obtained by removing up to two elements and then adding up to two elements.

12.5.2.1 Graph Coloring

We shall now exemplify simulated annealing using the *graph-coloring problem* already mentioned in Sect. 2.13. Recall that we are given an undirected graph $G = (V, E)$ and are looking for an assignment $c : V \rightarrow 1..k$ such that no two adjacent nodes are given the same color, i.e., $c(u) \neq c(v)$ for all edges $\{u, v\} \in E$. There is always a solution with $k = |V|$ colors; we simply give each node its own color. The goal is to minimize k. There are many applications of graph coloring and related problems. The most "classical" one is map coloring – the nodes are countries and edges indicate that countries have a common border, and thus those countries should not be rendered in the same color. A famous theorem of graph theory states that all maps (i.e., planar graphs) can be colored with at most four colors [265]. Sudoku puzzles are a well-known instance of the graph-coloring problem, where the player is asked to complete a partial coloring of the graph shown in Fig. 12.12 with the digits 1..9. We shall present two simulated-annealing approaches to graph coloring; many more have been tried.

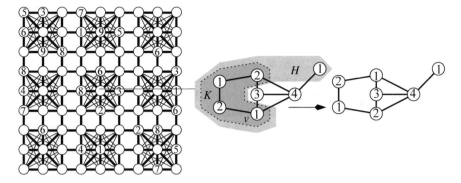

Fig. 12.12. The figure on the *left* shows a partial coloring of the graph underlying sudoku puzzles. The **bold** straight-line segments indicate cliques consisting of all nodes touched by the line. The figure on the *right* shows a step of Kempe chain annealing using colors 1 and 2 and a node v.

Kempe chain annealing. The obvious objective function for graph coloring is the number of colors used. However, this choice of objective function is too simplistic in a local-search framework, since a typical local move will not change the number of colors used. We need an objective function that rewards local changes that are "on a good way" towards using fewer colors. One such function is the sum of the squared sizes of the color classes. Formally, let $C_i = \{v \in V : c(v) = i\}$ be the set of nodes that are colored i. Then

$$f(c) = \sum_i |C_i|^2.$$

This objective function is to be maximized. Observe that the objective function increases when a large color class is enlarged further at the cost of a small color class. Thus local improvements will eventually empty some color classes, i.e., the number of colors decreases.

Having settled the objective function, we come to the definition of a local change or a neighborhood. A trivial definition is as follows: A local change consists in recoloring a single vertex; it can be given any color not used on one of its neighbors. Kempe chain annealing uses a more liberal definition of "local recoloring". Alfred Bray Kempe (1849–1922) was one of the early investigators of the four-color problem; he invented Kempe chains in his futile attempts at a proof. Suppose that we want to change the color $c(v)$ of node v from i to j. In order to maintain feasibility, we have to change some other node colors too: Node v might be connected to nodes currently colored j. So we color these nodes with color i. These nodes might, in turn, be connected to other nodes of color j, and so on. More formally, consider the node-induced subgraph H of G which contains all nodes with colors i and j. The connected component of H that contains v is the *Kempe chain K* we are interested in. We maintain feasibility by swapping colors i and j in K. Figure 12.12 gives an example. Kempe chain annealing starts with any feasible coloring.

*Exercise 12.25. Use Kempe chains to prove that any planar graph G can be colored with five colors. Hint: Use the fact that a planar graph is guaranteed to have a node of degree five or less. Let v be any such node. Remove it from G, and color $G - v$ recursively. Put v back in. If at most four different colors are used on the neighbors of v, there is a free color for v. So assume otherwise. Assume, without loss of generality, that the neighbors of v are colored with colors 1 to 5 in clockwise order. Consider the subgraph of nodes colored 1 and 3. If the neighbors of v with colors 1 and 3 are in distinct connected components of this subgraph, a Kempe chain can be used to recolor the node colored 1 with color 3. If they are in the same component, consider the subgraph of nodes colored 2 and 4. Argue that the neighbors of v with colors 2 and 4 must be in distinct components of this subgraph.

The penalty function approach. A generally useful idea for local search is to relax some of the constraints on feasible solutions in order to make the search more flexible and to ease the discovery of a starting solution. Observe that we have assumed so far that we somehow have a feasible solution available to us. However, in some situations, finding any feasible solution is already a hard problem; the eight-queens problem of Exercise 12.21 is an example. In order to obtain a feasible solution at the end of the process, the objective function is modified to penalize infeasible solutions. The constraints are effectively moved into the objective function.

In the graph-coloring example, we now also allow illegal colorings, i.e., colorings in which neighboring nodes may have the same color. An initial solution is generated by guessing the number of colors needed and coloring the nodes randomly. A neighbor of the current coloring c is generated by picking a random color j and a random node v colored j, i.e., $c(v) = j$. Then, a random new color for node v is chosen from all the colors already in use plus one fresh, previously unused color.

As above, let C_i be the set of nodes colored i and let $E_i = E \cap (C_i \times C_i)$ be the set of edges connecting two nodes in C_i. The objective is to minimize

$$f(c) = 2\sum_i |C_i| \cdot |E_i| - \sum_i |C_i|^2.$$

The first term penalizes illegal edges; each illegal edge connecting two nodes of color i contributes the size of the ith color class. The second term favors large color classes, as we have already seen above. The objective function does not necessarily have its global minimum at an optimal coloring. However, local minima are legal colorings. Hence, the penalty version of simulated annealing is guaranteed to find a legal coloring even if it starts with an illegal coloring.

Exercise 12.26. Show that the objective function above has its local minima at legal colorings. Hint: Consider the change in $f(c)$ if one end of a legally colored edge is recolored with a fresh color. Prove that the objective function above does not necessarily have its global optimum at a solution using the minimum number of colors.

Experimental results. Johnson et al. [169] performed a detailed study of algorithms for graph coloring, with particular emphasis on simulated annealing. We shall briefly

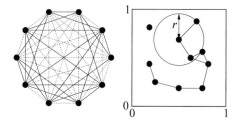

Fig. 12.13. *Left*: a random graph with 10 nodes and $p = 0.5$. The edges chosen are drawn solid, and the edges rejected are drawn dashed. *Right*: a random geometric graph with 10 nodes and range $r = 0.27$.

report on their findings and then draw some conclusions. Most of their experiments were performed on random graphs in the $G_{n,p}$ model or on random geometric graphs.

In the $G_{n,p}$ model, where p is a parameter in $[0, 1]$, an undirected random graph with n nodes is built by adding each of the $n(n-1)/2$ candidate edges with probability p. The random choices are independent for distinct edges. In this way, the expected degree of every node is $p(n-1)$ and the expected number of edges is $pn(n-1)/2$. For random graphs with 1000 nodes and edge probability 0.5, Kempe chain annealing produced very good colorings, given enough time. However, a sophisticated and expensive greedy algorithm, XRLF, produced even better solutions in less time. For very dense random graphs with $p = 0.9$, Kempe chain annealing performed better than XRLF. For sparser random graphs with edge probability 0.1, penalty function annealing outperformed Kempe chain annealing and could sometimes compete with XRLF.

Another interesting class of random inputs is *random geometric graphs*. Here, we choose n random, uniformly distributed points in the unit square $[0, 1] \times [0, 1]$. These points represent the nodes of the graph. We connect two points by an edge if their Euclidean distance is less than or equal to some given range r. Figure 12.13 gives an example. Such instances are frequently used to model situations where the nodes represent radio transmitters and colors represent frequency bands. Nodes that lie within a distance r from one another must not use the same frequency, to avoid interference. For this model, Kempe chain annealing performed well, but was outperformed by a third annealing strategy, called *fixed-K annealing*.

What should we learn from this? The relative performance of the simulated-annealing approaches depends strongly on the class of inputs and the available computing time. Moreover, it is impossible to make predictions about their performance on any given instance class on the basis of experience from other instance classes. So, be warned. Simulated annealing is a heuristic and, as for any other heuristic, you should not make claims about its performance on an instance class before you have tested it extensively on that class.

12.5.3 More on Local Search

We close our treatment of local search with a discussion of three refinements that can be used to modify or replace the approaches presented so far.

Threshold Acceptance. There seems to be nothing magic about the particular form of the acceptance rule used in simulated annealing. For example, a simpler yet also successful rule uses the parameter T as a threshold. New states with a value $f(x)$ below the threshold are accepted, whereas others are not.

Tabu Lists. Local-search algorithms sometimes return to the same suboptimal solution again and again – they cycle. For example, simulated annealing might have reached the top of a steep hill. Randomization will steer the search away from the optimum, but the state may remain on the hill for a long time. *Tabu search* steers the search away from local optima by keeping a *tabu list* of "solution elements" that should be "avoided" in new solutions for the time being. For example, in graph coloring, a search step could change the color of a node v from i to j and then store the tuple (v, i) in the tabu list to indicate that color i is forbidden for v as long as (v, i) is in the tabu list. Usually, this tabu condition is not applied if an improved solution is obtained by coloring node v with color i. Tabu lists are so successful that they can be used as the core technique of an independent variant of local search called *tabu search*.

Restarts. The typical behavior of a well-tuned local-search algorithm is that it moves to an area with good feasible solutions and then explores this area, trying to find better and better local optima. However, it might be that there are other, faraway areas with much better solutions. The search for Mount Everest illustrates this point. If we start in Australia, the best we can hope for is to end up at Mount Kosciuszko (altitude 2229 m), a solution far from optimal. It therefore makes sense to run the algorithm multiple times with different random starting solutions because it is likely that different starting points will explore different areas of good solutions. Starting the search for Mount Everest at multiple locations and in all continents will certainly lead to a better solution than just starting in Australia. Even if these restarts do not improve the average performance of the algorithm, they may make it more robust in the sense that it will be less likely to produce grossly suboptimal solutions.

12.5.4 Parallel Local Search

Local search is difficult to parallelize since, by definition, it performs one step after another. Of course, we can parallelize the operations in one step, for example by evaluating several members in the neighborhood $\mathcal{N}(x)$ in parallel. In Kempe chain annealing, several processors might explore different Kempe chains starting from different nodes. We can also try to perform more work in each step to reduce the number of steps, for example by using a larger neighborhood. We can also perform several independent local searches and then take the best result. In some sense, this is a parallelization of the restart strategy described in Sect. 12.5.3. A successful parallelization of a local search strategy might use parallelism on multiple levels. For example a few threads on each processor chip could evaluate an objective function in parallel, several of these thread groups working on different members of a large neighborhood, and multiple nodes of a distributed-memory machine work on independent local searches. However, for more scalable solutions, one should consider more advanced interactions between parallel solvers such as in the evolutionary algorithms described next.

12.6 Evolutionary Algorithms

Living beings are ingeniously adaptive to their environment, and master the problems encountered in their daily life with great ease. Can we somehow use the principles of life to develop good algorithms? The theory of evolution tells us that the mechanisms leading to this performance are *mutation*, *recombination*, and *survival of the fittest*. What could an evolutionary approach mean for optimization problems?

The genome describing an individual corresponds to the description of a feasible solution. We can also interpret infeasible solutions as dead or ill individuals. In nature, it is important that there is a sufficiently large *population* of genomes; otherwise, recombination deteriorates to incest, and survival of the fittest cannot demonstrate its benefits. So, instead of one solution as in local search, we now work with a pool of feasible solutions.

The individuals in a population produce offspring. Because resources are limited, individuals better adapted to the environment are more likely to survive and to produce more offspring. In analogy, feasible solutions are evaluated using a fitness function f, and fitter solutions are more likely to survive and to produce offspring. Evolutionary algorithms usually work with a solution pool of limited size, say N. Survival of the fittest can then be implemented as keeping only the N best solutions.

Even in bacteria, which reproduce by cell division, no offspring is identical to its parent. The reason is *mutation*. When a genome is copied, small errors happen. Although mutations usually have an adverse effect on fitness, some also improve fitness. Local changes in a solution are the analogy of mutations.

An even more important ingredient in evolution is *recombination*. Offspring contain genetic information from both parents. The importance of recombination is easy to understand if one considers how rare useful mutations are. Therefore it takes much longer to obtain an individual with two new useful mutations than it takes to combine two individuals with two different useful mutations.

We now have all the ingredients needed for a generic evolutionary algorithm; see Fig. 12.14. As with the other approaches presented in this chapter, many details need to be filled in before one can obtain an algorithm for a specific problem. The algorithm starts by creating an initial population of size N. This process should involve randomness, but it is also useful to use heuristics that produce good initial solutions.

Create an initial population $population = \{x^1, \ldots, x^N\}$
while not finished **do**
    if matingStep **then**
        select individuals x^1, x^2 with high fitness and produce $x' := mate(x^1, x^2)$
        else select an individual x^1 with high fitness and produce $x' = mutate(x^1)$
    $population := population \cup \{x'\}$
    $population := \{x \in population : x \text{ is sufficiently fit}\}$

Fig. 12.14. A generic evolutionary algorithm

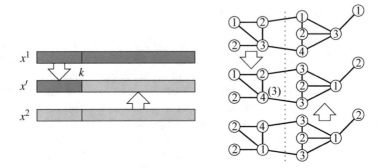

Fig. 12.15. Mating using crossover (*left*) and by stitching together pieces of a graph coloring (*right*).

In the loop, it is first decided whether an offspring should be produced by mutation or by recombination. This is a probabilistic decision. Then, one or two individuals are chosen for reproduction. To put selection pressure on the population, it is important to base reproductive success on the fitness of the individuals. However, it is usually not desirable to draw a hard line and use only the fittest individuals, because this might lead to too uniform a population and hence incest. For example, one can instead choose reproduction candidates randomly, giving a higher selection probability to fitter individuals. An important design decision is how to fix these probabilities. One choice is to sort the individuals by fitness and then to define the reproduction probability as some decreasing function of rank. This indirect approach has the advantage that it is independent of the objective function f and the absolute fitness differences between individuals, which are likely to decrease during the course of evolution.

The most critical operation is *mate*, which produces new offspring from two ancestors. The "canonical" mating operation is called *crossover*. Here, individuals are assumed to be represented by a string of n bits. An integer k is chosen. The new individual takes its first k bits from one parent and its last $n - k$ bits from the other parent. Figure 12.15 shows this procedure. Alternatively, one may choose k random positions from the first parent and the remaining bits from the other parent. For our knapsack example, crossover is a quite natural choice. Each bit decides whether the corresponding item is in the knapsack or not. In other cases, crossover is less natural or would require a very careful encoding. For example, for graph coloring, it would seem more natural to cut the graph into two pieces such that only a few edges are cut. Now one piece inherits its colors from the first parent, and the other piece inherits its colors from the other parent. Some of the edges running between the pieces might now connect nodes with the same color. This could be repaired using some heuristic, for example choosing the smallest legal color for miscolored nodes in the part corresponding to the first parent. Figure 12.15 gives an example.

Mutations are realized as in local search. In fact, local search is nothing but an evolutionary algorithm with population size 1.

The simplest way to limit the size of the population is to keep it fixed by removing the least fit individual in each iteration. Other approaches that provide room for different "ecological niches" can also be used. For example, for the knapsack problem, one could keep all Pareto-optimal solutions. The evolutionary algorithm would then resemble the optimized dynamic-programming algorithm.

12.6.1 Parallel Evolutionary Algorithms

In principle, evolutionary algorithms are easy to parallelize [14]. We simply perform multiple mating and mutation steps in parallel. This may imply that we have to work with a larger population than in a sequential algorithm. Rather than using a single large population, it then makes sense to work with multiple subpopulations. Each PE (or group of PEs) generally works on the local subpopulation. Subpopulations not only reduce communication and synchronization overhead but may also allow better solution quality. This is an effect also observed in nature – multiple subpopulations, for example on different islands, lead to higher biological diversity. This was already observed by Charles Darwin [82].

To avoid "incest", fit individuals are occasionally exchanged between the subpopulations. For example, each subpopulation can occasionally send one of its fittest individuals (or a mutation thereof) to a random other subpopulation. This approach guarantees a good balance between communication overhead and spreading of good solutions. It has been used very successfully for graph partitioning [277].

12.7 Implementation Notes

We have seen several generic approaches to optimization that are applicable to a wide variety of problems. When you face a new application, you are therefore likely to have a choice from among more approaches than you can realistically implement. In a commercial environment, you may have to home in on a single approach quickly. Here are some rules of thumb that may help:

- Study the problem, relate it to problems you are familiar with, and search for it on the web.
- Look for approaches that have worked on related problems.
- Consider black-box solvers.
- If the problem instances are small, systematic search or dynamic programming may allow you to find optimal solutions.
- If none of the above looks promising, implement a simple prototype solver using a greedy approach or some other simple, fast heuristic; the prototype will help you to understand the problem and might be useful as a component of a more sophisticated algorithm.
- Develop a local-search algorithm. Focus on a good representation of solutions and how to incorporate application-specific knowledge into the searcher. If you have a promising idea for a mating operator, you can also consider evolutionary algorithms. Use randomization and restarts to make the results more robust.

There are many implementations of linear-programming solvers. Since a good implementation is *very* complicated, you should definitely use one of these packages except in very special circumstances. The Wikipedia page "Linear programming" is a good starting point. Some systems for linear programming also support integer linear programming.

There are also many frameworks that simplify the implementation of local-search or evolutionary algorithms. Since these algorithms are fairly simple, the use of these frameworks is not as widespread as for linear programming. Nevertheless, the implementations available might have nontrivial built-in algorithms for dynamic setting of search parameters, and they might support parallel processing. The Wikipedia page "Evolutionary algorithm" contains pointers.

12.8 Historical Notes and Further Findings

We have only scratched the surface of (integer) linear programming. Implementing solvers, clever modeling of problems, and handling huge input instances have led to thousands of scientific papers. In the late 1940s, Dantzig invented the simplex algorithm [81]. Although this algorithm works well in practice, some of its variants take exponential time in the worst case. It is a well-known open problem whether some variant runs in polynomial time in the worst case. It is known, though, that even slightly perturbing the coefficients of the constraints leads to polynomial expected execution time [302]. Sometimes, even problem instances with an exponential number of constraints or variables can be solved efficiently. The trick is to handle explicitly only those constraints that may be violated and those variables that may be nonzero in an optimal solution. This works if we can efficiently find violated constraints or possibly nonzero variables and if the total number of constraints and variables generated remains small. Khachiyan [181] and Karmarkar [177] found polynomial-time algorithms for linear programming. There are many good textbooks on linear programming (e.g., [44, 98, 122, 240, 287, 321]).

Another interesting black-box solver is *constraint programming* [150, 204]. We hinted at the technique in Exercise 12.21. Here, we are again dealing with variables and constraints. However, now the variables come from discrete sets (usually small finite sets). Constraints come in a much wider variety. There are equalities and inequalities, possibly involving arithmetic expressions, but also higher-level constraints. For example, $allDifferent(x_1, \ldots, x_k)$ requires that $x_1, \ldots, x_k$ all receive different values. Constraint programs are solved using a cleverly pruned systematic search. Constraint programming is more flexible than linear programming, but restricted to smaller problem instances. Wikipedia is a good starting point for learning more about constraint programming.

13

Collective Communication and Computation

Counting votes after elections[1] is an early example of massively parallel computing. In a large country, there are millions and millions of ballots cast in thousands and thousands of polling places distributed all over the country. It is clearly a bad idea to ship all the ballot boxes to the capital and have them counted by a single clerk. Assuming the clerk can count one vote per second 24 h a day, counting 100 000 000 votes would take more than three years. Rather, the votes are counted in each station and the counts are aggregated in a hierarchy of election offices (e.g., polling place, city, county, state, capital). These counts are very compact and can be communicated by telephone in a few seconds. Overall, a well-organized counting process can yield a preliminary result in a few hours. We shall see that this is an example of a global reduction and that efficient parallel algorithms follow a very similar pattern.

Most parallel algorithms in this book follow the SPMD (single program multiple data) principle. More often than not, the high symmetry of these algorithms leads to highly regular interaction patterns involving all the PEs. This is a mismatch to the low-level primitives of our machine models such as point-to-point message exchange (Sect. 2.4.2) or concurrent access to memory locations (Sect. 2.4.1). Fortunately, these interaction patterns usually come from a small set of operations for which we can provide a library of efficient algorithms. Hence, these *collective communication operations* definitively belong to the basic toolbox of parallel algorithms and thus get their own chapter in this book. Figure 13.1 and Table 13.1 give an overview. Note that all bounds stated in this table are better than what one obtains using trivial algorithms. While the table gives asymptotic running times, in the following we shall look also at the constant factors involved in the terms depending on α and β. We shall call this the *communication time*. A bound of $a\alpha + b\beta$ will in all cases imply a running time of $a\alpha + b\beta + O(a + b)$.

---

[1] The illustration above shows a polling station in New York circa 1900. E. Benjamin Andrews, *History of the United States*, volume V. Charles Scribner's Sons, New York (1912).

© Springer Nature Switzerland AG 2019
P. Sanders et al., *Sequential and Parallel Algorithms and Data Structures*,
https://doi.org/10.1007/978-3-030-25209-0_13

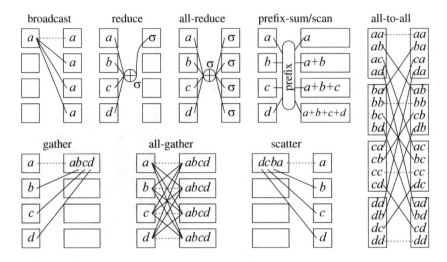

Fig. 13.1. Information flow (left to right) of collective communication operations. Note that the actual data flow might be different in order to improve performance.

Table 13.1. Collective communication operations and their asymptotic complexities; n = message size, p = #PEs, α = message startup latency, β = comm. cost per word

Name	# senders	# receivers	# messages	Computations?	Complexity	Section
Broadcast	1	p	1	no	$\alpha \log p + \beta n$	13.1
Reduce	p	1	p	yes	$\alpha \log p + \beta n$	13.2
All-reduce	p	p	p	yes	$\alpha \log p + \beta n$	13.2
Prefix sum	p	p	p	yes	$\alpha \log p + \beta n$	13.3
Barrier	p	p	0	no	$\alpha \log p$	13.4.2
Gather	p	1	p	no	$\alpha \log p + \beta pn$	13.5
All-gather	p	p	p	no	$\alpha \log p + \beta pn$	13.5
Scatter	1	p	p	no	$\alpha \log p + \beta pn$	13.5
All-to-all	p	p	p^2	no	$\log p(\alpha + \beta pn)$ or $\quad p(\alpha + \beta n)$	13.6

We begin with a group of operations with essentially the same complexity, and some closely related algorithms: sending a message to all PEs (*broadcast*, Sect. 13.1), combining one message from each PE into a single message using an associative operator (*reduce*, Sect. 13.2), a combination of reduce and subsequent broadcast (*all-reduce*), and computation of the partial sums $\sum_{i \le i_{\text{proc}}} x_i$, where, once more, the sum can be replaced by an arbitrary associative operator (*prefix sum* or *scan*, Sect. 13.3).

In Sect. 13.4, we discuss the *barrier* synchronization, which ensures that no PE proceeds without all the other PEs having executed the barrier operation. This operation is particularly important for shared memory, where it is used to ensure that no data is read by a consumer before being produced. On shared-memory machines, we also need further synchronization primitives such as *locking*, which are also ex-plained in Sect. 13.4.

The operations *gather* (concatenate p messages, one from each PE), *all-gather* (gather plus broadcast, also called *gossip* or *all-to-all broadcast*), and *scatter* (one PE sends one individual message to each PE) are much more expensive, since they scale only linearly with p and should be avoided whenever possible. Nevertheless, we sometimes need them, and Sect. 13.5 provides some interesting algorithms. Finally, in Sect. 13.6 we consider the most expensive operation, *all-to-all*, which delivers individual messages between all PEs.

We explain the collective operations first for the distributed-memory model. Then we comment on the differences for shared memory, where we have hardware sup-port for some of the operations. For example, concurrent reading works well because the hardware internally implements a kind of broadcast algorithm. In small shared-memory systems, constant factors may also be more important than asymptotic com-plexity. For example, p concurrent fetch-and-add instructions may often be faster than a call to a tree-based reduction algorithm. However, this may not apply for large p and large objects to be combined.

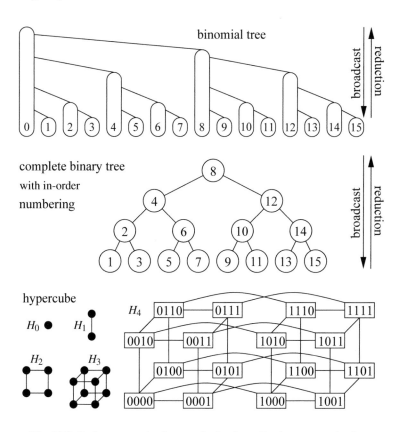

Fig. 13.2. Basic communication topologies for collective communications

All these algorithms are based on the three basic communication topologies *binomial tree*, *binary tree*, and *hypercube* shown in Fig. 13.2. The tree topologies are good for one-to-p patterns (broadcast and scatter) and p-to-one patterns (reduce and gather) and are introduced in Sect. 13.1 together with the broadcast operations. The hypercube introduced in Sect. 13.2.1 can in some cases accelerate p-to-p communication patterns and also leads to particularly simple algorithms.

13.1 Broadcast

In a broadcast, a root PE r wants to send a message m of size n to all PEs. Broadcasting is frequently needed in SPMD parallel programs to distribute parameters of the input or other globally needed values. Shared-memory machines with coherent caches support broadcasting by hardware to some extent. When one PE accesses a value, its cache line is copied to all levels of the cache hierarchy. Subsequently, other PEs accessing the same caches do not have to go down all the way to main memory. When all PEs access the same value, this hardware mechanism – with some luck[2] – will build a tree of caches and PEs in a similar fashion to the distributed-memory algorithms we shall see below. For example, the value may be copied from main memory to the L3 cache of all processor chips. From each L3 cache, it is then copied to the L2 and L1 caches of each core, and the hardware threads running on the same core need only to access their local L1 cache.

Going back to distributed memory, let us assume for now that $r = 0$ and that the PEs are numbered $0..p - 1$. Other arrangements are easy to obtain by renumbering the PEs, for example as $i_{\text{proc}} - r \bmod p$. A naive implementation uses a for-loop and $p - 1$ send operations. This strategy is purely sequential and needs time at least $\alpha(p - 1)$; it could be a bottleneck in an otherwise parallel program. A better strategy is to use a divide-and-conquer strategy. PE 0 sends x to PE $\lceil p/2 \rceil$ and delegates to it the task of broadcasting x to PEs $\lceil p/2 \rceil ..p - 1$ while PE 0 itself continues broadcasting to PEs $0.. \lceil p/2 \rceil - 1$. More generally, a PE responsible for broadcasting to PEs $i..j$ delegates to PE $\lceil (i + j)/2 \rceil$.

13.1.1 Binomial Trees

Suppose now that p is also a power of two. Then the resulting communication topology is a *binomial tree*. We have already seen binomial trees in the context of addressable priority queues (see Fig. 6.5 in Sect. 6.2.2). Fig. 13.2 draws them differently to visualize the communication algorithm. Edges starting further up are used earlier in the broadcast algorithm. Interestingly, the binary representation of the PE numbers tells us the role of a node in the tree: A PE number i with k trailing 0's indicates that this PE has k children with numbers $i + 1, i + 2, \ldots, i + 2^{k-1}$ and that PE $i - 2^k$ is its parent. This is so convenient that we also adopt it when p is not a power of two. We

---

[2] In the worst case, all PEs may try to read the value exactly at the same time and will then produce a lot of contention in trying to access the main memory.

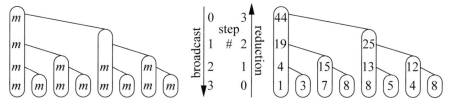

Fig. 13.3. Data flow for binomial tree broadcast (left) and reduction (right) with $p = 8$. The values at step i indicate the state of the computation at that step.

simply round up to the next power of two, use the binomial tree for that power of two and then drop the nodes with number $\geq p$. The resulting graph is still a tree, and it is easy to generalize the communication algorithms. Indeed, the following pseudocode works for any tree spanning all PEs:

> **Procedure** *Tree::broadcast(m)*
>   *receive(parent, m)*   *// does nothing on root*
>   **foreach** $c \in$ *children* **do** *send(c, m)*   *// largest first for binomial tree*

Figure 13.3 (left) shows the data flow for $p = 8$. It is very important that the messages to the children are sent in decreasing order of their size. Is tree broadcast a good algorithm? We first show an upper bound for binomial trees.

Theorem 13.1. *The broadcast algorithm for binomial trees needs communication time*

$$\lceil \log p \rceil (\alpha + \beta n).$$

Proof. We use induction on p. The base case $p = 1$ is easy, PE 0 is the only node. The receive operation does nothing and the set of children is empty. Hence, no communication is needed.

The induction step assumes that the claim is true for all $p \leq 2^k$ and shows that this implies that the claim also holds for all $p \leq 2^{k+1}$. For $2^k < p \leq 2^{k+1}$, the first send operation sends m to PE 2^k in time $\alpha + \beta n$. From then on, PEs 0 and 2^k execute the broadcast algorithm for a subtree of size $\leq 2^k$. Hence, by the induction hypothesis, they need time $k(\alpha + \beta n)$. Overall, the time is $(k+1)(\alpha + \beta n) = \lceil \log p \rceil (\alpha + \beta n)$. $\square$

This bound is in some sense the best possible: Since one communication operation can inform only a single PE about the content of message m, the number of PEs knowing anything about m can at most double with each communication. Hence, $\alpha \log p$ is a lower bound on the broadcasting time.

Exercise 13.1. Show that the running time of binomial tree broadcast becomes $\Omega(\alpha \log^2 p)$ if the order of the send operations is reversed, i.e., small children are served first.

Another obvious lower bound for broadcasting is βn – every nonroot PE must receive the entire message. However, binomial tree broadcast needs $\log p$ times more time. In fact, for long messages, there are better algorithms, approaching βn for large n.

13.1.2 Pipelining

There are two reasons why binomial trees are not good for broadcasting long messages. First, PE 0 sends the entire message to $\lceil \log p \rceil$ other PEs, so that it becomes a bottleneck. Hence, a faster algorithm should send only to a bounded number of other PEs. This problem is easy to fix – we switch to another tree topology where every PE has small outdegree. Outdegrees of one and two seem most interesting. For outdegree 1, the topology degenerates to a path, leading to communication time $\Omega(p\alpha)$. This is not attractive for large p. Hence, outdegree two – a binary tree – seems like a good choice. This is the smallest outdegree for which we can achieve logarithmic height of the tree.

Another problem is that any broadcasting algorithm which sends m as a whole will need time $\Omega(\log p(\alpha + \beta n))$. Hence, we should not immediately transfer the entire message. Rather, we should chop the message into smaller *packets*, which are sent independently. This way, broadcasting can spread its activity to all PEs much faster. Reinterpreting message m as an array of k packets of size $\lceil n/k \rceil$, we get the following algorithm that once more works for any tree topology:

> **Procedure** *Tree::pipelinedBroadcast*(m : *Array*$[1..k]$ **of** *Packet*)
>> **for** $i := 1$ **to** k **do**
>>> *receive*(*parent*, $m[i]$)
>>> **foreach** $c \in$ *children* **do** *send*(c, $m[i]$)

Figure 13.4 gives an example.

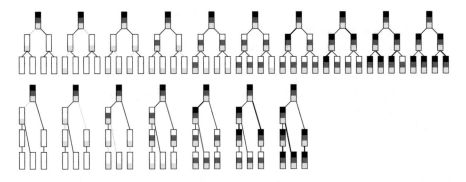

Fig. 13.4. Pipelined binary tree broadcast with $p = 7$ and $k = 3$. Top: 10 steps using a complete binary tree and half-duplex communication. Bottom: 7 steps using a skewed tree and full-duplex communication.

Lemma 13.2. *On a binary tree, the algorithm Tree::pipelinedBroadcast needs communication time*

$$T(L,n,k) \leq (2L + 3(k-1))\left(\alpha + \beta \left\lceil \frac{n}{k} \right\rceil\right), \tag{13.1}$$

where L is the length of the longest root to leaf path in the tree.

Proof. Sending or receiving one packet takes time $\alpha + \beta \lceil n/k \rceil$. It remains to show that after $2L + 3(k-1)$ such communication steps, every PE has received the entire message. A PE which has successfully received the first packet needs two steps to forward it to both of its neighbors. Hence, after $2L$ steps, every PE has received the first packet. An interior node of the tree can process one packet every three steps (receive, send left, send right). Hence, after $3(k-1)$ further steps, the last packet has arrived at the last PE.   □

It remains to determine k and L. If we use a perfectly balanced binary tree, we get $L = \lfloor \log p \rfloor$. Ignoring rounding issues, we can find an optimal value for k using calculus. We get $k = \sqrt{\beta n(2L-3)/3\alpha}$ if this value is ≥ 2. This yields the following bound.

Theorem 13.3. *Using pipelined broadcast on a perfectly balanced binary tree, we can obtain communication time*

$$T_2(k) = 3\beta n + 2\alpha \log p + O\left(\sqrt{\alpha \beta n \log p}\right) = O(\beta n + \alpha \log p).$$

Exercise 13.2. Prove Theorem 13.3.

13.1.3 Building Binary Trees

We have not yet explained how to efficiently build the topology for a perfectly balanced binary tree – each node (PE) needs to know the identity of its parent, its children, and the root. We describe a construction principle that is simple, is easy to compute, can also be used for reduction and prefix sums, and uses subtrees with contiguous numbers. The latter property might be useful for networks that exhibit a hierarchy reflected by the PE numbering. Figure 13.2 gives an example. Figure 13.5 gives pseudocode for a tree class defining a perfectly balanced binary tree whose nodes are numbered *in-order*, i.e., for any node v, the nodes in the subtree rooted at the left child of v have a smaller number than v and the nodes in the subtree rooted at the right child of v have a larger number than v. The constructor takes a user-defined root and calls a recursive procedure *buildTree* for defining a binary tree on PE numbers $a..b$. This procedure adopts the root x and parent y passed by the caller and calls itself to define the subtrees for PEs $a..x-1$ and $x+1..b$. The roots of the subtrees are placed in the middle of these subranges. Since PE i_{proc} needs only the local values for the parent, left child, and right child, only that branch of the recursion needs to be executed which contains i_{proc}. Thus, we get execution time $O(\log p)$ without any communication.

Exercise 13.3. Prove formally that the tree constructed in Fig. 13.5 has height $\leq \lceil \log p \rceil$ regardless of the root chosen.

Exercise 13.4. Design an algorithm that runs in time $O(\log p)$ and defines a balanced binary tree where the nodes are numbered layer by layer as in Sect. 6.1.

Class *InOrderTree*($r = \lceil p/2 \rceil : 1..p$)   // subclass of *Tree*
    leftChild, rightChild, parent, root : $1..p \cup \{\bot\}$
    root := r
    buildTree$(1, p, root, \bot)$

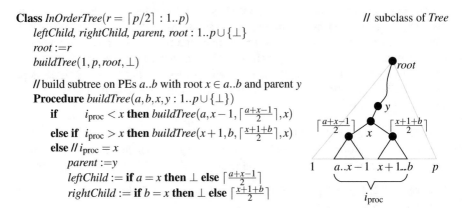

    // build subtree on PEs $a..b$ with root $x \in a..b$ and parent y
    Procedure *buildTree*$(a, b, x, y : 1..p \cup \{\bot\})$
        if $i_{\text{proc}} < x$ **then** *buildTree*$(a, x - 1, \lceil \frac{a+x-1}{2} \rceil, x)$
        else if $i_{\text{proc}} > x$ **then** *buildTree*$(x + 1, b, \lceil \frac{x+1+b}{2} \rceil, x)$
        else // $i_{\text{proc}} = x$
            parent := y
            leftChild := **if** $a = x$ **then** $\bot$ **else** $\lceil \frac{a+x-1}{2} \rceil$
            rightChild := **if** $b = x$ **then** $\bot$ **else** $\lceil \frac{x+1+b}{2} \rceil$

Fig. 13.5. Constructor of an in-order binary tree

13.1.4 *Faster Broadcasting

With binomial tree broadcasting we have a very simple algorithm that is optimal for small messages. Pipelined binary tree broadcasting is simple and asymptotically optimal for long messages also. However, it has several weaknesses that can be improved on.

First, we are not exploiting full-duplex communication. For *very* long messages, we can exploit full-duplex communication by using a *unary tree* – a path consisting of p nodes. The following *linear pipeline* algorithm exploits full-duplex communication. While receiving packet i, it forwards the previous packet $i - 1$ down the line:

    Procedure *Path::pipelinedBroadcast*(m : *Array*$[1..k]$ **of** *Packet*)
        receive(*parent*, $m[1]$)
        for $i := 2$ **to** k **do** *receive*(*parent*, $m[i]$) $\|$ *send*(*child*, $m[i - 1]$)
        send(*child*, $m[k]$)

The algorithm terminates in $p + k - 1$ steps. Optimizing for k as in Sect. 13.1.2 yields running time

$$T_1(k) = \beta n + p\alpha + O\left(\sqrt{\alpha\beta np}\right).$$

This is optimal for fixed p and $n \to \infty$.

The linear pipeline is much worse than pipelined binary tree broadcasting for large p unless n is extremely large. We can also use bidirectional communication for binary tree broadcasting, as in the following exercise.

Exercise 13.5. Adapt pipelined binary tree broadcasting so that it exploits full-duplex communication for communicating with one of its children and runs in time

$$T_2^* = 2\beta n + 2\alpha \log p + O\left(\sqrt{\alpha\beta n \log p}\right).$$

Hint: Use the same trick as for the linear pipeline. Figure 13.4 gives an example.

For large n, this is still a factor of two away from optimality. We seem to have the choice between two evils – a linear pipeline that does not scale with p and a binary tree that is slow for large n. This seems to be unavoidable for any tree that is not a path – the interior nodes have to do three communications for every packet and are thus overloaded, whereas the leaves do only one communication and are thus underloaded. Another trick solves the problem. By carefully scheduling communications, we can run *several* tree-based broadcast algorithms simultaneously. One approach uses $\log p$ spanning binomial trees embedded into a hypercube [171]. Another one uses just two binary trees [278]. The principle behind the latter algorithm is simple. By ensuring that an interior node in one tree is a leaf in the other tree, we cancel out the imbalance inherent in tree-based broadcasting. This algorithm is also easy to adapt to arbitrary values of p and to prefix sums and noncommutative reduction.

Another weakness of binary tree broadcasting can be seen in the example in Fig. 13.4. Some leaves receive the packets later than others. It turns out that one can reduce the latency of broadcasting by building the trees in such a way that all leaves receive the first message at the same time. This implies that, towards the right, the tree becomes less and less deep. Interestingly, one can obtain such a tree by modifying the procedure *buildTree* in Fig. 13.5 to place the root of a subtree not in the middle but according to the *golden ratio* $1 : (1 + \sqrt{5})/2$. Fig. 13.4 gives an example where all leaves receive a packet at the same time. When an interconnection network does not support arbitrary concurrent communications, one can adapt the tree structure so that not too many concurrent communications use the same wire. For example, Fig. 13.6 suggests how to embed a spanning binary tree of depth $\log p + O(1)$ into two-dimensional square meshes such that at most two tree edges are embedded into connections between neighboring PEs.

Exercise 13.6. (Research problem) Is there an embedding which uses any edge of the mesh for only a single tree edge?

Exercise 13.7. Construct a binary tree of depth $O(\log p)$ such that only a constant number of tree edges are embedded into connections between neighboring PEs of a $p = k \times 2^k$ mesh. Hint: Use horizontal connections spanning 2^i PEs in layer i.

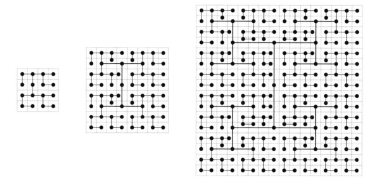

Fig. 13.6. *H-tree* embedding of a binary trees into 2D meshes.

13.2 Reduction

Reduction is important in many parallel programs because it is a scalable way to aggregate partial results computed on different PEs into a global result.

Given a message m_i on each PE i, a reduction computes $\bigotimes_{i<p} m_i$, where $\otimes$ can be any associative operator, i.e., $\forall a, b, c : (a \otimes b) \otimes c = a \otimes (b \otimes c)$. Some algorithms also require $\otimes$ to be commutative, and sometimes it is convenient if we have a neutral element (e.g., 0 for $\otimes = +$). In practice, the most frequent operators are $+$, min, and max, which are all commutative and have a neutral element.

Exercise 13.8. Explain how to compute both the minimum and the location of the minimum using a reduction. Is the resulting operation commutative?

Exercise 13.9. Show that floating-point addition is, strictly speaking, *not* associative. What are the consequences of treating it like an associative operation?

When long messages are involved in a reduction, we shall assume that they represent vectors of smaller objects and thus the reduction can be computed componentwise. This will be important in order to use pipelining techniques.

Shared-memory machines have limited support for reduction in the form of fetch-and-add instructions. However, when all processors are adding to a global value at the same time, software implementations along the lines of what is presented below may still be faster.

Mathematically speaking, associativity is the key to reducing in parallel. When we write $\bigoplus_{i<p} m_i$, this is conventionally interpreted as

$$((\cdots ((m_0 \otimes m_1) \otimes m_2) \cdots) \otimes m_{p-2}) \otimes m_{p-1}$$

which looks inherently sequential. However, associativity allows us to rewrite the sum as any binary tree whose interior nodes are labeled with $\otimes$ and whose ith leaf is m_i.

Parallel reduction algorithms thus use trees much like the broadcasting algorithms we have seen in Sect. 13.1. Indeed, that similarity goes much further. The algorithms we propose here are basically obtained by running a broadcasting algorithm "backwards". The leaves of the tree send their values to their parents, which add them up before forwarding the partial sum to their parents. This works even for noncommutative operators if the nodes of the tree are numbered in-order.

For binomial trees, we have already seen the resulting code in Sect. 2.4.2. Figure 13.3 gives an example and illustrates the symmetry with respect to broadcasting. Note that compared to broadcasting, the "running backwards" rule means that we also reverse the order in which we receive from the children – this time from left to right. This is important to achieve the same communication time as for the broadcasting algorithm.

For binary trees, we get the following pseudocode (for the nonpipelined case). To simplify some special cases, we use a neutral element **0**:

Procedure *InOrderTree::reduce(m)*
  $x := \mathbf{0};$ *receive(leftChild, x)*
  $z := \mathbf{0};$ *receive(rightChild, z)*
  send$(parent, x \otimes m \otimes z)$ *// root* returns result

We leave formulating pseudocode for pipelined tree reduction as an exercise but prove its complexity here, which is the same as for pipelined broadcasting (Lemma 13.2).

Lemma 13.4. *Pipelined binary tree reduction needs communication time*

$$T^{\otimes}(L,n,k) \le (2L + 3(k-1))\left(\alpha + \beta \left\lceil \frac{n}{k} \right\rceil\right), \qquad (13.2)$$

where k is the number of packets and L is the length of the longest root-leaf path in the tree.

Proof. We have to show that $2L + 3(k-1)$ packet communication steps are needed. Reduction activity propagates bottom up, one level every two steps. Hence, the root performs its first addition after $2L$ steps. An interior node of the graph (e.g., the right child of the root) can process one packet every three steps (receive left, receive right, send sum). Hence, after $3(k-1)$ further steps, the last packet has arrived at the root. □

Since the execution time is the same as for broadcasting, we can also optimize k in the same way and obtain the following corollary.

Corollary 13.5. *Using pipelined reduction on a perfectly balanced binary tree, we can obtain communication time*

$$T_2^{\otimes}(k) = 3\beta n + 2\alpha \log p + O\left(\sqrt{\alpha \beta n \log p}\right) = O(\beta n + \alpha \log p).$$

All the optimizations in Sect. 13.1.4 also transfer to reduction, except that using $\log p$ spanning trees [171] does not work for noncommutative reduction – it seems impossible to have in-order numberings of all the trees involved. However, this works for the two trees used in [278].

13.2.1 All-Reduce and Hypercubes

Often, *all* PEs (rather than only one root PE) need to know the result of a reduction. This *all-reduce* operation can be implemented by a reduction followed by a broadcast. For long messages and for shared-memory machines (where broadcast is supported by the hardware) this is also a good strategy. However, for short messages we can do better reducing the latency from $2\alpha \log p$ to $\alpha \log p$. We show this for the case where $p(= 2^d)$ is a power of two.

In this case, all-to-all communication patterns can often be implemented using a *hypercube*: A d-dimensional hypercube is a graph $H_d = (0..2^d - 1, E)$ where $(x,y) \in E$ if and only if the binary representations of x and y differ in exactly one bit. Edges corresponding to changing bit i are the edges along *dimension i*. Figure 13.2 shows the hypercubes for $d \in 0..4$. The hypercube communication pattern considers the PEs as nodes of a hypercube. It iterates through the dimensions, and in iteration i communicates along edges of dimension i.

For the all-reduce problem, we get the following simple code:

for $i := 0$ **to** $d - 1$ **do**
    $send(i_{\text{proc}} \oplus 2^i, m) \parallel receive(i_{\text{proc}} \oplus 2^i, m')$ // Or, for short:
    $m := m \otimes m'$ // $m\otimes = m@(i_{\text{proc}} \oplus 2^i)$

Understanding how and why a hypercube algorithm works often involves loop invariants dealing with i-dimensional *subcubes*. All the 2^i nodes sharing the most significant bits $i..d-1$ in their number form an i-dimensional hypercube. For all-reduction, the loop invariant says that after iteration i, all $i+1$-dimensional subcubes have computed an all-reduce within that subcube. Figure 13.7 gives an example.

13.3 Prefix Sums

A *prefix sum* or *scan* operation is a generalization of a reduction where we are interested not only in the overall sum but also in all the prefixes of the sum. More precisely, PE i wants to compute $\otimes_{i \leq i_{\text{proc}}} m_i$. A variant of this definition computes the *exclusive prefix sum* $\otimes_{i < i_{\text{proc}}} m_i$ on PE i.

Prefix sums with $\otimes = +$ are frequently used for distributing work between PEs; see also Sect. 14.2. We have seen an example for a different operator in Sect. 1.1.1, where it was used for computing carry lookaheads in parallel addition. Prefix sums are equally important for shared-memory and distributed-memory algorithms, and there is no hardware support for them on shared-memory machines.

Here is a very simple hypercube algorithm for computing prefix sums which just adds two lines to the all-reduce algorithm presented in Sect. 13.2.1.

$x := m;$ **invariant** x *is the prefix sum in current subcube*
for $i := 0$ **to** $d - 1$ **do**
    $send(i_{\text{proc}} \oplus 2^i, m) \parallel receive(i_{\text{proc}} \oplus 2^i, m')$
    $m := m \otimes m'$ // update overall sum in subcube
    if i_{proc} **bitand** 2^i **then** $x := m' \otimes x$ // update prefix sum in subcube

Figure 13.7 gives an example. Note that this algorithm computes *both* the prefix sum *and* the overall sum on each PE. This is handy, since in many applications we actually need both values, for example for distributed-memory quicksort (Sect. 5.7.1). We obtain the following result:

Theorem 13.6. *The hypercube algorithm computes a prefix sum plus all-reduce in communication time*

$$\log p(\alpha + n\beta).$$

Hypercube algorithms only work when p is a power of two. Also, for long messages, it is problematic that all PEs are active in every step and hence we cannot use

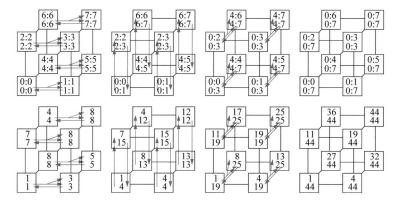

Fig. 13.7. A hypercube prefix sum (top values)/all-reduce (bottom values) for $p = 8$. The *top* row shows the general computation pattern, with $x : y$ as an abbreviation for $\oplus_{i=x}^{y} m_i$. The *bottom* row shows a concrete numeric example.

pipelining. In order to obtain a fast algorithm for arbitrary p and for long messages, we have therefore developed an algorithm based on in-order binary trees [198]. Figure 13.8 gives such an algorithm. For simplicity, we use neutral elements and do not show the code for pipelining. The algorithm consists of two phases. First, there is an upward phase, which works in the same way as reduction. Then comes a downward phase, which resembles a broadcast but computes the prefix sum and sends different data to the left and the right child. We exploit the fact that the PEs are numbered in-order, i.e., each PE i_{proc} is the root of a subtree containing all the PEs in a range $a..b$ of integers. During the upward phase, PE i_{proc} receives the sum x of the elements $a..i_{\text{proc}} - 1$ from its left subtree. The downward phase is implemented in such a way that PE i_{proc} receives the sum ℓ of the elements on PEs $0..a - 1$. Hence, PE i_{proc} can compute its local prefix sum as $\ell + x + m$. The left subtree is numbered $a..i_{\text{proc}} - 1$, so that i_{proc} forwards ℓ there. The right subtree is numbered $i_{\text{proc}} + 1..b$, so that i_{proc} sends the local prefix sum $\ell + x + m$. Figure 13.9 gives an example.

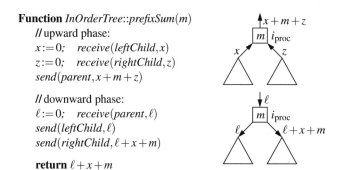

Function *InOrderTree::prefixSum(m)*
    // upward phase:
    $x := 0$;   *receive(leftChild, x)*
    $z := 0$;   *receive(rightChild, z)*
    send(parent, x + m + z)

    // downward phase:
    $\ell := 0$;   *receive(parent, ℓ)*
    send(leftChild, ℓ)
    send(rightChild, ℓ + x + m)

    return $\ell + x + m$

Fig. 13.8. Prefix sum using an *InOrderTree*

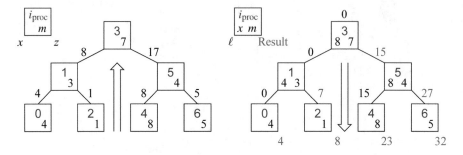

Fig. 13.9. Tree-based prefix sum computation for $p = 7$ over the sequence $\langle 4, 3, 1, 7, 8, 4, 5 \rangle$. See also Fig. 13.8.

Since the communication operations are exactly the same as in a reduction followed by a broadcast, we can adopt the strategy for pipelining and the analysis presented in the previous sections and get obtain the following corollary.

Corollary 13.7. *Using pipelined prefix sums on binary trees, we can obtain communication time*

$$T_2(k) = 6\beta n + 4\alpha \log p + O\left(\sqrt{\alpha \beta n \log p} \right) = O(\alpha n + \beta \log p).$$

The optimizations in Sect. 13.1.4 also transfer to prefix sums except that we cannot use $\log p$ spanning trees simultaneously [171]. We need the same in-order numbering of all trees involved in order to compute prefix sums, even for commutative operations. Using the result for two compatibly numbered trees [278], we obtain execution time $2\beta n + 4\alpha \log p + O\left(\sqrt{\alpha \beta n \log p} \right)$.

13.4 Synchronization

In a synchronization event, different threads inform each other that they have reached a certain location in their code. Thus synchronization is communication even if no additional information is exchanged. We have already introduced shared-memory locks in Chap. 2. In Sect. 13.4.1, we shall fill in some details. We discuss global synchronization of all PEs in Sect. 13.4.2.

13.4.1 Locks

In Sect. 2.5, we learned about binary locks. A lock is represented by a memory cell $S[i]$ which takes on values 0 and 1. A thread acquires this lock by changing the value from 0 to 1 and releases the lock by changing it back to 0. In order to make sure that only one thread can acquire the lock, the variable is set using a CAS instruction:

repeat until $CAS(i, desired := 0, 1)$

This kind of lock is called a *spin lock* because the thread trying to acquire it simply waits in a loop (spin) while repeatedly checking whether the lock is available. If many threads contend for the same lock, a basic spin lock may become inefficient, and a more careful implementation is called for:

> **Procedure** $lock(i : \mathbb{N})$ // acquire lock in memory cell $S[i]$
> $desired = 0 : \mathbb{N}$
>   **loop**
>   **if** $S[i] = 1$ **then** *backoff*   // wait a bit
>   **else if** $CAS(i, desired, 1)$ **then return**

We have made two improvements here. First, we attempt the expensive CAS instruction only when an ordinary memory read tells us that we "might" be successful. Second, we do not spin in a tight loop gobbling up a lot of hardware resources but instead call a procedure *backoff* that uses various measures to save resources. First of all, *backoff* should call the machine instruction `pause`, which tells the processor to yield resources to another thread on the same core. In addition, *backoff* may explicitly wait before even attempting to read $S[i]$ again in order to avoid situations where many threads wait for $S[i] = 0$ and each of them executes an expensive CAS when $S[i]$ becomes 0. One such adaptive strategy is *exponential backoff* – the backoff period is multiplied by a constant in every loop iteration until it hits a maximum in $\Theta(p)$. Thus, even if all PEs do little else but contend for the same lock, we can achieve a constant success probability for the CAS instruction. After a successful lock, the backoff period is reduced by the same factor. Waiting can be done with a for-loop that does nothing other than execute the `pause` instruction. Unlocking a spin lock is very simple – simply set $S[i] := 0$. No CAS instruction is necessary.[3]

Distinguishing readers and writers is slightly more complicated but can also be implemented with the spin-locking idea. Now the lock variable $S[i]$ is set to $p+1$ if a writer has exclusive access, where p is an upper bound on the number of threads. Otherwise, $S[i]$ gives the number of readers. Locking for writing has to wait for $S[i] = 0$. Locking for reading has to wait for $S[i] \leq p$ and it then increments $S[i]$. Unlocking from reading means decrementing $S[i]$ atomically.

Sometimes locks are held for a very long time. For example, a thread holding a lock may be waiting for I/O or a user interaction. Then a more complicated kind of lock makes sense, where the runtime system, in cooperation with the operating system, suspends threads that are waiting for a lock. An *unlock* then has to activate the first thread in the queue of waiting threads.

---

[3] This is true at least on the x86 architecture. Other architectures may need additional memory fence operations: see also Sect. B.3.

13.4.2 Barrier Synchronization

When a PE in an SPMD program calls the procedure *barrier*, it waits until all other PEs have called this routine. A barrier is thus a global synchronization of all PEs. This is needed when a parallel computation can only proceed if all PEs have finished a computation, for example updating a shared data structure.

The behavior of a barrier is entailed by calling an all-reduce with an empty (or dummy) operand.

Corollary 13.8. *A barrier needs communication time* $O(\alpha \log p)$.

On distributed-memory machines, there is little else to say. On shared-memory machines, barriers are so important that it is worth looking at some implementation details affecting the constant factors involved. Figure 13.10 gives pseudocode. Here, we adopt the basic idea to implement an all-reduce and choose a combination of a binomial tree reduce and a (hardware supported) broadcast. The idea is to replace a send operation by a write operation to a single memory cell (*readyEpoch*) that no other PE ever writes to. A receive operation becomes waiting for this memory cell to change. If there were only a single call to the procedure *barrier*, it would suffice to use a simple flag variable. However, in general, we would need an additional mechanism to reset these flags, which costs additional time. We circumvent this complication by using counters rather than flags.

$epoch = 0 : \mathbb{N}$ // how often was *barrier* called locally?
$readyEpoch = 0 : \mathbb{N}$ // how often did my subtree finish a barrier?
Procedure *barrier*
    $epoch{+}{+}$
    for $i := 0$ **to** $|childrenInBinomialTree|$ **do**   // that number should be precomputed
        while $readyEpoch @ (i_{\text{proc}} + 2^i) \neq epoch$ **do** *backoff*
    $readyEpoch := epoch$
    while $readyEpoch @ 0 \neq epoch$ **do** *backoff*   // implicit broadcast

Fig. 13.10. SPMD code for a shared-memory barrier.

13.4.3 Barrier Implementation

In this section we provide a C++ implementation of the binomial tree barrier and benchmark it against the simple folklore barrier shown in Listing 13.1. The latter implementation uses a PE counter c and an *epoch* counter that are aligned with cache-line boundaries (64 bytes on x86) to avoid false sharing. The barrier wait method accepts the ID of PE *iPE* and the total number of PEs p. In this implementation, only p is used but for the sake of generality we stick with this interface to be compatible with other barrier implementations. Each PE atomically increments the counter c with the last PE (that passes the barrier) incrementing the *epoch* and resetting the PE counter c (lines 6–8). All other PEs wait for this epoch transition in line 10. Note that the *epoch* variable is declared with the `volatile` storage class (see Sect. B.3) to prevent caching it into a local stack variable or a register through the compiler. The barrier semantics (proceed if all PEs have finished a computation) also requires a CPU memory fence because otherwise the processor is, in general, allowed to reorder computations including loads/stores before and after the barrier, a contradiction to the semantics. For this purpose we have inserted a CPU memory fence (line 12). On x86 architectures, this fence is not required, because the code flow includes an atomic operation on counter c (line 6), which is an implicit CPU memory fence on x86.

Listing 13.1. A simple barrier in C++

```
class SimpleBarrier {                                                    1
public:                                                                  2
  SimpleBarrier(): c(0), epoch(0) {}                                     3
  void wait(const int /* iPE */, const int p) {                         4
    register const int startEpoch = epoch;                              5
    if(c.fetch_add(1) == p−1) {                                         6
      c = 0;                                                            7
      ++epoch;                                                          8
    } else {                                                            9
      while(epoch == startEpoch) backoff();                            10
    }                                                                  11
    atomic_thread_fence(memory_order_seq_cst); // not required on x86  12
  }                                                                    13
protected:                                                             14
  atomic<int> c __attribute__((aligned(64)));                          15
  char pad[64 − sizeof(atomic<int>)];                                  16
  volatile int epoch __attribute__((aligned(64)));                     17
};//SPDX−License−Identifier: BSD−3−Clause; Copyright(c) 2018 Intel Corporation   18
```

The implementation of the binomial tree barrier is shown in Listing 13.2. It stores the PE-local *epoch* and the number of children *numChildren* in the binomial tree in the array *peData*. The number of children is precomputed in the recursive function *initNumChildren* (lines 8–17) called in the constructor. The items in the arrays *peData* and *readyEpoch* are padded to avoid false sharing (lines 6, 36, and 37). The *readyEpoch* array is initialized in line 21. Note that the actual value is accessed by the *first* member of the *paddedInt* C++ pair class.

The barrier *wait* function increments the local PE's *epoch* and caches it together with *numChildren* into registers in lines 24 and 25. In the following `for` loop the PE waits until all its children (if any) have reached *myEpoch*. The reference to *readyEpoch*@$(iPE + 2^i)$ is precomputed (line 27) before the polling loop. When all children are ready, the PE's *readyEpoch* is updated with *myEpoch* (line 30). The following CPU memory fence is required to prevent possible CPU memory reordering of this update (and, in general, other operations before or after the barrier). The reference e to *readyEpoch*@(0) is precomputed in line 32 before the loop that polls e until it reaches the value of *myEpoch*.

Listing 13.2. An implementation of a binomial tree barrier in C++

```
class BinomialBarrier {                                                       1
  BinomialBarrier();                                                          2
  struct EpochNumChildren {                                                   3
    EpochNumChildren() : epoch(0), numChildren(0) {}                          4
    int epoch, numChildren; // co−locate to avoid additional cache miss       5
    char padding[64 − 2*sizeof(int)];                                         6
  };                                                                          7
  template <class It>                                                         8
  void initNumChildren(It begin, int size) {                                  9
    if(size < 2) return;                                                      10
    for(int i = 1; i < size; i *= 2) {                                        11
      ++(begin−>numChildren);                                                 12
      int child = i, childSize = i;                                           13
      if(size < child + childSize) childSize = size − child;                  14
      initNumChildren(begin + child, childSize);                             15
    }                                                                         16
  }                                                                           17
public:                                                                       18
  BinomialBarrier(int p) : readyEpoch(p), peData(p) {                         19
    initNumChildren(peData.begin(), p);                                       20
    for(auto && e : readyEpoch) e.first = 0;                                  21
  }                                                                           22
  void wait(const int iPE, const int /* p */) {                              23
    register const int myEpoch = ++(peData[iPE].epoch);                       24
    register const int numC = peData[iPE].numChildren;                        25
    for(int i=0; i < numC; ++i) {                                            26
      auto & e = readyEpoch[iPE + (1<<i)].first;                             27
      while(e != myEpoch) backoff();                                          28
    }                                                                         29
    readyEpoch[iPE].first = myEpoch;                                          30
    atomic_thread_fence(memory_order_seq_cst);                               31
    auto & e = readyEpoch[0].first;                                           32
    while(e != myEpoch) backoff();                                            33
  }                                                                           34
private:                                                                      35
  typedef std::pair<volatile int, char [64−sizeof(int)]> paddedInt;          36
  vector<paddedInt> readyEpoch;                                               37
  vector<EpochNumChildren> peData;                                            38
};//SPDX−License−Identifier: BSD−3−Clause; Copyright(c) 2018 Intel Corporation  39
```

To benchmark the barrier implementations, we chose a diffusion-like application that iteratively averages neighboring entries of an array:

Procedure *diffuse(a : Array* $[0..n+1]$ **of** *double, k : int)*
    b : Array $[0..n+1]$ **of** *double;*
    for $i := 1$ **to** k **step** *2* **do**
        for $j := 1$ *to* n **do**$\|$
            $b[i] := (a[i-1] + a[i] + a[i+1])/3.0;$
        barrier;
        for $j := 1$ *to* n **do**$\|$
            $a[j] := (b[j-1] + b[j] + b[j+1])/3.0;$
        barrier;

Table 13.2 shows running times and speedups using 64 and 128 threads. We used $k = 10^8 p/n$ iterations, i.e., we kept the number of arithmetic operations per thread fixed. The tests were done on the four-socket machine described in Sect. B. If the barrier synchronization overhead is significant compared with the amount of computation (small values of n/p), we observe speedups of up to 2.35 compared with the simple barrier. We also tested the standard barrier implementation from the POSIX PThread library. Unfortunately, it does not scale at all and did not finish even after several hours. The reason is that it uses an exclusive lock to manage the internal structure for every barrier synchronization; see github.com/lattera/glibc/blob/master/nptl/pthread_barrier_wait.c. On the other hand, we did experiments with the OpenMP barrier implementation of the Intel® C++ Compiler (version 18.0.0), which shows similar performance than our binomial tree barrier. It turns out that it uses a similar algorithm.

Table 13.2. Running times (in milliseconds, median of nine runs) of the barrier benchmark with 64 and 128 threads.

p	n/p	simple barrier	binomial tree barrier	speedup
64	100	6 826	3 702	1.84
64	1 000	895	580	1.54
64	10 000	335	318	1.05
64	100 000	282	279	1.01
64	1 000 000	2 371	2 341	1.01
128	100	8 012	3 413	2.35
128	1 000	1465	790	1.85
128	10 000	632	571	1.11
128	100 000	3 006	2 917	1.03
128	1 000 000	3 945	3 897	1.01

13.5 (All)-Gather/Scatter

13.5.1 (All)-Gather

Given a local message m on each PE, a gather operation moves all these messages $m@0, m@1, \ldots, m@(p-1)$ to PE 0 (or some other specified PE). All-gather moves these messages to *all* PEs. The all-gather operation is also called *gossiping* or *all-to-all broadcast*. These operations are sometimes needed to assemble partial results

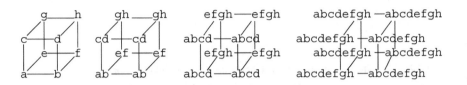

Fig. 13.11. Hypercube all-gather on eight PEs.

into a global picture. Whenever possible, gathers should be avoided since they can easily become a performance bottleneck. In a naive implementation of gather, all PEs send their message directly to PE 0. This takes communication time $(p-1)(\alpha + n\beta)$. We cannot do much about the term involving β, since PE 0 needs to receive all the messages in some way or the other. However, we can reduce the number of startup latencies. We simply use the binomial tree reduction algorithm presented in Sect. 13.2 using the concatenation of messages as the operator (see Fig. 13.12). This is correct since concatenation is associative. However, we have to redo the algorithm analysis, since the communicated messages have nonuniform length.

Theorem 13.9. *Gather using binomial tree reduction needs communication time*

$$\lceil \log p \rceil \alpha + (p-1)n\beta.$$

Proof. Counting from the bottom, in level i of the reduction tree, messages of length $2^i n$ are sent. This takes time $\alpha + 2^i n\beta$. Summing over all levels we get

$$\sum_{i=0}^{\lceil \log p \rceil - 1} \alpha + 2^i n\beta = \lceil \log p \rceil \alpha + n\beta \sum_{i=0}^{\lceil \log p \rceil - 1} 2^i = \lceil \log p \rceil \alpha + (p-1)n\beta. \qquad \square$$

Using the hypercube algorithm for all-reduce in Sect. 13.2.1 instead, we get an all-gather with the same complexity (when p is a power of two). Figure 13.11 gives an example.

Corollary 13.10. *All-gather using hypercube all-reduce needs communication time* $\alpha \log p + (p-1)n\beta.$

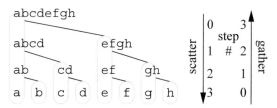

Fig. 13.12. Binomial tree gather and scatter for $p = 8$.

13.5.2 Scatter

When PE 0 (or some other specified PE) has messages $m_0, \ldots, m_{p-1}$ and delivers them so that PE i gets message m_i, this is called a *scatter* operation. This operation can be useful for distributing data evenly to all PEs. As with gather, the presence of a scatter operation may be an indicator of a program with poor scalability.

Scatter and gather and broadcast and reduction are, in a certain sense, duals of each other. Indeed, by slightly modifying the binomial tree broadcast algorithm in Sect. 13.1.1, we can get an efficient scatter algorithm: When sending to a subtree containing the nodes $a..b$, the concatenation of messages $m_a,\ldots,m_b$ is sent to the root of that subtree. Figure 13.12 gives an example. An analysis analogous to that in Theorem 13.9 yields execution time

$$\lceil \log p \rceil \alpha + (p-1)n\beta.$$

13.6 All-to-All Message Exchange

The most general communication operation is when every PE i has a message m_{ij} for every other PE j. We have seen this pattern for several sorting algorithms (bucket sort, radix sort, sample sort, and multiway mergesort) and for bulk operations on hash tables. We first look at the case where all messages have the same size n. In Sect. 13.6.1, we shall see how to achieve this task with direct delivery of data in communication time

$$T_{\text{all}\rightarrow\text{all}}(n) \le (p-1)(n\beta + \alpha). \tag{13.3}$$

For small messages, indirect delivery is better. Using a hypercube algorithm, we obtain

$$T_{\text{all}\rightarrow\text{all}}(n) \le \log p \left(n\frac{p}{2}\beta + \alpha \right), \tag{13.4}$$

see Sect. 13.6.2. In Sect. 13.6.3 we consider the more general case when messages are allowed to have different sizes. Let $h = \max_i \max(\sum_j |m_{ij}|, \sum_j |m_{ji}|)$ denote the maximum amount of data to be sent or received by a PE. We shall see a simple algorithm that needs communication time

$$T^*_{\text{all}\rightarrow\text{all}}(h) \le 2T_{\text{all}\rightarrow\text{all}}\left(\frac{h}{p} + 2p\right). \tag{13.5}$$

13.6.1 Uniform All-to-All with Direct Data Delivery

At first glance, all-to-all communication looks easy – simply send each message to its destination. However, our model of communication allows only one message to be sent at a time, and senders and receivers have to agree in which order the messages are sent. To keep the notation short, we abbreviate $i_{\text{proc}} \in 0..p-1$ to i in this section.

There is a particularly simple solution if p is a power of two: In step $k \in 1..p-1$, PE i communicates with PE $i \oplus k$. Since

$$(i \oplus k) \oplus k = i \oplus (k \oplus k) = i,$$

the senders and receivers agree on their communication partner. Moreover, every pair of PEs communicates in some step – PEs i and j communicate in step $i \oplus j$ (since $i \oplus (i \oplus j) = (i \oplus i) \oplus j = j$). We obtain an algorithm that needs communication time $(p-1)(\alpha + n\beta)$. This is optimal if messages are to be sent directly. Moreover, in each step the PEs are paired and the partners in each pair exchange their messages. This is referred to as the *telephone model* of point-to-point communication.

With full-duplex communication, we may also arrange the PEs into longer cycles (note that pairs are cycles of length two). The following code implements this idea:

for $k := 1$ **to** $p-1$ **do**
    $send((i+k) \bmod p, m_{i,(i+k) \bmod p}) \parallel receive((i-k) \bmod p, m_{(i-k) \bmod p,i})$

Exercise 13.10. Show the correctness of this solution.

We next give a solution in the telephone model for general p. We first show how to achieve the communication task for odd p with p steps and then derive a solution for even p with $p-1$ steps. The solution also applies to speed-dating and speed chess.

So assume p is odd. We first observe that the sequence $2r \bmod p$, $r \in 0..p-1$, first enumerates the even numbers $0, 2, \ldots p-1$ and then the odd numbers $1, 3, \ldots$, $p-2$ less than p. Consider any round r and let $k = 2r \bmod p$. We pair PEs i and j such that $i + j = k \bmod p$. Then $j = k - i \bmod p$ or

$$j = \begin{cases} k-i & \text{if } i \leq k \\ p+k-i & \text{if } i > k. \end{cases}$$

In this way, PE i will send to any PE (including itself) exactly once, and if PE i sends to j, j will send to i, i.e., we are within the telephone model. In each step, one PE communicates superfluously with itself. This PE has number $k/2$ if k is even and number $(p+k)/2$ if k is odd. Since $k = 2r$ if k is even and $k = 2r - p$ if k is odd, the idle PE has number r. In order to have this nice correspondence between round number and number of idle PE, we have k enumerate the numbers less than p in the order $0, 2, 4, \ldots, p-1, 1, 3, \ldots, p-2$.

The algorithm for even p is only slightly more complicated. We essentially run the algorithm for the case $p-1$. The only modification is that the "idle" PE communicates with PE $p-1$. We obtain the following protocol:

$$p' := p - isEven(p)$$
for $r := 0$ **to** $p' - 1$ **do**
$\quad k := 2r \bmod p'$
$\quad j := (k - i) \bmod p'$
$\quad$**if** $isEven(p) \wedge i = r \quad$ **then** $j := p - 1$
$\quad$**if** $isEven(p) \wedge i = p - 1 \quad$ **then** $j := r$
$\quad send(j, m_{ij}) \parallel receive(j, m_{ji})$

Figure 13.13 gives an example of the resulting communication pattern for this *1-factor algorithm*. The resulting communication time is $(p - isEven(p))(\alpha + n\beta)$.

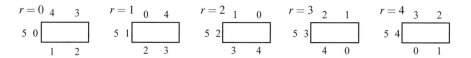

Fig. 13.13. Speed-dating for $p = 6$. PEs 0 to 4 sit around a rectangular table. Initially, PE 0 sits on one of the short sides of the table and the other PEs sit in counter-clockwise order along the long sides of the table. PE 5 sits separately next to PE 0. The PEs sitting on the long sides of the table partner with their counterpart on the other side of the table. In each round, PEs 0 to 4 move clockwise by one position. The index of the partner grows by two modulo $p - 1$ in each round. For example, the partners of PE 3 are 2, 4, 1, 5, 0 in the five rounds.

13.6.2 Hypercube All-to-All

For small n, direct data delivery is wasteful since the communication time will be dominated by startup overheads. Once more, we can reduce the number of startup overheads to $\log p$ using a hypercube algorithm when p is a power of two. The following pseudocode defines the algorithm:

$M := \{m_{i_{\mathrm{proc}} j} : j \in 0..p - 1\}$ *// messages i_{proc} has to deliver*
for $k := d - 1$ **downto** 0 **do**
$\quad M_s := \{m_{ij} \in M : i_{\mathrm{proc}}$ **bitand** $2^k \neq j$ **bitand** $2^k\}$*// messages for other k-cube*
$\quad send(i \oplus 2^k, M_s) \parallel receive(i \oplus 2^k, M_r)$
$\quad M := M_r \cup M \setminus M_s$

Each PE maintains a set M of messages it has to deliver. This time, we iterate from $d - 1$ downward because this simplifies formulating the loop invariant: After iteration k, all messages destined for a k-dimensional subcube are located somewhere in that subcube. Thus, the messages $m_{ij} \in M_s$ to be sent in each iteration are those where the kth bit of the recipient j differs from the kth bit of the local PE i_{proc}. Figure 13.14 gives an example. The algorithm sends and receives $p/2$ messages in each iteration. Thus, its communication complexity is

$$\log p \left(n \frac{p}{2} \beta + \alpha \right).$$

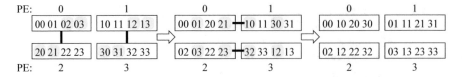

Fig. 13.14. Hypercube all-to-all with $p = 4$. The digit pair ij is a shorthand for m_{ij}. The boxes contain the message set in the current step. The shaded messages are moved in the next step along the communication links given by the bold lines.

This is good for small n but wasteful for large n since a message is moved $\log p/2$ times on average.

There are compromises between moving the data with p startups and moving it a logarithmic number of times with a logarithmic number of startups.

*Exercise 13.11. Suppose $p = k^d$ for integers k and d. Design an algorithm that delivers all messages using communication time $\leq kd\alpha + dnp\beta$. Hint: Arrange the PEs as a d-dimensional grid of side length k. Generalize the hypercube algorithm for this case. Use the direct delivery algorithm within each dimension. Generalize further for the case where $p = k_1 \cdot k_2 \cdots k_d$. Give an algorithm that needs communication time $\alpha \sum_i k_i + \beta dnp$.

13.6.3 All-to-All with Nonuniform Message Sizes

We now consider the case where the messages m_{ij} to be delivered have arbitrary sizes. A good way to measure the difficulty of such a problem is to look for the *bottleneck* communication volume – what is the maximum amount of data to be sent or received by any PE? In the full-duplex model of point-to-point communication, this is

$$h = \max_i \max \left(\sum_j |m_{ij}|, \sum_j |m_{ji}| \right).$$

We shall therefore also call this problem an *h-relation*. The uniform all-to-all algorithms we have seen so far can become quite slow if we use them naively by padding all messages to the maximum occurring message size. Even optimized implementations that deliver only the data actually needed could become slow because PEs currently delivering short messages will have to wait for those delivering long messages. We therefore need new algorithms.

An elegant solution is to reduce the all-to-all problem with variable message lengths to two all-to-all problems with a uniform message length. We chop each message m_{ij} into p pieces m_{ijk} of size $\lceil |m_{ij}|/p \rceil$ and send piece m_{ijk} from PE i to PE j via PE k. For the first uniform data exchange, PE i combines all messages m_{ijk}, $0 \leq j \leq p - 1$, into a single message m_{i*k} and sends it to PE k. The message consists of p fields specifying where the pieces begin plus the p pieces itself. Thus,

$$|m_{i*k}| = p + \sum_j |m_{ijk}| \leq p + \sum_j \left\lceil \frac{|m_{ij}|}{p} \right\rceil \leq p + \sum_j \left(\frac{|m_{ij}|}{p} + 1 \right) \leq \frac{h}{p} + 2p.$$

Note that this bound is independent of the individual message sizes and depends only on the global values h and p. Thus, the first data exchange can be performed in time $T_{\text{all}\to\text{all}}(h/p+2p)$ using any uniform all-to-all operation.

For the second data exchange, PE k combines the pieces $m_{ijk}, 0 \le i \le p-1$, into a message m_{*jk} and sends it to PE j. We have

$$|m_{*jk}| = p + \sum_i |m_{ijk}| \le p + \sum_i \left\lceil \frac{|m_{ij}|}{p} \right\rceil \le p + \sum_i \left(\frac{|m_{ij}|}{p} + 1 \right) \le \frac{h}{p} + 2p.$$

Once more, this can be done with a uniform all-to-all with message size $h/p+2p$. Figure 13.15 gives an example. Overall, assuming the regular all-to-all is implemented using direct data delivery, we get a total communication cost

$$T^*_{\text{all}\to\text{all}}(h) \le 2T_{\text{all}\to\text{all}}\left(\frac{h}{p}+2p^2\right) \approx 2p\alpha + (2h+4p)\beta.$$

This is not quite what we may want. For large h this is a factor of 2 away from the lower bound $h\beta$ that we may hope to approximate. For $h \ll p$, the overhead for rounding and communicating message sizes is an even worse penalty. There are approaches to improve the situation, but we are not aware of a general solution – routing h-relations in a practically and theoretically satisfactory way is still a research problem.

For example, there is an intriguing generalization of the graph theoretical model discussed in Sect. 13.6.1. Suppose the messages are partitioned into packets of uniform size. Now h measures the maximum communication cost in number of packets. We can model the communication problem as a bipartite graph $H = (\{s_1,\ldots,s_p\} \cup \{r_1,\ldots,r_p\}, E)$. Node s_i models the sending role of PE i and node r_i models its receiving role. A packet to be sent from PE i to PE j is modeled as an edge (s_i, r_j). Since messages can consist of multiple packets, *parallel edges* are allowed, i.e., we are dealing with a *multigraph*. Assuming packets are to be delivered directly, scheduling communication of an h-relation is now equivalent to coloring the edges of H. Consider a coloring $\chi : E \to 1..k$ where no two incident edges have the same color. Then all the packets corresponding to edges with color c can be delivered in a single step. Conversely, we can obtain a coloring of H for any schedule for the h-relation: If a packet x is delivered in step c, we can assign color c to the corresponding edges in H.

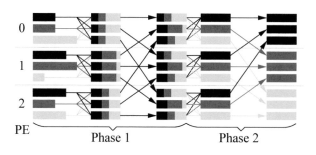

Fig. 13.15. Two-phase nonuniform all-to-all for $p = 3$.

Exercise 13.12. Prove this equivalence formally.

The maximum degree of H is h. It is known that a bipartite multigraph with maximum degree h can be edge-colored with h colors [188]; the maximum degree is clearly also a lower bound. Moreover, such a coloring can be found in polynomial time – actually, in time $O(|E|\log h)$ [77]. There are also fast parallel algorithms [202]

The discussion in the preceding paragraph suggests that delivering an h-relation may be possible in time close to $h\beta$ for large h. However, it is not so clear how to use an algorithm based on edge coloring in practice, since we would need to compute the coloring in parallel and highly efficiently.

13.7 Asynchronous Collective Communication

Sometimes one wants to overlap a collective communication with other computations. For example, the double counting termination detection protocol used in Sect. 14.5.3 requires a reduction operation running concurrently with the application. Most of the collective communication operations described here can be adapted to an asynchronous scenario. We must take care, however, to use asynchronous message send operations that do not have to wait for the actual delivery. We should also be aware that additional delays may be introduced by PEs that do not immediately react to incoming messages. One way to avoid such delays is to run the asynchronous operations using separate threads. Another useful concept is *active messages* that trigger a predefined behavior on the receiving side. For example, an asynchronous broadcast might trigger a transfer of the message to the children in a tree data structure. One can also emulate active messages by using a message handler within the application that can handle all possible asynchronous events by calling appropriate callback functions.

14

Load Balancing

Building a house comprises a large number of individual tasks such as digging a hole, laying the foundations, putting up walls, installing windows, wiring, and pipes, etc. Looking more closely, tasks consist of subtasks. For example, to put up a wall, you have to lay a large number of bricks. If you want to build fast, many workers have to cooperate. You have to plan which tasks can be done in parallel, what their dependencies are, and what resources (qualified workers, raw material, tools, space, ...) are needed. It is often difficult to estimate in advance how long a certain task is going to take. For example, raw materials may be delayed, a worker may fall ill, or a subtask may never have been done before. Hence, plans have to be revised as new information becomes available.

A parallel computation is quite similar to the construction site example above. When designing a parallel algorithm, we identify *tasks* (often also called *jobs*) that can be done in parallel. These tasks are assigned to PEs (workers) with possibly varying capabilities (qualifications of workers). The tasks may depend on each other, and they may require further resources such as memory or communication channels. We may or may not know how long a task is going to take (or what other resource requirements it has). *Load balancing* or *scheduling* is the process of assigning resources to tasks in order to make the overall computation efficient. Since the parallel algorithms in this book are fairly simple, we can usually get away with load-balancing algorithms that are also fast and simple and do not have to deal with all complications of the general problem. Nevertheless, we need several kinds of nontrivial load balancing techniques, which we describe in this chapter.

---

[0] The photograph shows a construction site in Hong Kong (Ding Yuin Shan www.flickr.com/photos/90461913@N00/5363967194/in/photostream/).

© Springer Nature Switzerland AG 2019
P. Sanders et al., *Sequential and Parallel Algorithms and Data Structures*,
https://doi.org/10.1007/978-3-030-25209-0_14

14.1 Overview and Basic Assumptions

Except in Sect. 14.6, we assume identical PEs and a set of m independent tasks. A justification for the independence assumption is that parallel algorithms often take dependencies into account explicitly. They split their computation into multiple phases so that the load balancer only has to deal with the independent tasks of a single phase. Section 14.2 deals with the case where we (pretend to) know exactly the work involved in each task. Then we can assign jobs very efficiently using prefix sums. The following three sections deal with the case where the size of the jobs is known only roughly or not at all. A straightforward solution is to choose a *master*, treat the other PEs as *workers* and let the master assign the jobs to the workers; see Sect. 14.3. However, this *master–worker scheme* does not scale well. When the number of jobs is large, Sect. 14.4 shows that it is quite effective to simply distribute the jobs randomly. Finally, Sect. 14.5 introduces a highly scalable dynamic load-balancing algorithm that can deal with widely varying and completely unknown task sizes provided that tasks can be split into subtasks when necessary. The idea is that idle PEs *steal* work from randomly chosen busy PEs. If an idle PE wants to steal work from a busy PE, the task at the busy PE is split into two subtasks, and the idle PE takes over one of the subtasks.

14.1.1 Independent Tasks

Many of the algorithms described below can be understood and analyzed as assigning m independent tasks to p PEs. Let t_j denote the size (or execution time) of task $j \in [m]$. We use the following notation in the discussion of the algorithms: the total execution time $T := \sum_i t_i$, the maximum task size $\hat{t} := \max_j t_j$, the average task size $\bar{t} := T/m$, and the *load* ℓ_i of PE i – the total size of the tasks assigned to PE i. The goal is to minimize the maximum load $\hat{\ell} := \max_i \ell_i$ (see also Sect. 12.2). Figure 14.1 illustrates these quantities.

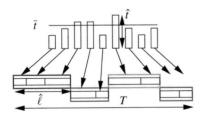

Fig. 14.1. An example of the quantities $\bar{t}$, $\hat{t}$, $\hat{\ell}$, and T that characterize sets of independent tasks. The large horizontal rectangles indicate PEs.

Table 14.1 compares the scalability of the load balancing algorithms presented below for independent jobs in the distributed-memory model, assuming that jobs are fully characterized by their ID number. We also ignore the cost for collecting the result. Clearly, it is easier to balance the load if there is more total work to be distributed. We want the algorithms to guarantee $T_{\text{par}} \leq (1 + \varepsilon)T/p$, i.e., optimal

Table 14.1. Scalability of four load-balancing algorithms for independent tasks. The second column shows the minimum required total work $T = \sum_j t_j$ for the algorithm to guarantee $T_{par} \leq (1+\varepsilon)T/p$. The startup overhead α is treated as a variable.

Algorithm	Sect.	$T = \Omega(\cdots)$	Remarks
prefix sum	14.2	$\dfrac{p}{\varepsilon}(\hat{t} + \alpha \log p)$	known task sizes
master–worker	14.3	$\dfrac{p}{\varepsilon} \cdot \dfrac{\alpha p}{\varepsilon} \cdot \dfrac{\hat{t}}{\bar{t}}$	bundle size $\sqrt{\dfrac{m\alpha}{\hat{t}}}$
randomized static	14.4	$\dfrac{p}{\varepsilon} \cdot \dfrac{\log p}{\varepsilon} \cdot \hat{t}$	randomized
work stealing	14.5	$\dfrac{p}{\varepsilon}(\hat{t} + \alpha \log p)$	randomized

balance up to a factor $1+\varepsilon$. We list for each algorithm the minimum total work required for this guarantee.

The prefix sum method is our "gold standard" – a simple method that is hard to improve upon if we know the task sizes. The other methods do not require task sizes but are more expensive in most cases. Work stealing comes within a constant factor of the gold standard in expectation. Randomized static load balancing behaves well when the tasks are very fine-grained, i.e., $\hat{t}/\varepsilon \ll \alpha$. Furthermore, randomized load balancing is very simple and can sometimes be done without explicitly moving the data. The master–worker scheme is the "ugly duckling" in this asymptotic comparison. It needs work quadratic in p, whereas the other methods only need work $p \log p$ and there are additional factors that can be big for large values of α, $1/\varepsilon$, or $\hat{t}/\bar{t}$. We include the master–worker scheme because it is simple and very popular in practice.

*Exercise 14.1. Prove the bounds given in Table 14.1 using the results discussed in Sections 14.2–14.5.

14.1.2 A Toy Example – Computing the Mandelbrot Set

We shall use a popular example from recreational mathematics as an example in Sect.s 14.4 and 14.5. In part, the discussion is similar to that in the book [281] including some of the images.

For every complex number $c \in \mathbb{C}$, we define the sequence $z_c(m) : \mathbb{N} \to \mathbb{C}$ as $z_c(0):=0$ and $z_c(m+1):=z_c(m)^2 + c$. The Mandelbrot set M is then defined as

$$M := \{c \in \mathbb{C} : z_c(m) \text{ is bounded}\} .$$

We want to know which points from a quadratic subset $A = \{z_0 + x + iy : x,y \in [0,a_0]\}$ of the complex plane belong to M. We shall approximate this question in two respects. First, we sample $n \times n$ grid points from this set (see Fig. 14.2): $c(j,k):=z_0 + (j+ik)a_0/n$ with $j,k \in 0..n-1$. Second, we consider only the first $\hat{t}$ members of the sequence $z_{c(j,k)}$. We exploit the mathematical fact that z_c is unbounded if $|z_c(t)| > 2$ for some t. Conversely, if $|z_c(t)| \leq 2$ for all $t \in 1..\hat{t}$, we assume that $c \in M$. Figure 14.2 shows an approximation to M.

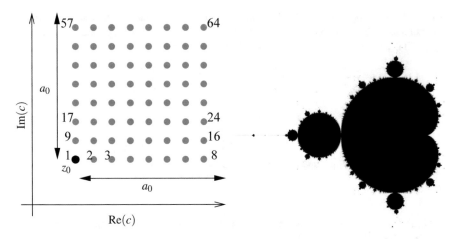

Fig. 14.2. *Left*: The points of the complex plane considered by our algorithm, and how they are numbered. *Right*: Approximation to the Mandelbrot set M.

There are quite sophisticated algorithms for computing the above approximation (e.g., [251]). We shall consider only a very simple method. We have $n \times n$ independent tasks. To simplify the description, we shall enumerate the tasks from 1 to $m = n^2$ as given in Fig. 14.2. The task for $c(j,k)$ iterates through $z_{c(j,k)}(t)$ and stops when this value becomes larger than 2 or when $t = \hat{t}$. Note that we have no a priori knowledge of how long a task will take.

14.1.3 Example – Parallel Breadth-First Search

An example, where we have a good estimate for the job size is parallel BFS in graphs (Sect. 9.2.4). In each iteration of the BFS algorithm, each node in the queue Q represents a job whose execution time is (roughly) proportional to the degree of the node. If some nodes have very high degree, we may have to design the implementation so that jobs are splittable, i.e., multiple PEs cooperate in exploring the edges out of a node.

14.2 Prefix Sums – Independent Tasks with Known Size

Suppose we have to assign m independent tasks with known sizes (execution times) t_j, $1 \le j \le m$. We first assume that tasks can be split between PEs and later remove this assumption. If tasks can be split, a simple idea works. We visualize task j as an interval of length t_j, arrange the intervals next to each other on a line segment of length $T = \sum_i t_i$, and cut the line segment into pieces of equal length, one for each PE. To get closer to an actual implementation, we assume from now on that the t_j's are integers and that we want to split only at integer positions.

Let $s = \lceil T/p \rceil$. We view the line segment as consisting of T unit intervals numbered starting at 1 and make PE i responsible for the unit intervals numbered $(i-1) \cdot s + 1$ to $i \cdot s$. Task j corresponds to the unit intervals numbered $T_j + 1$ to $T_j + t_j$, where $T_j = \sum_{k<j} t_k$ denotes the exclusive prefix sum up to task j. In this way, PEs $\lfloor T_j/s \rfloor + 1$ to $\lceil (T_j + t_j)/s \rceil$ are collectively responsible for task j. Note that $(i-1)s + 1 \leq T_j + 1$ iff $i \leq T_j/s + 1$ iff $i \leq \lfloor T_j/s \rfloor + 1$, where the last "iff" uses the fact that i is an integer. Similarly, $i \cdot s \geq T_j + t_j$ iff $i \geq (T_j + t_j)/s$ iff $i \geq \lceil (T_j + t_j)/s \rceil$.

Exercise 14.2. Show that task j is split into at most $\lceil t_j/s \rceil + 1$ pieces.

To compute the assignment in parallel, we essentially have to compute the prefix sums T_j. We can reduce this computation to local operations plus a prefix sum over one value per PE. This global prefix sum is discussed in Sect. 13.3. We use Brent's principle (see Sect. 2.10). Suppose PE i is responsible for assigning tasks $a_i..b_i$. PE i first locally calculates the local sum $s_i := \sum_{j \in a_i..b_i} t_j$. Then the global prefix sum $S_i := \sum_{k<i} s_k$ is computed. In a second pass over the local data, PE i computes $T_j = S_i + \sum_{a_i \leq k < j} t_k$ for $j \in a_i..b_i$. Overall, this takes time $O(\alpha \log p + \max_i(b_i - a_i))$ on a distributed-memory machine. On top of that, we may need time to move the tasks to the PEs they are assigned to. This depends on the size of the task descriptions, of course. We can, however, exploit the fact that consecutive tasks are moved to consecutive PEs.

Atomic Tasks. When tasks are nonsplittable, we face the scheduling problem already discussed in Sect. 12.2. "Atomic" is a synonym for "nonsplittable". While the basic greedy algorithm is inherently sequential, we can obtain a scalable parallel algorithm with similar performance guarantees by interpreting the prefix sums calculated for the splittable case differently. We can simply assign all of task j to PE $\lfloor T_j/s \rfloor + 1$ (see Fig. 14.3 for an example), i.e., instead of assigning the task to an interval of PEs starting at PE $\lfloor T_j/s \rfloor + 1$, we assign the entire task to this PE. This leads to good

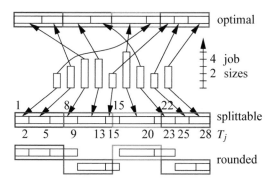

Fig. 14.3. Assigning 9 tasks to 4 PEs. We have $T = 28$ and $s = 28/4 = 7$. The yardstick on the right indicates the job sizes. Three assignments are shown. *Top*: optimal, with $\hat{\ell} = 7$. *Middle*: using prefix sums assuming jobs can be split, $\hat{\ell} = 7$. *Bottom*: using prefix sums and rounding, $\hat{\ell} = 9$. Note that the first three jobs are assigned to PE 1 since $\lfloor T_3/s \rfloor + 1 = \lfloor 5/7 \rfloor + 1 = 1$.

load balance when the task sizes are much smaller than the contingent s of each PE. Indeed, prefix-sum-based assignment represents a two-approximation algorithm for the problem of assigning tasks to identical PEs.

Exercise 14.3. Prove the above claim. Hint: Recall that ℓ_i denotes the load of PE i and that $\hat{\ell} = \max_i \ell_i$ and $\hat{t} = \max_j t_j$. Show the following three facts and use them to prove that the prefix-sum-based assignment is within a factor two of optimal:

(a) $\hat{\ell} \leq s + \hat{t}$ for the prefix-sum-based algorithm.
(b) $\hat{\ell} \geq \hat{t}$ for any schedule of atomic tasks.
(c) $\hat{\ell} \geq s$ for any schedule.
(d) If $\hat{t} \leq \varepsilon s$, the prefix-sum-based assignment is within a factor of $1 + \varepsilon$ of optimal.

14.3 The Master–Worker Scheme

One of the most widely used dynamic load-balancing algorithms uses a dedicated *master* PE to deal out tasks to *worker* PEs. Initially, the master has all tasks, and all workers are idle. An idle worker sends a work request to the master. The master waits for requests and sends one task to the requesting PE. When all tasks have been sent, the master informs requesting PEs about this. Figure 14.4 illustrates this scheme. The master–worker scheme is easy to implement even with a rudimentary communication infrastructure. This scheme may also be a good choice if the tasks are generated by a sequential (legacy) algorithm or are read from a centralized I/O device. Another advantage is that even *without* knowing the task sizes, good load balance is achieved – disregarding the scheduling overhead, one will get the same balancing quality as with the prefix-sum-based algorithm presented in the previous section.

Load balancing can be further improved if at least an estimate of the task sizes is available. It is then a good idea to deal out large tasks first. Exercise 12.8 mentions an approximation ratio of $4/3$ in a similar situation.

Unsurprisingly, the main disadvantage of the master–worker scheme is that it does not scale to large p. Unless the average task size is much larger than αp, the startup overheads alone will already limit performance. If there is a large number

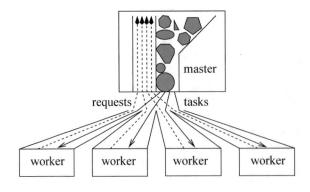

Fig. 14.4. The master–worker load-balancing scheme

of small tasks, one can mitigate this effect by dealing out *task bundles* – multiple tasks at a time. Unfortunately, this introduces a trade-off between load imbalance and scheduling overhead. Let us analyze this for a fixed bundle size b and small task descriptions. The master has to spend time $\geq \alpha m/b$ on sending out task bundles. The last bundle sent out might incur work $b\hat{\iota}$. Hence the parallel execution time is at least $\alpha m/b + b\hat{\iota}$. Using calculus, we can determine the optimal bundle size as $b = \sqrt{\alpha m/\hat{\iota}}$. Note that this consideration is only useful if we have a reasonable estimate of $\alpha/\hat{\iota}$. A somewhat better but more complicated strategy is to change the bundle size dynamically – beginning with a large bundle size and decreasing it as the computation progresses. It can be argued that a near-optimal strategy is to try constructing a bundle involving total work $W/(pC) + W_{\min}$ where W is an estimate of the total work not yet handed out and where C and $W_{\min}$ are tuning parameters that depend on the accuracy of the work estimation [29].

**Exercise 14.4. Investigate whether using this strategy can improve the scalability of the master–worker scheme in shown in Table 14.1 so that the factor $1/\varepsilon^2$ is replaced by $\frac{1}{\varepsilon} \log \frac{1}{\varepsilon}$.

A more sophisticated approach to making the master–worker scheme more scalable is to make it hierarchical – an overall master deals out task bundles to submasters, which deal out smaller bundles to subsubmasters, etc. Unfortunately, this destroys the main advantage of the basic scheme – its simplicity. In particular, we get tuning parameters that are harder to control. When you are in the situation that a basic scheme does not work well. Thinking about clean new strategies is often better than making the basic scheme more and more complicated. For example, one could consider the relaxed FIFO queue described in Sect. 3.7.2 as a truly scalable means of producing and consuming tasks.

Shared-Memory Master–Worker. If the tasks are stored in an array $t[1..m]$, a simple version of the master–worker scheme can be implemented without an explicit PE for the master. We simply use a shared counter to indicate the next task to be delegated. This also has limited scalability for large p, but the constant factors are much better than in the distributed-memory case.

The Mandelbrot Example. Dealing out individual tasks is not very scalable here, in particular, if most of the area under consideration is not in the Mandelbrot set M. In that case, most of the iterations may terminate after a small number of iterations. Bundling multiple tasks is easy in the sense that we can describe them succinctly as a range of task numbers, which we can easily convert into the starting number z.

Exercise 14.5. Develop a formula $c(i)$ that computes the parameter c of the sequence z_c from a task number i.

Furthermore, we can derive an upper bound on the execution time of a task from the maximum iteration count $\hat{\iota}$. However, estimating the size of k tasks as taking $k\hat{\iota}$ iterations may grossly overestimate the actual execution time, whereas using more aggressive bounds may be wrong when all tasks in a bundle correspond to members of M. We conclude that using the master–worker scheme with many PEs can cause major headaches even for trivial applications.

14.4 (Randomized) Static Load Balancing

A very simple and seemingly bad way to do load balancing is to simply assign a similar number of jobs to each PE without even considering their size. Figure 14.5 illustrates three different such static assignment strategies, which we shall now discuss for the Mandelbrot example. The most naive way to implement this idea is to map n/p consecutively numbered tasks to each PE or, even simpler, to assign a fixed number of rows to each PE. This makes displaying the results easy but looking at Fig. 14.2, one can see that this is a bad idea – some rows have many more elements of M than others, leading to bad load balance. A slightly more sophisticated approach is *round-robin* scheduling – task j is mapped to PE $j \bmod p$. For the Mandelbrot example, this is likely to work well. However, we get no performance guarantee in general. Indeed, for every fixed way to map tasks to PEs, there will be distributions of task sizes that lead to bad load balance.

Once more, randomization allows us to solve this problem. One can show that, for arbitrary distributions of task sizes, a random assignment of tasks to PEs leads to good load balance with high probability if certain conditions are met.

Theorem 14.1 ([268, 269]). *Suppose m independent tasks of size t_1, ..., t_m are mapped to random PEs. Let $\hat{t} := \max_j t_j$ and $T := \sum_j t_j$. Then the expected maximum load $E[\hat{\ell}]$ of a PE is no worse than for the case where we randomly map $\lceil T/\hat{t} \rceil$ tasks of size $\hat{t}$. In particular, for any constant $\varepsilon > 0$,*

$$E[\hat{\ell}] \leq (1+\varepsilon)\frac{T}{p} \quad if \quad \frac{T}{\hat{t}} \geq \frac{c}{\varepsilon^2}p\log p \quad for some appropriate constant c.$$

This is a very good bound if $\hat{t}$ is not too much larger than the average task size and if we can live with a large value of the imbalance ε. Decreasing the imbalance requires increasing the input size as $1/\varepsilon^2$.

One strength of randomized static load balancing is that a random distribution of the input is often enough to enforce a random distribution of the work *in every phase of a computation*. For example, consider the distributed-memory BFS discussed in Sect. 9.2.3. By assigning the nodes to random PEs, we ensure good expected load balance in every layer of the BFS algorithm which considers a sufficiently large number of edges, namely $\Omega(\Delta p\log p)$, where Δ is the maximum degree of the graph.

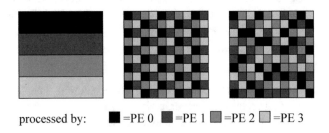

processed by: ■=PE 0   ■=PE 1   ■=PE 2   □=PE 3

Fig. 14.5. Block decomposition (*left*), round-robin (*middle*), and random decomposition (*right*) of a two-dimensional range

14.4.1 *Using Pseudorandom Permutations to Avoid Communication

In some situations, randomized static load balancing can be implemented essentially without communication (beyond broadcasting a description of the input). Consider the Mandelbrot example. Initially, we broadcast the parameters n, z_0, and a_0 describing the input. From these parameters and a task index $j \in 1..n$ we can reconstruct the tasks without communication. Hence, each PE needs only to know the indices of the tasks mapped to it. At first glance, this requires computing a (pseudo)random mapping $f : 1..n \to 1..p$ and communicating every i to PE $f(i)$. We can avoid the communication by using a (pseudo)random *permutation* $\pi : 1..n \to 1..n$ rather than a random function and making PE i responsible for tasks $\pi(j)$ for $j \in (i-1)\lceil n/p \rceil + 1..i\lceil n/p \rceil$. Effectively, this means that tasks are mapped using the inverse permutation π^{-1} of π. Since the inverse of a random permutation is also random, this means that the jobs are mapped using a random permutation rather than a random mapping. This is not the same. For example, a random permutation maps at most $\lceil n/p \rceil$ tasks to each PE, whereas a random mapping will, in general, allocate tasks less evenly. It can be shown that such a random-permutation mapping is at least as good as a plain random mapping [170, 268].

Although random permutations can be generated in parallel [270], this also requires communication. However, there are simple algorithms for computing *pseudorandom* permutations directly. *Feistel* permutations [237] are one possible approach. Assume for simplicity that $\sqrt{n}$ is an integer[1] and represent j as $j = j_a + j_b\sqrt{n}$. Now consider the mapping

$$\pi_k((j_a, j_b)) = (j_b, j_a + f_k(j_b) \bmod \sqrt{n}),$$

where $f_k : 1..\sqrt{n} \to 1..\sqrt{n}$ is a hash function.

Exercise 14.6. Prove that π_k is a permutation.

It is known that a permutation $\pi(x) = \pi_1(\pi_2(\pi_3(\pi_4(x))))$ built by chaining four Feistel permutations is "pseudorandom" even in a cryptographic sense if the f_k's are sufficiently random. The same holds if the innermost and outermost permutations are replaced by even simpler permutations [237]. This approach allows us to derive a pseudorandom permutation of $1..n$ from a small number of hash functions $f_k : 1..\sqrt{n} \to 1..\sqrt{n}$. When $\sqrt{n} = O(n/p)$, it is even feasible to use tables of random numbers to implement f_k.

14.5 Work Stealing

The load-balancing algorithms we have seen so far are simple but leave a lot to be desired. prefix-sum-based scheduling works only when we know the task sizes. The

---

[1] For general n, we can round n up to the next square number and generate corresponding empty jobs.

master–worker scheme suffers from its centralized design and the fact that task bundles may commit too much work to a single PE. Randomized static load balancing is not dynamic at all.

We now present a simple, fully distributed dynamic load-balancing algorithm that avoids these shortcomings and, in consequence, guarantees much better performance. This approach can even handle dependent jobs and fluctuating resource availability [20]. However, to keep things simple and understandable, we describe the approach for independent computations and refer the reader to the literature and Sect. 14.6 for more.

14.5.1 Tree-Shaped Computations

We use a simple but rather abstract model of the application – *tree-shaped computations* [273]. Initially, all the work is subsumed in a *root* task J_{root}. We can perform two basic operations on a task – *working* on it sequentially, and *splitting* it. When sequential processing finishes a task, we are left with an *empty* task J_0. Splitting replaces a task by two tasks representing the work of the parent task. Very little is assumed about the outcome of a task split, in particular, the two subproblems may have different sizes. One subproblem may even be empty. In that case, the parent task is *atomic*, i.e., unsplittable. Initially, the root task resides on PE 1. The computation terminates when all remaining tasks are empty. Figure 14.6 illustrates the concept. A tree-shaped computation can be represented by a binary tree. Each node of this tree consists of some (possibly zero) sequential computation, followed by a split (for internal nodes) or by a transition to an empty task (for leaves).

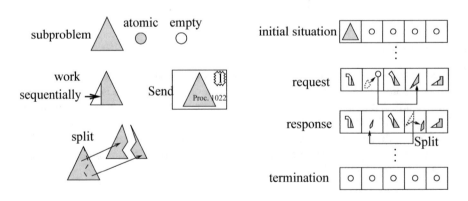

Fig. 14.6. Tree-shaped computations

Tree-shaped computations subsume many applications. For example, when applied to the independent job model, we obtain a fully distributed multilevel generalization of the master–worker scheme where every PE is both a master and a worker. We can also handle applications such as the tree exploration described in Sect. 12.4.2 for the knapsack problem, where we do not have a fixed set of jobs up front.

In order to analyze the performance of a load-balancing algorithm for tree-shaped computations, we need to model the cost of splitting and moving tasks. We also need some way to measure the performance of the splitting operation. This performance is expressed by a bound t_{atomic} on the size of an atomic task and by the *splitting depth* h, which is the maximum number of subsequent splits needed to arrive at an atomic task. In other words, h is a bound on the height of the task tree, disregarding splits of atomic tasks.

Example: Independent tasks. These can be translated into tree-shaped computations as follows: Let the range $a..b$ represent the task of processing the (atomic) tasks numbered a to b. Then the root problem is $J_{\text{root}} = 1..m$. Splitting $a..b$ yields $a..\lfloor(a+b)/2\rfloor$ and $\lfloor(a+b)/2\rfloor+1..b$. Atomic tasks have size at most $t_{\text{atomic}} = \hat{t}$, and the splitting depth is $h = \lceil\log m\rceil$.

14.5.2 The Randomized Work-Stealing Algorithm

We now describe a simple and efficient load-balancing algorithm for tree-shaped computations. This algorithm maintains the invariant that each PE works on exactly one task. Initially, this means that PE 1 works on the root task J_{root} while all other PEs have an empty task $J_\emptyset$.

Exercise 14.7. Work out the details of an optional fast initialization for the work-stealing algorithm that first broadcasts J_{root} and then locally splits it $\lfloor\log p\rfloor$ or $\lceil\log p\rceil$ times, each time throwing away either the first or the second subproblem. Afterwards, the locally present tasks should still represent the entire computation.

After initialization, a busy PE (one working on a nonempty task) works on it sequentially. An idle PE i, i.e., one with an empty task, sends a work request to another, random PE and waits for a reply. This is repeated until PE i receives a nonempty task. A PE receiving a work request splits its local task, keeps one part, and sends the other part to the requesting PE. Idle PEs or PEs working on an atomic task may not actually call the general splitting routine and also need not send a full-blown task description. Rather, they can simply respond with a rejection. However, we prefer to interpret this rejection as a compact representation of an empty task in order to underline the simplicity of the algorithm.

We still need to discuss one issue – how can we make a PE working sequentially respond to requests? On a distributed-memory machine, one solution is to periodically check for incoming messages during sequential work. On a shared-memory machine, the task description could be a concurrent data structure that allows concurrent splitting and sequential work. For example, our "packaging" of independent tasks into tree-shaped computations could implement the required operations by simple atomic updates of the range $a..b$ of unprocessed tasks. Working sequentially means, most of the time, working on task $a-1$. When this task is finished, a is atomically incremented. Work stealing atomically updates a to $\lfloor(a+b)/2\rfloor+1$. The same approach can also be used on a distributed-memory machine if, on each node, we keep a separate thread for handling incoming messages.

We shall not give a complete analysis of the asynchronous work-stealing algorithm described above, but instead analyze a related synchronous algorithm. Nevertheless, this analysis still illuminates the essential ideas behind the general analysis. We consider a *random shift* algorithm. This algorithm works in synchronized *rounds* of duration Δ, large enough to allow sending a request, splitting a subproblem, and sending one piece back to the requestor. A PE i that is idle at the beginning of a round sends a request to PE $(i+k \bmod p)+1$, where $k \in 1..p-1$ is a uniformly distributed random value that is the *same* on all PEs. The receiving PE answers this request by splitting its local subproblem and sending one of the pieces to the requesting PE. Note that each PE receives at most one request per round, since all PEs use the same k. We make the further simplifying assumption that busy PEs that do *not* receive a request in a round are in no way delayed by the load-balancing process. In other words, in a round an unfinished nonatomic task is either split or completed, or its unfinished work is reduced by Δ. If the task is split, the unfinished work is distributed over the two subtasks. An atomic task is either finished in a round or its unfinished work is reduced by Δ. The task tree has depth h and hence at most 2^h leaves. Each leaf stands for an atomic task, and atomic tasks require at most work t_{atomic}.

Theorem 14.2. *For[2] $h \geq \log p$ and any constant $d > 0$, the random shift algorithm has an execution time bounded by*

$$\frac{T_{\text{seq}}}{p} + (3ch+3)\Delta + t_{\text{atomic}} = \frac{T_{\text{seq}}}{p} + O(t_{\text{atomic}} + h\Delta)$$

with probability at least $1 - p^{-d}$. Here, $c = 2 + 2(d+1)\ln 2$.

Proof. Let

$$k = \left\lceil \frac{T_{\text{seq}}}{p\Delta} + 3ch \right\rceil + 1. \tag{14.1}$$

We shall show that, with probability at least $1 - p^{-d}$, all unfinished tasks are atomic after k rounds. Assume this is the case. Then the parallel execution time is at most $k\Delta + \lceil t_{\text{atomic}}/\Delta \rceil \Delta \leq (k+1)\Delta + t_{\text{atomic}}$.

Suppose there are m_i idle PEs at the beginning of round i. Each idle PE issues a split request, and each such request may hinder a PE that is working on a task. Moreover, up to m_{i+1} PEs may become idle during that period and hence not do the full amount Δ of work. We conclude that in round i, at least $p - 2m_i - m_{i+1}$ PEs do Δ units of useful sequential work each. Let $M := \sum_{1 \leq i \leq k} m_i$. Since the useful sequential work accounted for in this way cannot be more than the sequential execution time, we obtain

$$T_{\text{seq}} \geq \sum_{i=1}^{k-1}(p - 2m_i - m_{i+1})\Delta \geq (k-1)p\Delta - 3\Delta M.$$

Thus

_____

[2] Note that the case $h < \log p$ is not interesting, since the number of nontrivial tasks generated is at most 2^h and hence some PEs will always stay idle.

$$k \leq 1 + \frac{T_{\text{seq}} + 3\Delta M}{p\Delta}. \tag{14.2}$$

In combination with (14.1), we obtain $M \geq chp$.

Consider any nonempty task t after round k. Let the random variable $X_i \in \{0,1\}$ express the event that the ancestor of t was split in round i. Then t is the result of $X := \sum_i X_i$ splits. If $X \geq h$, t is atomic. We have $\text{prob}(X_i = 1) = m_i/(p-1)$ since in round i, m_i split requests are issued and for each split request there is one choice of k that will lead to a split of the ancestor of t. Thus $\text{E}[X] = \sum_i m_i/(p-1) \geq M/p$.

We want to use Chernoff bounds (see Sect. A.3) to bound the probability that X is less than h. For this, we require that the X_i's are independent. They are not, because m_i and hence the probability that X_i is equal to 1 depend on the preceding rounds. However, for any fixed choice of the m_i's, the X_i's are independent (they depend only on the random shift values). Hence, we may apply Chernoff bounds if we afterwards take the worst case over the choices of m_i's. We shall see that only the sum of the m_i's is relevant and hence we are fine.

Let $\varepsilon = 1 - 1/c$. Then $(1 - \varepsilon)\text{E}[X] \geq 1/c \cdot M/p \geq h$, and hence the probability that X is fewer than h is at most $\gamma := e^{-(1-1/c)^2 ch/2}$. The probability that *any* task is split less than h times is then bounded by $2^h \gamma$. Recall that there are at most 2^h leaf tasks. We show that $2^h \gamma \leq p^{-d}$ for our choice of c. For this it suffices to have (take natural logarithms on both sides)

$$-h \ln 2 + \frac{1}{2}\left(1 - \frac{1}{c}\right)^2 ch \geq d \ln p.$$

Since $h \geq \log p = (\ln p)/\ln 2$, this holds true for the choice $c = 2 + 2(d+1)\ln 2$. Namely,

$$\frac{1}{2}\left(1 - \frac{1}{c}\right)^2 ch \geq \frac{1}{2}(c-2)h \geq (d+1)h\ln 2 \geq d\ln p + h\ln 2. \qquad \square$$

Let us apply the result to allocating m independent tasks in order to establish the entry in Table 14.1. We have $h = \lceil \log m \rceil$, $t_{\text{atomic}} = \hat{t}$ and $\Delta = \text{O}(\alpha)$. Thus, Theorem 14.2 yields a parallel execution time $T/p + \text{O}(\hat{t} + \alpha \log m)$. If $m = \Omega(p^2)$, the term $T/p = \hat{t}m/p$ asymptotically dominates the term $\alpha \log m$. Then

$$\frac{T}{p} + \text{O}(\hat{t}) \leq (1+\varepsilon)\frac{T}{p} \quad \Leftrightarrow \quad \text{O}(\hat{t}) \leq \varepsilon\frac{T}{p} \quad \Leftrightarrow \quad T = \Omega(p\hat{t}/\varepsilon).$$

If $m = \text{O}(p^2)$, we have $\log m = \text{O}(\log p)$ and hence

$$\frac{T}{p} + \text{O}(\hat{t} + \alpha \log p) \leq (1+\varepsilon)\frac{T}{p} \quad \Leftrightarrow \quad T = \Omega\left(\frac{p}{\varepsilon}(\hat{t} + \alpha \log p)\right).$$

14.5.3 Termination Detection

The randomized work-stealing algorithm as described above does not detect when all PEs have become idle – they will simply keep sending work requests in vain.

One solution is that a PE i receiving a task t is responsible for reporting back to the PE from which t was received when all the work in t has been performed. In order to do this, PE i has to keep track of how many subtasks have been split away from t. Only when termination responses for all these subtasks have been received and the remaining work has been performed sequentially, can PE i send the termination response for t. Ultimately, PE 1 will receive termination responses for the subtasks split away from J_{root}. At this point it can broadcast a termination signal to all PEs. The overhead for this protocol does not affect the asymptotic cost of load balancing – the termination responses mirror the pieces of work sent. The final broadcast adds a latency $O(\alpha \log p)$. This termination detection protocol can also be used to perform final result calculations.

Double Counting Termination Detection. Even faster termination detection is possible if we adapt a general purpose termination detection protocol suitable for asynchronous message exchange. Mattern [209] describes such a method. We count the number of received and completed tasks locally. On PE 1, we initially set the receive counter to 1 for the root problem. All other counters are initially 0. We repeatedly run an asynchronous reduction algorithm to determine the global sums of these two counters (see also Sect. 13.7). The corresponding messages are sent when a PE is idle and has received the messages from its children in the reduction tree. If the global sums of sent and completed subproblems are equal *and* the next cycle yields exactly the same global sums, then the system has globally terminated. If we replace the linear-latency summation algorithms given in [209] with our logarithmic algorithms, we get overall latency $O(\log p)$ for the termination detection.

14.5.4 Further Examples: BFS and Branch-and-Bound

We discuss two further examples here.

Breadth-First Search. On a shared-memory machine, we can use work stealing as a load balancer in each iteration of BFS.[3] The simple variant in Sect. 9.2.1 that parallelizes over the nodes in the queue could view each node in the queue as an independent job. We get $h = \lceil \log n \rceil$ and t_{atomic} proportional to the maximum node degree.

The algorithm for handling high-degree nodes in Sect. 9.2.4 could be adapted for work stealing by making tasks splittable even when they only work on a single node. In principle, an atomic task could simply represent the operations performed on a single edge. This might be too fine-grained, though, since there is some overhead for reserving an edge for sequential processing. We can amortize this overhead by always assigning a bundle of k edges to a PE. Theorem 14.2 tells us that this has little effect on the load-balancing quality as long as $k = O(\log n)$.

---

[3] On distributed memory, global work stealing would require more communication than we can afford. We could use randomized static load balancing to assign graph nodes to compute nodes. Cores within a compute node could use work stealing as described here.

Branch-and-Bound for the Knapsack Problem. Let n be the number of items. The algorithm backtracks through a binary tree of depth n that enumerates all solutions that cannot be excluded using upper and lower bounds (see Sect. 12.4.2). A task is a subtree of this tree. By always splitting off the highest unexplored subtree of a task, we obtain a splitting depth of n. For describing a split off at level k of the tree, it suffices to transfer a bit vector of length k indicating which decisions have been made up to level k.

14.6 Handling Dependencies

A quite general way to specify a parallel computation is to view it as a directed acyclic graph (DAG) $G = (V, E)$. Tasks are nodes of this graph and edges express *dependencies*, i.e., an edge (u, v) indicates that v can only be executed after u has been completed. As before, let t_v denote the execution time of task v. This is a natural generalization of our model of independent tasks considered in Sect. 14.1.1 and also in Sect. 12.2. We have already seen task DAGs for integer multiplication (Fig. 1.5) and for tiled computation on arrays (Fig. 3.2). Even disregarding data flow between the tasks, it is already an interesting question how one should assign the tasks to the PEs. It is **NP**-hard to find a schedule that minimizes the parallel execution time (the *makespan*). On the other hand, any "reasonable" scheduling algorithm achieves a two-approximation.

In order to explain this statement in more detail, we need to introduce some terminology. A schedule x assigns each task j to a PE $p_j \in 1..p$ and specifies a starting time $s_j \geq 0$. A schedule is feasible if the task execution times on any particular processor do not overlap (i.e., $\forall i, j : p_i = p_j \Rightarrow s_j \notin [s_i, s_i + t_i))$ and no task starts before it is ready for execution (i.e., $\forall j : \forall (i, j) \in E : s_j \geq s_i + t_i$). The makespan is the last finishing time, $\max_j(s_j + t_j)$, of a job. We are looking for a schedule with minimum

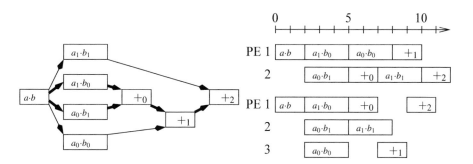

Fig. 14.7. Two optimal schedules for the recursive multiplication algorithm of Sect. 1.4 for two-digit numbers. One schedule is for two PEs and the other for three PEs. It is assumed that a recursive call and an addition/shift cost 2 units of time while a multiplication costs 3 units. The thick edges indicate the critical paths (of length 11). A schedule drawn as above with one line per PE is also known as a *Gantt chart*.

makespan. A trivial lower bound for the makespan is the average work $\sum_j t_j/p$. Another lower bound is the length $\sum_{j \in P} t_j$ of any path P in G. The *critical path length* is the maximum such length over all paths in G. Figure 14.7 gives an example.

Theorem 14.3. *Consider any schedule that never leaves a PE idle when a task is ready for execution. Then its makespan is at most the average work plus the critical path length. This is a two-approximation of the optimal schedule.*

Proof. Let $G = (V, E)$ be the scheduling problem and let T denote the makespan of the schedule. Partition $[0, T]$ into (at most $2|V|$) intervals $I_1, \ldots, I_k$ such that jobs start or finish only at the beginning or end of an interval. Call an interval *busy* if all p processors are active during that interval and call it *idle* otherwise. Then T is the total length of the busy intervals plus the total length of the idle intervals. The total length of the busy intervals is at most $\sum_j t_j$. Now consider any path P through G and any idle interval. Since the schedule leaves no ready job idle, some job from P must be executing during the interval or all jobs on P must have finished before the interval. Thus the length of P is bounded by the total length of the idle intervals.

Since both the average work and the critical path length are lower bounds for the makespan, their sum must be a two-approximation of the makespan.   □

A more careful analysis yields an approximation ratio of $2 - 1/p$. Improving upon this seems difficult. The only known better bounds increase the constant factor in the $1/p$ term [120]. We view this as a good reason to stick to the simple schedules characterized above. In particular, Theorem 14.3 applies even when the execution times are unknown and G unfolds with the computation. We only have to make sure that idle PEs find ready jobs efficiently. All the load-balancing algorithms described in this section can be adapted for this purpose.

Master–Worker. The master is informed about finished tasks. When a task becomes ready, it is inserted into the queue of tasks that can be handed out to idle PEs.

Randomized static. Each PE executes the ready jobs assigned to it.

Work stealing. Multithreaded computations [20] define a computation DAG implicitly by spawning tasks and waiting for them to finish. It can be shown that randomized work stealing leads to asymptotically optimal execution time. Compared with our result for tree-shaped computations, this is a more general result but also a constant factor worse with respect to the T/p term. Also, in practice, we might observe this constant factor because a multithreaded computation needs to generate the entire computation graph whereas tree-shaped computations only split work when this is actually needed.

A

Mathematical Background

A.1 Mathematical Symbols

$\{e_0,\ldots,e_{n-1}\}$: set containing elements e_0, ..., e_{n-1}.

$\{e : P(e)\}$: set of all elements that fulfill the predicate P.

$\langle e_0,\ldots,e_{n-1}\rangle$: sequence consisting of elements e_0, ..., e_{n-1}.

$\langle e \in S : P(e)\rangle$: subsequence of all elements of sequence S that fulfill the predicate P.

$|x|$: the absolute value of x, for a real number x.

$\lfloor x \rfloor$: the largest integer $\leq x$, for a real number x.

$\lceil x \rceil$: the smallest integer $\geq x$, for a real number x.

$[a,b] := \{x \in \mathbb{R} : a \leq x \leq b\}$.

$i..j$: abbreviation for $\{i, i+1, \ldots, j\}$.

A^B: the set of all functions mapping B to A.

$A \times B$: the set of ordered pairs (a,b) with $a \in A$ and $b \in B$.

$\bot$: an undefined value.

$(+/-)\infty$: (plus/minus) infinity.

$\forall x : P(x)$: For *all* values of x, the proposition $P(x)$ is true.

$\exists x : P(x)$: There *exists* a value of x such that the proposition $P(x)$ is true.

$\mathbb{N}$: nonnegative integers; $\mathbb{N} = \{0, 1, 2, \ldots\}$.

$\mathbb{N}_+$: positive integers; $\mathbb{N}_+ = \{1, 2, \ldots\}$.

© Springer Nature Switzerland AG 2019
P. Sanders et al., *Sequential and Parallel Algorithms and Data Structures*,
https://doi.org/10.1007/978-3-030-25209-0

$\mathbb{Z}$: integers.

$\mathbb{R}$: real numbers.

$\mathbb{R}_{>0}/\mathbb{R}_{\geq 0}$: positive/nonnegative real numbers.

$\mathbb{Q}$: rational numbers.

$|, \&, \ll, \gg, \oplus$: bitwise OR, bitwise AND, shift left, shift right, and exclusive OR (XOR) respectively.

$\sum_{i=1}^{n} a_i = \sum_{1 \leq i \leq n} a_i = \sum_{i \in 1..n} a_i := a_1 + a_2 + \cdots + a_n$.

$\prod_{i=1}^{n} a_i = \prod_{1 \leq i \leq n} a_i = \prod_{i \in 1..n} a_i := a_1 \cdot a_2 \cdots a_n$.

$n! := \prod_{i=1}^{n} i$, the *factorial* of n.

$H_n := \sum_{i=1}^{n} 1/i$, the nth *harmonic number* (see (A.13)).

$a \cdot b := \sum_{i=1}^{n} a_i b_i$ is the dot product of vectors $a = (a_1, \ldots, a_n)$ and $b = (b_1, \ldots, b_n)$.

$\log x$: the logarithm to base two of x, $\log_2 x$, for $x > 0$.

$\log^* x$, for $x > 0$: the smallest integer k such that $\underbrace{\log(\log(\ldots \log(x) \ldots))}_{k\text{-fold application}} \leq 1$.

$\ln x$: the (natural) logarithm of x to base $e = 2.71828\ldots$.

$\mu(s,t)$: the shortest-path distance from s to t; $\mu(t) := \mu(s,t)$.

div: integer division; $m \operatorname{div} n := \lfloor m/n \rfloor$.

mod : modular arithmetic; $m \bmod n = m - n(m \operatorname{div} n)$.

$a \equiv b \bmod m$: a and b are congruent modulo m, i.e., $a + im = b$ for some integer i.

$\prec$: some ordering relation. In Sect. 9.3, it denotes the order in which nodes are marked during depth-first search.

1, 0: the Boolean values "true" and "false".

Σ^*: The set $\{\langle a_1, \ldots, a_n \rangle : n \in \mathbb{N}, a_1, \ldots, a_n \in \Sigma\}$ of all *strings* or *words* over Σ. We usually write $a_1 \ldots a_n$ instead of $\langle a_1, \ldots, a_n \rangle$.

$|x|$: The number n of characters in a word $x = a_1 \ldots a_n$ over Σ; also called the length of the word.

A.2 Mathematical Concepts

antisymmetric: A relation $R \subseteq A \times A$ is *antisymmetric* if for all a and b in A, $a \, R \, b$ and $b \, R \, a$ implies $a = b$.

associative: An operation $\otimes$ with the property that $(x\otimes y)\otimes z = x\otimes(y\otimes z)$ for all x, y, and z.

asymptotic notation:

$$O(f(n)) := \{g(n) : \exists c > 0 : \exists n_0 \in \mathbb{N}_+ : \forall n \geq n_0 : g(n) \leq c\cdot f(n)\}.$$
$$\Omega(f(n)) := \{g(n) : \exists c > 0 : \exists n_0 \in \mathbb{N}_+ : \forall n \geq n_0 : g(n) \geq c\cdot f(n)\}.$$
$$\Theta(f(n)) := O(f(n))\cap\Omega(f(n)).$$
$$o(f(n)) := \{g(n) : \forall c > 0 : \exists n_0 \in \mathbb{N}_+ : \forall n \geq n_0 : g(n) \leq c\cdot f(n)\}.$$
$$\omega(f(n)) := \{g(n) : \forall c > 0 : \exists n_0 \in \mathbb{N}_+ : \forall n \geq n_0 : g(n) \geq c\cdot f(n)\}.$$

See also Sect. 2.1.

concave: A function f is concave on an interval $[a,b]$ if

$$\forall x,y \in [a,b] : \forall t \in [0,1] : f(tx+(1-t)y) \geq tf(x)+(1-t)f(y),$$

i.e., the function graph is never below the line segment connecting the points $(x,f(x))$ and $(y,f(y))$.

convex: A function f is convex on an interval $[a,b]$ if

$$\forall x,y \in [a,b] : \forall t \in [0,1] : f(tx+(1-t)y) \leq tf(x)+(1-t)f(y),$$

i.e., the function graph is never above the line segment connecting the points $(x,f(x))$ and $(y,f(y))$.

equivalence relation: a transitive, reflexive, and symmetric relation.

field: a set of elements (with distinguished elements 0 and 1) that support addition, subtraction, multiplication, and division by nonzero elements. Addition and multiplication are associative and commutative, and have neutral elements analogous to 0 and 1 for the real numbers. The most important examples are $\mathbb{R}$, the real numbers; $\mathbb{Q}$, the rational numbers; and $\mathbb{Z}_p$, the integers modulo a prime p.

iff: abbreviation for "if and only if".

lexicographic order: the canonical way of extending a total order on a set of elements to tuples, strings, or sequences over that set. We have $\langle a_1,a_2,\ldots,a_k\rangle < \langle b_1,b_2,\ldots,b_\ell\rangle$ if and only if there is an $i \leq \min\{k,\ell\}$ such that $\langle a_1,a_2,\ldots,a_{i-1}\rangle = \langle b_1,b_2,\ldots,b_{i-1}\rangle$ and $a_i < b_i$ or if $k < \ell$ and $\langle a_1,a_2,\ldots,a_k\rangle = \langle b_1,b_2,\ldots,b_k\rangle$. An equivalent recursive definition is as follows: $\langle\rangle < \langle b_1,b_2,\ldots,b_\ell\rangle$ for all $\ell > 0$; for $k > 0$ and $\ell > 0$, $\langle a_1,a_2,\ldots,a_k\rangle < \langle b_1,b_2,\ldots,b_\ell\rangle$ if and only if $a_1 < b_1$ or $a_1 = b_1$ and $\langle a_2,\ldots,a_k\rangle < \langle b_2,\ldots,b_\ell\rangle$.

linear order: (also total order) a reflexive, transitive, antisymmetric, and total relation. Linear orders are usually denoted by the symbol $\leq$. For $a \leq b$, one also writes $b \geq a$. The strict linear order $<$ is defined by $a < b$ if and only if $a \leq b$ and

$a \neq b$. The relation $<$ is transitive, irreflexive ($a < b$ implies $a \neq b$), and total in the sense that for all a and b either $a < b$ or $a = b$ or $a > b$. A typical example is the relation $<$ for real numbers.

linear preorder: (also linear quasi-order) a reflexive, transitive, and total relation. The symbols $\leq$ and $\geq$ are also used for linear preorders. Note that there can be distinct elements a and b with $a \leq b$ and $b \leq a$. The strict variant $<$ is defined as $a < b$ if $a \leq b$ and not $a \geq b$. An example is the relation $R \subseteq \mathbb{R} \times \mathbb{R}$ defined by $x \, R \, y$ if and only if $|x| \leq |y|$.

median: an element with rank $\lceil n/2 \rceil$ among n elements.

multiplicative inverse: If an object x is multiplied by a *multiplicative inverse* x^{-1} of x, we obtain $x \cdot x^{-1} = 1$ – the neutral element of multiplication. In particular, in a *field*, every element except 0 (the neutral element of addition) has a unique multiplicative inverse.

prime number: An integer n, $n \geq 2$, is a prime if and only if there are no integers $a, b > 1$ such that $n = a \cdot b$.

rank: Let $\leq$ be a linear preorder on a set $S = \{e_1, \dots, e_n\}$. A one-to-one mapping $r : S \to 1..n$ is a *ranking function* for the elements of S if $r(e_i) < r(e_j)$ whenever $e_i < e_j$. If $\leq$ is a linear order, there is exactly one ranking function.

reflexive: A relation $R \subseteq A \times A$ is reflexive if $a \, R \, a$ for all $a \in A$.

relation: a set of ordered pairs R over some set A. Often we write relations as infix operators; for example, if $R \subseteq A \times A$ is a relation, $a \, R \, b$ means $(a, b) \in R$.

symmetric relation: A relation $R \subseteq A \times A$ is *symmetric* if for all a and b in A, $a \, R \, b$ implies $b \, R \, a$.

total order: a synonym for linear order.

total relation: A relation $R \subseteq A \times A$ is *total* if for all a and b in A, either $a \, R \, b$ or $b \, R \, a$ or both. If a relation R is total and transitive, then the relation $\sim_R$ defined by $a \sim_R b$ if and only if $a \, R \, b$ and $b \, R \, a$ is an equivalence relation.

transitive: A relation $R \subseteq A \times A$ is *transitive* if for all a, b, and c in A, $a \, R \, b$ and $b \, R \, c$ imply $a \, R \, c$.

A.3 Basic Probability Theory

Probability theory rests on the concept of a *sample space* $\mathscr{S}$. For example, to describe the rolls of two dice, we would use the 36-element sample space $\{1, \dots, 6\} \times \{1, \dots, 6\}$, i.e., the elements of the sample space (also called elementary events or simply events) are the pairs (x,y) with $1 \leq x,y \leq 6$ and $x,y \in \mathbb{N}$. Generally, a sample space is any nonempty set. In this book, all sample spaces are finite.[1] In a *random experiment*, any element of $s \in \mathscr{S}$ is chosen with some elementary *probability* p_s, where $\sum_{s \in \mathscr{S}} p_s = 1$. The function that assigns to each event s its probability p_s is called a *distribution*. A sample space together with a probability distribution is called a *probability space*. In this book, we use *uniform distributions* almost exclusively; in this case $p_s = p = 1/|\mathscr{S}|$. Subsets $\mathscr{E}$ of the sample space are called *events*. The probability of an *event* $\mathscr{E} \subseteq \mathscr{S}$ is the sum of the probabilities of its elements, i.e., $\mathrm{prob}(\mathscr{E}) = |\mathscr{E}|/|\mathscr{S}|$ in the uniform case. So the probability of the event $\{(x,y) : x+y = 7\} = \{(1,6),(2,5),\ldots,(6,1)\}$ is equal to $6/36 = 1/6$, and the probability of the event $\{(x,y) : x+y \geq 8\}$ is equal to $15/36 = 5/12$.

A *random variable* is a mapping from the sample space to the real numbers. Random variables are usually denoted by capital letters to distinguish them from plain values. For our example of rolling two dice, the random variable X could give the number shown by the first die, the random variable Y could give the number shown by the second die, and the random variable S could give the sum of the two numbers. Formally, if $(x,y) \in \mathscr{S}$, then $X((x,y)) = x$, $Y((x,y)) = y$, and $S((x,y)) = x+y = X((x,y))+Y((x,y))$.

We can define new random variables as expressions involving other random variables and ordinary values. For example, if V and W are random variables, then $(V+W)(s) = V(s)+W(s)$, $(V \cdot W)(s) = V(s) \cdot W(s)$, and $(V+3)(s) = V(s)+3$.

Events are often specified by predicates involving random variables. For example, $X \leq 2$ denotes the event $\{(1,y),(2,y) : 1 \leq y \leq 6\}$, and hence $\mathrm{prob}(X \leq 2) = 1/3$. Similarly, $\mathrm{prob}(X+Y = 11) = \mathrm{prob}(\{(5,6),(6,5)\}) = 1/18$.

Indicator random variables are random variables that take only the values 0 and 1. Indicator variables are an extremely useful tool for the probabilistic analysis of algorithms because they allow us to encode the behavior of complex algorithms into simple mathematical objects. We frequently use the letters I and J for indicator variables. Indicator variables and events are in a one-to-one correspondance. If $\mathscr{E}$ is an event, then $I_{\mathscr{E}}$ with $I_{\mathscr{E}}(s) = 1$ if and only if $s \in \mathscr{E}$ is the corresponding indicator variable. If an event is specified by a predicate P, one sometimes writes $[P]$ for the corresponding indicator variable, i.e., $[P](s) = 1$ if $P(s)$ and $[P](s) = 0$ otherwise.

The *expected value* of a random variable $Z : \mathscr{S} \to \mathbb{R}$ is

$$E[Z] = \sum_{s \in \mathscr{S}} p_s \cdot Z(s) = \sum_{z \in \mathbb{R}} z \cdot \mathrm{prob}(Z = z), \tag{A.1}$$

i.e., every sample s contributes the value of Z at s times its probability. Alternatively, we can group all s with $Z(s) = z$ into the event $Z = z$ and then sum over the $z \in \mathbb{R}$.

In our example, $E[X] = (1+2+3+4+5+6)/6 = 21/6 = 3.5$, i.e., the expected value of the first die is 3.5. Of course, the expected value of the second die is also 3.5. For an indicator random variable I we have

---

[1] All statements made in this section also hold for countable infinite sets, essentially with the same proofs. Such sample spaces are, for example, needed to model the experiment "throw a die repeatedly until the value six occurs".

$$E[I] = 0 \cdot \text{prob}(I = 0) + 1 \cdot \text{prob}(I = 1) = \text{prob}(I = 1).$$

Sometimes we are more interested in a random variable Z and its behavior than in the underlying probability space. In such a situation, it suffices to know the range $Z[\mathscr{S}]$ of Z and the induced probabilities $\text{prob}(Z = z)$, $z \in Z[\mathscr{S}]$. We refer to the function $z \mapsto \text{prob}(Z = z)$ defined on $Z[\mathscr{S}]$ as the *distribution of Z*. Two random variables X and Y with the same distribution are called *identically distributed*.

For a random variable Z that takes only values in the natural numbers, there is a very useful formula for its expected value:

$$E[Z] = \sum_{k \geq 1} \text{prob}(Z \geq k), \quad \text{if } Z[\mathscr{S}] \subseteq \mathbb{N}. \tag{A.2}$$

This formula is easy to prove. For $k, i \in \mathbb{N}$, let $p_k = \text{prob}(Z \geq k)$ and $q_i = \text{prob}(Z = i)$. Then $p_k = \sum_{i \geq k} q_i$ and hence

$$E[Z] = \sum_{z \in Z[\mathscr{S}]} z \cdot \text{prob}(Z = z) = \sum_{i \in \mathbb{N}} i \cdot \text{prob}(Z = i) = \sum_{i \in \mathbb{N}} \sum_{1 \leq k \leq i} q_i = \sum_{k \geq 1} \sum_{i \geq k} q_i = \sum_{k \geq 1} p_k.$$

Here, the next to last equality is a change of the order of summation.

Often we are interested in the expectation of a random variable that is defined in terms of other random variables. This is particulary easy for sums of random variables due to the *linearity of expectations* of random variables: For any two random variables V and W,

$$E[V + W] = E[V] + E[W]. \tag{A.3}$$

This equation is easy to prove and extremely useful. Let us prove it. It amounts essentially to an application of the distributive law of arithmetic. We have

$$E[V + W] = \sum_{s \in \mathscr{S}} p_s \cdot (V(s) + W(s))$$
$$= \sum_{s \in \mathscr{S}} p_s \cdot V(s) + \sum_{s \in \mathscr{S}} p_s \cdot W(s)$$
$$= E[V] + E[W].$$

As our first application, let us compute the expected sum of two dice. We have

$$E[S] = E[X + Y] = E[X] + E[Y] = 3.5 + 3.5 = 7.$$

Observe that we obtain the result with almost no computation. Without knowing about the linearity of expectations, we would have to go through a tedious calculation:

$$E[S] = 2 \cdot \tfrac{1}{36} + 3 \cdot \tfrac{2}{36} + 4 \cdot \tfrac{3}{36} + 5 \cdot \tfrac{4}{36} + 6 \cdot \tfrac{5}{36} + 7 \cdot \tfrac{6}{36} + 8 \cdot \tfrac{5}{36} + 9 \cdot \tfrac{4}{36} + \ldots + 12 \cdot \tfrac{1}{36}$$
$$= \frac{2 \cdot 1 + 3 \cdot 2 + 4 \cdot 3 + 5 \cdot 4 + 6 \cdot 5 + 7 \cdot 6 + 8 \cdot 5 + \ldots + 12 \cdot 1}{36} = 7.$$

Exercise A.1. What is the expected sum of three dice?

We shall now give another example with a more complex sample space. We consider the experiment of throwing n balls into m bins. The balls are thrown at random and distinct balls do not influence each other. Formally, our sample space is the set of all functions f from $1..n$ to $1..m$. This sample space has size m^n, and $f(i)$, $1 \le i \le n$, indicates the bin into which the ball i is thrown. All elements of the sample space are equally likely. How many balls should we expect in bin 1? We use W to denote the number of balls in bin 1. To determine $E[W]$, we introduce indicator variables I_i, $1 \le i \le n$. The variable I_i is 1 if ball i is thrown into bin 1 and is 0 otherwise. Formally, $I_i(f) = 0$ if and only if $f(i) \ne 1$. Then $W = \sum_i I_i$. We have

$$E[W] = E\left[\sum_i I_i\right] = \sum_i E[I_i] = \sum_i \mathrm{prob}(I_i = 1),$$

where the second equality is the linearity of expectations and the third equality follows from the I_i's being indicator variables. It remains to determine the probability that $I_i = 1$. Since the balls are thrown at random, ball i ends up in any bin[2] with the same probability. Thus $\mathrm{prob}(I_i = 1) = 1/m$, and hence

$$E[I] = \sum_i \mathrm{prob}(I_i = 1) = \sum_i \frac{1}{m} = \frac{n}{m}.$$

Products of random variables behave differently. In general, we have $E[X \cdot Y] \ne E[X] \cdot E[Y]$. There is one important exception: If X and Y are *independent*, equality holds. Random variables $X_1, \ldots, X_k$ are independent if and only if

$$\forall x_1, \ldots, x_k : \mathrm{prob}(X_1 = x_1 \wedge \cdots \wedge X_k = x_k) = \prod_{1 \le i \le k} \mathrm{prob}(X_i = x_i). \qquad (A.4)$$

As an example, when we roll two dice, the value of the first die and the value of the second die are independent random variables. However, the value of the first die and the sum of the two dice are not independent random variables.

Exercise A.2. Let I and J be independent indicator variables and let $X = (I+J)$ mod 2, i.e., X is 1 if and only if I and J are different. Show that I and X are independent, but that I, J, and X are dependent.

We will next show

$$E[X \cdot Y] = E[X] \cdot E[Y] \quad \text{if } X \text{ and } Y \text{ are independent.}$$

We have

---

[2] Formally, there are exactly m^{n-1} functions f with $f(i) = 1$.

$$E[X] \cdot E[Y] = \left(\sum_x x \cdot \mathrm{prob}(X = x) \right) \cdot \left(\sum_y y \cdot \mathrm{prob}(X = y) \right)$$

$$= \sum_{x,y} x \cdot y \cdot \mathrm{prob}(X = x) \cdot \mathrm{prob}(X = y)$$

$$= \sum_{x,y} x \cdot y \cdot \mathrm{prob}(X = x \wedge Y = y)$$

$$= \sum_z \sum_{x,y \text{ with } z = x \cdot y} z \cdot \mathrm{prob}(X = x \wedge Y = y)$$

$$= \sum_z z \cdot \sum_{x,y \text{ with } z = x \cdot y} \mathrm{prob}(X = x \wedge Y = y)$$

$$= \sum_z z \cdot \mathrm{prob}(X \cdot Y = z)$$

$$= E[X \cdot Y].$$

How likely is it that a random variable will deviate substantially from its expected value? *Markov's inequality* gives a useful bound. Let X be a nonnegative random variable and let c be any constant. Then

$$\mathrm{prob}(X \geq c \cdot E[X]) \leq \frac{1}{c}. \tag{A.5}$$

The proof is simple. We have

$$E[X] = \sum_{z \in \mathbb{R}} z \cdot \mathrm{prob}(X = z)$$

$$\geq \sum_{z \geq c \cdot E[X]} z \cdot \mathrm{prob}(X = z)$$

$$\geq c \cdot E[X] \cdot \mathrm{prob}(X \geq c \cdot E[X]),$$

where the first inequality follows from the fact that we sum over a subset of the possible values and X is nonnegative, and the second inequality follows from the fact that the sum in the second line ranges only over z such that $z \geq c E[X]$.

Much tighter bounds are possible for some special cases of random variables. The following situation arises several times in the book. We have a sum $X = X_1 + \cdots + X_n$ of n independent indicator random variables $X_1, \ldots, X_n$ and want to bound the probability that X deviates substantially from its expected value. In this situation, the following variant of the *Chernoff bound* is useful. For any $\varepsilon > 0$, we have

$$\mathrm{prob}(X < (1 - \varepsilon)E[X]) \leq e^{-\varepsilon^2 E[X]/2}, \tag{A.6}$$

$$\mathrm{prob}(X > (1 + \varepsilon)E[X]) \leq \left(\frac{e^\varepsilon}{(1 + \varepsilon)^{(1 + \varepsilon)}} \right)^{E[X]}. \tag{A.7}$$

A bound of the form above is called a *tail bound* because it estimates the "tail" of the probability distribution, i.e., the part for which X deviates considerably from its expected value.

Let us see an example. If we throw n coins and let X_i be the indicator variable for the ith coin coming up heads, $X = X_1 + \cdots + X_n$ is the total number of heads. Clearly, $E[X] = n/2$. The bound above tells us that $\text{prob}(X \le (1 - \varepsilon)n/2) \le e^{-\varepsilon^2 n/4}$. In particular, for $\varepsilon = 0.1$, we have $\text{prob}(X \le 0.9 \cdot n/2) \le e^{-0.01 \cdot n/4}$. So, for $n = 10\,000$, the expected number of heads is 5000 and the probability that the sum is less than 4500 is smaller than e^{-25}, a very small number.

Exercise A.3. Estimate the probability that X in the above example is larger than 5050.

If the indicator random variables are independent and identically distributed with $\text{prob}(X_i = 1) = p$, X is *binomially distributed*, i.e.,

$$\text{prob}(X = k) = \binom{n}{k} p^k (1-p)^{(n-k)}. \tag{A.8}$$

Exercise A.4 (balls and bins continued). As above, let W denote the number of balls in bin 1. Show that

$$\text{prob}(W = k) = \binom{n}{k} \left(\frac{1}{m}\right)^k \left(1 - \frac{1}{m}\right)^{(n-k)},$$

and then attempt to compute $E[W]$ as $\sum_k \text{prob}(W = k)k$.

A.4 Useful Formulae

We shall first list some useful formulae and then prove some of them:

- *A simple approximation to the factorial:*

$$\left(\frac{n}{e}\right)^n \le n! \le n^n \quad \text{or, more precisely} \quad e\left(\frac{n}{e}\right)^n \le n! \le (en)\left(\frac{n}{e}\right)^n. \tag{A.9}$$

- *Stirling's approximation to the factorial:*

$$n! = \left(1 + O\left(\frac{1}{n}\right)\right) \sqrt{2\pi n} \left(\frac{n}{e}\right)^n. \tag{A.10}$$

- *An approximation to the binomial coefficients:*

$$\binom{n}{k} \le \left(\frac{n \cdot e}{k}\right)^k. \tag{A.11}$$

- *The sum of the first n integers:*

$$\sum_{i=1}^{n} i = \frac{n(n+1)}{2}. \tag{A.12}$$

- *The harmonic numbers*:

$$\ln n \leq H_n = \sum_{i=1}^{n} \frac{1}{i} \leq \ln n + 1. \tag{A.13}$$

- *The geometric series*:

$$\sum_{i=0}^{n-1} q^i = \frac{1-q^n}{1-q} \quad \text{for } q \neq 1 \text{ and } \quad \sum_{i \geq 0} q^i = \frac{1}{1-q} \quad \text{for } |q| < 1. \tag{A.14}$$

$$\sum_{i \geq 0} 2^{-i} = 2 \quad \text{and} \quad \sum_{i \geq 0} i \cdot 2^{-i} = \sum_{i \geq 1} i \cdot 2^{-i} = 2. \tag{A.15}$$

- *Jensen's inequality*:

$$\sum_{i=1}^{n} f(x_i) \leq n \cdot f\left(\frac{\sum_{i=1}^{n} x_i}{n}\right) \tag{A.16}$$

for any concave function f. Similarly, for any convex function f,

$$\sum_{i=1}^{n} f(x_i) \geq n \cdot f\left(\frac{\sum_{i=1}^{n} x_i}{n}\right). \tag{A.17}$$

- *Approximations to the logarithm following from the Taylor series expansion*:

$$x - \frac{1}{2}x^2 \leq \ln(1+x) \leq x - \frac{1}{2}x^2 + \frac{1}{3}x^3 \leq x. \tag{A.18}$$

A.4.1 Proofs

For (A.9), we first observe that $n! = n(n-1)\cdots 1 \leq n^n$. For the lower bound, we recall from calculus that $e^x = \sum_{i \geq 0} x^i / i!$ for all x. In particular, $e^n \geq n^n / n!$ and hence $n! \geq (n/e)^n$.

We now come to the sharper bounds. Also, for all $i \geq 2$, $\ln i \geq \int_{i-1}^{i} \ln x \, dx$, and therefore

$$\ln n! = \sum_{2 \leq i \leq n} \ln i \geq \int_{1}^{n} \ln x \, dx = \left[x(\ln x - 1)\right]_{x=1}^{x=n} n(\ln n - 1) + 1.$$

Thus

$$n! \geq e^{n(\ln n - 1) + 1} = e(e^{\ln n - 1})^n = e(n/e)^n.$$

For the upper bound, we use $\ln(i-1) \leq \int_{i-1}^{i} \ln x \, dx$ and hence $(n-1)! \leq \int_{1}^{n} \ln x \, dx = e(n/e)^n$. Thus $n! \leq (en)(n/e)^n$.

Equation (A.11) follows almost immediately from (A.9). We have

$$\binom{n}{k} = \frac{n(n-1)\cdots(n-k+1)}{k!} \leq \frac{n^k}{(k/e)^k} = \left(\frac{n \cdot e}{k}\right)^k.$$

Equation (A.12) can be computed by a simple trick:

$$1+2+\ldots+n = \frac{1}{2}((1+2+\ldots+n-1+n)+(n+n-1+\ldots+2+1))$$
$$= \frac{1}{2}((n+1)+(2+n-1)+\ldots+(n-1+2)+(n+1))$$
$$= \frac{n(n+1)}{2}.$$

The sums of higher powers are estimated easily; exact summation formulae are also available. For example, $\int_{i-1}^{i} x^2\, dx \leq i^2 \leq \int_{i}^{i+1} x^2\, dx$, and hence

$$\sum_{1 \leq i \leq n} i^2 \leq \int_{1}^{n+1} x^2\, dx = \left[\frac{x^3}{3}\right]_{x=1}^{x=n+1} = \frac{(n+1)^3 - 1}{3}$$

and

$$\sum_{1 \leq i \leq n} i^2 \geq \int_{0}^{n} x^2\, dx = \left[\frac{x^3}{3}\right]_{x=0}^{x=n} = \frac{n^3}{3}.$$

For (A.13), we also use estimation by integral. We have $\int_{i}^{i+1}(1/x)\,dx \leq 1/i \leq \int_{i-1}^{i}(1/x)\,dx$, and hence

$$\ln n = \int_{1}^{n} \frac{1}{x}\,dx \leq \sum_{1 \leq i \leq n} \frac{1}{i} \leq 1 + \int_{1}^{n} \frac{1}{x}\,dx = 1 + \ln n.$$

Equation (A.14) follows from

$$(1-q) \cdot \sum_{0 \leq i \leq n-1} q^i = \sum_{0 \leq i \leq n-1} q^i - \sum_{1 \leq i \leq n} q^i = 1 - q^n.$$

If $|q| < 1$, we may let n pass to infinity. This yields $\sum_{i \geq 0} q^i = 1/(1-q)$. For $q = 1/2$, we obtain $\sum_{i \geq 0} 2^{-i} = 2$. Also,

$$\sum_{i \geq 1} i \cdot 2^{-i} = \sum_{i \geq 1} 2^{-i} + \sum_{i \geq 2} 2^{-i} + \sum_{i \geq 3} 2^{-i} + \ldots$$
$$= (1 + 1/2 + 1/4 + 1/8 + \ldots) \cdot \sum_{i \geq 1} 2^{-i}$$
$$= 2 \cdot 1 = 2.$$

For the first equality, observe that the term 2^{-i} occurs in exactly the first i sums on the right-hand side.

Equation (A.16) can be shown by induction on n. For $n = 1$, there is nothing to show. So assume $n \geq 2$. Let $x^* = \sum_{1 \leq i \leq n} x_i / n$ and $\bar{x} = \sum_{1 \leq i \leq n-1} x_i / (n-1)$. Then $x^* = ((n-1)\bar{x} + x_n)/n$, and hence

$$\sum_{1\leq i\leq n} f(x_i) = f(x_n) + \sum_{1\leq i\leq n-1} f(x_i)$$

$$\leq f(x_n) + (n-1)\cdot f(\bar{x}) = n\cdot\left(\frac{1}{n}\cdot f(x_n) + \frac{n-1}{n}\cdot f(\bar{x})\right)$$

$$\leq n\cdot f(x^*),$$

where the first inequality uses the induction hypothesis and the second inequality uses the definition of concavity with $x = x_n$, $y = \bar{x}$, and $t = 1/n$. The extension to convex functions is immediate, since convexity of f implies concavity of $-f$.

B

Computer Architecture Aspects

In Sect. 2.2, we introduced several basic models of parallel computing, and then in Sects. 2.4.3 and 2.15 hinted at aspects that make practical parallel computing more complex (see Fig. 2.6 for an overview). However, in order to understand some aspects of our practical examples, one needs a little more background. The purpose of this appendix is to fill this gap in a minimalistic way – learning about these aspects from sources dedicated to them might still be a good idea.

We discuss a concrete example – the machine used to run our examples of shared-memory programs. This example will nevertheless lead to general insights, as the most relevant aspects of its architecture have been stable for more than a decade and essentially also apply to processors from other vendors. For more details of computer architecture in general, see the textbook by Hennessy and Patterson [149]. Details of the x86 architecture can be found in the Architecture Reference Manual [161].

B.1 Cores and Hardware Threads

We programmed our shared-memory examples on a machine with four Intel Xeon E7-8890 v3 processors (previously codenamed Haswell-EX). Each of these processors has 18 *cores*. A processor core has one or several arithmetical and logical units for executing instructions and a *pipeline* for controlling the execution of an instruction stream. The instructions come from up to two *hardware threads*. Each hardware thread has a dedicated set of registers for storing operands. Most other resources of a core are shared by the two threads.[1] Since x86 is a *CISC* (complex instruction set computer) architecture, it has a large range of instructions. Instructions can perform a mix of memory access and arithmetic operations and their numbers of operands and

---

[1] The hope is that the two hardware threads of a core will substantially increase the performance of the core. Because each hardware thread has its own set of registers, switching between them incurs almost no cost. *Software threads* are managed by the operating system. Activating a software thread requires it to be made into a hardware thread; in particular, the registers must be loaded. The switch therefore incurs substantial cost.

© Springer Nature Switzerland AG 2019
P. Sanders et al., *Sequential and Parallel Algorithms and Data Structures*,
https://doi.org/10.1007/978-3-030-25209-0

encoding lengths vary. Since CISC instructions are difficult to process in a pipeline, they are first translated to *microinstructions* that are similar to the simple machine instructions used in the machine model in Sect. 2.2 (RISC – reduced instruction set computer).

The pipeline processes micro-instructions in up to 19 *stages*, where each stage is responsible for some small substep of the execution (such as decoding, fetching operands, . . .). Thus, each stage becomes very fast, allowing high clock frequencies. Additional *instruction parallelism* is introduced because our machine is *superscalar*, i.e., up to eight instructions can be in the same pipeline stage at the same time. The hardware (assisted by the compiler) automatically extracts instruction parallelism from a sequential program in order to keep this pipeline reasonably well filled. In particular, the instructions might be executed in a different order than that specified by the machine program as long as this does not change the semantics of the (sequential) program. Thus, a core can execute several instructions in every clock cycle.

There are many reasons why the number eight is rarely reached; including data dependencies between instructions, large memory access latencies (cache misses), and conditional branches. Thus, a typical sequential program will most of the time use only a small fraction of the available arithmetical units or slots in the execution pipeline. This waste can be reduced using the two hardware threads sharing these resources.[2] Thus, in some codes with many high-latency operations, the two hardware threads do twice the work that a single thread could do. Usually however, the performance gain is smaller and can even be negative, for example with highly optimized numerical libraries that are carefully tuned to avoid cache misses and to use all the available arithmetical units. Other examples are codes that suffer from software overhead of additional threads (additional work for parallelization, scheduling, and locking). *Thus the PEs we are talking about in this book are either hardware threads or dedicated cores – whatever gives better overall performance.*

B.2 The Memory Hierarchy

Each core of our machine has a 32 KB level-one (L1) data cache; this cache is shared by the hardware threads supported by the core. There is a separate L1 cache of the same size that stores machine instructions. The content of the cache is managed in *blocks* (aka *cache lines*) of 64 bytes whose address starts at a position divisible by the block size – these addresses are said to be *aligned* to the cache line size.

Each core also has a larger (L2) cache of size 256 KB for data and instructions (a *unified cache*). The reason for this division between L1 and L2 is that the L1 cache is made as fast as possible almost without regard for cost per bit of memory, whereas the L2 cache uses a more compact design at the price of higher access latencies and lower throughput. Indeed, even using the same technology, a large cache will inevitably have larger access latency than a small one, ultimately because the speed of light is limited.

---

[2] Some architectures support four or eight hardware threads per core.

Using an even more compact technology, all cores on a processor chip share a large L3 cache of size 45 MB. Many concurrent memory operations on threads on the same chip can be performed within the L3 cache without having to go through the main memory. This is one reason why many shared-memory programs scale well as long as they run on a single chip but scale worse (or not at all) when running on multiple chips.

Our machine has 128 GB of main memory attached to each processor chip. For uniformity, let us call this level four of the memory hierarchy. The processor chips are connected by a high-speed interconnect interface. On our machine, every chip has a dedicated link to every other chip.[3] Any thread can transparently access the main memory of every chip. However, nonlocal accesses will incur higher latency and yield lower overall bandwidth. This effect is called *Non Uniform Memory Access* (NUMA) and processor chips (or *sockets*) are therefore also called *NUMA nodes*[4].

B.3 Cache Coherence Protocols

We first review how a typical sequential cache replacement strategy works. When a core accesses the data at address i, it looks for the cache line b containing that data in the data caches. Suppose the highest[5] level of the hierarchy containing b is j. Then b is copied to the caches at levels $j-1$ down to 1 and then accessed from level 1. The reason for this strategy is that the main cost is for reading the data from level j and that having copies in the higher levels makes future accesses cheaper. A consequence of moving a cache line into a cache is that another block may have to be evicted to make room. If this block has been modified since the last access, it is written to the next lower layer of the memory hierarchy.

It can happen that the data being accessed straddles two cache lines despite being smaller than the block size. Such *nonaligned* operations cause additional costs by requiring both blocks to be moved. Moreover, certain operations cause additional overheads for nonaligned accesses or do not work at all (e.g., 16-byte CAS). Hence, an important performance optimization involves avoiding unaligned accesses by being careful with the data layout.

Assuming a *write-back cache*, write operations go to the L1 cache. When the cache line being accessed is already present in the L1 cache, this is easy. Otherwise, some other cache line may need to be evicted in order to make room. Moreover, the accessed block first has to be read from the lower levels of the memory hierarchy.

[3] On larger or cheaper machines, a more sparse network might be used, e.g., a ring of four chips or a mesh network.
[4] Identifiying sockets with NUMA nodes is an oversimplification because chips or multichip modules on a single processor socket might also define multiple NUMA nodes.
[5] The L1 cache is the highest level and the main memory is the lowest level. Instead of "highest level" one may also say "closest level", i.e., the level closest to the processing unit.

This is necessary in order to avoid additional bookkeeping about which parts of what cache block contain valid data.[6]

In a shared-memory parallel computer, things get more complicated when several PEs access the same memory block. Suppose block b is written by PE i. Some other PE j may have a copy of b in a local cache. This copy is no longer valid, and thus has to be *invalidated*. The inter-PE communication needed for invalidation and rereading invalidated copies causes significant overhead and is one of the main reasons for the limited scalability of shared-memory programs. Note that this overhead is incurred even when PEs i and j access *different* memory locations in the same cache line. This effect is called *false sharing* and has to be avoided by careful memory layout. For example, we should usually allocate a full cache line for a lock variable or use the remainder of that cache line for data that is accessed only by a thread that owns that lock. The technique of making a data structure larger than necessary is called *padding*. More generally, whenever we have the situation that multiple threads are trying to write to the same cache line b at the same time, performance can go down significantly. We call this situation *contention* on b.

We can see that accessing memory on a real-world machine is far away from the idealized view of an instantaneous, globally visible effect. The *cache coherence* mechanism of the hardware can only provide an approximation of the idealized view to the application programs. The possible deviations from the idealized view are defined in the *memory consistency model* or *memory model* of the machine. Coping with these deviations is a major challenge for writers of parallel shared-memory programs. Unfortunately, memory models vary between different architectures. Here we describe the memory model of the x86 architecture and hint at differences in other architectures.

The compiler or the hardware can reorder the memory access operations of a thread in order to keep the pipelines filled. Additionally, in order to improve memory performance, some write operations are delayed by buffering the data in additional *memory buffers*. Within a sequential program, this is done in such a way that the outcome of the computation is not changed. However, in a concurrent program this can lead to problems. Thus, the first thing is to instruct the compiler to abstain from undesired reorderings. In C++ the storage class `volatile` ensures that a variable is always loaded from memory before using it. It is also ensured that the compiler does not reorder accesses to volatile variables. In order to also exclude improper reordering of other memory accesses the statement

```
atomic_signal_fence(memory_order_seq_cst);
```

defines a *memory fence* for the compiler – all reads and writes before the fence have to be compiled to happen before all reads and writes after the fence. Once the compiler is tamed, the x86 architecture guarantees that the read operations of a thread appear to all other threads in the right order. The same is true for the write operations. However, the x86 hardware may reorder reads with older writes to different

---

[6] In order to avoid unnecessary overheads when the user knows that the entire cache line will be written in the near future, one can use *write combining* aided by *nontemporal* write instructions. An example in the case of sorting can be found in [279].

locations. The C++ command `atomic_thread_fence()` can be used to also guarantee the ordering between read and write operations.

Note that it may still happen that the operations of different threads can be mixed arbitrarily and may also appear to different threads in different orders.

B.4 Atomic Operations

We have already explained the compare-and-swap (CAS) operation in Sect. 2.4.1. On the x86 architecture it works exactly as described there and is available for accesses to 32-, 64-, or aligned 128-bit data. Some other architectures have CAS operations up to only 64 bits or with slightly weaker guarantees – they may sometimes fail even though the actual value is equal to the desired value. This sometimes requires additional check loops.

The x86 architecture and others also offer atomic fetch-and-add/subtract/min/-max/and/ or/xor instructions. Although it is easy to implement them using CAS (see also Sect. 2.4.1), using the built-in operation may be faster, in particular when several threads contend for operating on the same variable.

B.5 Hardware Transactional Memory

Beginning with the Haswell microarchitecture, Intel x86 processors have supported an implementation of restricted hardware transactional memory called *Transactional Synchronization Extensions*.[7] The programmer can enclose a critical section using the machine instructions `XBEGIN` and `XEND`. There is an important difference compared to the simple mechanism described in Sect. 2.4.1. A transaction *t* may fail, i.e., the hardware notices that another thread has accessed a cache line touched by *t* in such a way that *t* does not appear to be executed atomically.[8] In that case, the transaction is *rolled back*, i.e., all its effects are undone and an *abort handler* is called. It is possible to retry a transaction. However, it may happen that transactions keep failing without the system making progress. Therefore, the programmer has to provide some ultimate fallback mechanism that is guaranteed to work. For example, using locks or atomic operations. The listing at the end of Sect. 4.6.3 gives an example. Hardware transactions are attractive from the point of view of performance when an abort is unlikely, i.e., when there is little contention on the pieces of memory accessed by the transaction.

B.6 Memory Management

In reality, memory is not the simple one-dimensional array we have seen for the RAM model in Sect. 2.2. First of all, a process sees logical addresses in *virtual*

---

[7] IBM had already introduced a similar mechanism with the POWER8 architecture.

[8] There are further reasons why a transaction may fail, e.g., that too many cache lines have been touched.

memory instead of physical addresses. Logical addresses are mapped to physical memory by the hardware in cooperation with the operating system. Pages of virtual memory (address ranges beginning at a multiple of the page size which is a power of 2) are mapped to physical memory by changing the most significant digits of the addresses using a combination of hardware buffers (translation lookaside buffers – TLBs) and a translation data structure managed by the operating system.

Besides the overhead in virtual address translation, it matters which socket a piece of virtual memory is mapped to. This has to be taken into account when allocating memory. In LINUX, a useful mechanism is that a thread allocating memory will by default allocate it on its local socket. Other placement policies are possible. For example, sometimes it makes sense to map blocks round-robin, i.e., the ith block of an array is mapped to socket i mod P, where P is the number of socket.

B.7 The Interconnection Network

The processors of a parallel machine are connected through an interconnection network. When the number of processors is small, a complete network with all point-to-point links is feasible. When the number of processors is larger, a sparse network must be used. Figure B.1 shows some typical interconnection networks. Two-dimensional and three-dimensional meshes have the advantage that they can be built arbitrarily large with bounded wire length. This is also possible for *torus* networks, which consist of ring interconnections in every direction; Fig. B.1 shows a physical layout with only short connections. The log p-dimensional mesh is better known under the name "*hypercube*" (see also Fig. 13.2). A good approximation to a complete interconnection network can be achieved using hierarchical networks such as the *fat tree* [200]. In this book, we have consciously avoided topology-dependent algorithms in order to keep things simple and portable. However, we sometimes point out opportunities for adapting algorithms to the network topology. We gave a nontrivial example at the end of Sect. 13.1.4.

In sparse interconnection networks, arbitrary point-to-point communication has to be realized by sending messages along paths through the network. Some of these

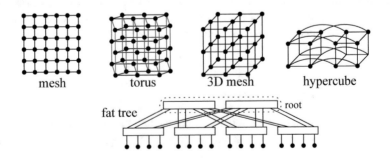

Fig. B.1. Common interconnection networks. The fat tree is an example of a *multistage network*, connecting 16 PEs using two layers of 8-way switches.

paths will have to share the same wire and thus cause *contention*. Carefully designed routing algorithms can keep this contention small. For example, in a hypercube with p nodes, the maximum length of a path between two nodes is $\Theta(\log p)$ and hence, the minimum delivery time is $\Theta(\log p)$. There are routing algorithms for the hypercube that guarantee that p messages with distinct sources and distinct destinations can be delivered in time $O(\log p)$, thus achieving the minimum. We refer our readers to the textbook [198] for a discussion of routing algorithms.

Processors contain dedicated subprocessors for efficient I/O and communication. Using *direct memory access (DMA)*, they move data between memory and interface hardware controlling disk and network access. Therefore, it is possible to overlap communication, I/O, and computation.

B.8 CPU Performance Analysis

Code optimization requires profiling for two reasons. First, algorithm analysis makes asymptotic statements and ignores constant factors. Second, our models of sequential and, even more so, of parallel computers are only approximations of real hardware. Profiling provides detailed information about the execution of a program. Modern processors have dedicated hardware performance-monitoring units (PMUs) to support profiling. PMUs count performance events (CPU cycles, cache misses, branch mis-predictions, etc.) and also map them to particular instructions.

PMUs support hundreds of very detailed performance events[9], which can be programmed using profiling software (CPU profilers). The open-source Linux perf profiler[10] has good support for common CPU architectures. The most advanced profiler for the Intel CPU architecture is the Intel VTune Amplifier.[11] Profilers usually offer a basic analysis of program hotspots (where most of the cycles are spent). Advanced analysis uses fine-grained performance events to identify bottlenecks in the processor which are responsible for incurring many cycles during instruction execution [333].

Reading performance counters introduces a penalty itself, and therefore most profilers use statistical sampling of hardware performance events to minimize the side effects on program execution. If mapping to instructions is not required (for example in an initial phase of performance analysis or when the bottlenecks in the *code* are well understood), then PMUs can be used in counting mode where they are read only when required, for example, before and after program execution to count the total number of events. The Processor Counter Monitor[12] and Linux perf support this kind of lightweight analysis.

For the analysis of performance issues in very short actions (for example for real-time processing in car engines or the response time of graphical user interfaces),

---

[9] For Intel processors, the events are listed in the software developer manuals; see www.intel.com/sdm.
[10] perf.wiki.kernel.org
[11] software.intel.com/en-us/intel-vtune-amplifier-xe
[12] github.com/opcm/pcm/

sampling or counting performance events is not appropriate, because of the high relative overhead. Recently, Intel Processor Trace (PT) has been introduced to address the analysis of such performance issues related to responsiveness. PT allows one to trigger collection of a full instruction execution trace together with timestamps. Performance analysis based on PT is available in Intel VTune Amplifier and Linux perf.

B.9 Compiler

If not stated otherwise we used GNU C++ Compiler (g++) version 4.7.2 to compile our examples of shared-memory programs. The specific compiler options can be found in Makefiles at `github.com` under `basic-toolbox-sample-code/basic-toolbox-sample-code/`.

C

Support for Parallelism in C++

C++11 (a version of the C++ programming language standard) extends C++ by new language constructs for native support of parallelism through multithreading in shared-memory systems. The new constructs hide the implementation details of thread management. They provide basic locks, generic atomic operations, and support for asynchronous tasks. The C++14 version of the standard adds a new lock type for shared access. In our shared-memory implementation examples, we use only constructs available in C++11.

In the following sections we give a short introduction to the most important C++11/14 classes for multithreading.[1]

The parallel-programming support in C++ is not only rather recent but also fairly low-level. Hence, there are also tools outside the C++ standard that can be used for parallel programming with C++. Section C.5 introduces some frequently used ones.

C.1 "Hello World" C++11 Program with Threads

Listing C.1 shows a minimalistic program that demonstrates basic management of threads. It spawns the maximum number of threads supported by the hardware. These threads execute a user-defined worker function. The parameters of the worker function are passed in the constructor of the thread (the current thread count). Each C++ thread is scheduled for execution on one of the hardware threads available in the system. The C++ thread interface does not provide any methods to control the scheduling such that execution of C++ threads might be delayed and/or several C++ threads might need to share the same hardware thread [2]. In our simple example, each worker thread just prints its thread identifier passed as parameter to the C++ standard output (cout). Since the cout object is a shared resource, concurrent access to it might

---

[1] For an exhaustive overview, see the standard document [162] and the online C++ reference at en.cppreference.com/w/. A list of textbooks is available at isocpp.org/get-started.

[2] The only available scheduling call is a *yield*(), function giving a hint to deschedule the thread.

© Springer Nature Switzerland AG 2019   455
P. Sanders et al., *Sequential and Parallel Algorithms and Data Structures*,
https://doi.org/10.1007/978-3-030-25209-0

Listing C.1. Threading "hello world" example

```cpp
#include <iostream>
#include <vector>
#include <thread>
#include <mutex>

using namespace std;

mutex m;

void worker(int iPE) {
  m.lock();
  cout << "Hello_from_thread_"<< iPE << endl;
  m.unlock();
}

int main() {
  vector<thread> threads(thread::hardware_concurrency());
  int i = 0;
  for(auto & t: threads) t = thread(worker, i++);
  for(auto & t: threads) t.join();
  return 0;
}//SPDX−License−Identifier: BSD−3−Clause; Copyright(c) 2018 Intel Corporation
```

jumble the characters. To avoid this, we protect the access using an exclusive lock (mutex) provided by C++11. The main thread waits for completion of every worker thread by calling their *join* function.

Since spawning and joining threads are expensive operations involving operating system calls, it does not pay off to have a separate C++ thread for each small work item. For robust multithreaded applications, it is advisable to create all required threads at once (a *thread pool*) and to pass the work to them as needed using the load-balancing methods described in Chap. 14.

C.2 Locks

We introduced binary locks in Sect. 2.5. The corresponding C++11 class is *mutex* (mutual exclusion lock). More advanced lock classes provide the ability to allow timeouts (*recursive_timed_mutex*), to acquire a lock multiple times (*recursive_mutex*), or to distinguish between readers and writers (*shared_timed_mutex*). To make the usage of locks less error-prone, C++11 provides several helper classes. The class *lock_guard* acquires a C++11 mutex in its constructor and automatically releases it in its destructor when the execution leaves the scope where the guard was created. The function *lock* is passed multiple mutexes and locks all of them in such a way that no deadlocks can occur.

C.3 Asynchronous Operations

The classes *promise* and *future* provide mechanisms for thread interaction that are slightly higher-level than the basic locking mechanisms we use in this book. Roughly, one thread can produce a value, whereas others can wait for this value to become available. One useful variant of this method is an asynchronous function call that returns a *future* object so that at a later point, a thread can wait for the function call to be completed.

C.4 Atomic Operations

C++11 provides abstractions of atomic operations (Sect B.4) on shared-memory parallel computers. These operations include atomic load and store, exchange, strong and weak compare-and-exchange (equivalent to compare-and-swap (CAS); see Sect B.4), fetch-and-add, fetch-and-substract, fetch-and-or, fetch-and-and, and fetch-and-xor on C++ built-in integers and char types of length 8, 16, 32, and 64 bits. An atomic version of C++ *bool* exists too. *atomic_flag* is a Boolean variable supporting atomic *test_and_set* and *clear* operations that can easily be used to implement a simple spin lock.

C++11 atomic operations have (*_explicit*) versions that allow one to specify the memory-ordering guarantee (Sect. B.3) around an atomic operation. Their range is very rich, such that developers who are new to memory ordering should be very careful when choosing relaxed guarantees. A wrong memory ordering is a latent bug that is very hard to discover. The default and the safest memory-ordering type is *memory_order_seq_cst* (sequential consistency), which is recommended for beginners and is also relatively fast on common x86 architectures. This ordering provides many guarantees: No reads or writes can be reordered, all writes in other threads that access the same atomic variable are visible in the current thread, all writes in the current thread are visible in other threads that access the same atomic variable and writes that carry a dependency into the atomic variable become visible in other threads that access the same atomic variable, all threads observe all modifications in the same order. To enforce a required memory ordering between arbitrary operations (including nonatomic ones), C++ offers the *atomic_thread_fence* function. The weaker *atomic_signal_fence* prevents only reordering of the operations by the compiler but not by the processor. See also Sect. B.3.

The first versions of the C++11 compilers did not always implement the atomic operations listed above using the fastest available CPU instruction. Sometimes the heaviest and slowest CPU instruction was used. Therefore we recommend to use the latest compiler version.

Although the set of atomic operations and types provided in the C++11 standard is very rich, it still represents the lowest common denominator of the vendor-specific processor capabilities. For example, the 128-bit CAS operation which was very useful in Sect 4.6.3 is supported by the x86 architecture but is not part of the C++ standard.

C.5 Useful Tools

OpenMP. This is a compiler extension for shared-memory parallel programming; see also [64] and `openmp.org`. The basic idea behind OpenMP is that one annotates a sequential program with *compiler pragma*s that help the compiler to parallelize it. OpenMP supports SPMD programming, local (*private*) and global (*shared*) variables, and parallel loops. OpenMP supports locking directly, but can also be used together with other libraries. For example, one can use the class *std :: mutex* from the C++-standard.

Task-parallel programming. Since version 3.0, OpenMP has supported task-parallel programming. However, in the first implementations, performance was not very good. The Intel tools Cilk Plus[3] and Threading Building Blocks (Intel TBB[4]) give better performance using the work-stealing load balancer described in Sect. 14.5. However, algorithms using task-parallel programming sometimes do not scale beyond one processor chip, since memory access locality is not very good.

Software libraries. Good software libraries considerably ease the life of a software designer. A large variety of libraries is available for C++ and some of them exploit parallelism. Often, using these parallelized libraries can be the key to parallelizing an application. Important examples are libraries for linear algebra and parallel implementations of the the C++ standard template library STL. Note that STL not only supports classical algorithms like sorting, merging, selection, or random permutation, but also a comprehensive set of seemingly simple operations such as *for_each*. If those are also parallelized (perhaps using dynamic load balancing) this blurs the distinction between libraries and parallelization tools like OpenMP or task parallel programming – we can express many parallel algorithms as a set of STL calls. For example this has been done for a minimum spanning tree algorithm [246]. MC-STL [298] is a good parallelization of the STL and is part of the GNU C++ distribution.

C.6 Memory Management

The C++ memory allocation function (*new*) calls the underlying operating system allocator (e.g. *malloc* on Linux). The standard memory allocators are general-purpose and are not optimized for maximum scalability. Typically, small-sized allocations are serviced from a process-local heap protected by a per-process exclusive lock, which leads to scalability bottlenecks. Larger allocations are requested directly from the operating system which also involves locking of operating system memory structures responsible for bookkeeping (i.e., virtual memory page tables). Another issue is that, for security reasons, all allocated memory has to be initialized by the operating system before it can be given to an application. In Sect. 5.13, we saw an

---

[3] `www.cilkplus.org`
[4] `www.threadingbuildingblocks.org`

example where this initialization turned out to run sequentially – introducing a major scalability roadblock in our sample sort implementation. Thus it can be much faster to reuse memory rather than to allocate and deallocate it over and over again. To facilitate such reuse, Intel Threading Building Blocks and Boost libraries provide user-space memory pools that have standard allocation and deallocation interfaces (see *tbb* :: *memory_pool*[5] and *boost* :: *pool*[6]).

There are also libraries that replace the standard C++ allocators by more scalable implementations. The most known such allocators are Google's Thread Caching Malloc[7] and the Intel TBB scalable memory allocator. They provide per-thread heaps, avoiding global locks, and also automatically cache the memory in user pools for reuse. The TBB allocators additionally provide explicit interfaces for specific data structures that require scalable allocation. They also work with the C++ standard containers (*vector*, *stack*, etc.).

Sometimes there is a requirement to allocate memory on a boundary with a certain alignment. For example, the 16-byte CAS instruction of the x86 architecture requires 16-byte alignment. Also, to prevent false sharing (Sect. B.3), a data structure must begin at a fresh cache line. Common operating systems have custom allocators with support for alignment (*posix_memalign* on Linux and *_aligned_malloc* on Windows). Intel TBB provides a scalable allocator (*tbb* :: *cache_aligned_allocator*) which returns cache-line-aligned pieces of memory that can also be used with C++ containers.

As discussed in Sect. B.6, for performance reasons, the application might want to have control of the memory placement on specific sockets. Most operating systems have libraries and interfaces that support such control: For example, *libnuma* on Linux and *VirtualAllocExNuma* on Windows. See also the next section.

C.7 Thread Scheduling

By default, user threads have no guarantees about when and on what hardware threads or cores they will be executed. The operating system is allowed to deschedule and migrate them arbitrarily following some optimization goal (usually a heuristic). During its execution, a thread can be migrated from one hardware thread to another. In some cases thread migration can be very undesirable from a performance perspective: The migrated thread can no longer use its recently accessed cache lines (its *cache footprint*), it now has higher latency for accessing memory allocated on a different socket, etc. To prevent such migrations, the developer can *pin* a user thread to a set of hardware threads using the *pthread_setaffinity_np* call on Linux and *SetThreadGroupAffinity* on Windows.

On Linux, thread-pinning functions can be also used to control NUMA allocation. If a thread touches a virtual memory block which is not yet assigned to physical

---

[5] software.intel.com/en-us/blogs/2011/12/19/
scalable-memory-pools-community-preview-feature
[6] www.boost.org/doc/libs/1_48_0/libs/pool/doc/html/index.html
[7] goog-perftools.sourceforge.net/doc/tcmalloc.html

memory (lazy memory allocation), the default policy for Linux is to try to allocate the physical memory on the local socket.

D

The Message Passing Interface (MPI)

MPI[1] is a software library for message passing in clusters. It is the de facto standard for high-performance computing. It was initially developed for Fortran and C in 1994. This lineage still shows in the function interfaces – the handling of data types is rather low-level. On the other hand, MPI offers a quite complete set of collective communication operations (see Sect D.3) that are missing from most alternatives. MPI is the result of a standardization process that is a careful compromise between performance, portability, and generality. Most of the functionality needed in this book is already in the MPI-1 standard. Therefore this is also the focus of this appendix. MPI-2, from 1997, adds high-performance I/O and one-sided communication – a way to get some of the functionality one uses in shared-memory programs, albeit with some performance overhead. MPI-3, from 2012, adds nonblocking collective operations.

D.1 "Hello World" and What Is an MPI Program?

An MPI program is an ordinary "sequential" (Fortran), C, or C++ program that includes mpi.h and calls the MPI functions declared there. This program is executed in parallel on all PEs of our parallel machine, i.e., MPI programs follow the SPMD approach to parallel programming. Listing D.1 shows a minimalistic example.

The call *MPI_Init* initializes the library. In particular, it initializes a global variable *MPI_COMM_WORLD* that stores a *communicator* object describing the set of PEs available to the program. The procedures *MPI_Comm_size* and *MPI_Comm_rank* extract the total number of PEs p and the PE number (from $0..p-1$). Communicators are also passed to all MPI functions to define the context of the communication. In particular, it is possible to define communicators spanning only a subset of the PEs in *MPI_COMM_WORLD*. In our simple example, each PE outputs p and its processor number i. The overall output of our program is not

---
[1] www.mpi-forum.org

© Springer Nature Switzerland AG 2019
P. Sanders et al., *Sequential and Parallel Algorithms and Data Structures*,
https://doi.org/10.1007/978-3-030-25209-0

uniquely defined. The characters or lines of output coming from different PEs may
be jumbled or the output may be written to one file for each PE.

Listing D.1. MPI hello world example

```
#include <iostream>
#include <mpi.h>

int main(int argc, char** argv)
{ int p, i;
 MPI_Init(&argc, &argv);
 MPI_Comm_size(MPI_COMM_WORLD, &p);
 MPI_Comm_rank(MPI_COMM_WORLD, &i);
 std::cout << "PE " << i << " out of " << p << std::endl;
 MPI_Finalize();
}
```

In MPI terminology, a PE is called a *process* and, indeed, an MPI process often
corresponds to a process in the host operating system. Whether each node of the
cluster runs one or several processes depends on how the MPI program is launched.
One important option is one process per core (or hardware thread) of the node. This is
convenient because then MPI takes care of all the parallelism. At the other extreme,
we might have only one process per node. In order to exploit the actual parallelism
of the machine, the node should then run a multithreaded shared-memory parallel
program. In that case it is advisable that, at any time, only a single thread per node
makes MPI calls.[2] If the multithreaded and message-passing parts of the program
work well together, this option may lead to better performance at the price of a more
complicated program. The middle ground may also make sense. For example, one
could run one MPI process on each socket, thus explicitly taking NUMA effects into
account.

To actually start our example program, the simplest case is when it runs on the
cores of a single machine. For example, under Linux with OpenMPI v3.0.0 (www.
open-mpi.org/) one would compile the program with the command line

```
mpic++ example.cpp -o example
```

where mpic++ is a script that calls the GNU compiler with appropriate parameters.
To then run the program using four processes one uses another script

```
mpirun -np 4 example
```

On a supercomputer, starting the program is a bit more complicated. One typically
writes a configuration file describing which program to call, how many nodes should
be used and how many processes run on each node. Then one passes this configa-
ration file to a job scheduler that allocates the appropriate ressources and starts the
program on all nodes.

---

[2] If MPI calls are made concurrently, some MPI implementations do not work at all and
others have performance problems.

D.2 Point-to-Point Communication

Supposing message m is an array of k integers, our pseudocode operation $send(i,m)$ can be translated into the MPI call

$MPI\_Send(\&m,k,MPI\_INT,i,t,MPI\_COMM\_WORLD)$

where the integer t is a *message tag* that helps the receiver to distinguish different types of messages. The destination PE i is a rank within the global communicator MPI_COMM_WORLD. Other communicators can also be used which can encode subsets of PEs – for example the rows and columns in Sect. 5.2. Further data types can be used by replacing the constant MPI_INT by another predefined constant such as $MPI\_CHAR/SHORT/LONG/FLOAT/DOUBLE$. User-defined data types are also possible.

There are several variants of send operations – *Send/Ssend/Isend/Issend/Bsend*. When the ordinary send returns, the message buffer m can be reused (and overwritten) without affecting the delivery of the message. *Ssend* guarantees in addition that the receiver has begun to actually receive the message. The *nonblocking* or *immediate* operations *Isend* and *Issend* are a little more complicated. They have an additional return parameter that returns a *request object*. Their buffer can only be reused when an additional (blocking) operation waiting for the request to finish has been called. The advantage of nonblocking operations is that they return immediately. Thus, several communications can be initiated together. One can also use this feature to overlap communication and internal work. Function *MPI_Bsend* also returns immediately *and* guarantees that the message buffer can be immediately removed. The disadvantage of this convenience is that the user has to supply additional buffer memory (using the operation *MPI_Buffer_attach*) and that this causes additional copy operations, which incur some overhead.

A receive operation matching the above call is

$MPI\_Recv(\&m,k,MPI\_INT,j,t,MPI\_COMM\_WORLD,\&status)$

where j either specifies the PE from which the caller expects to receive a message or is equal to MPI_ANY_SOURCE. In the latter case, a message from any sender can be received. Similarly, t specifies the expected tag or MPI_ANY_TAG. The parameter k specifies the allocated length of the message buffer. This buffer may be longer then the message actually received. The actual length of the received message can be read from the status variable (which has type MPI_Status) using the operation MPI_Get_count. A status object has fields $status.MPI\_TAG$ and $status.MPI\_SOURCE$ that tell the tag and sender, respectively, of the received message. Sometimes the receiver does not have a useful upper bound on the length of the message to be received. In that case, the operation MPI_Probe can be called first which delivers a status that tells the message length (and its tag and source PE). There is also a nonblocking receive operations MPI_Irecv.

Finally, there is an operation $MPI\_Sendrecv$ that corresponds to our pseudocode operations $send(\cdots) \parallel receive(\cdots)$.

To work with the nonblocking operations *Isend* and *Irecv*, one additionally needs operations *MPI_Wait/Waitany/Waitall* and *MPI_Test/Testany/Testall*. These operations are passed the request objects returned by *Isend* and *Irecv*. The wait operations block until the specified requests have finished. The test operations do not block, and thus allow us to perform computations while communication operations are executed in the background.

D.3 Collective Communication

Table D.1. Collective communication operations in MPI.

Our name	MPI name	See also Sect.
broadcast	MPI_Bcast	13.1
reduce	MPI_Reduce	13.2
all-reduce	MPI_Allreduce	13.2
prefix sum	MPI_Scan	13.3
barrier	MPI_Barrier	13.4.2
gather	MPI_Gather(v)	13.5
all-gather	MPI_Allgather(v)	13.5
scatter	MPI_Scatter(v)	13.5
all-to-all	MPI_Alltoall(v)	13.6

MPI supports all the collective communication operations discussed in Chap. 13. Starting with MPI 3.0, this includes the asynchronous ones presented in Sect. 13.7. We view this as a major strength of MPI in particular in comparison with other frameworks for parallel processing. However, one should not assume that all MPI implementations implement all collective operations efficiently. Careful profiling and occasional manual reimplementations of the required operations are therefore important for achieving good performance in practice. Table D.1 summarizes the available collective operations. The collective operations with irregular message size have names ending with v. These expect the receiver of a message to specify the length of that message. This often implies that the message lengths have to be transferred in a separate operation.

An example call for a collective operation is

MPI_Reduce(&c, &sum, 1, MPI_INT, MPI_SUM, 0, MPI_COMM_WORLD)

which will perform a sum-reduction of the values of the local variable c. The second to last parameter specifies that the overall result will be stored in the variable *sum* at PE 0. MPI supports a number of further predefined reduction operations besides *MPI_SUM*. User-defined operations are also possible.

E

List of Commercial Products, Trademarks and Software Licenses

The following list includes the names of commercial products and trademarks mentioned in the book.

- Microsoft® Windows® (Windows)
- Oracle® Java®
- IBM® RS/6000®
- IBM® Power®
- IBM® POWER8®
- IBM® Blue Gene®/Q
- Intel® Core™
- Intel® Xeon®
- Intel® Pentium®
- Intel® Threading Building Blocks (Intel TBB)
- Intel® Cilk™ Plus
- Intel® VTune™ Amplifier
- Intel® Processor Trace (Intel PT)
- Intel® Transactional Synchronization Extensions (Intel TSX)
- Wikipedia®
- OpenMP®

The listings in the book are distributed under the Open Source BSD-3-Clause license.

E.1 BSD 3-Clause License

Redistribution and use in source and binary forms, with or without modification, are permitted provided that the following conditions are met:

- Redistributions of source code must retain the above copyright notice, this list of conditions and the following disclaimer.

© Springer Nature Switzerland AG 2019
P. Sanders et al., *Sequential and Parallel Algorithms and Data Structures*,
https://doi.org/10.1007/978-3-030-25209-0

References

[1] E. H. L. Aarts and J. Korst. *Simulated Annealing and Boltzmann Machines*. John Wiley & Sons Ltd., 1989.

[2] J. Abello, A. L. Buchsbaum, and J. R. Westbrook. A functional approach to external graph algorithms. *Algorithmica*, 32(3):437–458, 2002.

[3] I. Abraham, D. Delling, A. V. Goldberg, and R. F. F. Werneck. A hub-based labeling algorithm for shortest paths in road networks. In *10th Symposium on Experimental Algorithms (SEA)*, volume 6630 of *LNCS*, pages 230–241. Springer, 2011.

[4] W. Ackermann. Zum hilbertschen Aufbau der reellen Zahlen. *Mathematische Annalen*, 99:118–133, 1928.

[5] G. M. Adel'son-Vel'skii and E. M. Landis. An algorithm for the organization of information. *Soviet Mathematics Doklady*, 3:1259–1263, 1962.

[6] A. Aggarwal and J. S. Vitter. The input/output complexity of sorting and related problems. *Communications of the ACM*, 31(9):1116–1127, 1988.

[7] A. V. Aho, J. E. Hopcroft, and J. D. Ullman. *The Design and Analysis of Computer Algorithms*. Addison-Wesley, 1974.

[8] A. V. Aho, B. W. Kernighan, and P. J. Weinberger. *The AWK Programming Language*. Addison-Wesley, 1988.

[9] R. K. Ahuja, R. L. Magnanti, and J. B. Orlin. *Network Flows*. Prentice Hall, 1993.

[10] R. K. Ahuja, K. Mehlhorn, J. B. Orlin, and R. E. Tarjan. Faster algorithms for the shortest path problem. *Journal of the ACM*, 3(2):213–223, 1990.

[11] M. Ajtai, J. Komlós, and E. Szemerédi. An O($n \log n$) sorting network. In *15th ACM Symposium on Theory of Computing (STOC)*, pages 1–9, 1983.

[12] Y. Akhremtsev and P. Sanders. Fast parallel operations on search trees. In *23rd IEEE Conference on High Performance Computing (HIPC)*, pages 291–300, 2016.

[13] M. Akra and L. Bazzi. On the solution of linear recurrence equations. *Computational Optimization and Applications*, 10(2):195–210, 1998.

[14] E. Alba and M. Tomassini. Parallelism and evolutionary algorithms. *IEEE Transactions on Evolutionary Computation*, 6(5):443–462, 2002.

© Springer Nature Switzerland AG 2019
P. Sanders et al., *Sequential and Parallel Algorithms and Data Structures*,
https://doi.org/10.1007/978-3-030-25209-0

[15] N. Alon, M. Dietzfelbinger, P. B. Miltersen, E. Petrank, and E. Tardos. Linear hash functions. *Journal of the ACM*, 46(5):667–683, 1999.

[16] A. Andersson, T. Hagerup, S. Nilsson, and R. Raman. Sorting in linear time? *Journal of Computer and System Sciences*, pages 74–93, 1998.

[17] F. Annexstein, M. Baumslag, and A. Rosenberg. Group action graphs and parallel architectures. *SIAM Journal on Computing*, 19(3):544–569, 1990.

[18] D. L. Applegate, R. E. Bixby, V. Chvátal, and W. J. Cook. *The Traveling Salesman Problem: A Computational Study*. Princeton University Press, 2007.

[19] L. Arge, M. T. Goodrich, M. Nelson, and N. Sitchinava. Fundamental parallel algorithms for private-cache chip multiprocessors. In *20th ACM Symposium on Parallelism in Algorithms and Architectures (SPAA)*, pages 197–206, 2008.

[20] N. S. Arora, R. D. Blumofe, and C. G. Plaxton. Thread scheduling for multiprogrammed multiprocessors. In *10th ACM Symposium on Parallel Algorithms and Architectures (SPAA)*, pages 119–129, 1998.

[21] J. Arz, D. Luxen, and P. Sanders. Transit node routing reconsidered. In *12th Symposium on Experimental Algorithms (SEA)*, volume 7933 of *LNCS*, pages 55–66. Springer, 2013.

[22] G. Ausiello, P. Crescenzi, G. Gambosi, V. Kann, A. Marchetti-Spaccamela, and M. Protasi. *Complexity and Approximation: Combinatorial Optimization Problems and Their Approximability Properties*. Springer, 1999.

[23] M. Axtmann, T. Bingmann, P. Sanders, and C. Schulz. Practical massively parallel sorting. In *27th ACM Symposium on Parallelism in Algorithms and Architectures, (SPAA)*, pages 13–23, 2015.

[24] M. Axtmann, A. Wiebigke, and P. Sanders. Lightweight MPI communicators with applications to perfectly balanced quicksort. In *32nd IEEE International Parallel and Distributed Processing Symposium (IPDPS)*, pages 254–265, 2018.

[25] M. Axtmann, S. Witt, D. Ferizovic, and P. Sanders. In-place parallel super scalar samplesort (IPSSSSo). In *25th European Symposium on Algorithms (ESA)*, pages 9:1–9:14, 2017. full paper at arXiv:1705.02257 [cs.DC].

[26] D. A. Bader and G. Cong. Fast shared-memory algorithms for computing the minimum spanning forest of sparse graphs. *Journal of Parallel and Distributed Computing*, 66:1366–1378, 2006.

[27] M. Bader. *Space-filling Curves – An Introduction with Applications in Scientific Computing*, volume 9 of *Texts in Computational Science and Engineering*. Springer, 2012.

[28] T. Balyo and P. Sanders. HordeSat: A massively parallel portfolio SAT solver. In *18th Conference on Theory and Applications of Satisfiability Testing (SAT)*, volume 9340 of *LNCS*, pages 156–172. Springer, 2015.

[29] H. Bast. Scheduling at twilight the easy way. In *19th Symposium on Theoretical Aspects of Computer Science (STACS)*, volume 2285 of *LNCS*, pages 166–178. Springer, 2002.

[30] H. Bast, D. Delling, A. Goldberg, M. Müller-Hannemann, T. Pajor, P. Sanders, D. Wagner, and R. F. Werneck. Route planning in transportation networks. In

L. Kliemann and P. Sanders, editors, *Algorithm Engineering*, volume 9220 of *LNCS*, pages 19–80. Springer, 2016.

[31] H. Bast, S. Funke, P. Sanders, and D. Schultes. Fast routing in road networks with transit nodes. *Science*, 316(5824):566, 2007.

[32] H. Bast and T. Hagerup. Fast parallel space allocation, estimation and integer sorting. *Information and Computation*, 123(1):72–110, 1995.

[33] K. E. Batcher. Sorting networks and their applications. In *AFIPS Spring Joint Computing Conference*, pages 307–314, 1968.

[34] R. Bayer and E. M. McCreight. Organization and maintenance of large ordered indexes. *Acta Informatica*, 1(3):173–189, 1972.

[35] A. Beckmann, R. Dementiev, and J. Singler. Building a parallel pipelined external memory algorithm library. In *23rd IEEE International Symposium on Parallel and Distributed Processing (IPDPS)*, 2009.

[36] R. Beier and B. Vöcking. Random knapsack in expected polynomial time. *Journal of Computer and System Sciences*, 69(3):306–329, 2004.

[37] R. Belazzougui, F. C. Botelho, and M. Dietzfelbinger. Hash, displace, and compress. In *17th European Symposium on Algorithms (ESA)*, volume 5757 of *LNCS*, pages 682–693. Springer, 2009.

[38] R. Bellman. On a routing problem. *Quarterly of Applied Mathematics*, 16(1):87–90, 1958.

[39] M. A. Bender, E. D. Demaine, and M. Farach-Colton. Cache-oblivious B-trees. In *41st IEEE Symposium on Foundations of Computer Science (FOCS)*, pages 399–409, 2000.

[40] M. A. Bender, M. Farach-Colton, G. Pemmasani, S. Skiena, and P. Sumazin. Lowest common ancestors in trees and directed acyclic graphs. *Journal of Algorithms*, 57(2):75–94, 2005.

[41] J. L. Bentley and M. D. McIlroy. Engineering a sort function. *Software Practice and Experience*, 23(11):1249–1265, 1993.

[42] J. L. Bentley and T. A. Ottmann. Algorithms for reporting and counting geometric intersections. *IEEE Transactions on Computers*, C-28(9):643–647, 1979.

[43] J. L. Bentley and R. Sedgewick. Fast algorithms for sorting and searching strings. In *8th ACM Symposium on Discrete Algorithms (SODA)*, pages 360–369, 1997.

[44] D. Bertsimas and J. N. Tsitsiklis. *Introduction to Linear Optimization*. Athena Scientific, 1997.

[45] T. Bingmann, T. Keh, and P. Sanders. A bulk-parallel priority queue in external memory with STXXL. In *14th Symposium on Experimental Algorithms (SEA)*, volume 9125 of *LNCS*, pages 28–40. Springer, 2015.

[46] G. E. Blelloch, D. Ferizovic, and Y. Sun. Just join for parallel ordered sets. In *28th ACM Symposium on Parallelism in Algorithms and Architectures (SPAA)*, pages 253–264, 2016.

[47] G. E. Blelloch, C. E. Leiserson, B. M. Maggs, C. G. Plaxton, S. J. Smith, and M. Zagha. A comparison of sorting algorithms for the connection machine

CM-2. In *3rd ACM Symposium on Parallel Algorithms and Architectures (SPAA)*, pages 3–16, 1991.

[48] M. Blum, R. W. Floyd, V. R. Pratt, R. L. Rivest, and R. E. Tarjan. Time bounds for selection. *Journal of Computer and System Sciences*, 7(4):448, 1972.

[49] N. Blum and K. Mehlhorn. On the average number of rebalancing operations in weight-balanced trees. *Theoretical Computer Science*, 11:303–320, 1980.

[50] Boost.org. Boost C++ Libraries. www.boost.org.

[51] O. Borůvka. O jistém problému minimálním. *Práce Moravské Přírodovědecké Společnosti*, 3:37–58, 1926. In Czech.

[52] F. C. Botelho, R. Pagh, and N. Ziviani. Practical perfect hashing in nearly optimal space. *Information Systems*, 38(1):108–131, 2013.

[53] G. S. Brodal. Worst-case efficient priority queues. In *7th ACM-SIAM Symposium on Discrete Algorithms (SODA)*, pages 52–58, 1996.

[54] G. S. Brodal and J. Katajainen. Worst-case efficient external-memory priority queues. In *6th Scandinavian Workshop on Algorithm Theory (SWAT)*, volume 1432 of *LNCS*, pages 107–118. Springer, 1998.

[55] G. S. Brodal, J. L. Träff, and C. D. Zaroliagis. A parallel priority queue with constant time operations. *Journal of Parallel and Distributed Computing*, 49(1):4–21, 1998.

[56] N. G. Bronson, J. Casper, H. Chafi, and K. Olukotun. A practical concurrent binary search tree. *ACM SIGPLAN Notices*, 45(5):257–268, 2010.

[57] M. R. Brown and R. E. Tarjan. Design and analysis of a data structure for representing sorted lists. *SIAM Journal of Computing*, 9:594–614, 1980.

[58] R. Brown. Calendar queues: A fast O(1) priority queue implementation for the simulation event set problem. *Communications of the ACM*, 31(10):1220–1227, 1988.

[59] A. Buluc, H. Meyerhenke, I. Safro, P. Sanders, and C. Schulz. Recent advances in graph partitioning. In L. Kliemann and P. Sanders, editors, *Algorithm Engineering*, volume 9220 of *LNCS*, pages 117–158. Springer, 2014.

[60] C. K. Caldwell and Y. Cheng. Determining Mills' constant and a note on Honaker's problem. *Journal Integer Sequences*, 8(4):Article 05.4.1, 2005.

[61] J. L. Carter and M. N. Wegman. Universal classes of hash functions. *Journal of Computer and System Sciences*, 18(2):143–154, 1979.

[62] S. K. Cha, S. Hwang, K. Kim, and K. Kwon. Cache-conscious concurrency control of main-memory indexes on shared-memory multiprocessor systems. In *27th Conference on Very Large Data Bases (VLDB)*, pages 181–190, 2001.

[63] D. Chakrabarti, Y. Zhan, and C. Faloutsos. R-MAT: A recursive model for graph mining. In *SIAM International Conference on Data Mining (SDM)*, pages 442–446, 2004.

[64] R. Chandra. *Parallel Programming in OpenMP*. Morgan Kaufmann, 2001.

[65] B. Chazelle. A minimum spanning tree algorithm with inverse-Ackermann type complexity. *Journal of the ACM*, 47:1028–1047, 2000.

[66] B. Chazelle and L. J. Guibas. Fractional cascading: I. A data structuring technique. *Algorithmica*, 1(2):133–162, 1986.

[67] B. Chazelle and L. J. Guibas. Fractional cascading: II. Applications. *Algorithmica*, 1(2):163–191, 1986.

[68] J.-C. Chen. Proportion extend sort. *SIAM Journal on Computing*, 31(1):323–330, 2001.

[69] Y. Cheng. Explicit estimate on primes between consecutive cubes. *Rocky Mountain J. Math.*, 40:1, 117–153, 2010.

[70] J. Cheriyan and K. Mehlhorn. Algorithms for dense graphs and networks. *Algorithmica*, 15(6):521–549, 1996.

[71] B. V. Cherkassky, A. V. Goldberg, and T. Radzik. Shortest path algorithms: Theory and experimental evaluation. *Mathematical Programming*, 73:129–174, 1996.

[72] Y.-J. Chiang, M. T. Goodrich, E. F. Grove, R. Tamassia, D. E. Vengroff, and J. S. Vitter. External-memory graph algorithms. In *6th ACM-SIAM Symposium on Discrete Algorithms (SODA)*, pages 139–149, 1995.

[73] E. G. Coffman, M. R. Garey, and D. S. Johnson. Approximation algorithms for bin packing: A survey. In D. Hochbaum, editor, *Approximation Algorithms for NP-Hard Problems*, pages 46–93. 1996.

[74] D. Cohen-Or, D. Levin, and O. Remez. Progressive compression of arbitrary triangular meshes. In *IEEE Conference on Visualization (VIS)*, pages 67–72, 1999.

[75] R. Cole. Parallel merge sort. *SIAM Journal on Computing*, 17(4):770–785, 1988.

[76] R. Cole, P. N. Klein, and R. E. Tarjan. Finding minimum spanning forests in logarithmic time and linear work using random sampling. In *8th ACM Symposium on Parallel Algorithms and Architectures (SPAA)*, pages 243–250, 1996.

[77] R. Cole, K. Ost, and S. Schirra. Edge-coloring bipartite multigraphs in $O(E \log D)$ time. *Combinatorica*, 21(1):5–12, 2000.

[78] S. A. Cook. *On the Minimum Computation Time of Functions*. PhD thesis, Harvard University, 1966.

[79] S. A. Cook. The complexity of theorem proving procedures. In *3rd ACM Symposium on Theory of Computing (STOC)*, pages 151–158, 1971.

[80] A. Crauser, K. Mehlhorn, U. Meyer, and P. Sanders. A parallelization of Dijkstra's shortest path algorithm. In *23rd Symposium on Mathematical Foundations of Computer Science (MFCS)*, number 1450 in LNCS, pages 722–731. Springer, 1998.

[81] G. B. Dantzig. Maximization of a linear function of variables subject to linear inequalities. In T. C. Koopmans, editor, *Activity Analysis of Production and Allocation*, pages 339–347. Wiley, 1951.

[82] C. Darwin. *The Origin of Species by Means of Natural Selection: or, the Preservation of Favoured Races in the Struggle for Life*. John Murray, London, 1859.

[83] A. Davidson, D. Tarjan, M. Garland, and J. D. Owens. Efficient parallel merge sort for fixed and variable length keys. In *Innovative Parallel Computing (InPar)*, pages 1–9. IEEE, 2012.

[84] A. De, P. P. Kurur, C. Saha, and R. Saptharishi. Fast integer multiplication using modular arithmetic. *SIAM Journal on Computing*, 42(2):685–699, 2013.

[85] M. de Berg, M. van Kreveld, M. Overmars, and O. Schwarzkopf. *Computational Geometry: Algorithms and Applications*. Springer, 2nd edition, 2000.

[86] J. Dean and S. Ghemawat. MapReduce: simplified data processing on large clusters. *Communications of the ACM*, 51(1):107–113, January 2008.

[87] R. Dementiev, L. Kettner, J. Mehnert, and P. Sanders. Engineering a sorted list data structure for 32 bit keys. In *6th Workshop on Algorithm Engineering and Experiments (ALENEX)*, pages 142–151, 2004.

[88] R. Dementiev, L. Kettner, and P. Sanders. STXXL: standard template library for XXL data sets. *Softw., Pract. Exper.*, 38(6):589–637, 2008.

[89] R. Dementiev and P. Sanders. Asynchronous parallel disk sorting. In *15th ACM Symposium on Parallelism in Algorithms and Architectures (SPAA)*, pages 138–148, 2003.

[90] R. Dementiev, P. Sanders, D. Schultes, and J. Sibeyn. Engineering an external memory minimum spanning tree algorithm. In *Exploring New Frontiers of Theoretical Informatics*, volume 155 of *IFIPAICT*, pages 195–208. Springer, 2004.

[91] L. Devroye. A note on the height of binary search trees. *Journal of the ACM*, 33(3):289–498, 1986.

[92] R. B. Dial. Shortest-path forest with topological ordering. *Communications of the ACM*, 12(11):632–633, 1969.

[93] M. Dietzfelbinger, T. Hagerup, J. Katajainen, and M. Penttonen. A reliable randomized algorithm for the closest-pair problem. *Journal of Algorithms*, 25(1):19–51, 1997.

[94] M. Dietzfelbinger, A. Karlin, K. Mehlhorn, F. Meyer auf der Heide, H. Rohnert, and R. E. Tarjan. Dynamic perfect hashing: Upper and lower bounds. *SIAM Journal of Computing*, 23(4):738–761, 1994.

[95] M. Dietzfelbinger and C. Weidling. Balanced allocation and dictionaries with tightly packed constant size bins. *Theoretical Computer Science*, 380(1–2):47–68, 2007.

[96] E. W. Dijkstra. A note on two problems in connexion with graphs. *Numerische Mathematik*, 1(1):269–271, 1959.

[97] E. A. Dinic. Economical algorithms for finding shortest paths in a network. In Y. Popkov and B. Shmulyian, editors, *Transportation Modeling Systems*, pages 36–44, 1978.

[98] W. Domschke and A. Drexl. *Einführung in Operations Research*. Springer, 2007.

[99] J. R. Driscoll, N. Sarnak, D. D. Sleator, and R. E. Tarjan. Making data structures persistent. *Journal of Computer and System Sciences*, 38(1):86–124, 1989.

[100] M. Drmota and W. Szpankowski. A master theorem for discrete divide and conquer recurrences. *Journal of the ACM*, 60(3):16:1–16:49, 2013.

[101] S. Edelkamp and A. Weiss. BlockQuicksort: Avoiding branch mispredictions in quicksort. In *24th European Symposium on Algorithms (ESA)*, volume 57 of *LIPIcs*, pages 38:1–38:16, 2016.

[102] A. Elmasry, J. Katajainen, and M. Stenmark. Branch mispredictions don't affect mergesort. In *11th Symposium on Experimental Algorithms (SEA)*, volume 7276 of *LNCS*, pages 160–171. Springer, 2012.

[103] J. Fakcharoenphol and S. Rao. Planar graphs, negative weight edges, shortest paths, and near linear time. *Journal of Computer and System Sciences*, 72(5):868–889, 2006.

[104] L. K. Fleischer, B. Hendrickson, and A. Pınar. On identifying strongly connected components in parallel. In *Workshop on Solving Irregularly Structured Problems in Parallel (IPDPS Workshops)*, number 1800 in LNCS, pages 505–511. Springer, 2000.

[105] R. Fleischer. A tight lower bound for the worst case of Bottom-Up-Heapsort. *Algorithmica*, 11(2):104–115, 1994.

[106] R. Floyd. Assigning meaning to programs. In J. Schwarz, editor, *Mathematical Aspects of Computer Science*, pages 19–32. AMS, 1967.

[107] R. W. Floyd and R. L. Rivest. Expected time bounds for selection. *Communications of the ACM*, 18(3):165–172, 1975.

[108] L. R. Ford. Network flow theory. Technical Report P-923, Rand Corporation, Santa Monica, California, 1956.

[109] D. Fotakis, R. Pagh, P. Sanders, and P. Spirakis. Space efficient hash tables with worst case constant access time. *Theory of Computing Systems*, 38(2):229–248, 2005.

[110] W. D. Frazer and A. C. McKellar. Samplesort: A sampling approach to minimal storage tree sorting. *Journal of the ACM*, 17(3):496–507, 1970.

[111] E. Fredkin. Trie memory. *Communications of the ACM*, 3(9):490–499, 1960.

[112] M. L. Fredman. On the efficiency of pairing heaps and related data structures. *Journal of the ACM*, 46(4):473–501, 1999.

[113] M. L. Fredman, J. Komlós, and E. Szemerédi. Storing a sparse table with $O(1)$ worst case access time. *Journal of the ACM*, 31(3):538–544, 1984.

[114] M. L. Fredman, R. Sedgewick, D. D. Sleator, and R. E. Tarjan. The pairing heap: A new form of self-adjusting heap. *Algorithmica*, 1:111–129, 1986.

[115] M. L. Fredman and R. E. Tarjan. Fibonacci heaps and their uses in improved network optimization algorithms. *Journal of the ACM*, 34(3):596–615, 1987.

[116] M. Frigo, C. E. Leiserson, H. Prokop, and S. Ramachandran. Cache-oblivious algorithms. In *40th IEEE Symposium on Foundations of Computer Science (FOCS)*, pages 285–298, 1999.

[117] M. Fürer. Faster integer multiplication. *SIAM Journal on Computing*, 39(3):979–1005, 2009.

[118] H. N. Gabow. Path-based depth-first search for strong and biconnected components. *Information Processing Letters*, 74(3–4):107–114, 2000.

[119] E. Gamma, R. Helm, R. Johnson, and J. Vlissides. *Design Patterns: Elements of Reusable Object-Oriented Software*. Addison-Wesley, 1995.

[120] D. Gangal and A. Ranade. Precedence constrained scheduling in $(2 - 7/(3p - 1))$ optimal. *Journal of Computer and System Sciences*, 74(7):1139–1146, 2008.

[121] M. R. Garey and D. S. Johnson. *Computers and Intractability: A Guide to the Theory of NP-Completeness*. W. H. Freeman, 1979.

[122] B. Gärtner and J. Matoušek. *Understanding and Using Linear Programming*. Springer, 2006.

[123] R. Geisberger, P. Sanders, D. Schultes, and C. Vetter. Exact routing in large road networks using contraction hierarchies. *Transportation Science*, 46(3):388–404, 2012.

[124] J. Giacomoni, T. Moseley, and M. Vachharajani. Fastforward for efficient pipeline parallelism: A cache-optimized concurrent lock-free queue. In *13th ACM SIGPLAN Symposium on Principles and Practice of Parallel Programming (PPoPP)*, pages 43–52, 2008.

[125] P. B. Gibbons. A more practical PRAM model. In *1st ACM Symposium on Parallel Algorithms and Architectures (SPAA)*, pages 158–168, 1989.

[126] P. B. Gibbons, Y. Matias, and V. Ramachandran. The QRQW PRAM: Accounting for contention in parallel algorithms. In *5th ACM-SIAM Symposium on Discrete Algorithms (SODA)*, pages 638–648, 1994.

[127] GMP (GNU Multiple Precision Arithmetic Library). `http://gmplib. org/`.

[128] A. V. Goldberg. Scaling algorithms for the shortest path problem. *SIAM Journal on Computing*, 24(3):494–504, 1995.

[129] A. V. Goldberg. A simple shortest path algorithm with linear average time. In *9th European Symposium on Algorithms (ESA)*, number 2161 in LNCS, pages 230–241. Springer, 2001.

[130] A. V. Goldberg and C. Harrelson. Computing the shortest path: A^* meets graph theory. In *16th ACM-SIAM Symposium on Discrete Algorithms (SODA)*, pages 156–165, 2005.

[131] G. H. Gonnet and R. Baeza-Yates. *Handbook of Algorithms and Data Structures: In Pascal and C*. Addison-Wesley, 2nd edition, 1991.

[132] M. T. Goodrich. Randomized Shellsort: A simple data-oblivious sorting algorithm. *Journal of the ACM*, 58(6):27:1–27:26, 2011.

[133] M. T. Goodrich. Zig-zag sort: A simple deterministic data-oblivious sorting algorithm running in $O(n \log n)$ time. In *46th ACM Symposium on Theory of Computing (STOC)*, pages 684–693, 2014.

[134] G. Graefe. A survey of B-tree locking techniques. *ACM Transactions on Database Systems (TODS)*, 35(3):16, 2010.

[135] G. Graefe and P.-A. Larson. B-tree indexes and CPU caches. In *17th International Conference on Data Engineering (ICDE)*, pages 349–358. IEEE, 2001.

[136] R. L. Graham. Bounds on multiprocessing timing anomalies. *SIAM Journal of Applied Mathematics*, 17(2):416–429, 1969.

[137] R. L. Graham, D. E. Knuth, and O. Patashnik. *Concrete Mathematics*. Addison-Wesley, 2nd edition, 1994.

[138] J. F. Grantham and C. Pomerance. Prime numbers. In K. H. Rosen, editor, *Handbook of Discrete and Combinatorial Mathematics*, chapter 4.4, pages 236–254. CRC Press, 2000.

[139] R. Grossi and G. Italiano. Efficient techniques for maintaining multi-dimensional keys in linked data structures. In *26th International Colloquium on Automata, Languages and Programming (ICALP)*, volume 1644 of *LNCS*, pages 372–381. Springer, 1999.

[140] B. Haeupler, S. Sen, and R. E. Tarjan. Rank-pairing heaps. In *17th European Symposium on Algorithms (ESA)*, volume 5757 of *LNCS*, pages 659–670, 2009.

[141] S. Halperin and U. Zwick. Optimal randomized EREW PRAM algorithms for finding spanning forests and for other basic graph connectivity problems. In *7th ACM-SIAM Symposium on Discrete Algorithms (SODA)*, pages 438–447, 1996.

[142] Y. Han and M. Thorup. Integer sorting in $O(n\sqrt{\log\log n})$ expected time and linear space. In *42nd IEEE Symposium on Foundations of Computer Science (FOCS)*, pages 135–144, 2002.

[143] G. Handler and I. Zang. A dual algorithm for the constrained shortest path problem. *Networks*, 10(4):293–309, 1980.

[144] T. D. Hansen, H. Kaplan, R. E. Tarjan, and U. Zwick. Hollow heaps. *ACM Transactions on Algorithms (TALG)*, 13(3):42:1–42:27, 2017.

[145] J. Hartmanis and J. Simon. On the power of multiplication in random access machines. In *5th IEEE Symposium on Foundations of Computer Science (FOCS)*, pages 13–23, 1974.

[146] D. Harvey and J. van der Hoeven. Integer multiplication in time $O(n\log n)$. https://hal.archives-ouvertes.fr/hal-02070778, Mar. 2019.

[147] M. Held and R. Karp. The traveling-salesman problem and minimum spanning trees. *Operations Research*, 18(6):1138–1162, 1970.

[148] M. Held and R. Karp. The traveling-salesman problem and minimum spanning trees, part II. *Mathematical Programming*, 1:6–25, 1971.

[149] J. L. Hennessy and D. A. Patterson. *Computer Architecture: A Quantitative Approach*. Morgan Kaufmann, 5th edition, 2011.

[150] P. V. Hentenryck and L. Michel. *Constraint-Based Local Search*. MIT Press, 2005.

[151] M. Herlihy and N. Shavit. *The Art of Multiprocessor Programming, Revised Reprint*. Elsevier, 2012.

[152] C. A. R. Hoare. An axiomatic basis for computer programming. *Communications of the ACM*, 12(10):576–585, 1969.

[153] C. A. R. Hoare. Proof of correctness of data representations. *Acta Informatica*, 1(4):271–281, 1972.

[154] R. D. Hofstadter. Metamagical themas. *Scientific American*, 248(2):16–22, 1983.

[155] S. Hong, N. C. Rodia, and K. Olukotun. On fast parallel detection of strongly connected components (SCC) in small-world graphs. In *ACM Conference on*

High Performance Computing, Networking, Storage and Analysis (SC), pages 92:1–92:11, 2013.

[156] J. E. Hopcroft and J. D. Ullman. Set merging algorithms. *SIAM Journal on Computing*, 2(4):294–303, 1973.

[157] P. Høyer. A general technique for implementation of efficient priority queues. In *3rd Israeli Symposium on Theory of Computing and Systems (ISTCS)*, pages 57–66, 1995.

[158] L. Hübschle-Schneider and P. Sanders. Communication efficient algorithms for top-k selection problems. In *30th IEEE International Parallel and Distributed Processing Symposium (IPDPS)*, pages 659–668, 2016.

[159] S. Huddlestone and K. Mehlhorn. A new data structure for representing sorted lists. *Acta Informatica*, 17(2):157–184, 1982.

[160] J. Iacono. Improved upper bounds for pairing heaps. In *7th Scandinavian Workshop on Algorithm Theory (SWAT)*, volume 1851 of *LNCS*, pages 32–45. Springer, 2000.

[161] ©Intel. *Intel® 64 and IA-32 Architectures Optimization Reference Manual*, 2016. Order No. 248966-032.

[162] ISO/IEC. C++ programming language standard. Technical Report 14882:2014, ISO/IEC, 2014.

[163] A. Itai, A. G. Konheim, and M. Rodeh. A sparse table implementation of priority queues. In *8th International Colloquium on Automata, Languages and Programming (ICALP)*, volume 115 of *LNCS*, pages 417–431. Springer, 1981.

[164] J. Jájá. *An Introduction to Parallel Algorithms*. Addison-Wesley, 1992.

[165] V. Jarník. O jistém problému minimálním (Z dopisu panu O. Borůvkovi). *Práce Moravské Přírodovědecké Společnosti*, 6:57–63, 1930. In Czech.

[166] K. Jensen and N. Wirth. *Pascal User Manual and Report. ISO Pascal Standard*. Springer, 1991.

[167] jgrapht.org. JGraphT Java Graph Library. http://jgrapht.org.

[168] T. Jiang, M. Li, and P. Vitányi. Average-case complexity of Shellsort. In *26th International Colloquium on Automata, Languages and Programming (ICALP)*, number 1644 in LNCS, pages 453–462. Springer, 1999.

[169] D. S. Johnson, C. R. Aragon, L. A. McGeoch, and C. Schevon. Optimization by simulated annealing: Experimental evaluation; part II, graph coloring and number partitioning. *Operations Research*, 39(3):378–406, 1991.

[170] N. L. Johnson, S. Kotz, and A. W. Kemp. *Univariate Discrete Distributions*. Wiley, 1989.

[171] S. L. Johnsson and C. T. Ho. Optimum broadcasting and personalized communication in hypercubes. *IEEE Transactions on Computers*, 38(9):1249–1268, 1989.

[172] L. V. Kalé and A. B. Sinha. Information sharing mechanisms in parallel programs. In *8th International Parallel Processing Symposium (IPPS)*, pages 461–468. IEEE, 1994.

[173] K. Kaligosi and P. Sanders. How branch mispredictions affect quicksort. In *14th European Symposium on Algorithms (ESA)*, volume 4168 of *LNCS*, pages 780–791. Springer, 2006.

[174] H. Kaplan and R. E. Tarjan. New heap data structures. Technical Report TR-597-99, Princeton University, 1999.

[175] A. Karatsuba and Y. Ofman. Multiplication of multidigit numbers on automata. *Soviet Physics Doklady*, 7(7):595–596, 1963.

[176] D. Karger, P. N. Klein, and R. E. Tarjan. A randomized linear-time algorithm for finding minimum spanning trees. *Journal of the ACM*, 42(1):321–329, 1995.

[177] N. Karmarkar. A new polynomial-time algorithm for linear programming. *Combinatorica*, 4(4):373–395, 1984.

[178] J. Katajainen and B. B. Mortensen. Experiences with the design and implementation of space-efficient deques. In *5th Workshop on Algorithm Engineering (WAE)*, volume 2141 of *LNCS*, pages 39–50. Springer, 2001.

[179] I. Katriel, P. Sanders, and J. L. Träff. A practical minimum spanning tree algorithm using the cycle property. In *11th European Symposium on Algorithms (ESA)*, number 2832 in LNCS, pages 679–690. Springer, 2003.

[180] H. Kellerer, U. Pferschy, and D. Pisinger. *Knapsack problems*. Springer, 2004.

[181] L. Khachiyan. A polynomial time algorithm in linear programming (in Russian). *Soviet Mathematics Doklady*, 20(1):191–194, 1979.

[182] T. Kieritz, D. Luxen, P. Sanders, and C. Vetter. Distributed time-dependent contraction hierarchies. In *9th Symposium on Experimental Algorithms (SEA)*, volume 6049 of *LNCS*, pages 83–93. Springer, 2010.

[183] V. King. A simpler minimum spanning tree verification algorithm. *Algorithmica*, 18(2):263–270, 1997.

[184] S. Knopp, P. Sanders, D. Schultes, F. Schulz, and D. Wagner. Computing many-to-many shortest paths using highway hierarchies. In *9th Workshop on Algorithm Engineering and Experiments (ALENEX)*, pages 36–45, 2007.

[185] D. E. Knuth. *The Art of Computer Programming – Sorting and Searching*, volume 3. Addison-Wesley, 2nd edition, 1998.

[186] D. E. Knuth. *MMIXware: A RISC Computer for the Third Millennium*, volume 1750 of *LNCS*. Springer, 1999.

[187] P. Konecny. Introducing the Cray XMT. In *Cray User Group meeting (CUG)*, 2007.

[188] D. König. Über Graphen und ihre Anwendung auf Determinantentheorie und Mengenlehre. *Mathematische Annalen*, 77(4):453–465, 1916.

[189] R. E. Korf. Depth-first iterative-deepening: An optimal admissible tree search. *Artificial Intelligence*, 27(1):97–109, 1985.

[190] B. Korte and J.Vygen. *Combinatorial Optimization: Theory and Algorithms*. Springer, 2000.

[191] J. Kruskal. On the shortest spanning subtree of a graph and the traveling salesman problem. *Proceedings of the American Mathematical Society*, 7(1):48–50, 1956.

[192] V. Kumar, A. Grama, A. Gupta, and G. Karypis. *Introduction to Parallel Computing. Design and Analysis of Algorithms*. Benjamin/Cummings, 1994.

[193] A. LaMarca and R. E. Ladner. The influence of caches on the performance of heaps. *ACM Journal of Experimental Algorithmics (JEA)*, 1:4:1–4:32, 1996.

[194] E. L. Lawler, J. K. Lenstra, A. H. G. Rinooy Kan, and D. B. Shmoys. *The Traveling Salesman Problem*. Wiley, 1985.

[195] LEDA (Library of Efficient Data Types and Algorithms). www.algorithmic-solutions.com.

[196] L. Q. Lee, A. Lumsdaine, and J. G. Siek. *The Boost Graph Library: User Guide and Reference Manual*. Addison-Wesley, 2002.

[197] P. L. Lehman and S. B. Yao. Efficient locking for concurrent operations on b-trees. *ACM Transactions on Database Systems (TODS)*, 6(4):650–670, 1981.

[198] T. Leighton. *Introduction to Parallel Algorithms and Architectures*. Morgan Kaufmann, 1992.

[199] N. Leischner, V. Osipov, and P. Sanders. GPU sample sort. In *24th IEEE International Parallel and Distributed Processing Symposium (IPDPS)*, 2010. see also arXiv:0909.5649.

[200] C. E. Leiserson. Fat-trees: universal networks for hardware-efficient super-computing. *IEEE Transactions on Computers*, 100(10):892–901, 1985.

[201] LEMON. LEMON C++ Graph Library. http://lemon.cs.elte.hu.

[202] G. Lev, N. Pippenger, and L. Valiant. A fast parallel algorithm for routing in permutation networks. *IEEE Transactions on Computing*, 30(2):93–100, 1981.

[203] L. Levin. Universal search problems (in Russian). *Problemy Peredachi Informatsii*, 9(3):265–266, 1973.

[204] I. Lustig and J.-F. Puget. Program does not equal program: Constraint programming and its relationship to mathematical programming. *Interfaces*, 31(3):29–53, 2001.

[205] T. Maier and P. Sanders. Dynamic Space Efficient Hashing. In *25th European Symposium on Algorithms (ESA)*, volume 87 of *LIPIcs*, pages 58:1–58:14, 2017.

[206] T. Maier, P. Sanders, and R. Dementiev. Concurrent hash tables: Fast *and* general?(!). *CoRR*, arXiv:1601.04017 [cs.DS], 2016. Short version in *Principles and Practice of Parallel Processing (PPoPP)*, 2016.

[207] S. Martello and P. Toth. *Knapsack Problems – Algorithms and Computer Implementations*. Wiley, 1990.

[208] C. Martínez and S. Roura. Optimal sampling strategies in Quicksort and Quickselect. *SIAM Journal on Computing*, 31(3):683–705, 2002.

[209] F. Mattern. Algorithms for distributed termination detection. *Distributed Computing*, 2(3):161–175, 1987.

[210] C. McDiarmid. Concentration. In M. Habib, C. McDiarmid, and J. Ramirez-Alfonsin, editors, *Probabilistic Methods for Algorithmic Discrete Mathematics*, pages 195–247. Springer, 1998.

[211] C. McGeoch, P. Sanders, R. Fleischer, P. R. Cohen, and D. Precup. Using finite experiments to study asymptotic performance. In *Experimental Algorithmics*

– *From Algorithm Design to Robust and Efficient Software*, volume 2547 of *LNCS*, pages 93–126. Springer, 2002.

[212] MCSTL: The Multi-Core Standard Template Library. `http://algo2.iti.uni-karlsruhe.de/singler/mcstl/`.

[213] K. Mehlhorn. A faster approximation algorithm for the Steiner problem in graphs. *Information Processing Letters*, 27(3):125–128, Mar. 1988.

[214] K. Mehlhorn. Amortisierte Analyse. In T. Ottmann, editor, *Prinzipien des Algorithmenentwurfs*, pages 91–102. Spektrum Lehrbuch, 1998.

[215] K. Mehlhorn and U. Meyer. External-memory breadth-first search with sublinear I/O. In *10th European Symposium on Algorithms (ESA)*, volume 2461 of *LNCS*, pages 723–735. Springer, 2002.

[216] K. Mehlhorn and S. Näher. Bounded ordered dictionaries in $O(\log\log N)$ time and $O(n)$ space. *Information Processing Letters*, 35(4):183–189, 1990.

[217] K. Mehlhorn and S. Näher. Dynamic fractional cascading. *Algorithmica*, 5(2):215–241, 1990.

[218] K. Mehlhorn and S. Näher. *The LEDA Platform for Combinatorial and Geometric Computing*. Cambridge University Press, 1999.

[219] K. Mehlhorn, S. Näher, and P. Sanders. Engineering DFS-based graph algorithms. *arXiv preprint arXiv:1703.10023*, 2017.

[220] K. Mehlhorn, V. Priebe, G. Schäfer, and N. Sivadasan. All-pairs shortest-paths computation in the presence of negative cycles. *Information Processing Letters*, 81(6):341–343, 2002.

[221] K. Mehlhorn and P. Sanders. Scanning multiple sequences via cache memory. *Algorithmica*, 35(1):75–93, 2003.

[222] K. Mehlhorn and S. Saxena. A still simpler way of introducing the interior-point method for linear programming. *Computer Science Review*, 22:1–11, 2016.

[223] K. Mehlhorn and M. Ziegelmann. Resource constrained shortest paths. In *8th European Symposium on Algorithms (ESA)*, volume 1879 of *LNCS*, pages 326–337. Springer, 2000.

[224] R. Mendelson, R. E. Tarjan, M. Thorup, and U. Zwick. Melding priority queues. In *9th Scandinavian Workshop on Algorithm Theory (SWAT)*, volume 3111 of *LNCS*, pages 223–235. Springer, 2004.

[225] *Meyers Konversationslexikon*. Bibliographisches Institut, 1888.

[226] B. Meyer. *Object-Oriented Software Construction*. Prentice-Hall, second edition, 1997.

[227] U. Meyer. Average-case complexity of single-source shortest-path algorithms: lower and upper bounds. *Journal of Algorithms*, 48(1):91–134, 2003. preliminary version in SODA 2001.

[228] U. Meyer and P. Sanders. Δ-stepping: A parallelizable shortest path algorithm. *Journal of Algorithms*, 49(1):114–152, 2003.

[229] U. Meyer, P. Sanders, and J. Sibeyn, editors. *Algorithms for Memory Hierarchies*, volume 2625 of *LNCS Tutorial*. Springer, 2003.

[230] G. L. Miller and J. H. Reif. Parallel tree contraction and its application. In *26st IEEE Symposium on Foundations of Computer Science (FOCS)*, pages 478–489, 1985.

[231] B. M. E. Moret and H. D. Shapiro. An empirical analysis of algorithms for constructing a minimum spanning tree. In *2nd Workshop on Algorithms and Data Structures (WADS)*, volume 519 of *LNCS*, pages 400–411. Springer, 1991.

[232] R. Morris. Scatter storage techniques. *Communications of the ACM*, 11(1):38–44, 1968.

[233] S. S. Muchnick. *Advanced Compiler Design and Implementation*. Morgan Kaufmann, 1997.

[234] I. Müller, P. Sanders, A. Lacurie, W. Lehner, and F. Färber. Cache-efficient aggregation: Hashing is sorting. In *ACM SIGMOD Conference on Management of Data*, pages 1123–1136, 2015.

[235] I. Müller, P. Sanders, R. Schulze, and W. Zhou. Retrieval and perfect hashing using fingerprinting. In *13th Symposium on Experimental Algorithms (SEA)*, volume 8504 of *LNCS*, pages 138–149. Springer, 2014.

[236] S. Näher and O. Zlotowski. Design and implementation of efficient data types for static graphs. In *10th European Symposium on Algorithms (ESA)*, volume 2461 of *LNCS*, pages 748–759. Springer, 2002.

[237] M. Naor and O. Reingold. On the construction of pseudorandom permutations: Luby-Rackoff revisited. *Journal of Cryptology*, 12(1):29–66, 1999.

[238] G. Navarro. *Compact Data Structures – A Practical Approach*. Cambridge University Press, 2016.

[239] G. Nemhauser and Z. Ullmann. Discrete dynamic programming and capital allocation. *Management Science*, 15(9):494–505, 1969.

[240] G. Nemhauser and L. Wolsey. *Integer and Combinatorial Optimization*. Wiley, 1988.

[241] J. Nešetřil, H. Milková, and H. Nešetřilová. Otakar Borůvka on minimum spanning tree problem: Translation of both the 1926 papers, comments, history. *Discrete Mathematics*, 233:3–36, 2001.

[242] K. S. Neubert. The Flashsort1 algorithm. *Dr. Dobb's Journal*, pages 123–125, February 1998.

[243] J. v. Neumann. First draft of a report on the EDVAC. Technical report, University of Pennsylvania, 1945.

[244] J. Nievergelt and E. Reingold. Binary search trees of bounded balance. *SIAM Journal on Computing*, 2(1):33–43, 1973.

[245] K. Noshita. A theorem on the expected complexity of Dijkstra's shortest path algorithm. *Journal of Algorithms*, 6(3):400–408, 1985.

[246] V. Osipov, P. Sanders, and J. Singler. The filter-Kruskal minimum spanning tree algorithm. In *10th Workshop on Algorithm Engineering and Experiments (ALENEX)*, pages 52–61, 2009.

[247] A. Pagh, R. Pagh, and M. Ružić. Linear probing with constant independence. *SIAM Journal on Computing*, 39(3):1107–1120, 2009.

[248] R. Pagh and F. Rodler. Cuckoo hashing. *Journal of Algorithms*, 51(2):122–144, 2004.

[249] M. Patrascu and M. Thorup. On the k-independence required by linear probing and minwise independence. In *37th International Colloquium on Automata, Languages and Programming (ICALP, Part I)*, volume 6198 of *LNCS*, pages 715–726, 2010.

[250] W. J. Paul, P. Bach, M. Bosch, J. Fischer, C. Lichtenau, and J. Röhrig. Real PRAM programming. In *8th Euro-Par*, volume 2400 of *LNCS*, pages 522–531. Springer, 2002.

[251] H.-O. Peitgen and P. H. Richter. *The Beauty of Fractals*. Springer-Verlag, 1986.

[252] W. W. Peterson. Addressing for random access storage. *IBM Journal of Research and Development*, 1(2), Apr. 1957.

[253] S. Pettie. Towards a final analysis of pairing heaps. In *46th IEEE Symposium on Foundations of Computer Science (FOCS)*, pages 174–183, 2005.

[254] S. Pettie and V. Ramachandran. An optimal minimum spanning tree algorithm. In *27th International Colloquium on Automata, Languages and Programming (ICALP)*, volume 1853 of *LNCS*, pages 49–60. Springer, 2000.

[255] J. Pinkerton. *Voyages and Travels*, volume 2. 1808.

[256] P. J. Plauger, A. A. Stepanov, M. Lee, and D. R. Musser. *The C++ Standard Template Library*. Prentice-Hall, 2000.

[257] R. C. Prim. Shortest connection networks and some generalizations. *Bell Systems Technical Journal*, 36(6):1389–1401, 1957.

[258] W. Pugh. Skip lists: A probabilistic alternative to balanced trees. *Communications of the ACM*, 33(6):668–676, 1990.

[259] M. Pătrașcu and M. Thorup. The power of simple tabulation hashing. *Journal of the ACM*, 59(3):14:1–14:50, 2012.

[260] M. Rahn, P. Sanders, and J. Singler. Scalable distributed-memory external sorting. In *26th IEEE International Conference on Data Engineering (ICDE)*, pages 685–688, 2010.

[261] S. Rajasekaran and J. H. Reif. Optimal and sublogarithmic time randomized parallel sorting algorithms. *SIAM Journal on Computing*, 18(3):594–607, 1989.

[262] O. Ramaré and Y. Saouter. Short effective intervals containing primes. *Journal on Number Theory*, 98(1):10–33, 2003.

[263] A. Ranade, S. Kothari, and R. Udupa. Register efficient mergesorting. In *7th Conference on High Performance Computing (HIPC)*, volume 1970 of *LNCS*, pages 96–103. Springer, 2000.

[264] J. H. Reif. Depth-first search is inherently sequential. *Information Processing Letters*, 20(5):229–234, 1985.

[265] N. Robertson, D. P. Sanders, P. Seymour, and R. Thomas. Efficiently four-coloring planar graphs. In *28th ACM Symposium on Theory of Computing (STOC)*, pages 571–575. ACM Press, 1996.

[266] G. Robins and A. Zelikwosky. Improved Steiner tree approximation in graphs. In *11th ACM-SIAM Symposium on Discrete Algorithms (SODA)*, pages 770–779, 2000.

[267] S. Roura. Improved master theorems for divide-and-conquer recurrences. *Journal of the ACM*, 48(2):170–205, 2001.

[268] P. Sanders. *Lastverteilungsalgorithmen für parallele Tiefensuche*. PhD thesis, University of Karlsruhe, 1996.

[269] P. Sanders. On the competitive analysis of randomized static load balancing. In S. Rajasekaran, editor, *First Workshop on Randomized Parallel Algorithms*, Honolulu, Hawaii, 16 April, 1996. http://algo2.iti.kit.edu/sanders/papers/rand96.pdf.

[270] P. Sanders. Random permutations on distributed, external and hierarchical memory. *Information Processing Letters*, 67(6):305–310, 1998.

[271] P. Sanders. Randomized priority queues for fast parallel access. *Journal Parallel and Distributed Computing, Special Issue on Parallel and Distributed Data Structures*, 49(1):86–97, 1998.

[272] P. Sanders. Fast priority queues for cached memory. *ACM Journal of Experimental Algorithmics*, 5, 2000.

[273] P. Sanders. Randomized receiver initiated load balancing algorithms for tree shaped computations. *The Computer Journal*, 45(5):561–573, 2002.

[274] P. Sanders, S. Lamm, L. Hübschle-Schneider, E. Schrade, and C. Dachsbacher. Efficient random sampling – parallel, vectorized, cache-efficient, and online. *ACM Transactions on Mathematical Software*, 44(3), 2018.

[275] P. Sanders, S. Schlag, and I. Müller. Communication efficient algorithms for fundamental big data problems. In *IEEE Conference on Big Data*, pages 15–23, 2013.

[276] P. Sanders and D. Schultes. Highway hierarchies hasten exact shortest path queries. In *13th European Symposium on Algorithms (ESA)*, volume 3669 of *LNCS*, pages 568–597. Springer, 2005.

[277] P. Sanders and C. Schulz. Distributed evolutionary graph partitioning. In *14th Workshop on Algorithm Engineering and Experiments (ALENEX)*, pages 16–29. SIAM, 2012.

[278] P. Sanders, J. Speck, and J. L. Träff. Two-tree algorithms for full bandwidth broadcast, reduction and scan. *Parallel Computing*, 35(12):581–594, 2009.

[279] P. Sanders and J. Wassenberg. Engineering a multi-core radix sort. In *17th Euro-Par*, volume 6853 of *LNCS*, pages 160–169. Springer, 2011.

[280] P. Sanders and S. Winkel. Super scalar sample sort. In *12th European Symposium on Algorithms (ESA)*, volume 3221 of *LNCS*, pages 784–796. Springer, 2004.

[281] P. Sanders and T. Worsch. *Parallele Programmierung mit MPI – ein Praktikum*. Logos Verlag Berlin, 1997. ISBN 3-931216-76-4.

[282] R. Santos and F. Seidel. A better upper bound on the number of triangulations of a planar point set. *Journal of Combinatorial Theory Series A*, 102(1):186–193, 2003.

[283] N. Satish, M. Harris, and M. Garland. Designing efficient sorting algorithms for manycore GPUs. In *23rd IEEE International Symposium on Parallel and Distributed Processing (IPDPS)*, 2009.

[284] R. Schaffer and R. Sedgewick. The analysis of heapsort. *Journal of Algorithms*, 15(1):76–100, 1993.

[285] A. Schönhage. Storage modification machines. *SIAM Journal on Computing*, 9(3):490–508, 1980.

[286] A. Schönhage and V. Strassen. Schnelle Multiplikation großer Zahlen. *Computing*, 7(3–4):281–292, 1971.

[287] A. Schrijver. *Combinatorial Optimization (3 Volumes)*. Springer Verlag, 2003.

[288] R. Sedgewick. Analysis of Shellsort and related algorithms. In *4th European Symposium on Algorithms (ESA)*, volume 1136 of *LNCS*, pages 1–11. Springer, 1996.

[289] R. Sedgewick and P. Flajolet. *An Introduction to the Analysis of Algorithms*. Addison-Wesley, 1996.

[290] R. Seidel and C. R. Aragon. Randomized search trees. *Algorithmica*, 16(4–5):464–497, 1996.

[291] R. Seidel and M. Sharir. Top-down analysis of path compression. *SIAM Journal of Computing*, 34(3):515–525, 2005.

[292] J. Sewall, J. Chhugani, C. Kim, N. Satish, and P. Dubey. PALM: Parallel Architecture-Friendly Latch-Free Modifications to B+ Trees on Many-Core Processors. *PVLDB*, 4(11):795–806, 2011.

[293] M. Sharir. A strong-connectivity algorithm and its applications in data flow analysis. *Computers and Mathematics with Applications*, 7(1):67–72, 1981.

[294] J. C. Shepherdson and H. E. Sturgis. Computability of recursive functions. *Journal of the ACM*, 10(2):217–255, 1963.

[295] J. Shun and G. E. Blelloch. Phase-concurrent hash tables for determinism. In *26th ACM Symposium on Parallelism in Algorithms and Architectures (SPAA)*, pages 96–107, 2014.

[296] J. Shun, G. E. Blelloch, J. T. Fineman, and P. B. Gibbons. Reducing contention through priority updates. In *25th ACM Symposium on Parallelism in Algorithms and Architectures (SPAA)*, pages 152–163, 2013.

[297] J. Shun, G. E. Blelloch, J. T. Fineman, P. B. Gibbons, A. Kyrola, H. V. Simhadri, and K. Tangwongsan. The problem based benchmark suite. In *24th ACM Symposium on Parallelism in Algorithms and Architectures (SPAA)*, pages 68–70, 2012. http://www.cs.cmu.edu/~pbbs.

[298] J. Singler, P. Sanders, and F. Putze. MCSTL: The multi-core standard template library. In *13th Euro-Par*, volume 4641 of *LNCS*, pages 682–694. Springer, 2007.

[299] M. Sipser. *Introduction to the Theory of Computation*. MIT Press, 1998.

[300] D. D. Sleator and R. E. Tarjan. A data structure for dynamic trees. *Journal of Computer and System Sciences*, 26(3):362–391, 1983.

[301] D. D. Sleator and R. E. Tarjan. Self-adjusting binary search trees. *Journal of the ACM*, 32(3):652–686, 1985.

[302] D. Spielman and S.-H. Teng. Smoothed analysis of algorithms: why the simplex algorithm usually takes polynomial time. *Journal of the ACM*, 51(3):385–463, 2004.

[303] M. Stephan and J. Docter. Jülich Supercomputing Centre. JUQUEEN: IBM Blue Gene/Q Supercomputer System at the Jülich Supercomputing Centre. *Journal of Large-Scale Research Facilities*, A1, 1 2015.

[304] A. Stivala, P. J. Stuckey, M. G. de la Banda, M. Hermenegildo, and A. Wirth. Lock-free parallel dynamic programming. *Journal of Parallel and Distributed Computing*, 70(8):839–848, 2010.

[305] G. L. Taboada, S. Ramos, R. R. Expósito, J. Touriño, and R. Doallo. Java in the high performance computing arena: Research, practice and experience. *Science of Computer Programming*, 78(5):425–444, 2013.

[306] R. E. Tarjan. Depth first search and linear graph algorithms. *SIAM Journal on Computing*, 1(2):146–160, 1972.

[307] R. E. Tarjan. Efficiency of a good but not linear set union algorithm. *Journal of the ACM*, 22(2):215–225, 1975.

[308] R. E. Tarjan. Shortest paths. Technical report, AT&T Bell Laboratories, 1981.

[309] R. E. Tarjan. Amortized computational complexity. *SIAM Journal on Algebraic and Discrete Methods*, 6(2):306–318, 1985.

[310] R. E. Tarjan and U. Vishkin. An efficient parallel biconnectivity algorithm. *SIAM Journal on Computing*, 14(4):862–874, 1985.

[311] M. Thorup. Undirected single source shortest paths in linear time. *Journal of the ACM*, 46(3):362–394, 1999.

[312] M. Thorup. Even strongly universal hashing is pretty fast. In *11th ACM-SIAM Symposium on Discrete Algorithms (SODA)*, pages 496–497, 2000.

[313] M. Thorup. Compact oracles for reachability and approximate distances in planar digraphs. *Journal of the ACM*, 51(6):993–1024, 2004.

[314] M. Thorup. Integer priority queues with decrease key in constant time and the single source shortest paths problem. *Journal of Computer and System Sciences*, 69(3):330–353, 2004.

[315] M. Thorup and U. Zwick. Approximate distance oracles. *Journal of the ACM*, 52(1):1–24, 2005.

[316] A. Toom. The complexity of a scheme of functional elements realizing the multiplication of integers. *Soviet Mathematics Doklady*, 150(3):496–498, 1963.

[317] P. Tsigas and Y. Zhang. A simple, fast parallel implementation of quicksort and its performance evaluation on SUN enterprise 10000. In *11th Euromicro Conference on Parallel, Distributed and Network-Based Processing (PDP)*, pages 372–381. IEEE, 2003.

[318] Unknown. *Der Handlungsreisende – wie er sein soll und was er zu thun hat, um Auftraege zu erhalten und eines gluecklichen Erfolgs in seinen Geschaeften gewiss zu sein – Von einem alten Commis-Voyageur.* 1832.

[319] L. Valiant. A bridging model for parallel computation. *Communications of the ACM*, 33(8):103–111, 1990.

[320] P. van Emde Boas. Preserving order in a forest in less than logarithmic time. *Information Processing Letters*, 6(3):80–82, 1977.

[321] R. Vanderbei. *Linear Programming: Foundations and Extensions*. Springer, 2001.

[322] P. J. Varman, S. D. Scheufler, B. R. Iyer, and G. R. Ricard. Merging multiple lists on hierarchical-memory multiprocessors. *Journal on Parallel & Distributed Computing*, 12(2):171–177, 1991.

[323] V. Vazirani. *Approximation Algorithms*. Springer, 2000.

[324] C. Vetter. Fast and exact mobile navigation with OpenStreetMap data. Master's thesis, KIT, 2010.

[325] J. Vuillemin. A data structure for manipulating priority queues. *Communications of the ACM*, 21(4):309–314, 1978.

[326] L. Wall, T. Christiansen, and J. Orwant. *Programming Perl*. O'Reilly, 3rd edition, 2000.

[327] J. Wassenberg, W. Middelmann, and P. Sanders. An efficient parallel algorithm for graph-based image segmentation. In *13th Conference on Computer Analysis of Images and Patterns (CAIP)*, pages 1003–1010, 2009.

[328] I. Wegener. BOTTOM-UP-HEAPSORT, a new variant of HEAPSORT beating, on an average, QUICKSORT (if *n* is not very small). *Theoretical Computer Science*, 118(1):81–98, 1993.

[329] I. Wegener. *Complexity Theory: Exploring the Limits of Efficient Algorithms*. Springer, 2005.

[330] R. Wickremesinghe, L. Arge, J. S. Chase, and J. S. Vitter. Efficient sorting using registers and caches. *ACM Journal of Experimental Algorithmics*, 7(9), 2002.

[331] R. Wilhelm and D. Maurer. *Compiler Design*. Addison-Wesley, 1995.

[332] J. W. J. Williams. Algorithm 232: Heapsort. *Communications of the ACM*, 7(6):347–348, 1964.

[333] A. Yasin. A top-down method for performance analysis and counters architecture. *IEEE International Symposium on Performance Analysis of Systems and Software (ISPASS)*, pages 35–44, 2014.

[334] W. Zhou. A practical scalable shared-memory parallel algorithm for computing minimum spanning trees. Master's thesis, Karlsruhe Institute of Technology, 2017.

[335] A. L. Zobrist. A new hashing method with application for game playing. Technical Report TR88, Computer Sciences Department, University of Wisconsin, Madison, 1970.

Index

p, **38**
@, 45
15-puzzle, 375

Aarts, E. H. L., 384
(a,b)-tree, *see under* sorted sequence
Abello, J., 354
Ackermann, W., 342
Ackermann function (inverse), 342
active message, 418
addition, **2**
address, 33
Adel'son-Vel'skii, G. M., 256
adjacency array, *see under* graph
adjacency list, *see under* graph
adjacency matrix, *see under* graph
adjacent, **69**
Aggarwal, A., 188
aggregation, 135
Aho, A. V., 256
Ahuja, R. K., 313, 355
Ajtai, M., 210
Akhremtsev, Y., 253
Akra, M., 56
al-Khwarizmi, Muhammad ibn Musa, **1**, 10
ALD, *see under* shortest path
algorithm, **1**
algorithm analysis, *see also* running time, **52**, 52
  amortized, 82, 221, 248, 315
    accounting method, **99**
    binary counter, 101
    deamortization, 101

general definition, **102**
operation sequence, **102**
parallel, 106
potential method, **99**
token, **99**
unbounded array, 97
universality of potential method, **105**
approximation algorithm, 367
average case, **57**, 121, 158, 166, 169, 179, 184, 185, 209, 237, 311, 317, 371
global, **57**
master theorem, **54**, 161
parallel, **62**
randomized, **63**, 95, 166, 169, 179, 190
recursion, 13, 21, 53, 161
recursive, 13, 18
smoothed analysis, 392
sum, 7, **53**
worst case, 169
algorithm design, 1
  "make the common case fast", 97
  algebraic, 13, 124, 127, 156, 263, 268
  black-box solvers, 359, 374, 391
  certificate, **48**, 51, 71, 292
  deterministic, **66**, 154
  diversification, 365
  divide-and-conquer, **11**, 49, 54, 160
    building a heap, 216
    mergesort, 160
    MSD radix sort, 184
    multiplication, 11
    multiway mergesort, 188
    parallel, 396

© Springer Nature Switzerland AG 2019
P. Sanders et al., *Sequential and Parallel Algorithms and Data Structures*,
https://doi.org/10.1007/978-3-030-25209-0

parallel multiplication, 16
quicksort, 168, 178
dynamic programming, **368**, 391
  Bellman–Ford algorithm, 318
  changing money, 371
  edit distance, 372
  knapsack, 369, 371
  matrix products, chained, 371
  parallel, 372
  principle of optimality, **368**, 372
  shortest paths, 303
evolutionary algorithm, **389**, 391
  parallel, 391
greedy, 156, **365**, 387, 391
  changing money, 371
  cycle detection, 71
  Dijkstra's algorithm, 307
  Jarník–Prim algorithm, 337
  knapsack, 363, 365
  Kruskal's algorithm, 338
  machine scheduling, 367
  parallel, 294, 297, 368
local search, **379**, 391
  hill climbing, 379
  relaxing constraints, 386
  restarts, 388
  simplex algorithm, 380
  simulated annealing, 382
  tabu search, 388
  threshold acceptance, 387
lookup table, 315
multilevel algorithm, **93**
parallel recursion, 16
portfolio approach, 365
preprocessing, 49, 154
random sampling, 189, 354
randomized, **63**, 131, 209, 256, 345, 391
  Las Vegas, **67**, 122, 168, 178
  load balancing, 426
  Monte Carlo, **67**, 96, 156
randomized algorithm
  load balancing, 429
recursion, 11, 13, 73, 160, 168, 173, 178,
    184, 216, 282, 373
result checking, 10, **48**, 156, 309
systematic search, **373**, 376, 391
  constraint programming, 374, **392**
  ILP solving, 376
  iterative deepening, 375

knapsack, 373
  parallel, 376
  use of sorting, 49, 153–155, 210, 265, 363
algorithm engineering, **1**, 9, 14, 17, 130,
    147, 170, 189, 208, 209, 254, 310,
    322, 387, 391
alignment, 11, 254, 448, 449, 459
all-gather, 159, 191, **412**, 464
  on hypercube, 412
all-pairs shortest path, *see under*
    shortest path
all-reduce, 138, **403**, 464
  on hypercube, 404
all-to-all, 80, 136, 192, **413**, 464
  MPI, 198
  nonuniform, 138, 193, 416
    two-phase algorithm, 416
  on hypercube, 413
all-to-all broadcast, **412**
allocate, 33
Alon, N., 150
Amdahl's law, 63
amortized, *see under* algorithm analysis
analysis, *see* algorithm analysis
ancestor, **73**
AND, **30**
Andersson, A, 210
antisymmetric, **436**
Applegate, D. L., 352
approximation algorithm, 333, **366**, 424, 433
approximation ratio, **366**
Aragon, C. R., 256, 386
arbitrage, 320
Arge, L., 80, 208
arithmetical unit, 447
arithmetics, 30
Arora, N. S., 113, 428, 434
array, **33**, 81
  access [·], **97**
  associative, 117
    build, **118**
    find, **118**
    forall, **118**
    insert, **118**
    remove, **118**
  circular, **108**, 110, 313
  growing, 97
  popBack, **97**
  pushBack, **97**

reallocate, 97
shrinking, 97
size, **97**
sorting, 170
unbounded, **97**, 261
assertion, **46**
assignment, 34
associative operator, 402
asymptotic, 15, **27**, 31
atomic operation, 39, 451, 455, 457
hashing, 118
atomic task, 428
atomic variable, 457
Ausiello, G., 75
average case, *see under* running time
AVL tree, *see under* sorted sequence
AWK, 118
Axtmann, M., 159, 177, 210

B (block size), 32
B-tree, 255
backoff, 407
Bader, D. A., 354
Balyo, T., 365
bandwidth, 32
barrier, 193, 195, 203, **408**, 464
binomial tree, 408
base, **2**
Bast, H., 210
batch, 118
Batcher, K. E., 210
Bayer, R., 255
Bazzi, L., 56
Beckmann, A., 226
Beier, R., 371
Bellman, R., 318
Bellman–Ford algorithm, *see under*
sorted path
Bender, M. A., 257, 336
Bentley, J. L., 173, 209
Bertsekas, D. P., 392
best case, *see under* running time
best-first branch-and-bound, 212
BFS, *see under* graph
bin packing, 234, 368
binary heap, *see under* priority queue
binary operation, 30
binary search, *see under* searching, 163

binary search tree, *see under*
sorted sequence
binary tree, 396, 398
distributed construction, 399
Bingmann, T., 159, 210, 226
binomial coefficient, **443**
binomial heap, *see under* priority queue
binomial tree, **222**, 396, 412, 413
broadcast, 396
bisection method, **51**
bit operation, **30**
Bixby, E. E., 352
Blelloch, G. E., 139, 191, 253, 329, 354
block, *see* memory block
BlueGene/Q, 200
Blum, N., 209, 257
Blumofe, R. D., 113, 428, 434
Boolean formula, 364
Boolean value, **32**
Boost, 78
Bellman–Ford algorithm, 331
Dijkstra's algorithm, 331
graph, 268
graph traversal, 298, 331
MST, 354
union–find, 354
boost, 459
Borůvka, O., 348
Botelho, F., 150
bottleneck shortest path, 333
bottom-up heap operation, 216
bounded array, **81**
branch, 30, 448
branch prediction, 209, 254
branch-and-bound, 212, 373
parallel, 377
branch-and-cut, 376
Brent's principle, **62**, 423
Bro Miltersen, P., 150
broadcast, 158, 175, 191, **396**, 464
asynchronous, 418
binomial tree, **396**
by hardware, 396
by two trees, 401
hypercube algorithm, 401
linear pipeline, 400
lower bound, 397
naive, 396
on a mesh, 401

pipelined, 398
using a tree, 397
Brodal, G. S., 226, 230
Bronson, N. G., 252
Brown, M. R., 116
Brown, R., 231
BSP, *see under* machine model
Buchsbaum, A., 354
bucket, 189
bucket sort, *see under* sorting
bulk operation
    FIFO, 110
    hashing, 136
    priority queue, 226
    sorted sequence, 253
Buluc, A., 266

C, 32
C++, 23, 32, 37, 78, 114, 148, 208, 229, 255,
        267, 331, 354, 455
C++11/14, 455
cache, 31
    block, 448
    coherence, 449
    footprint, 459
    instruction, 448
    levels, 448
    limited associativity, 207
    replacement strategy, 449
    unified, 448
    write-back, 449
cache hierarchy, 396
cache line, 448
cache miss, 448
cache-oblivious, 230, 257
Caldwell, C. K., 150
calendar queue, *see under* priority queue
call by reference, **35**
call by value, **35**
carry, 2
Carter, J., 150
CAS, **39**, 139, 144, 177, 206, 328, 406, 451,
        457, 459
    16 byte, 141
cascading cut, 223
Casper, J., 252
casting out nines, 10
Cayley, A., 268
census, 153

certificate, *see* algorithm design
certification
    MST, 336
    strongly connected components, 292
certifying algorithm, **48**
Cha, S. K., 252
Chandra, R., 458
changing money, 371
characteristic function, 75
Chase, S., 208
Chazelle, B., 258, 354
checksum, 10
Cheng, Y., 150
Cheriyan, J., 299
Cherkassky, B., 331
Chernoff bound, 137, 190, 203, 227, 431,
        442
chess, 117
Chiang, Y-J, 294
child, **73**
Chvátal, V., 352
Cilk, 78, 458
CISC, 447
class, 32, 33, **37**
clique, *see under* graph
clock cycle, 32
clustering, 333
Coffman, E. G., 234
Cohen-Or, D., 269
coherence, 449
Cole, R., 354, 418
collective communication, **393**, 464
    asynchronous, 418, 464
collision, 119
combinatorial search, 117
communication network, 41
communication pattern
    one-to-p, 396
    p-to-one, 396
    p-to-p, 396
communication time, **393**
communication topology, 396
communication volume, 80
communicator, 461
commutative operator, 402
compare-and-swap, **39**
comparison, **30**
    three-way, 49, 168, 169
    two-way, 51

comparison-based algorithm, 49, 165
competitive ratio, **368**
compiler, 7, 32, 80, 117, 208, 454
   symbol table, 117
compiler pragma, 458
complex number, 37, 154
complexity, *see also* running time, 31
complexity theory, 74
composite data structure, 33
composite type, 32
computation, model of, 30
computer architecture, 447
concave function, 312, **437**
conditional branch instruction, 209
conditional statement, **34**
cone, 381
Cong, G., 354
congruent, 436
consistency, 450
constant, **30**
constant factor, 27, 31
constraint, 359
constraint programming, *see under*
   algorithm design, systematic search
contention, 40, 135, 146, 450, 453
contract, **47**
contraction, 93
convex, **437**
convex polytope, 380
Cook, S. A., 76
Cook, W. J., 23, 352
cooling schedule, 384
coprocessor, 44
core, 43, 447
correctness, **46**
cost vector, 359
counting votes, 393
Crauser, A., 330
critical item, 363
critical path length, 434
critical section, **39**
crossover operation, 390
C#, 32
cuneiform script, **81**
cycle, **70**
   Hamiltonian, **70**, 74
   simple, **70**
   testing for, **71**

DAG, *see* graph, directed, acyclic
Dantzig, G. B., 359
Darwin, C., 391
data dependency, 31, 448
data distribution
   block cyclic, 82
   blocked, 82
   cyclic, 82
   round robin, 82
data parallelism, 174
data structure, **vii**
   circular array, 110
   (a,b)-tree, 238
   abstraction, 37
   augmenting search trees, 250
   binary search tree, 235
   deque, 107
   FIFO queue, 107
      distributed-memory, 112
      parallel, 110
      relaxed, 110
      shared-memory, 110
   graph, 259–269
   hash table, 117–151
   implementations, 78
   invariant, *see under* invariant
   linked list, 86
   lock, *see* lock
   lock-free, 46
   non-blocking, 46
   persistent, 258
   priority queue, 211–231
      addressable, 218
      binary heap, 213
      bucket queue, 313
      external, 224
      Fibonacci heap, 221
      pairing heap, 220
      parallel, 226
      radix heap, 313
   red–black tree, 255
   skip list, 257
   sorted sequence, 233–258
      parallel, 252
   stack, 106
   unbounded array, 97
   union–find, 340
   wait-free, 46
data stucture

priority queue
    monotone, 313
data type, *see* type
database, 120, 135, 140, 235, 255
database join, 117
Davidson, A., 208
de Berg, M., 355
De, A., 24
deadlock, **46**, 456
Dean, J., 138
decision problem, 75
declaration, **32**, 35
    implicit, 35
decrement ($--$), **34**
degree, **69**
Delaunay triangulation, 355
Demaine, E. D., 257
Dementiev, R., 139, 208, 209, 226, 257, 344
deque, **107**, 116
    first, **107**
    last, **107**
    popBack, **107**
    pushFront, **107**
    pushBack, **107**
    pushFront, **107**
dereference, 33
descendant, **73**
design by contract, **47**
deterministic algorithm, *see under*
    algorithm design
Devroye, L., 237
dictionary, **117**, 153
diet problem, 359
Dietzfelbinger, M., 150
digit, **1**
digraph, *see* graph, directed
Dijkstra's algorithm, *see under*
    shortest path
Dijkstra, E., 307, 337
discrete-event simulation, 212
disk, *see* hard disk
dispose, 33
distributed memory, **41**
distributed system, 44
div, 30
divide-and-conquer, *see under*
    algorithm design
division (integer), 10
DMA, 453

dopar, 45
dot product, **358**
Driscoll, J., 258
Drmota, M., 56
dynamic programming, *see under*
    algorithm design
dynamic tree, 339

edge, **68**
    associated information, 259
    backward, 271, 284
    contraction, 299
    cost, **70**
    cross, 271, 284, 286
    crossing, 71
    forward, 271, 284
    parallel, 259, 268
    reduced cost, *see also* node potential, 320
    tree, 271, 284
    weight, **70**, 259
edge coloring, 417
edge contraction, 345
edge query, 260, 263
edgeArray, **260**
edit distance, 372
efficiency, **62**, *see* running time
Eiffel, 77
eight-queens problem, 374, 386
element, **33**, 153
embarassingly parallel, **45**
empty sequence ⟨⟩, **34**
equals ($=$), **30**
equivalence relation, **437**
Eratosthenes, 37
event, **439**
evolutionary algorithm, *see under*
    algorithm design
exchange argument, 335, 364
exclusive OR ($\oplus$), **30**
execution time, *see* running time
execution trace, 454
existence problem, **357**
expected value, 57, **439**
exponential backoff, 407
exponential search, 51
external memory, *see also* machine model
    building heap, 217
    lower bound, 188
    merging, 187

MST, 344
parallel, 80
parallel disks, 189
priority queue, **224**
queue, 109
scanning, 187
semiexternal algorithm, 345
sorting, 187, 189, 208
stack, 109

factorial, 443
Fakcharoenphol, J., 332
false, **30**
false sharing, **450**, 459
Farach-Colton, M., 257, 336
fast memory, 32
fat-tree, 452
Feistel permutation, 427
fence, 409, 450, 457
Ferizovic, D., 253
ferry connections, 333
fetch-and-add, **40**, 294, 402, 425, 451, 457
Fibonacci, L., 221
Fibonacci heap, *see under* priority queue
field (algebraic), **124**, **437**
field (of variable), 33
FIFO queue, **107**, 273
    distributed-memory, 112
    epoch, 112
    external-memory, 109
    first, **107**
    parallel, 110
    popFront, **107**
    pushBack, **107**
    relaxed, 110
    shared-memory, 110
    using circular array, 108
    using two stacks, 107
file, 34
filing card, 233
Flajolet, P., 56
Fleischer, R., 230
floating-point, **30**, 77, 315
flow, 361
Floyd, R. W., 80, 181, 209
"folklore" (result), 116
for, **35**
Ford, L. R., Jr., 318
forest, **71**

Fortran, 78
Fotakis, D., 150
Fredkin, E., 257
Fredman, M. L., 150, 221, 230
frequency allocation, 387
Frigo, M., 230
Fürer, M., 24
full-duplex, 414
function object, 148
function pointer, 208
future, 457

Gabow, H., 299
Gärtner, B., 392
Gangal, D, 434
garbage collection, **79**
Garey, M. R., 75, 234
Garland, M., 208
gather, 191, **412**, 464
    binomial tree, 412
    naive, 412
Geisberger, R., 325
generic methods, 357
generic programming, **37**, 229, 268
genome, 389
geometric series, *see under* sum, 444
geometry, 381
Ghemawat, S., 138
Giacomoni, J., 110
Gibbons, P. B., 40, 191
GMP, 23
Goldberg, A., 317, 325, 331, 332
golden ratio, 401
Goodrich, M. T., 80, 210, 294
gossiping, **412**
Graefe, G., 252, 255
Graham, R. L., 56, 80, 367
Grama, A., 62
Grantham, J. F., 150
graph, **68**
    2-edge-connected components, 292
    adjacency array, **261**
    adjacency list, **261**
    adjacency matrix, **263**
        undirected, 263
    average degree, 347
    BFS, **272**, 302, 432
        implementation, 298
        load balancing, 422

biconnected components, 293, 299
bidirected, **69**, 259, 262
bipartite, **48**, 268, 417
breadth-first search, *see* BFS
Cayley, 268
citation network, 259
clique, **74**, 75
coloring, 48, **74**, 75, 384, 386
    fixed-*K* annealing, 387
    Kempe chain annealing, 385
    penalty function annealing, 386
    XRLF greedy algorithm, 387
communication network, 271
complete, 74
component, **70**
compression, 269
connected components, **70**, 274
construction, 260
conversion, 260, 261
counting paths, 264
cut, 264, 334
cycle detection, 262
DAG, *see* graph, directed, acyclic (DAG)
dense, 263
depth-first search, *see* DFS
DFS, 271, **282**, 319
    backtrack, 282
    init, 282
    root, 282
    traverseNonTreeEdge, 282
    traverseTreeEdge, 282
diameter, **321**
directed, **68**
    acyclic (DAG), **70**, 71, 72, 285, 433
dynamic, 260, 261
ear decomposition, 299
edge, *see under* edge
edge sequence, 260, 338
exploration, *see* graph traversal
face, 268
grid, 264
hypergraph, 268
input, 260
interval graph, 154, 264
Kempe chain, 385
layer, 272
LEMON, 229, 268
linked edge objects, 262

minimum spanning tree, *see*
    MST
MST, *see* MST
multigraph, 259, 268
navigation, 260, 261
negative cycle, *see under* shortest path
network design, 333
node, *see* node
open ear decomposition, 293
output, 260
parallel
    Bellman–Ford, 328
    BFS, 280, 432
    coloring, 295, 297
    Dijkstra's algorithm, 330
    independent sets, 295
    representation, 265
    shortest paths, 328
    shortest paths in DAGs, 307
    topological sorting, 294
partitioning, **265**
    by sorting, 266
    by space-filling curve, 266
    evolutionary, 391
    parallel, 391
    web graph, 266
planar, **71**, 268
    4-coloring, 384
    5-coloring, 385
    embedding, 299
    testing planarity, 299
random, 320, 387
random geometric graph, 387
representation, **259**
reversal information, 260
SCC, *see* graph, strongly connected
    components
shortest path, *see* shortest path
    Δ-stepping, 330
shrunken graph, 286
sparse, 261
static, 261
Steiner tree, 351
    2-approximation, 351
street network, 71
strongly connected components, **70**, 271,
    286
    certificate, 292
    closed, 287

implementation, 298
invariant, 288
more algorithms, 299
open, 287, 288
subgraph (induced), **69**
topological sorting, 285, 294, 306
transitive closure, 274
traversal, **271**
triconnected components, 299
undirected, **69**
parallel traversal, 295
vertex, *see* node
visitor, 298, 331
graph partitioning, *see under*
graph
graph500 benchmark, 299
greater than ($>$), **30**
greedy algorithm, *see under* algorithm
design
grid network, 416
Grossi, R., 257
group, 268
grouping, 154
growth rate, 27
Guibas, L. J., 258
Gupta, A., 62

h-relation, 80, **416**
H-tree, 401
Haeupler, B., 230
Hagerup, T., 210
half-space, 380
Halperin, S., 354
Hamilton, W. R., 70
Han, Y., 210, 231
handle, **33**, 86, 212, 234
Handler, G., 332
hard disk, 32
hardware thread, 447, 448
harmonic numbers, 444
harmonic sum, *see under* sum
Harrelson, C., 325
Harris, M., 208
Hartmanis, J, 80
Harvey, D., 24
hash function, 118
hash table, *see* hashing
hashing, **117**, 154
available implementations, 147

closed, **128**
concurrent, 138
insertOrUpdate, **134**
large elements, 148
large keys, 148
linear probing, 119, 128
cyclic, 130
find, **128**
insert, **128**
parallel, 138
remove, **129**
unbounded, 130
open, **128**
perfect, 131
perfect (dynamic), 133
permutation invariant, 156
realistic analysis, 123
shared memory, 138
shared memory implementation, 141
universal, 122
bit strings, 123
by integer multiplication, 127
by shifting, 127
by table lookup, 127
simple linear, 127
using bit matrix multiplication, 126
using scalar products, 124
universal family, 123
unrealistic analysis, 121
update, **134**
use of, 154, 156, 167, 260
with chaining, 119, **120**
average case, 122
fat, **148**
find, **120**
implementation, 147
insert, **120**
parallel, 140
remove, **120**
slim, **148**
unbounded, 122
heap property, **214**, *see also*
priority queue
heapsort, *see under* sorting
Held, M., 353
Held–Karp lower bound, *see under*
MST
Hennessy, J. L., 24, 447
Herlihy, M., 110

heuristic, 60, 297
high-performance computing, 44
hill climbing, *see under* algorithm design,
    local search
H_n, *see* sum, harmonic
Ho, C. T., 401, 403, 406
Hoare, C. A. R., 80
Hollerith, H., 153
Hopcroft, J., 256, 343
Huddlestone, S., 116, 256
Hübschle-Schneider, L., 181, 182, 228
Hwang, S, 252
hyper-threading, 146
hypercube, 396, 401, **403**, 412, 413, 415,
    452
  prefix sum, 404
  subcube, 404
hyperplane, 380
Høyer, P., 230

I/O step, 32
Iacono, J., 230
IBM, 153
IEEE floating-point, 77
if, 34
iff, **437**
ILP, *see* linear program, integer
imperative programming, 32
implementation note, 31
incident, **69**
increment (++), **34**
incumbent, 374
indentation, 34
independent random variable, **441**
independent set, 94, 95, **296**
index, 33, 81
indicator random variable, 57, 170
inequality
  Chernoff, **442**
  Jensen's, **444**
  Markov's, 68, **442**
infinity (∞), **33**, 77
initialization, 32
inlining, **35**
input, 30
input size, 26, 29
inserting into a sequence, 82
insertion sort, *see under* sorting
instance, 26

instruction, **30**, 31
instruction parallelism, 448
integer, **32**
integer arithmetics, **1**
internal memory, 32
invariant, **46**, 288
  data structure invariant, 47, 48, 86, 213,
    218, 238, 249, 256, 314, 340
  loop invariant, **47**, 49, 129, 157
inverse, **438**
isoefficiency function, **62**, 162
Itai, A., 257
Italiano, G., 257
item, 86
iteration, 35
iterative deepening search, 375
iterator, *see under* STL

Jájá, J., 80
Jarník, V., 337
Jarník–Prim algorithm, *see under*
  MST
Java, 23, 32, 37, 79, 115, 149, 208, 256, 331
  deque, 115
  hashCode, 149
  hashMap, 149
  linked list, 115
  memory management, 115
  PriorityQueue, 229
  SortedMap, 256
  SortedSet, 256
  sorting, 208
  stack, 115
  TreeMap, 256
  TreeSet, 256
  vector, 115
JDSL
  graph traversal, 298
Jensen's inequality, 444
Jensen, K., 32
JGraphT, 229, 268, 298, 331, 354
  graph, 229, 268
  MST, 354
  union–find, 354
Jiang, T., 209
job, 419
Johnson, D. S., 75, 234, 386
Johnson, N. L., 427
Johnsson, S. L., 401, 403, 406

join
  hash-, 120
  parallel, 140, 165
jump, 30

Kaligosi, K., 209
Kaplan, H., 230
Karatsuba, A., 13
Karger, D., 354
Karlin, A., 150
Karmarkar, N., 362
Karp, R., 353
Karypis, G., 62
Katajainen, J., 116, 226
Katriel, I., 354
Keh, T., 226
Kellerer, H., 357
Kemp, A. W., 427
Kempe, A. B., 385
Kempe chain, *see under* graph
Kettner, L., 208, 257
key, 118, 153, 211
Khachiyan, L., 362
Kieritz, T., 329
King, V., 354
Klein, P. N., 354
knapsack, **74**, 301
knapsack problem, 357
  2-approximation (*round*), 366
  as an ILP, 362
  average case, 371
  branch-and-bound algorithm, 373
  dynamic programming, 369
    by profit, 371
  evolutionary algorithm, 390
  fractional, 363, 374
  fractional solver, 364
  greedy algorithm, 365
  local search, 379
  parallel, 428, 433
  simulated annealing, 384
  use of, 357
knot, 81
Knuth, D. E., 56, 80, 149, 209
Komlós, J., 150, 210
Konecny, P., 207
Konheim, A. G., 257
König, D., 418
Korf, R. E., 375

Korst, J., 384
Korte, B., 355
Kosaraju, S. R., 299
Kothari, S., 208
Kotz, S., 427
Kruskal, J., 338
Kumar, V., 62
Kurur, P. P., 24

Ladner, R. E., 230
LaMarca, A., 230
Lamm, S., 191
Landis, E. M., 256
Larsen, P.-A., 255
Las Vegas algorithm, *see under*
    algorithm design, randomized
latency, 32
Lawler, E. L., 352, 355
leading term, 28
leaf, **73**
LEDA, 23, 78, 298
  Bellman–Ford algorithm, 331
  bounded stack, 115
  Dijkstra's algorithm, 331
  graph, 267
  graph traversal, 298
  h_array, 149
  list, 115
  map, 149
  MST, 354
  node_pq, 331
  priority queue, 229
  queue, 115
  sortseq, 256
  stack, 115
  static graph, 267
  union–find, 354
Lee, L. W., 268
Lee, M., 78
left-to-right maximum, **58**, 64, 170, 312
Lehman, P.L., 252
Leighton, T., 80, 405, 453
Leischner, N, 208
Leiserson, C. E., 230
LEMON, 229, 268, 354
  graph traversal, 298, 331
  MST, 354
  union–find, 354
Lenstra, J. K., 352, 355

less than ($<$), **30**
Lev, G., 418
Levenshtein distance, 372
Levin, D., 269
Levin, L., 76
lexicographic order, 154, **437**
Li, M., 209
linear algebra, 263, 381
linear order, 154, 332, **437**
linear preorder, **438**
linear program (LP), **359**
  fractional solution, 363
  integer (ILP), 360, **362**
    0-1 ILP, 363
    0-1 ILP, 376
    branch-and-cut, 376
    knapsack, 362
    pigeonhole principle, 365
    set covering, 364
  maximum flow, 362
  minimum-cost flow, 362
  mixed integer (MILP), **362**
  relaxation of ILP, 363
  rounding, 363
  shortest path, 360
  simplex algorithm, 380
  smoothed analysis, 392
  solver, 392
  strict inequality, 381
  tight inequality, 381
linearity of expectations, 57, 121, 123, 170,
    347, **440**
Linux, 196
list, 34, 81, 120, 261
  blocked, 109, 162, 185
  bulk insert, **162**
  circular, 221, 261
  concat, 89, 91
  concatenate, 86, 90
  doubly linked, **86**, 234
  dummy item, **87**, 261
  empty, 87
  find, 88, 91
  findNext, 89, 91
  first, 89, 91
  head, 89, 91
  insert, **88**, 89, 91
  interference between ops., 90
  invariant, 86

isEmpty, 89, 91
last, 89, 91
linked, 86
makeEmpty, 89, 91
memory management, 87, 89
move item, **88**
popBack, 89
popFront, 89, 91
pushBack, 89, 91
pushFront, 89, 91
remove, 87, 89, 91
rotation, 89
singly linked, **90**, 147
size, **90**
sorting, 162
splice, **86**, 90
swapping sublists, **89**
list ranking, **91**
  by doubling, **92**
  by independent set removal, 93
load balancing, 45, 63, 112, 174, 377, **419**,
    456
  by prefix sums, **422**
  dependent tasks, 433
  master–worker, 421, **424**
    by fetch-and-add, 425
    hierarchical, 425
  randomized static, 421, 426
  scalability comparison, 420
  work stealing, 112, 377, 421, **427**, 458
    randomized, 429
load instruction, **30**
local search, *see under* algorithm design
  parallel, 388
locate, *see under* sorted sequence
lock, **46**, 406, 451, 455, 456
  binary, **46**
  queue, 407
  spin, 407
  timeout, 456
lock-free, 46
logarithm, **436**
logical operations, **30**
loop, **34**, 53
loop fusion, 7
loop invariant, *see under* invariant, 404
lower bound, 367
  "breaking", 182
  broadcast, 397

element uniqueness, **167**
external sorting, 188
minimum, **167**
pairing heap priority queue, 230
sorting, **165**
lower-order term, **28**
lowest common ancestor, 336
LP, *see* linear program
Lucas, É., 107
Lumsdaine, A., 268
Lustig, I. J., 392
Luxen, D., 329

M (size of fast memory), 32
machine instruction, *see* instruction
machine model, 27, **29**
    accurate, 31
    BSP, 80, 281
    complex, 31
    distributed memory, **41**
    external memory, 31
    parallel, 30, **38**
    PEM, 80
    RAM, **29**, 32
    real, 31
    sequential, 29
    shared memory, **38**
    simple, 31
    von Neumann, 29
    word, 210
machine program, **30**, 32
machine scheduling, **366**
    online algorithm, **367**
    shortest-queue algorithm, **367**
machine word, 29, 31
Magnanti, R. L., 355
Maier, T., 139, 150
main memory, 449
makespan, 366, 433
Mandelbrot set, 421, 425
map coloring, 384
MapReduce, 138
Markov, A., 68
Markov's inequality, *see under* inequality
Martello, S., 357
Martinez, C., 209
master, 424

master theorem, *see under* algorithm analysis
master–worker, *see under* load balancing
Matias, Y., 40, 191
mating, 390
Matoušek, J., 392
matrix, 263
matrix products, chained, 371
Mattern, F., 432
Mauer, D., 80
maximization problem, **357**
maximum flow, 362
McCreight, E. M., 255
McDiarmid, C. J. H., 96
McGeoch, L. A., 386
McIlroy, M. D., 209
median, 178, *see also* selection, **438**
Mehlhorn, K., 78, 116, 150, 256–258, 298, 299, 313, 321, 330, 332, 352
Mehnert, J., 257
member variable, **37**
memcpy, 115
memory
    allocator, 458
    initialization, 458
memory access, 31
memory block, **32**
memory cell, *see also* machine word, 29
memory fence, 407, 450
memory hierachy, 448
memory management, **33**, **451**, 458
memory model, 450
memory ordering, 457
memory size, 31
Mendelson, R., 231
mergesort, *see under* sorting
merging, 159, 160, 371
    external, 187
    multiway, **188**
    parallel, 162, 372
mesh, 449, 452
Meyer auf der Heide, F., 150
Meyer, B., 77
Meyer, U., 299, 317, 330
Meyerhenke, H., 266
Michel, L., 392
microinstruction, 448
minimization problem, **357**

minimum spanning forest, *see* MST
minimum spanning tree, *see* MST
mobile device, 43
mod, 30
modulo, 10, **436**
Monte Carlo algorithm, *see under* algorithm design, randomized
Moret, B., 353
Morris, R., 149
Morton-ordering, 266
Moseley, T., 110
most significant distinguishing index, 314
move-to-front, 60
MPI, 198, **461**
msd, *see* most significant distinguishing index
MST, **333**
    2-approximation of TSP, 352
    Borůvka's algorithm, 348
    bottleneck paths, 336
    certification, 336
    clustering, 333, 355
    cut property, **335**, 339
    cycle property, 335, 354
    Euclidean, 355
    external memory, 344
    Held–Karp lower bound, 353
    Jarník–Prim algorithm, **337**
    maximum-cost spanning tree, 334
    parallel
        Kruskal, 355
    semiexternal Kruskal algorithm, 345
    streaming algorithm, 339
    uniqueness conditions, 336
    use of, 333, 351, 355
Müller, I., 80, 135
multicore processor, **43**
multigraph, 259, 268, 417
multikey quicksort, 173
multilevel algorithm, 93, 266
multiplication (integer)
    Karatsuba, **13**
    refined, 17
    recursive, **11**
    school method, 1, **6**
    use of, 1
multithreading, **43**, 455

Musser, D. R., 78
mutation, 389
mutex, 456

Nadathur, S., 208
Näher, S., 78, 257, 258, 263, 298
Naor, M., 427
Nemhauser, G., 370, 376
network, 44, *see also* graph, 452
    communication network, 68
    contention, 453
    design, 333
    fat-tree, 452
    hypercube, 452
    mesh, 452
    torus, 452
Neubert, K. S., 210
Nilsson, S., 210
node, **68**
    active, 283
    associated info., 259
    depth, 72, 272
    dfsNum, 284
    finishing time, 284
    interior, **73**
    marked, 282
    numbering, 260
    ordering relation ($\prec$), 284
    potential, **320**, 324, 353
    reached, 272, 308
    representative, 274, 288
    scanned, 307
NodeArray, **260**, 267
non-blocking, 46
nonblocking, 463
Noshita, K., 311
NOT, **30**
NP, **73**
NP-complete, **74**
NP-hard, **76**, 363, 433
NUMA, 193, 200, 201, **449**, 462
    allocation, 459
numeric type, 33

O($\cdot$), **27**
o($\cdot$), **27**
object, 33
object-oriented, **37**
objective function, **357**

of (in type declaration), 33, 34, 37
Ofman, Y., 13
$\Omega(\cdot)$, **27**
$\omega(\cdot)$, **27**
online algorithm, 61, **367**
OpenMP, 78, 458
optimization, **357**
optimization problem, 77, **357**
OR, **30**
Orlin, J. B., 313, 355
Osipov, V., 208, 354
Ost, K., 418
oversampling, 189
owner computes, **83**, 205, 265, 280

P, **73**
packet, 398
padding, 450
page, **452**
Pagh, A., 151
Pagh, R., 150, 151
pair, **33**
pairing heap, *see under* priority queue
parallel, 3, 15, 24, 38, 44, 61, 77–80, 106,
       109, 118, 134, 155, 158, 162, 173,
       180, 186, 226, 252, 265, 294, 368, 372
   addition, 3
   array processing, 82
   DAG traversal, 294
   divide and conquer, 16
   evolutionary algorithm, 391
   following pointers, 92
   graph construction, 265
   graph partitioning, 391
   graph representation, 265
   hardware multiplication, 24
   hashing, 134
   independent sets, 295
   knapsack problem, 428, 433
   linear programming, 362
   local search, 388
   loop, 458
   machine model, **38**
   merging, 162
   multiplication, 15
   partitioning, 175
   recursion, 16
   sampling, 175
   sorted sequence, 252
   sorting, *see* sorting
   parallel assignment, 34
   parallel external memory, 80
   parallel processing, 31, 189, 392
parameter, **35**
   actual, 35
   formal, 35
parameterized class, **37**
parent, **73**
Pareto, V., 370
Pareto-optimal, 370, 391
parser, 73
partition, 340
partitioning, 83
Pascal, 32
Patashnik, O., 56, 80
path, **70**
   simple, **70**
Pătraşcu, M., 138, 151
Patterson, D. A., 24, 447
Paul, W. J., 207
pause instruction, 407
PE, **38**, 448
Peitgen, H.-O., 422
Pemmasani, G., 336
perf profiler, 453
performance analysis, 453
Perl, 118
permutation, 58, 154, 156
   random, 59, 64
persistent data structure, 258
Peru, 81
Peterson, W. W., 128
Petrank, E., 150
Pettie, S., 230, 354
Pferschy, U., 357
pigeonhole principle, 364
pipe, 110
pipeline, 447
pipeline stage, 448
pipelining, 9, 398
Pippenger, N., 418
Pisinger, D., 357
pivot, **168**, 189
   selection, 171, 209
Plauger, J.J., 78
Plaxton, C. G., 113, 428, 434
point-to-point communication, 41
pointer, **32**

polynomial, 28, *see also under*
    running time, 156
polytope, 380
Pomerance, C., 150
population, 389
postcondition, **47**
potential function, *see* node, potential
powers (of numbers), 47
PRAM, 80, 95
  aCRQW, **40**, 140
  CRCW, **38**, 205, 294, 330
  CREW, **38**
  CRQW, **40**
  EREW, **38**
  QRQW, **40**
Pratt, V. R., 209
precedence relation, 68
precondition, **47**
predecessor, **82**, 86
prefix sum, 6, 94, 112, 176, 186, 282, 294,
    368, 372, **404**, 422, 464
  by two trees, 406
  exclusive, **404**
  on hypercube, 404
  on tree, 405
Priebe, V., 321
Prim, R. C., 337
Prim's algorithm, *see* MST, Jarník–Prim alg.
prime number, 37, 124, 156, **438**
  abundance, 125
primitive operation
  full adder, 2
  product, 2
principle of optimality, **368**, 372
priority queue, **211**
  addressable, **212**, 218, 309
  binary heap, **213**, 310
    addressable, 213, 217
    bottom-up *deleteMin*, 230
    building, 216
    bulk insertion, 217
    deleteMin, **215**
    insert, **215**
    invariant, 213
    siftDown, **215**
    siftUp, **215**
  binomial heap, 222
  bounded, 213
  bucket queue, 231, **313**

  invariant, 314
  calendar queue, 231
  decrease key, **212**, 310
  deleteMin, **212**
  double-ended, 246
  external, 224
  fat heap, 231
  Fibonacci heap, **221**, 310, *see also*
    priority queue, heap-ordered forest
    decreaseKey, **223**
    deleteMin, **221**
    item, **221**
    rank, **221**
  heap-ordered forest, **218**
    cut, **218**
    decreaseKey, **218**
    deleteMin, **218**
    insert, **218**
    invariant, 218
    link, **218**
    merge, **218**
    new tree, **218**
    remove, **218**
  hollow heap, 231
  insert, **212**
  integer, 229, 231, **313**
  item, 218
  memory management, 229
  merge, **212**
  minimum, **212**, 214, 218
  monotone, 212, 231, 309, **313**
  naive, 213, 310
  pairing heap, **220**, *see also*
    priority q., heap-ordered forest
    complexity, 230
    three-pointer items, 220
    two-pointer items, 220
  parallel, 226
    bulk operation, 226
  radix heap, **313**
    base b, 316
  remove, **212**
  thin heap, 230
  unbounded, 213
  use of, 157, 188, 209, 212, 213, 309, 345
priority update principle, 329
probability, **439**
probability space, 58, **439**
problem instance, 26

procedure, **35**
process, 462
processor core, 447
profiling, 195, 453
profit vector, *see* cost vector
program, **30**
program analysis, *see* algorithm analysis
programming language, 32, 34, 80
  functional, 162
  logical, 162
programming model, *see* machine model
Prokop, H., 230
pseudo-polynomial algorithm, 371
pseudocode, **32**, 77
pseudorandom permutation, 427
Puget, J.-F., 392
Pugh, W., 257
Putze, F., 197

quartile, *see also* selection, 178
queue, 34, 262, *see also* FIFO
  parallel, 109
quickselect, *see under* selection
quicksort, *see under* sorting
quipu, **81**

radix sort, *see under* sorting
Radzik, T., 331
Rahn, M, 210
Rajasekaran, S, 210, 265
RAM model, *see under* machine model
Ramachandran, S., 230
Ramachandran, V., 40, 354
Raman, R., 210
Ramaré, O., 150
Ranade, A., 208, 434
random access, 118
random experiment, **439**
random number, **64**
random source, 77
random variable, 57, **439**
  independent, **441**
  indicator, **439**
  product, **441**
randomized algorithm, *see under*
    algorithm design; algorithm analysis
rank, 158, **438**
ranking, 158
Rao, S., 332

realloc, 115
receive, 463
recombination, 389
record, *see* composite type
recurrence relation, 13, 21, 50, **53**, 80
recursion, **35**, *see also under*
    alg. design; alg. analysis
  elimination, 173, 229
red–black tree, *see under* sorted sequence
reduction, **76**, 159, 228, 393, **402**, 464
  asynchronous, 418, 432
  by two trees, 403
  on vectors, 205
  pipelined, 403
  shared-memory, 402
  tree, 402
reflexive, **438**
register, **29**, 31, 43, 447
Reif, J. H., 210, 265, 299
Reingold, O., 427
relation, **438**
  antisymmetric, **436**
  equivalence, **437**
  reflexive, **438**
  symmetric, **438**
  total, **438**
  transitive, **438**
relaxation, 386, *see also under*
    linear program
remainder, **30**
Remez, O., 269
removing from a sequence, 82
repeat, **34**
resource, 419
result checking, *see under* algorithm design
return, 35
Richter, P. H., 422
ring, 449
Rivest, R. L., 181, 209
road map, 68
Robertson, N., 384
Robins, G., 352
Rodeh, M., 257
Rodler, F., 150
roll back, 451
root, *see under* tree
root PE, 396
round-robin, 426
Roura, S., 56, 209

routing, 453
run, *see under* sorting
running time, 26, **31**, 35, *see also*
    algorithm analysis, 52
  average case, **26**, 57
  best case, **26**, 31
  polynomial, **74**
  worst case, **26**
Ružić, M., 151

Sabela, R., 79
Saha, C., 24
sample space, **438**
Sanders, D. P., 384
Sanders, P., 80, 135, 139, 150, 177, 181, 182,
    191, 192, 197, 208–210, 226, 228,
    230, 253, 257, 298, 325, 330, 344,
    354, 355, 365, 391, 401, 403, 406,
    421, 426–428, 450
Santos, R., 269
Saptharishi, R., 24
Sarnak, N., 258
SAT solver, 364
satisfiability problem, **74**
satisfiable, 364
scalability, 62
  of load balancing, 420
scan, **404**
scatter, **413**, 464
  binomial tree, 413
Schäfer, G., 321
Schaffer, R., 230
scheduling, 212, 301, **366**, *see also under*
    load balancing
  threads, 459
Scheufler, S. D., 204
Schevon, C., 386
Schirra, S., 418
Schlag, S., 80
Schönhage, A., 24
Schrijver, A., 392
Schultes, D., 325, 344
Schulz, C., 391
search problem, **357**
search tree, *see* sorted sequence
searching, **233**, *see also* sorted sequence
  binary search, **49**, 77, 154, 189, 239
  dynamic, **59**
  exponential, 51, 77

linear, 59
range, 154
shortest path, *see under* shortest path
Sedgewick, R., 56, 173, 209, 230
Seidel, R., 256, 269, 342
selection, **178**
  deterministic, 209
  from sorted sequences, 163, 202
  parallel, 180
  quickselect, **178**, 205
  streaming, **180**
  using two pivots, 181
self-loop, **69**
semicolon (in pseudocode), 34
Sen, S., 230
send, 463
  asynchronous, **41**, 226, 418
sentinel, **88**, 147, 157, 162, 229
sequence, **34**, 81, 154
  overview of operations, 113
  space efficiency, 113
sequential consistency, 457
series, *see* sum
server, 44
set, **34**
set covering, 364
Seymour, P., 384
Shapiro, H. D., 353
shared memory, **38**
Sharir, M., 299
Shavit, N., 110
Shell sort, *see under* sorting
Shepherdson, J. C., 29, 79
shift, **30**
Shmoys, D. B., 352
shortest path, **301**
  acyclic, 302
  ALD (average linear Dijkstra), **317**, 332
  all-pairs, 301
  all-pairs with negative costs, 320
  arbitrary edge costs, 318
  as a linear program, 360
  A*-search, 324
  Bellman–Ford algorithm, 318
    refined, 331
  bidirectional search, 322
  bottleneck, 333, 355
  constrained, 332, 372
  correctness criterion, 305

DAG, 306
Dijkstra's algorithm, 226, **307**
  invariant, 313
  edge relaxation, **304**
  geometric, 332
  goal-directed search, 324
  hierarchical search, 325, 329
  integer edge cost, 302
  linear average time, 317
  multicriteria, 332
  negative cycle, 303
  nonnegative edge cost, 302
  parallel
    hierarchical search, 329
  parent pointer, 304
  public transportation, 307
  query, 322
  relaxing of edges, **304**
  single-source, 301
  subpath, 303
  tentative distance, 304
  tree, 304
  uniqueness, 304
  unit edge cost, 302
  use of, 301, 320
shrunken graph, 286
Shun, J., 139, 329, 354
Sibeyn, J., 344
sibling, **73**
sibling pointer, 221
Siek, J. G., 268
sieve of Eratosthenes, 37
SIMD, **43**, 147
Simon, J., 80
simplex algorithm, *see under*
    linear programming
simulated annealing, *see under*
    algorithm design, local search
Singler, J., 197, 208, 210, 226, 354
Sipser, M., 75
Sivadasan, N., 321
Skiena, S., 336
Sleator, D., 116, 230, 256, 258, 339
slow memory, 32
snow plow heuristic, 209
socket, **449**, 452
solution
  feasible, **357**
  potential, **357**

sorted sequence, 49, **233**
  (a, b)-tree, **238**
    split (node), **240**
    amortized update cost, **248**
    augmentation, **250**
    balance, **243**
    build/rebuild, **246**
    concatenation, **246**
    fusing, **244**
    height, **239**
    insert, **239**
    invariant, **238**
    item, **240**
    locate, **240**
    parent pointer, **250**
    reduction, **250**
    remove, **243**
    removing a range, **248**
    splitter, **239**
    splitting, **247**
  adaptable, 256
  AVL tree, 256
  binary search tree, **235**
    degenerate, 237
    expected height, 237
    implicit, 238
    insert, 237
    locate, **236**
    perfect balance, 237
    rotation, 238, 245
    selection, 251
  cache-oblivious, 257
  concatenation, 253
  concurrent access, 252
  finger search, 251
  first, **234**, 246
  insert, **233**
  integer, 257
  last, **234**, 246
  locate, **233**, 235
  merging, 251
  navigation, 234
  persistent, 258
  pred, **234**
  randomized search tree, 256
  range searching, 246
  red–black tree, 245
  remove, **233**
  skip list, 257

sparse table, 257
splay tree, 256
split, 253
strings, 257
succ, **234**
trie, 257
use of, **234, 235**
weight-balanced tree, 257
sorting, **153**
  almost sorted inputs, 158
  bottom-up heapsort, 230
  bucket, **183**
  by ranking, 158
  comparison-based, 183
  dynamic, 157
  external, **187**
  flash, 210
  heapsort, 212, **217**
  in-place, **156**, 170
  insertion, 53, **157**, 162
  large elements, 207
  list, 162
  lower bound, 182
  mechanical, 153
  mergesort, **160**, 208
  multiway merge, **188**
  numbers, 183, 207, 261
  parallel, 189
    bucket, 186
    by merging, 162
    by multiway merging, 201
    distr. memory multiway merging, 204
    distributed-memory mergesort, 165
    distributed-memory quicksort, 174
    fast inefficient, **158**, 192, 206
    log latency, 205
    MPI, 198
    MSD radix, 187
    overview, 155
    radix, 187
    sample (implementation), **193**
    shared-memory quicksort, 177
  quicksort, **168**, 208, 209, 237
  radix, **183**
    LSD, **184**
    MSD, **184**, 187, 207
  random numbers, **184**
  run formation, 164, 187, 209
  sample, **189**

selection, **156**, 212
Shell sort, 209
small inputs, 157, 167
small subproblems, 171
stable algorithm, **183**
strings, 173, 183
use of, 49, 153–155, 210, 265, 345, 363
word model, 210
source node, **69**
space-filling curve, 266
  Morton ordering, 266
  z-order, 266
span, 16, **63**, 95
Speck, J., 401, 403, 406
speed of light, 448
speedup, **62**
  superlinear, 378
spellchecking, 210
Spielmann, D., 392
spin lock, 407
Spirakis, P., 150
splitter, 189, 236
SPMD, **44**, 393, 458, 461
stack, 34, 36, 106, 107
  bounded, 108
  external-memory, 109
  pop, **107**
  push, **107**
  top, **107**
  unbounded, 108
statement, 34
static array, **33**, 81
statistics, 178
stencil, 85
Stepanov, A. A., 78
Stirling's approximation, 166, 185, **443**
Stivala, A., 139
STL, 18, 78, 255
  deque, 115
  iterator, 115, 208
  list, 115
  map, 255
  multimap, 255
  multiset, 255
  parallel, 458
  priority_queue, 229
  set, 255
  sort, 208
  stack, 115

unordered_map, 148
unordered_multimap, 148
unordered_multiset, 148
unordered_set, 148
store instruction, **30**
Strassen, V., 24
streaming algorithm, 180, 339
string, 34, 81, 154, 436
striping, 209
struct, *see* composite type
Stuckey, P. J., 139
Sturgis, H. E., 29, 79
STXXL, 208, 226, 229
subcube, 404
subroutine, **35**
successor, **82**, 86
succinct data structure, 151
Sudoku, 384
sum, 80, *see also under* algorithm analysis
  estimation by integral, 445
    geometric, 54, **444**
    harmonic, 59, 64, 126, 170, 312, 347, 436, **444**
Sumerian, 81
Sun, Y., 253
superscalar, 448
survival of the fittest, 389
swap, 34
sweep-line algorithm, 235
symmetric, **438**
synchronization, **406**
syntax, 32
Szemerédi, E., 150, 210
Szpankowski, W., 56

table, 81
tablet, 81
Taboada, L., 79
tabu list, *see* tabu search
tabu search, *see under* algorithm design, local search
tabulation hashing, 138
tail bound, **442**
tail recursion, *see* recursion, elimination
Tardos, E., 150
target node, **69**
Tarjan, D., 208
Tarjan, R. E., 116, 150, 209, 221, 230, 231, 256, 258, 299, 313, 331, 339, 342, 354

task, 45, 419
  atomic, 423, 428
  based programming, 458
  bundle, 425
  DAG, 433
  dependencies, 420
  graph, 16
  independent, 420, 429
  parallelism, 162, 174
  splitting, 428
Taylor series, 444
TBB, 78, 145, 458, 459
telephone book, 153
telephone model, 414
template programming, 37, 208
Teng, S. H., 392
termination, **48**, 50
termination detection, 418, 431
$\Theta(\cdot)$, **27**
Thomas, R., 384
Thompson, K., 373
Thorup, M., 138, 147, 151, 210, 231, 332
thread, **43**, 45, 447
  pinning, 459
thread pool, 456
threshold acceptance, *see under* algorithm design, local search
tie breaking, 202
tiling, **84**
time, *see* running time
time forward processing, 294
time step, **31**
Toom, A., 23
topological sorting, *see under* graph
torus, 452
total order, 154, **438**
Toth, P., 357
tournament tree, 209
Tower of Hanoi, 107
Träff, J. L., 226, 354, 401, 403, 406
transaction, **39**
  hardware support, 141
  roll back, 451
transactional memory, **451**
transitive, **438**
translation, 33–35, 37
traveling salesman problem, **74**, 75, 77, **352**
  2-exchange, 380

3-exchange, 380
Held–Karp lower bound, 353
hill climbing, 380
tree, **71**, 236
  binary, 428
  binomial, **222**, *see* binomial tree
  depth, 72
  dynamic, 339
  expression tree, **73**
  H, 401
  height, **72**
  implicitly defined, 213
  in-order numbering, 399, 402
  interior node, **73**
  ordered, **73**, **235**
  reduction, 402
  representation, 221
  root, **72**
  sorting tree, **165**
  traversal, 73
tree-shaped computation, **428**
triangle inequality, **352**, 380
trie, *see under* sorted sequence
triple, **33**
true, **30**
truth value, **30**
Tsigas, P., 177
Tsitsiklis, J. N., 392
TSP, *see* traveling salesman problem
tuple, **33**, 154
type, **32**

Udupa, R., 208
Ullman, J. D., 256, 343
Ullmann, Z., 370
unary operation, 30
unbounded array, 82, **97**
undefined value (⊥), **33**
uniform memory, 29
union–find, 340
  path compression, 341
  union by rank, 341
universe ($\mathcal{U}$), 357
upper bound, *see* worst case

Vöcking, B., 371
Vachharajani, M., 110
Valiant, L. G., 80, 418
van der Hoeven, J., 24

van Emde Boas layout, 257
van Emde Boas, P., 257
van Hentenryck, P., 392
van Kreveld, M., 355
Vanderbei, R. J., 392
variable, **32**, 359
Varman, P. J., 204
Vazirani, V., 355
vector (in C++), 114
verification, 47, 158
vertex, *see* node
Vetter, C., 329
virtual memory, **452**, 458
  allocation strategy, 452
Vishkin, U., 299
visitor, *see under* graph, *see under*
  graph
Vitányi, P., 209
Vitter, J. S., 188, 208
volatile, 409
von Neumann, J., 29
von Neumann machine, *see under*
  machine model
Vuillemin, J., 222
Vygen, J., 355

wait-free, 46
Wassenberg, J., 355, 450
weak scaling, 195
Wegener, I., 75, 230
Wegman, M., 150
Weidling, C., 150
Westbrook, J., 354
while, 35
Wickremesinghe, R., 208
Wilhelm, R., 80
Williams, J. W. J., 213
Winkel, S., 192, 209, 230
Wirth, N., 32
witness, *see* algorithm design, certificate
Wolsey, L., 376
word, *see* machine word, 436
work, **63**, 95
work stealing, *see under* load balancing
worker, 424
Worsch, T., 421
worst case, *see under* running time
write combining, **450**
write contention, **40**

XOR (⊕), **30**, 315

Yao, S. B., 252
Yasin, A., 453

z-order, 266
Zang, I., 332
Zaroliagis, C., 226

Zelikowski, A., 352
Zhang, Y., 177
Zhou, W., 354
Ziegelmann, M., 332
Ziviani, N., 150
Zlotowski, O., 263
Zobrist, A. L., 127, 151
Zwick, U., 231, 354

Printed in the United States
By Bookmasters